HANDBOOK OF
ADVANCED WASTEWATER TREATMENT

SECOND EDITION

Russell L. Culp
George Mack Wesner
Gordon L. Culp

D0592725

Van Nostrand Reinhold Environmental Engineering Series

VAN NOSTRAND REINHOLD COMPANY

NEW YORK   CINCINNATI   ATLANTA   DALLAS   SAN FRANCISCO
LONDON   TORONTO   MELBOURNE

Van Nostrand Reinhold Company Regional Offices:
New York   Cincinnati   Atlanta   Dallas   San Francisco

Van Nostrand Reinhold Company International Offices:
London   Toronto   Melbourne

Library of Congress Catalog Card Number: 77-24483
ISBN: 0-442-21784-6

Manufactured in the United States of America

Published by Van Nostrand Reinhold Company
135 West 50th Street, New York. N.Y. 10020

Published simultaneously in Canada by Van Nostrand Reinhold Ltd.

15 14 13 12 11 10 9 8 7 6 5 4 3 2

Library of Congress Cataloging in Publication Data

Culp, Russell L   1916–
    Advanced wastewater treatment.

(Van Nostrand Reinhold environmental engineering series)
    Includes bibliographies and index.
    1. Sewage—Purification.   I. Wesner, George Mack,
joint author.   II. Culp, Gordon L., joint author.
III. Title.
TD745.C8   1977   628′.3   77-24483
ISBN 0-442-21784-6

Van Nostrand Reinhold Environmental Engineering Series

Van Nostrand Reinhold Environmental Engineering Series

THE VAN NOSTRAND REINHOLD ENVIRONMENTAL ENGINEER-ING SERIES is dedicated to the presentation of current and vital information relative to the engineering aspects of controlling man's physical environment. Systems and subsystems available to exercise control of both the indoor and outdoor environment continue to become more sophisticated and to involve a number of engineering disciplines. The aim of the series is to provide books which, though often concerned with the life cycle—design, installation, and operation and maintenance—of a specific system or subsystem, are complementary when viewed in their relationship to the total environment.

Books in the Van Nostrand Reinhold Environmental Engineering Series include ones concerned with the engineering of mechanical systems designed (1) to control the environment within structures, including those in which manufacturing processes are carried out, (2) to control the exterior environment through control of waste products expelled by inhabitants of structures and from manufacturing processes. The series will include books on heating, air conditioning and ventilation, control of air and water pollution, control of the acoustic environment, sanitary engineering and waste disposal, illumination, and piping systems for transporting media of all kinds.

Preface

The first edition of *Advanced Wastewater Treatment*, which was published in 1971, was the first book available in this relatively new field of engineering endeavor. It has found wide acceptance and use as a reference by design engineers and pollution control authorities. It is used as a text or reference in graduate and undergraduate environmental engineering courses in universities throughout the world. The book is also now available in a Japanese-language translation.

The book was prepared originally in recognition of the fact that the control of water pollution in the United States and many other countries required treatment techniques far more efficient and reliable than the conventional processes of the past. At that time, there was a substantial gap between available, proven technology and that which had actually been brought to bear by the technical and political forces seeking to control pollution. One purpose of the book was to present the basic principles, engineering design information, and actual operating experiences related to treatment techniques which were relatively new to the wastewater treatment field, with the hope that it would assist in closing the then existing gap in practical application of new technology.

Since 1971, advanced wastewater treatment processes have been employed to a much greater extent than predicted by the most optimistic forecasts of that time. They have been used in a wide variety of ways to upgrade the quality of effluents, improve stream conditions, facilitate wastewater reuse, and improve the reliability of treatment plants. The national goal of zero discharge of pollutants by 1983 established in Public Law 92-500 gives added impetus to advanced wastewater treatment (AWT). Although this goal has yet to be quantitatively defined, there is little doubt that Public Law 92-500 reflects a national philosophy which will stimulate the use of advanced wastewater treatment techniques.

Since the original publication, there has been much progress in refining

AWT processes and in developing new methods, a great deal more experience has been gained in the design and operation of AWT plants, and some material in the book has become outdated. In recognition of this situation, a second edition has been prepared. It has been expanded from 300-odd to over 500 pages. A third author, George Mack Wesner, has joined the authors of the first edition. Dr. Wesner's training and experience in this field broadens the base of information which has been incorporated in the revision and expansion of the text. The discussion of treatment methods, as before, is restricted to those designed to remove pollutants normally remaining after conventional secondary treatment.

The major changes in the second edition are the expansion of design examples and case histories; the addition of new chapters on biological nitrogen removal, selective ion exchange, breakpoint chlorination, disinfection, chemical sludge handling, land treatment, demineralization, and estimating costs; and new information on the use and regeneration of powdered activated carbon.

Special thanks are expressed to the consulting engineers, equipment manufacturers, and government officials who were so generous in supplying illustrative material for the text. The encouragement offered by our families and colleagues throughout the preparation of this second edition was essential to its completion.

RUSSELL L. CULP
GEORGE MACK WESNER
GORDON L. CULP

August 11, 1976

Contents

HANDBOOK OF
ADVANCED WASTEWATER TREATMENT

I

Purpose
and benefits
of advanced
wastewater treatment

PURPOSE

Advanced wastewater treatment technology is designed to
remove pollutants which are not adequately removed by
conventional secondary treatment processes, previously
considered "complete" processes. These pollutants may include soluble inorganic compounds such as phosphorus or
nitrogen, which may support algal growths in receiving
waters; organic materials contributing biochemical oxygen
demand (BOD), chemical oxygen demand (COD), color,
taste, and odor; bacteria; viruses; colloidal solids contributing turbidity; or soluble minerals which may interfere with
subsequent reuse of the wastewater. The purpose of advanced waste treatment may be to alleviate pollution of a receiving watercourse or to provide a water quality adequate
for reuse, or both. The advanced waste treatment process
may be used following or in conjunction with the conventional secondary process, or it may replace secondary
treatment entirely.

1

Increasing population and increasing water use has already created, in many locations, pollution problems which cannot be adequately solved by secondary treatment. It is inevitable that the number of these instances will increase in the future. It is also inevitable that the deliberate reuse of treated wastewaters will be required in order to meet future water demands. Indirect water reuse is already commonly practiced, with some estimates indicating that 40 percent of the United States population is using water that has been used at least once before for domestic or industrial purposes. This indirect reuse will also increase in the future. All of these factors indicate that use of advanced wastewater treatment techniques will become increasingly common.

PUBLIC ATTITUDE AND NATIONAL SIGNIFICANCE

The public attitude toward pollution control, which bordered on apathy during the first half of the twentieth century, has undergone drastic change in the early 1970s as part of the surge in public concern for the quality of the environment. The strong desire of the people for adequate pollution control programs is reflected in the overwhelming margins by which very large pollution control bond issues have carried. For example, the voters of the State of New York approved a $1 billion bond issue for pollution control in 1968 by a margin of four to one. The people of St. Louis approved a $95 million pollution bond issue by a five-to-one margin. These results reflect the intense public desire to improve our environment. Sincere and concerted public concern will be required over a long period of time to make the necessary changes in society to bring about significant improvements in our environment. Much more than clever technological advances will be needed. Major changes in our political, social, legal, and economic approaches to pollution control will be required.

All municipal wastewater could be completely eliminated as a source of pollution in the United States and converted to a quality adequate to provide a valuable water resource for many types of reuse at a national annual cost of only about one dollar per person per month.

CHARACTERISTICS OF SECONDARY EFFLUENTS

It is only a question of time before conventional secondary processes operating at their highest efficiency will be inadequate in a significant portion of the United States. Materials present in effluent from a properly operating secondary plant which may be of concern can be placed in the general categories of soluble organic compounds, soluble inorganic compounds, particulate solid material, and pathogenic organisms.

Organic Compounds

Efficient secondary processes employ biological treatment to remove essentially all of the soluble, biologically degradable organic material in municipal wastewaters. A portion of the soluble organics removed are converted to biological organic cell material which in turn, can exert oxygen demand in the effluent. Generally, the net removal of biodegradable organics is on the order of 90 percent. The remaining degradable organics will exert a demand on the oxygen resources of the receiving body of water which may or may not have an adverse effect, depending on the assimilative capacity of that body of water.

Nondegradable organics are, of course, not removed by secondary processes using biodegradation techniques. These organics can cause taste and odor problems in downstream water supplies. They also impart a color to the effluent which may make it unsuitable for many reuse applications and may make the receiving stream aesthetically unacceptable for recreation. In some cases, they may cause objectionable tastes in fish residing in the receiving watercourse. They may also pass through downstream water treatment plants or react with disinfectants added, causing as yet unknown long term physiological effects on downstream water users. In some cases, they may cause foam in a receiving stream, although the introduction of biodegradable detergents has done much to reduce this problem.

Inorganic Compounds

Phosphorus and nitrogen are two key elements required by algae for growth which are not significantly removed by conventional secondary processes. Phosphates also may interfere with the coagulation processes used in downstream water treatment plants. A major source of phosphorus is the phosphate builders used in modern detergents. The growth of algae in a receiving body of water may create aesthetically unacceptable conditions for recreation, may create taste and odor problems in downstream water supplies, may cause operating problems in downstream water filtration plants, and may create a significant oxygen demand during nighttime hours or after death of the algae. There remains some debate on the minimum phosphorus and nitrogen concentrations which will support objectionable algal growths. However, removal of phosphorus to a concentration of about 0.1 mg/l has reduced algal growths to an insignificant level in a reservoir made up solely of effluent from the South Lake Tahoe wastewater reclamation plant.

During use of water in a municipality, the mineral quality of the water is altered. Inorganic salts containing calcium, magnesium, sodium, potassium, chlorides, sulfates, and phosphates are among those added. Normal water treatment practices at downstream locations do not remove these salts. As a result, the dissolved solids content increases as a fixed supply source passes through several users in series. It is generally agreed that 500 mg/l of dissolved solids is the upper limit for palatable water. Excessive dissolved solids concentrations can cause laxative action in the user, although no harmful permanent physiological effects are known. Dissolved solids concentrations can also adversely affect irrigation use, industrial use, or stock and wildlife watering. Calcium and magnesium contribute to downstream water hardness.

Particulate Solids

Although an efficient secondary plant removes 90–95 percent of the incoming suspended solids (SS), much poorer removals occur all too frequently during "upsets" of the secondary plant due to poor operation, hydraulic or organic overloads, or mechanical failures. The 90–95 percent removal is not adequate for many reuse applications. Suspended solids can interfere with disinfection of the effluent, leading to the discharge of pathogenic organisms. In cases of gross secondary plant failures and small receiving streams, sludge deposits may result which can exert long-term oxygen demands in addition to being aesthetically unacceptable. The historically inconsistent performance of secondary plants is a major weakness which can be overcome with proper application of advanced waste treatment techniques to remove all suspended solids.

Pathogenic Organisms

Secondary processes provide substantial reductions in incoming viral and bacterial concentrations, but it has been shown that both viruses and bacteria

TABLE 1-1 Water Quality at Various Stages in Treatment

Quality Parameter	Raw Wastewater	EFFLUENT	
		Primary	Secondary
BOD (mg/l)	300	100	30
COD (mg/l)	480	220	40
SS (mg/l)	230	100	26
Turbidity (JU)*	250	150	50
MBAS (mg/l)**	7	6	2.0
Phosphorus (mg/l)	12	9	6
Coliform	50	15	2.5
(MPN/100 ml)*	million	million	million

*JU = Jackson Units.
MBAS = Methylene blue active substance.
MPN = Most probable number.

Table 1-2 Anticipated Performance of Various Unit Process Combinations.

AWT Pretreatment[a]	AWT Process[b]	ESTIMATED AWT PROCESS EFFLUENT QUALITY						
		BOD (mg/l)	COD (mg/l)	Turb. (JU)	PO_4 (mg/l)	SS (mg/l)	Color (units)	NH_3-N (mg/l)
Preliminary[a]	C,S	50–100	80–180	5–20	2–4	10–30	30–60	20–30
	C,S,F	30–70	50–150	1–2	0.5–2	2–4	30–60	20–30
	C,S,F,AC	5–10	25–45	1–2	0.5–2	2–4	5–20	20–30
	C,S,NS,F,AC	5–10	25–45	1–2	0.5–2	2–4	5–20	1–10
Primary	C,S	50–100	80–180	5–15	2–4	10–25	30–60	20–30
	C,S,F	30–70	50–150	1–2	0.5–2	2–4	30–60	20–30
	C,S,F,AC	5–10	25–45	1–2	0.5–2	2–4	5–20	20–30
	C,S,NS,F,AC	5–10	25–45	1–2	0.5–2	2–4	5–20	1–10
High rate Trickling Filter	F	10–20	35–60	6–15	20–30	10–20	30–45	20–30
	C,S	10–15	35–55	2–9	1–3	4–12	25–40	20–30
	C,S,F	7–12	30–50	0.1–1	0.1–1	0–1	25–40	20–30
	C,S,F,AC	1–2	10–25	0.1–1	0.1–1	0–1	0–15	20–30
	C,S,NS,F,AC	1–2	10–25	0.1–1	0.1–1	0–1	0–15	1–10
Conventional Activated Sludge	F	3–7	30–50	2–8	20–30	3–12	25–50	20–30
	C,S	3–7	30–50	2–7	1–3	3–10	20–40	20–30
	C,S,F	1–2	25–45	0.1–1	0.1–1	0–1	20–40	20–30
	C,S,F,AC	0–1	5–15	0.1–1	0.1–1	0–1	0–15	20–30
	C,S,NS,F,AC	0–1	5–15	0.1–1	0.1–1	0–1	0–15	1–10

[a]Preliminary treatment—grit removal, screen chamber, Parshall flume, overflow.
[b]C,S—coagulation and sedimentation; F—mixed-media filtration; AC—activated carbon adsorption; NS—ammonia stripping. Lower effluent NH_3 value at 18°C; upper value at 13°C.

TABLE 1-3   Quality of Sewage from Domestic Use of Colorado River Water.

Constituent	Colorado River water (mg/l)	70 gpcd incr.	70 gpcd conc.	100 gpcd incr.	100 gpcd conc.	120 gpcd incr.	120 gpcd conc.
BOD	0	310	310	216	216	180	180
COD	0	475	475	330	330	280	280
TSS	0	360	360	250	250	210	210
TDS	750	450	1200	315	1065	260	1010
Set Sd.	0	13	13	9	9	7.5	7.5
P—Tot.	0	13	13	9	9	7.5	7.5
P—Ortho.	0	9	9	6	6	5	5
N—Tot. Org.	0	19	19	13	13	11	11
—NO$_3$	1.3	0.3	1.6	0.2	1.5	0.15	1.45
—NO$_2$	0	0.06	0.06	0.05	0.05	0.04	0.04
—NH$_3$	0	32	32	22	22	19	19
B	0	0.5	0.5	0.4	0.4	0.3	0.3
Na	111	64	175	45	156	38	149
% Na	—	—		—	—	—	
K	5	15	20	11	16	9	14
Ca	88	26	114	18	106	15	103
Mg	33	13	46	9	42	7.5	40
Cl	97	103	200	72	169	60	157
SO$_4$	327	40	367	27	354	22	349
HCO$_3$	148	130	278	90	238	75	223
CO$_3$	1		1	0	1	0	1
E.C.	1160	640	1800	450	1610	375	1535
Grease and float	0	130	130	90	90	75	75
Total heavy metals	0						
F	0.4		1.0		0.7		0.5
SiO$_2$	8		8		8		8
pH	8.3						
Temp (°F)	69						

Source: "Irvine Ranch Sewerage Survey," report for Irvine Ranch Water District, Brown & Caldwell, Engrs. (Nov. 1968).

are normally present in secondary effluents. Infectious hepatitis has been confirmed to be a waterborne viral disease with the virus capable of surviving ten weeks in clean water. Certainly, reuse of secondary effluent for many purposes is not acceptable because of the health hazard.

Wastewater Quality

Wastewater quality depends upon the chemical quality of the raw water supply, per capita water use, and the nature and quantity of materials discharged to sewers. Several illustrative examples follow.

Table 1-1 presents water quality data for raw wastewater, primary effluent, and secondary (activated sludge) effluent at South Lake Tahoe, which may be typical for soft water supply areas and one cycle of domestic use. Table 1-2 gives the anticipated performance of various AWT (advanced wastewater treatment) unit process combinations. Table 1-3 shows the quality of sewage resulting from domestic use of Coloraro River water. Note that the incremental concentrations for the constituents vary inversely with the per capita wastewater flow rates. Table 1-4 gives representative use increments for 22 U.S. cities.

In an article entitled, "Quality Considerations in Successive Water Use," in the August 1971 *Journal of the Water Pollution Control Federation*, Linstedt and co-workers present very detailed information on the Denver water supply. The differences in the city water supply and the wastewater effluent represent the use increment under circumstances at Denver.

The effect of several successive cycles of water reuse for municipal purposes is to reach an equilibrium effluent quality provided that freshwater makeup is provided. Studies made at a number of places indicate that two parts of makeup water must be added to one part of recycled reclaimed wastewater in order to prevent development of excessive concentrations of certain chemical constituents which are not completely removed in treatment. The use of $33\frac{1}{3}$ percent reclaimed wastewater is just a rough rule of thumb, however, and each individual set of circumstances must be analyzed where successive reuse is under consideration.

TABLE 1-4   Representative Use Increments for 22 U.S. Cities.
(*Neale, J. H.,* Advanced Waste Treatment by Distillation, *AWTR-7, USPHS Publication Number 999-WP-7, 1964.*)

	USE INCREMENT (mg/l)		
Constituent	Min.	Max.	Avg.
Sodium	8	101	66
Calcium	1	50	18
Magnesium	0	15	6
Ammonium	0	36	15
Chloride	6	200	74
Sulfate	12	57	28
Bicarbonate	−44	265	100
Nitrate	−5	26	10
Silica (SiO_3)	9	22	15
Phosphate (PO_4)	7	50	24
TDS	128	541	320
BOD	8	27	16
COD	22	159	87

TABLE 1-5   Raw Water and Waste Effluent Data.

Constituent	Supply Water Composite	Wastewater Effluent	Use Increment
Physical:			
SS (mg/l)	0.0	98	98.0
Turbidity (JTU)	0.7	45	44.0
Color (color units)	5.2	75	69.8
Odor (TON)	0.0	27.2	27.2
Microbiological (no./100 ml):			
Coliforms	0.1	160,000	160,000
Fecal coliforms	0.0	26,000	26,000
Fecal strep	0.0	2,000	2,000
Organic constituents (mg/l):			
CCE	0.059	2.478	2.419
CAE	0.091	1.185	1.094
MBAS	0.000	0.116	0.116
COD	0.0	62.0	62.0
BOD	0.0	24.0	24.0
Phenols	0.0	0.01	0.01
Nutrients (mg/l):			
Phosphate	0.04	8.7	8.7
Nitrate-N	0.03	Trace	0.0
Ammonia-N	0.03	17.5	17.5
Kjeldahl-N	0.1	28.2	28.1
Toxic chemicals (mg/l):			
Arsenic	<0.001	0.003	0.003
Barium	0.085	0.192	0.107
Cadmium	<0.001	<0.001	<0.001
Chromium (total)	0.039	0.050	0.011
Cyanide	<0.01	<0.01	—
Fluoride	0.91	1.09	0.18
Lead	0.030	0.082	0.052
Selenium	<0.001	<0.001	—
Silver	<0.001	0.008	0.008
Inorganic ions (major):			
Alkalinity ($CaCO_3$) (mg/l)	91.4	246	155
Calcium (mg/l)	25.6	62	36.4
Chloride (mg/l)	20.0	120	100
Hardness ($CaCO_3$) (mg/l)	82.7	199	116
Magnesium (mg/l)	7.3	10.6	3.3
Sulfate (mg/l)	41.1	168	127
TDS (mg/l)	123.7	480	356
Specific conductance (μmho)	201	1,030	829
Sodium (mg/l)	25.2	155	130

TABLE 1-5   (Continued)

Constituent	Supply Water Composite	Wastewater Effluent	Use Increment
Silica (mg/l)	4.6	10.1	5.5
Potassium (mg/l)			
pH	7.9	7.7	—
Trace elements (mg/l):			
Aluminum	0.11	0.16	0.05
Bromine	0.053	0.197	0.144
Cobalt	<0.001	<0.001	<0.001
Columbium	<0.001	0.001	0.001
Copper	0.036	0.070	0.034
Germanium	<0.001	<0.001	<0.001
Gold	<0.001	<0.001	<0.001
Iron (filtered)	0.273	3.00	2.73
Lanthanum	<0.001	<0.001	<0.001
Manganese	0.040	0.075	0.035
Molybdenum	0.079	0.100	0.021
Nickel	0.015	0.120	0.105
Rubidium	0.021	0.061	0.040
Silver	<0.001	0.008	0.008
Strontium	0.280	0.400	0.120
Tin	<0.001	0.003	0.003
Titanium	<0.001	0.124	0.124
Tantalum	<0.001	<0.001	<0.001
Tungsten	<0.001	0.147	0.147
Uranium	0.015	0.041	0.026
Yitrium	0.001	0.001	0.000
Vanadium	<0.001	<0.001	<0.001
Zinc	0.134	0.185	0.051
Zirconium	0.001	0.013	0.012

Source: Lindstet et al.

WATER REUSE

Unavoidable Water Reuse

The location of several cities on a single river, stream, or lake leads to unavoidable water reuse when each city uses the water body as a water supply and receiving body for wastewater. A recent study by the American Water Works Association[11] showed that, on the average, one gallon out of every 30 used for water supply had passed through the wastewater system of an upstream community for the 155 cities studied. Streams containing water used by many upstream industries, agriculturists, and communities serve as sources of recreation and water supply. Unavoidable water reuse is already a well established fact and accepted practice, although not often recognized as being so.

Intentional Water Reuse

The need for direct and deliberate reuse of reclaimed sewage effluents is increasing in many areas of the world. Reuse holds the key to the efficient and effective utilization of the limited freshwater resource by making available a new valuable source of water to augment existing supplies and a major source of supply for the future. It is generally recognized that we are not running out of water, but have reached the point where maximum utilization of the available supply through elimination of wasteful and reckless degradation of the resource quality must occur if we are to provide for the water needs of the future.

The time is rapidly approaching in the United States when the degree of wastewater treatment required in many areas for pollution control will be so costly that cities will not be able to afford the luxury of discarding once used water. In many other areas of the world, it is only a matter of time before serious consideration must be given to direct reclamation and reuse of sewage effluent to supplement inadequate potable water supplies.

To illustrate the expected need for extensive water reuse, we note that by 1980 the estimated available daily supply in the United States will be 515 billion gallons, if careful resource management is practiced and increased impoundment capacity is provided, but the demand will have increased to 650 billion gallons. To meet the total demand, it is apparent that there will have to be a considerable amount of reclamation and reuse of "secondhand" waters. At the present time it is estimated that the total of used municipal water is approaching 20 billion gallons daily (bgd) and that over 40 percent of this rapidly growing supply could readily be reclaimed for industrial use. There is the potential that water quality suitable for potable use can be produced from sewage effluents when there is sufficient economic justification. For example, complete utilization of available potable water supplies has forced the city of Windhoek, South West Africa, to turn to direct potable use of reclaimed wastewater.[5]

The increasing awareness of the value of reclaimed water as a potential resource is reflected by the following statement of policy made by the Water Pollution Control Federation in 1963: "Wastewater represents an increasing fraction of the nation's total water resource and is of such value that it might well be reclaimed for beneficial reuse through the restoration of an appropriate degree of quality. To this end, the development of methods for waste-water reclamation and criteria for such reuse should be encouraged."[6] For a time, the process of recovering fresh water from the ocean received much public attention. Interest has waned somewhat as hopes that the ocean can supply unlimited amounts of fresh water at a reasonable price have not been fulfilled. However, desalination may provide an important supplemental supply for communities situated on the seacoast. In contrast to municipal wastewater, which contains less than 0.1 percent of impurities that can be effectively removed with existing advanced wastewater treatment methods, seawater contains 3.5 percent of dissolved salts plus considerable organic matter—over 35 times as much foreign matter as secondary sewage effluent. For these reasons capital investment and unit costs for desalination of seawater exceed costs for wastewater reclamation. The benefits to be gained from advanced wastewater treatment and water reuse include not only supplementation of fresh-water supplies, but also the alleviation, in the same step, of the pollution problem. Reclamation of wastewaters eliminates the source of pollution and qualifies as complete pollution control.

The importance of wastewater reclamation and reuse in the United States received detailed consideration as early as 1947. At that time, a symposium on reclamation of sewage effluents was held at a joint meeting between the Federation of Sewage Works Associations and the American Water Works Association. It was concluded that the economics of water reuse hinged on the value of the reclaimed water itself, rather than on the value of any impurities extracted from sewage. Some concern about the need for greater water reuse was expressed through the 1950s, but as the decade came to a close, the lack of severe water shortages in the United States provided little stimulation to take concrete action to encourage extensive reuse.

In 1947, sewage effluents were being utilized at 135 locations in the United States, of which 124 were for irrigation; 10 for industrial cooling, boiler feedwater, quenching and process water; and in one instance for a skating pond. The volume of water reused amounted to 0.3 percent of the daily sewage flow in the United States. As of 1954, it was estimated[6] that utilization of sewage effluents in the United States did not exceed 1 percent of the total sewage flow. Reuse as irrigation water far exceeded that for other purposes.

Industrial Use

Industrial use of water in the United States has been estimated to be about seven times that of municipal use and about equal to the irrigation use. It has been estimated that by 1980, industrial water use will increase by 146 percent, municipal use by 68 percent, and irrigation by 18 percent over 1960 levels. It is apparent that as the demand for and cost of freshwater supplies increases, that wastewater reclamation will be an economically attractive alternative for industry in an increasing number of instances. The required quality of water varies widely depending upon the specific use involved. Required cooling water quality differs greatly from that for boiler feedwater, for example. Water quality exceeding drinking water standards is needed in some industrial applications, such as certain electronic manufacturing manufacturing operations, for high pressure boiler feedwater, and in beverage preparation. It is impossible to tabulate the specific criteria required for all potential industrial uses. When considering a specific instance of industrial use of reclaimed wastewater, it is necessary to determine the needed water quality and evaluate the economics of obtaining this quality from a wastewater stream as compared to alternative sources. Major items to be considered are the consistency in quality and quantity of each source.

Municipal wastewater treatment plant effluents have been used as industrial cooling water in several locations. The major considerations are that the water should not encourage corrosion, deposition of scale, delignification of wooden cooling towers, growth of microorganisms, and excessive foaming in cooling towers. Municipal wastewaters usually show less variation in temperature than do surface waters, although wastewaters are frequently warmer.

Use of secondary effluents as cooling water has proven satisfactory in some cases, although chemical coagulation and sedimentation of secondary effluent is often provided prior to use as cooling water. If lime is used as a coagulant, care must be taken to insure that the pH is adjusted to a satisfactory value (6.5–7.5) to avoid delignification of wooden packings in cooling towers and to avoid deposition of calcium carbonate in the cooling system.

Wastewater constituents which may prove troublesome in cooling water supplies but are readily controlled by advanced wastewater treatment processes include suspended solids, hardness agents, dissolved organics, dissolved oxygen or gases, algal nutrients, and slime producing organisms. Turbidity and hardness values of less than 50 mg/l are generally desirable. Chemical additives to inhibit scale formation and slime growth may be used in conjunction with the wastewater treatment process.

Almost all industries use boilers and consequently have the problem of adequate boiler feedwater. The subject of required quality of boiler feedwater is a complex one which has been discussed in many published papers. As the operating pressure of the boiler increases, the quality of boiler feedwater must improve. For boilers of very high pressure (1000 psi or more), all hardness must be removed, and dissolved solids should be as low as possible, preferably less than 0.5 mg/l. Carefully deionized and deaerated water is required in these cases. Silica is especially troublesome because it forms a hard scale in boilers and boiler tubes. For high pressure boilers, dissolved silica cannot exceed 0.2 mg/l, while 5 mg/l can be tolerated at pressures of 250-400 psi. No ammonia should be present because it damages copper parts. High pressure boilers require treatment steps beyond those normally used for advanced wastewater treatment purposes to remove soluble minerals of concern.

Bethlehem Steel Company has used Baltimore's wastewater for process purposes for over 20 years. Wolman[29] reported the desirable water characteristics for steel manufacture as follows: temperature below 75° F, chlorides below 175 mg/l, pH between 6.8 and 7.0, hardness below 50 mg/l, suspended matter below 25 mg/l, organic content as low as possible, and corrosion potential as low as possible. The copper and aluminum industries also have several process steps in which reclaimed wastewater would be of suitable quality.

The phosphorus removal provided by the processes described in this book will reduce the algal growths which could interfere with oil well repressuring using stored effluent. Filtration is required to ensure that suspended solids will not plug the injection well.

Industrial use of renovated municipal effluent for processing of foods or ice for human consumption will be limited by the same concerns discussed in a subsequent section of this chapter on domestic reuse. Freshwater sources will remain preferable for the foreseeable future for these purposes. There are innumerable other industrial applications where the processes described in this book are capable of providing adequate water quality.

Agricultural Use

Irrigation use is second only to industrial use of water in the United States and comprised about 43 percent of all water use in 1960 (compared to 50 percent by industry).[9a] As noted earlier, use of effluent for irrigation has been the major instance of wastewater reuse in the past in the United States. The chief concern has been to restrict the use for irrigation to prevent health hazards from effluent contact with crops directly consumed by humans. The use of advanced wastewater treatment techniques offers the potential for unrestricted direct reuse for irrigation. Unrestricted reuse following groundwater recharge is already an accepted practice. Recent studies by Bouwer[4] have shown that secondary effluent from Phoenix, Arizona, can be renovated adequately to permit unrestricted irrigation by groundwater recharge through the normally dry bed of the Salt River. The success of the Muskegon County, Michigan project, where 6300 acres are irrigated with secondary effluent, has sparked considerable interest in treatment of wastewaters through application to the land. Chapter 10 discusses this concept in detail.

The salt content of the irrigation water is of concern as well as the sanitary quality, but absolute limits of desirable salt content cannot be fixed because plants vary widely in their tolerance of salinity and soil types, climatic conditions, and irrigation practices. The relationships among ions may be significant, as illustrated by the antagonistic influence between calcium and sodium. The concentration of salts in most municipal effluents will not usually be high enough to cause immediate injury to crops. If leaching of the root zone does not take place, the salt concentration will increase until it reaches the limit of solubility of each salt. Water containing up to 2000 mg/l of dissolved solids is suitable for most plants except sensitive ones. Dissolved solids of less than 1000 mg/l are suitable for all types of plants provided drainage is good. Calcium and magnesium in proper proportion maintain soil in good condition, while the opposite is true when sodium predominates. When the percentage of sodium exceeds the desirable limit for a given soil (see Chapter 10), granular soil structures begin to break down when the soil is moistened. The soil pores eventually seal, resulting in a decrease in soil permeability.

Certain soluble salts can be harmful if present in excessive quantities. For example, 700–1500 mg/l of chlorides in the root zone can cause leaf burn and possibly death of the plants. Although most municipal wastewaters will be of adequate chemical quality for irrigation, careful evaluation of this aspect as well as of the potential health hazards is required. Toxic compounds originating from industrial wastes which may be discharged to the municipal sewer system should be considered.

Irrigation of golf courses and parks with secondary effluent has long been practiced in the Southwest. More advanced treatment may be preferable in many cases to minimize health hazards due to chance contact or by wind-blown spray striking drinking fountains. Piping carrying the effluent must be clearly marked to prevent crossconnection or drinking use. The need for careful control of these irrigated areas is apparent if direct contact with the effluent is to be prevented during irrigation periods.

Where soil and hydrologic conditions are proper, groundwater recharge through surface spreading (or infiltration-percolation) of secondary effluent offers an economical method of providing an effluent for unrestricted irrigation. As noted earlier, a project of this type has been found adequate for unrestricted irrigation reuse at Phoenix, Arizona.

Domestic Use

There are some significant differences of opinion in different parts of the world on the suitability of reuse of wastewaters for a potable water supply.

The city of Windhoek, South West Africa, is successfully recycling advanced wastewater treatment plant effluent directly to the inlet of a water treatment plant. However, it is the present position of the Environmental Protection Agency that renovated sewage should not in any case be used as a source of drinking water when other sources are available. There are a number of good reasons to delay the reuse of renovated wastewaters for potable supplies until it is absolutely necessary. The federal drinking water standards are based on the assumption that a sanitary survey has shown the raw water source is relatively unpolluted. These standards appear to be reliable for supplies taken from freshwater sources, since acute health effects have not yet been attributed to waters that have met these standards. However, these standards may not be adequate for reclaimed wastewaters because of the presence of trace organic materials such as pesticides, antibiotics, hormones, or trace materials from industrial wastes. Also, the potential for viral contamination has not been accurately determined over a long period for various advanced wastewater treatment unit process combinations.

An instance of drought forced recycle of trickling filter effluent through a 17 day retention pond directly to a softening plant intake in Chanute, Kansas, has received widespread attention.[16] Although no health problems have been traced to this direct reuse, the presence of soluble materials not removed by secondary treatment led to poor consumer acceptance due to foaming and a pale yellow color in the tap water. Had advanced wastewater treatment facilities been practical, there is little doubt that the residents of Chanute would have been willing to pay many times the cost of normal water and sewage treatment to obtain a satisfactory water from their wastewater. Despite the lack of health problems at Chanute with rather crude techniques and at Windhoek with advanced technology, it appears that the most prudent approach at the present is to use reclaimed wastewaters for the many nondrinking uses, where it is unquestionably of adequate quality, to increase the availability of freshwater supplies for drinking. Indirect reuse through groundwater recharge, either intentional or unintentional, and through effluent discharge to surface supplies used by other communities, is an accepted practice and will continue until intentional reuse becomes economical and of accepted safety.

Recreational Use

Recreational water used for sports involving contact with the water must be aesthetically acceptable, must not contain substances that are toxic upon ingestion or irritating to the skin, and must be reasonably free from pathogenic organisms. Limits of temperature, color, odor, pH, turbidity, and specific ions may be established to define the first two conditions, although many state agencies prefer to use qualitative rather than quantitative standards for all but health related conditions. The condition that the water be reasonably free of pathogenic organisms has been subjected to quantitative standards in most states, although the standards vary considerably. No definitive relationship between bacterial quality of recreational waters and incidence of related disease has been established. State standards vary from 50 to 3000 coliforms/100 ml based either on arithmetical mean, geometrical mean, or the median of monthly samples.[14] English studies have concluded that unless the water is so fouled as to be aesthetically revolting, public health standards are reasonably well met. Certainly, these conditions need not be approached in a reservoir made up solely of reclaimed wastewater and the most restrictive state standards of less than 2.2 coliforms/100 ml are easily met by the proper combination of the unit processes described in this book.

The suitability of reclaimed water for recreational use has been demonstrated by the Santee County, California, Water District. Treated wastewater is the principal supply for a series of recreational lakes. Secondary effluent from a conventional activated sludge plant passes through an infiltration area to the ponds. Careful monitoring of the bacteriological quality of the infiltration system effluent following chlorination show coliform concentrations of less than 2/100 ml. No virus has been isolated from the water entering the lakes. Based on the high quality of the reclaimed lake waters, the record of community health, and the absence of virus in the lake waters, permission was granted by the State Health Officer to allow establishment of a swimming program during the summer of 1965.[15] The subsequent success of the swimming program demonstrates the ability to treat wastewater to a degree that will meet public health requirements and will be accepted by the public.

In 1969, Indian Creek Reservoir in Alpine County, California, was approved for all water contact sports by all local and state regulatory agencies. This lake is filled with one billion gallons of reclaimed wastewater from the South Tahoe Public Utility District plant at South Lake Tahoe, California. Recreation activities include sailing, swimming, and trout fishing (see Figure 1-1). The lake has been successfully stocked with rainbow trout. In two summers of sampling, viruses have been isolated on several occasions from the secondary stage of treatment but never from the final chlorinated effluent. Bacteriological tests of water delivered to this reservoir are consistently negative for coliform bacteria.

It is difficult to establish separate criteria for noncontact recreation such as boating and aesthetic enjoyment, since contact recreation is frequently concurrent with these other activities. Noncontact recreation is adversely affected by problems easily overcome by advanced wastewater treatment; such problems include floating or visible solids, slime growths, algal mats or

Fig. 1-1   This recreational lake is composed solely of reclaimed water from the South Lake Tahoe, California, advanced wastewater treatment plant. In addition to boating and swimming, a rainbow trout fishery is supported by the lake.

blooms, discoloration, gas bubbles, turbidity, oil, and foaming. Indeed, it is more likely that the reservoir of renovated wastewater as shown in Figure 1-1 is in greater danger of being polluted by boating and shoreline activities than of being a source of aesthetic pollution.

Fishing is a recreational activity where water quality is of obvious importance. It is impossible to establish universal quality criteria for fish because the effects of harmful substances vary with species, size, and age of the fish. The effects also vary with the chemical composition of the water supply. Certain salts act synergistically while others act antagonistically. Dissolved oxygen, pH, free carbon dioxide, ammonia nitrogen, suspended solids, temperature, and toxic metals are the major parameters to be evaluated for the particular species of fish involved. Warm water species have been successfully supported in the Santee project reservoirs, while rainbow trout have been supported in the South Lake Tahoe effluent reservoir. No health hazards have been related to eating fish from these reservoirs made solely of reclaimed effluent.

Direct water reuse for recreational purposes has been proven practical by the Santee and South Lake Tahoe projects and offers an attractive means of meeting the increased demand for water based recreation in water-short areas.

UNIT PROCESSES FOR ADVANCED WASTEWATER TREATMENT

A study of the effluent quality produced by conventional secondary treatment processes will quickly reveal that such treatment methods do not remove many pollutants which may create a pollution problem or prevent reuse of the effluent. Should the presence of materials found in secondary effluent be objectionable because of the desire to reuse the water or the desire to alleviate pollution, the selection from among the appropriate advanced waste treatment unit processes must be made.

Among the many factors to be considered when designing an advanced waste treatment facility are the disposition or use of the final effluent and the related requirements for effluent quality, the nature of the materials to be removed to achieve the required quality, the problems associated with handling of the solids or waste liquids generated in the liquid treatment process, the potential for recovery and reuse of coagulants or other materials used in the treatment processes, the limitations imposed by the sewage collection system and available plant sites, the potential for creating air or land pollution in the process of treating wastewater, the demand for energy and other consumable resources, and overall economic feasibility.

The unit processes now being used for advanced waste treatment have generally been used for various industrial purposes and have been adapted to waste treatment plant design as the need for higher effluent quality has developed. The following sections of this book explore the design considerations for various unit processes for liquid treatment and the handling of the solids or waste liquids generated by treatment of the liquid phase. This latter point is of major importance, since it is obvious that the residues of the waste treatment cannot be discharged into a usable source if a net gain is to be achieved by the advanced waste treatment process. In many instances the disposal of these residues may be the major factor governing the selection of the liquid treatment process. Generalized cost estimates for various processes are also presented.

REFERENCES

1. Anonymous, "Progress Report of Committee on Quality Tolerances of Water for Industrial Purposes," *Journal New England Water Works Assoc.* Aug., 1958, p. 1021.

2. Argo, D. G., and Culp, G. L., "Heavy Metals Removal in Wastewater Treatment Processes," *Water and Sewage Works*, Aug., 1972, p. 62 and Sept., 1972, p. 128.

3. Besik, F., "Renovation of Domestic Wastewater," *Water and Sewage Works*, April, 1973, p. 78.

4. Bouwer, H., "Returning Wastes to the Land, A new Role for Agriculture," *Journal Soil and Water Conservation*, 23, Sept.–Oct., 1968.

5. Cillie, G. G.; Van Vuuren, L. R. J.; Stander, G. J.; and Kolbe, F. F., "The Reclamation of Sewage Effluents for Domestic Use," *Proceedings, Third International Conference on Water Pollution Research, Munich, Germany*. Water Pollution Control Federation, Washington, D.C., 1966.

6. Connell, C. H., and Forbes, M. C., "Once Used Municipal Water as Industrial Supply, in Retrospect and Prospect," *Water and Sewage Works*, 1964, p. 397.

7. Culp, G. L., et al., "Water Resource Preservation by Planned Recycling of Treated Wastewater," *Journal American Water Works Assoc.*, October, 1973, p. 641.

8. Culp, G. L., and Culp, R. L., *New Concepts in Water Purification*. Van Nostrand Reinhold, New York, 1974.

9. Culp, G. L., and Hamann, C. L., "Advanced Waste Treatment Process Selection," *Public Works*, March, April, May, 1974.

9a. Gloyna, E. F., et al., "A Report Upon Present and Prospective Means for Improved Reuse of Water," prepared for U.S. Senate Committee on National Water Resources (1960).

10. Graeser, H. J., "Dallas' Wastewater Reclamation Studies," *Journal American Water Works Assoc.*, October, 1971, pp. 634–640.

11. Haney, Paul D., "Water Reuse for Public Supply," *Journal American Water Works Assoc.*, 1969, p. 73.

12. Heaton, R. D., et al., "Progress Toward Successive Water Use in Denver," paper presented at the Water Pollution Control Federation Conference, Denver, Colorado, October, 1974.

13. Kearns, J. T., "Water Conservation and Its Application to New England," *Journal American Water Works Assoc.*, 1966, p. 1379

14. McKee, J. E., and Wolf, H. W., eds., *Water Quality Criteria*. Resources Agency of California, State Water Quality Control Board, Pub. No. 3-A, 1963.

15. Merrill, J. C., Jr., and Katko, A., "Reclaimed Wastewater for Santee Recreational Lakes," *Journal Water Pollution Control Federation*, 1966, p. 1310.

16. Metzler, D. F., et al., "Emergency Use of Reclaimed Water for Potable Supply at Chanute, Kansas," *Journal American Water Works Assoc.*, Aug., 1958, p. 1021.

17. Ongerth, H. J., et al., "Public Health Aspects of Organics in Water," *Journal American Water Works Assoc.*, July, 1973, p. 495.

18. Parkhurst, J. D., and Garrison, W. E., "Water Reclamation at Whittier Narrows," *Journal Water Pollution Control Federation*, 1963, p. 1094.

19. "Reuse of Effluents: Methods of Wastewater Treatment and Health Safeguards," World Health Organization Technical Report Series No. 517, 1973.

20. Rose, J. L., "Injection of Treated Wastewater into Aquifers," *Water and Waste Engineering*, 5:10, 1968, p. 40.

21. Shuval, H. I., and Gruener, N., "Health Considerations in Renovating Wastewater for Domestic Use," *Environmental Science and Technology*, July, 1973, p. 601.

22. Steffen, A. J., "Control of Water Pollution by Waste Utilization: The Role of the WPCF," *Water and Sewage Works*, 1964, p. 384.

23. Stephan, D. G., and Weinberger, L. W., "Water Reuse—Has it Arrived?", *Journal Water Pollution Control Federation*, 1968, p. 529.

24. Weber, W. J., Jr., *Physicochemical Processes for Water Quality Control*. Wiley-Interscience, New York, 1972.

25. Wesner, G. M., and Baier, D. C., "Injection of Reclaimed Wastewater into Confined Aquifers," *Journal American Water Works Assoc.*, March, 1970, p. 203.

26. Wesner, G. M., and Culp, R. L., "Wastewater Reclamation and Seawater Desalination," *Journal Water Pollution Control Federation*, 44:10, October, 1972, pp. 1932–1939.

27. Whetstone, G. A., *Reuse of Effluent in the Future With an Annotated Bibliography*. Texas Water Development Board Report 8, Austin, Texas, 1965.

28. Wilcox, L. V., "Agricultural Uses of Reclaimed Sewage Effluent," *Sewage Works Journal*, 1948, p. 24.

29. Wolman, A., "Industrial Water Supply From Processed Sewage Treatment Plant Effluent at Baltimore, Md.," *Sewage Works Journal*, 1948, p. 15.

30. Zemansky, G. M., "Removal of Trace Metals During Conventional Water Treatment," *Journal American Water Works Assoc.*, October, 1974, p. 606.

2

Chemical clarification

GENERAL CONSIDERATIONS

The term chemical clarification as used herein is a treatment process made up of three distinct operations: (1) coagulation, (2) flocculation, and (3) sedimentation. The terms "coagulation" and "flocculation" have often been used interchangeably to describe the process for removal of suspended material from wastewater. These and other terms used in this book to describe chemical clarification processes are defined as follows:

Coagulation: The process whereby chemicals are added to a wastewater resulting in a reduction of the forces tending to keep suspended particles apart. This process physically occurs in a rapid mix or flash mix basin.

Flocculation: The agglomeration of suspended material to form particles that will settle by gravity.

Sedimentation: The separation of suspended solids from wastewater by gravity.

Coagulants: Chemicals, such as alum, iron salts or lime, added, in relatively large concentrations, to reduce the forces tending to keep suspended particles apart.

Coagulation–Flocculation Aids: Materials used in relatively small concentrations which are added either to the coagulation and/or flocculation basins and may be classified as: (1) oxidants, such as chlorine or ozone; (2) weighting agents, such as bentonite clay; (3) activated silica; and (4) polyelectrolytes. Polyelectrolytes dissolved in water may ionize to have a positive, a negative or no charge and are, therefore, referred to respectively as cationic, anionic and non-ionic.

Chemical clarification may take place in separate coagulation, flocculation and sedimentation basins, or in single basin solids contact type units. There have also been some efforts to eliminate sedimentation basins and use filters for the removal of the flocculated material.

It is possible to remove, via the chemical clarification process, suspended and dissolved organic and inorganic materials from secondary treated wastewater. Substances that can be substantially removed or reduced by chemical clarification include the following:

1. Suspended organic and inorganic material. It is generally possible to remove material greater than one micron in size by chemical clarification. Tests at Orange County, California showed that it is possible to routinely remove COD-causing materials in secondary effluent larger than 0.45 micron; the COD in wastewater passing a 0.45 micron filter was not removed.
2. Dissolved phosphate can usually be reduced to less than 1 mg/l with alum, iron salts or lime. Certain polymers are also capable of removing phosphates from some wastewaters.
3. Some calcium, magnesium, silica and fluoride can be removed with lime. Higher removals of calcium and magnesium can be achieved with lime in wastewaters with high carbonate hardness content.
4. Some heavy metals can be removed. Lime is especially effective in precipitating cadmium, chromium, copper, nickel, lead and silver.
5. Reduction in bacteria and viruses can be achieved. Lime is particularly effective in reducing bacteria and virus levels because of high pH, usually 10.5 to 11.5, in the flocculation and sedimentation basins.

Coagulation, flocculation and sedimentation are imperfectly understood phenomena and much research has attempted to develop rational explanations. Several mechanisms take place in the clarification process including: (1) chemical and physical reactions with colloidal particles during coagulation, (2) physical enmeshment or entrapment of particles during the flocculation and sedimentation operations, and (3) adsorption. The relative importance of the various mechanisms in any chemical clarification systems are not known, but undoubtedly all play a role.

Although present day theory is an aid in understanding the process, chemical clarification remains as much an art as a science, and theory alone cannot be used in the design of treatment units. The proper chemicals, dosages, and physical conditions must be determined from bench and pilot scale tests and from experience. Variations in flow volume and wastewater characteristics are additional reasons which make it desirable to design as much flexibility as possible into coagulation, flocculation and sedimentation units.

The purpose of this chapter is to briefly describe the fundamental processes and then proceed to a more detailed discussion of design criteria and several case studies.

COAGULATION

Coagulation takes place in rapid mix, or flash mix, basins. All research agrees that coagulation reactions are very rapid, probably taking less than one second. Therefore, the primary function of a rapid mix basin is to disperse the coagulant so that it contacts all of the wastewater.

Rapid mixing basins for dispersion of coagulants are usually equipped with high-speed mixing devices designed to create velocity gradients of 300 fps/ft (feet per second per foot), or more, with detention times of 15 to 60 sec. Power requirements of mechanical mixers are 0.25–1 hp/mgd (horsepower per million gallons per day). A typical rapid mixing mechanism is shown in Figure 2-1.

Two theories have been advanced to explain the basic mechanisms involved in the stability and instability of colloid systems: (1) Chemical theory assumes that colloids are aggregates of definite chemical structural units, and proposes that coagulation occurs because of specific chemical reactions between colloidal particles and the chemical coagulant added. (2) Physical theory proposes that reduction of forces tending to keep colloids apart occurs through the reduction of electrostatic forces, such as the zeta potential. The two theories are not mutually exclusive and both must be employed to explain the operation of an effective chemical coagulation process. Coagulation theory is discussed in several publications.[1,2,3] Much of the available experience and published theory relates to water treatment and must be used with caution in application to wastewaters. The presence, and relatively high concentration, of organic material makes the coagulation of wastewater a complex process. Generally good coagulation, flocculation and sedimentation is difficult to obtain in wastewater treatment.

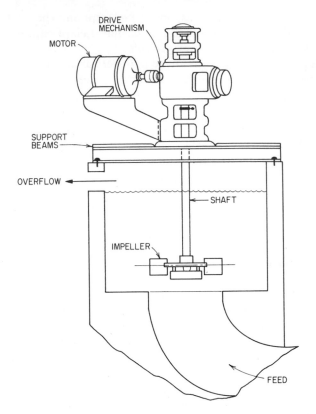

Fig. 2-1   Mechanical rapid mixing device. (*Courtesy Dorr-Oliver*)

The primary coagulants that have been used in wastewater treatment are: (1) lime, (2) alum, and (3) iron salts such as ferric chloride, ferric sulfate and ferrous sulfate. Some of the properties of these common coagulants are shown in Table 2-1. Chemical reactions of these coagulants with alkalinity and phosphate are given in this chapter in the form of chemical equations. Coagulation is a complex process and is undoubtedly far more complicated than these equations indicate; nevertheless, the equations do provide an estimate of the products of the reactions and the quantitative relationships.

Lime Coagulation

Lime is a term applied to a variety of forms of chemicals which are highly alkaline in character and contain principally calcium and oxygen, but also may contain magnesium. Table 2-1 summarizes information on the various

TABLE 2-1 Characteristics of Primary Coagulants

Common Name [Formula]	Molecular Weight	Available Forms	Bulk Density, lb/ft³	Specific gravity	Commercial Strength, percent	Solubility in Water, g/100 ml
Quicklime [CaO]	56	pebble crushed lump ground pulverized	55–65	3.2–3.4	70–96 CaO (below 88 can be poor quality)	reacts to form $Ca(OH)_2$ each lb of quicklime will form 1.16–1.32 lb of $Ca(OH)_2$, with 2–12% grit, depending on purity
Hydrated Lime [$Ca(OH)_2$]	74	powder (passes 200 mesh)	25–50	2.3–2.4	$Ca(OH)_2$ 82–98 CaO 62–74 (Std. 70)	0.18 at 0°C 0.16 at 20°C 0.15 at 30°C 0.077 at 100°C
Alum [$Al_2(SO_4)_3 \cdot 14H_2O$]	594	dry	62–68		17.1 Al_2O_3	
		liquid		1.3	8.3 Al_2O_3	8.7 @ 20°C
Ferric Chloride [$FeCl_3$]	162	liquid		1.4	40 $FeCl_3$	

forms of lime. Lime is defined by the National Lime Association as a general term which includes the chemical and physical types of quicklime, hydrated lime, and dolomitic lime. The two forms used most in wastewater treatment are:

1. Quicklime, CaO, usually a solid, white lump, results from the calcination of limestone at temperatures of about 1400–1700° F. It is shipped in different sizes, ranging from pulverized to 8 inches in diameter.
2. Hydrated lime, $Ca(OH)_2$, is a fine white powder obtained by treating quicklime with enough water to satisfy its chemical affinity for water.

As indicated in Table 2-1, lime may be shipped in a variety of containers. For most large plants, bulk shipments of quicklime are preferable. If bag shipment is used, the bags should be stored in a dry room on pallets rather than on concrete floors because moisture can be absorbed through the latter type of floor. Sixty to ninety days is the usual limit for storage in bags because of the gradual absorption of atmospheric moisture. In small plants where lime consumption is less than a carload a month, hydrated lime is preferred because of ease of handling. In dry storage, deterioration is not serious for periods up to one year.

The term *slaking* is used for the process of adding water to quicklime, or recalcined lime to produce a slurry of hydrated lime. Quick lime is slaked to produce hydrated lime as represented in the following reaction:

$$CaO + H_2O \longrightarrow Ca(OH)_2$$

Lime slaking and feeding systems are discussed later in this chapter.

Hydrated lime reacts with carbonate hardness and orthophosphate as follows:

$$Ca(OH)_2 + Ca(HCO_3)_2 \longrightarrow 2CaCO_3 \downarrow + 2H_2O$$
$$2Ca(OH)_2 + Mg(HCO_3)_2 \longrightarrow 2CaCO_3 \downarrow + Mg(OH)_2 \downarrow + 2H_2O$$
$$5Ca(OH)_2 + 3PO_4^{3-} \longrightarrow Ca_5OH(PO_4)_3 + 9OH^-$$
$$3Ca(OH)_2 + 2PO_4^{3-} \longrightarrow Ca_3(PO_4)_2 \downarrow + 6OH^-$$
$$4Ca(OH)_2 + 3PO_4^{3-} + H_2O \longrightarrow Ca_4H(PO_4)_3 \downarrow + 9OH^-$$

Several calcium orthophosphate salts are known to precipitate, but only hydroxylapatite, $Ca_5OH(PO_4)_3$, and tricalcium phosphate, $CA_3(PO_4)_2$, are believed by some researchers to form under treatment plant conditions.[4] However, other work[5] indicates that the calcium phosphate solid that is formed is octacalcium phosphate, $Ca_4H(PO_4)_3$.

Normally, 70–90 percent of the phosphorus in domestic sewage is in the form of orthophosphate or polyphosphate, which may hydrolyze to orthophosphate. Organic bound phosphorus makes up the rest of the total phosphorus. Although the orthophosphate may be precipitated, polyphosphate is probably removed by adsorption on the floc resulting from the precipitation of the orthophosphate.[6] Phosphorus may also be adsorbed on the surfaces of calcium carbonate particles. The removal of phosphorus by the addition of lime is probably achieved by several mechanisms operating simultaneously.

Magnesium precipitation begins at about pH 9.5 and is essentially complete at pH 11. Good clarification is usually not achieved until pH 11 is reached. Magnesium hydroxide is a gelatinous precipitate which will remove many colloidal particles as it settles. However, its gelatinous properties adversely affect sludge thickening and dewatering.

The lime dose for a particular wastewater could be calculated if the complete wastewater composition and all reactions were known. However, as indicated in the reactions given for phosphate precipitation, the stoichiometry is not completely known. Also, the reaction rates and the other mechanisms in the clarification process, such as enmeshment and adsorption, cannot be precisely defined. As a result, calculated dosages based on known wastewater characteristics, such as alkalinity, calcium, magnesium, phosphate, and pH, invariably give lower values than those found in jar tests or plant operation. An empirical relationship between lime dose required to reach pH 11 (the pH required for good clarification and phosphorus removal) and alkalinity was published in 1969.[7] Figure 2-2 is based on this relationship and later work in Orange County, California.

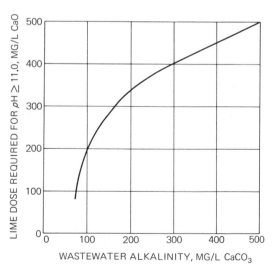

Fig. 2-2   Lime coagulation of wastewater.

The effect of chemical treatment on the mineral composition of the final effluent must be considered in some instances. The equations indicate that the addition of lime may actually effect a net reduction in total dissolved solids (TDS) concentration through precipitation of phosphorus, calcium and magnesium. However, in order to achieve good turbidity removal, it is necessary in most cases to add excess lime. Also, the efficiency of the sedimentation basin will affect the final effluent mineral concentration. It is often possible to achieve a 10–15 percent reduction in TDS with hard wastewater in an efficient chemical clarifier.

Lime Slaking and Feeding

Detailed information on the handling, application and storage of lime is available from the National Lime Association.[8]

Lime can be conveyed either mechanically, by screw conveyors or bucket conveyors, or pneumatically. Where humidity is high, air slaking of quicklime will take place in pneumatic conveyors by absorption of moisture from the atmosphere. Chutes and hopper sides for handling and storing bulk quicklime should have a minimum slope of 55 degrees to assure flow, since the angle of repose for quicklime is 30–40 degrees. Equipment for the mechanical handling of quicklime is generally constructed of mild steel, while pneumatic handling equipment is constructed of rubber and steel pipe. Steel bins and standard steel mechanical or air conveying equipment require no precautions against corrosion. Storage facilities for bulk quicklime should be airtight to prevent air slaking.

Lime slakers combine varying proportions of excess water and quicklime to produce a slurry of hydrated lime. Slakers of either detention or pug mill types are used to accomplish slaking in a plant operation. Heat is given off during slaking and the rise in temperature hastens the reaction. Each pound of calcium oxide gives off approximately 490 Btu during slaking. Thermostatic control of either type of slaker is commonly used. When a ratio of 3.5 lb of water per 1 lb of lime is used in detention slakers, and the lime contains 90 percent CaO, the temperature of the slaking mixture will rise about 115° F.

The detention slaker produces a creamy slurry of about 10 percent hydrated lime, while the pug mill type uses less water for slaking and produces a calcium hydroxide paste containing about 36 percent calcium hydroxide. This paste must be diluted as it leaves the slaking compartment to permit it to flow to the point of application. Pug mill slakers generally use a water to lime ratio of 2.5 to 1. The required slaking time varies with the source of lime. Fast slaking limes will complete the reaction, while poor quality limes may require up to 60 minutes and an external source of heat. Some investigators have reported that slaking characteristics of lime recovered and recalcined after being used for wastewater coagulation tend to deteriorate. However, this has not been observed in plant operation at South Tahoe.

Before selecting a slaker, it is advisable to determine the slaking time, best initial water temperature, and best weight ratio of water to lime for the lime to be used, according to the procedures for slaking tests recommended by the American Water Works Association. Detention slakers offer some advantages, especially in slaking slower slaking limes. Detention-type slakers will slake most limes if proper attention is given to the water to lime ratio and the temperature, and if hot water or steam are available, if needed, to raise the temperature of a slow slaking mixture. A vapor removal device is essential to prevent vapor rising into the feeder mechanism or into the slaking room. These devices usually consist of water jets which condense the vapors and wash the dust out of the air before discharge.

Lime feeders are either volumetric or gravimetric in their means of measuring the quantity of lime added. Gravimetric feeders are more accurate and are generally used in larger plants. In small plants volumetric feeders may be more economical. Both types of feeder can be arranged to feed in proportion to flow of wastewater. Lime is almost never fed in solution because of its low solubility in water. Quicklime is almost never applied dry and hydrated lime is rarely applied dry directly to the wastewater because it is more easily transported to the point of application as a slurry. "Dissolvers" for dry lime feeders are usually designed to provide 3–5 minutes detention when forming a 6 percent slurry (0.5 lb/gal).

Usually the slurry that is ready for use must be transferred at least a short distance to the rapid mix basin. It is here that serious problems from scaling can occur, and there is no foolproof solution. The following methods are used to avoid or lessen this:

1. The best method is to locate the feeder so that the slurry can flow by gravity a short distance directly into the rapid mix basin without handling. This may not be possible, particularly with complex existing plants.
2. Scale at the termination of the transfer line can be avoided by discharging the slurry through an air gap.
3. Heavy duty, flexible rubber hoses can be installed in lieu of metal pipe. Hose is easily disengaged and the scale broken by flexing the pipe and flushing. This method may require duplicate hose lines so that one is in use while the other is being cleaned.
4. Use of open troughs to convey lime suspensions instead of pipes, when possible, simplifies scale removal.

Alum Coagulation

Aluminum sulfate (alum) produced commercially for wastewater treatment has the formula $Al_2(SO_4)_3 \cdot 14H_2O$. Alum is produced by reacting bauxite ore or clays, which are rich in aluminum oxide, with sulfuric acid. Alum is available in either liquid or dry form and its characteristics are summarized in Table 2-1. The American Water Works Association Standard (AWWA B403-70) is a helpful guide for the use of alum in wastewater treatment.

The reactions of alum with alkalinity and phosphate in water are represented by the following equations:

$$Al_2(SO_4)_3 \cdot 14H_2O + 6HCO_3^- \longrightarrow 2Al(OH)_3\downarrow + 3SO_4^{2-} + 6CO_2 + 14H_2O$$

$$Al_2(SO_4)_3 \cdot 14H_2O + 2PO_4^{3-} \longrightarrow 2AlPO_4\downarrow + 2SO_4^{2-} + 14H_2O$$

The reactions that actually occur in water are more complex than these equations indicate. The equations do, however, indicate rough proportions and show that sulfate ion and carbon dioxide are added to the water.

According to these equations, 3.13 mg/l of alum per mg/l PO_4^{3-} is required to precipitate 1.29 mg/l $AlPO_4$; and each mg/l of alum decreases the HCO_3 concentration by 0.62 mg/l while precipitating 0.26 mg/l of $Al(OH)_3$. Also, each mg/l of alum added increases in SO_4^{2-} concentration by 0.48 mg/l and the CO_2 concentration by 0.44 mg/l. The Al:P mole ratio is 1:1 and the weight ratio is 0.87:1.

Stumm and Morgan[9] calculated solubilities of $AlPO_4$ at various pH values, neglecting the possible influences of complex formation, as follows:

pH	$AlPO_4$ solubility; mg/l
5	0.3
6	0.01
7	0.3

The hydrolysis of polyvalent metal ions in solution has been studied by several researchers. The hydrolysis products, as opposed to the metal ions themselves, are now believed to be of prime importance in coagulation. The hydrolysis of iron and aluminum ions to yield a variety of hydrolysis produces may be represented as follows:[1]

$$Al^{3+} + H_2O \rightleftharpoons AlOH^{2+} + H^+$$
$$Al^{3+} + 2H_2O \rightleftharpoons Al(OH)_2^+ + 2H^+$$
$$Al^{3+} + 3H_2O \rightleftharpoons Al(OH)_3 + 3H^+$$
$$Al^{3+} + 4H_2O \rightleftharpoons Al(OH)_4^- + 4H^+$$

There is general agreement that the settleable or filterable precipitates formed are hydrated complexes that have absorbed or chemically combined with impurities to be removed.

As a result of hydrolysis and other reactions, higher than stoichiometric quantities of alum are necessary in wastewater treatment. Jar tests with the wastewater to be treated are necessary to determine required alum dosages. The following ratios of alum (9.1 percent Al) are reported to be representative for alum treatment of municipal wastewater and may be used for initial estimating purposes:[10]

Percentage Phosphorus Reduction	Al:P	Weight Ratio Alum:P	Alum:PO_4^{3-}
75	1.2:1	13:1	4.2:1
85	1.5:1	16:1	5.3:1
95	2.0:1	22:1	7.2:1

Using these guidelines, the alum dose required to remove 95 percent of the phosphorus from wastewater containing 11 mg/l P (34 mg/l PO_4^{3-}) is about 240 mg/l.

Alum Storage and Feeding

Alum is available in either dry or liquid form. It is available dry in ground, powder, or lump forms shipped in bags, barrels, or carloads. The ground form is used for dry feeding; it weighs 60–75 lb/ft^3 and is only slightly hygroscopic. The liquid form is a 50 percent alum solution and is usually delivered in minimum loads of 4000 gal. The liquid form may be delivered by rail or truck. It must be stored and conveyed in corrosion resistant materials such as rubber lined steel, fiberglass, or stainless steel. The economic merits of dry versus liquid alum will depend primarily on local freight costs. The advantages of the liquid form are easier handling and feeding.

The following criteria are given as guidelines for alum storage and feeding:

Dry Alum

1. The storage space should be free of moisture, since dry alum becomes corrosive when it absorbs moisture.
2. Pallets should be provided for handling bags of alum. Ground and rice grades are the most commonly used grades in wastewater treatment.
3. Storage bins do not require agitation.
4. Respirators should be provided to personnel for protection against alum dust.

5. Storage bins may be constructed of mild steel or concrete. Dust collectors and vents should be provided.
6. Screw conveyors, pneumatic conveyors, or bucket elevators should be constructed of mild steel.
7. Storage bins should have at least an 8 hour capacity so that dry alum need not be added to the bin more than once per shift.
8. Bin bottoms should have a minimum slope of 60 degrees to prevent arching.
9. Facilities must be provided for isolating the feed equipment from the storage bin so that maintenance of feed equipment can be readily carried out.
10. Dissolving chambers must be corrosion resistant. Suitable materials include 316 stainless steel and fiberglass reinforced plastic. Dissolvers should be sized for a 6 percent alum solution (0.6 lb alum/gal water). The dissolving chamber should have a minimum detention time of 5 minutes at maximum feed rate and be equipped with water meters so that the alum to water ratio may be controlled.
11. Piping and valving for transport of dissolved alum must be corrosion resistant. Suitable materials include fiberglass, PVC, polyethylene, polypropylene and rubber lined pipe.

Liquid Alum

1. Storage tanks must be maintained at temperatures above 18°F to prevent crystallization. Tanks must be corrosion resistant. Suitable materials include fiberglass, 316 stainless steel, and steel lined with rubber or PVC.
2. Liquid alum should not be diluted in storage or feeding. Commercial liquid alum is commonly 8.3 percent Al_2O_3 or 48.3 percent $Al_2(SO_4)_3 \cdot 14H_2O$ (5.4 lb dry alum per gallon).
3. Hydraulic diaphragm pumps constructed of resistant material are preferable for feeding. Relief valves should be provided to protect the pump.
4. Piping material must be corrosion resistant; suitable materials include fiberglass, plastics and 316 stainless steel.

Ferric Chloride Coagulation

The reactions of ferric chloride with alkalinity and phosphate are represented by the following equations:

$$2FeCl_3 + 3Ca(HCO_3)_2 \longrightarrow 2Fe(OH)_3 \downarrow + 3CaCl_2 + 6CO_2$$
$$FeCl_3 + PO_4^{3-} \longrightarrow FePO_4 \downarrow + 3Cl^-$$

According to these equations 1 mg/l $FeCl_3$ reacts with 1.13 mg/l HCO_3 and produces 0.66 mg/l $Fe(OH)_3$; and 1 mg/l $FeCl_3$ reacts with 0.58 mg/l PO_4 to produce 0.93 mg/l of $FePO_4$. The reactions with iron are similar to those discussed for alum in that the actual reactions are more complex and the hydrolysis products are important in the clarification process.

Stumm and Morgan[9] calculated the solubility of $FePO_4$ as follows:

pH	Approximate $FePO_4$ solubility, mg/l
5	0.1
6	2
6.5	10

The experience with iron salts in AWT applications is quite limited and the actual dosage of iron required should be determined from jar tests. The small size of the $FePO_4$ particle requires a sufficient excess of ferric ion for the formation of a hydroxide precipitate, which probably enmeshes some of the $FePO_4$ particles and acts as an adsorbent for other phosphorus compounds. Experience has shown that efficient phosphorus removal requires the stoichiometric amount of iron (1.8 mg/l Fe per mg/l P) to be supplemented by at least 10 mg/l of iron for hydroxide formation. The iron requirements reported for municipal wastewaters to provide phosphorus reductions of 85–90 percent vary from 40 to 200 mg/l and the use of polymers is required in some cases.

Ferric Chloride Storage and Feeding

The following criteria are provided as guidelines for storing and feeding liquid ferric chloride:

1. Feeding equipment must be corrosion resistant. The criteria for liquid alum also apply to ferric chloride.
2. Storage tanks must be corrosion resistant. Suitable materials include fiberglass and rubber or plastic lined steel.
3. Temperatures which produce crystallization vary with the strength of solution purchased and must be carefully checked during design.
4. Piping should be rubber or Saran lined steel, hard rubber, fiberglass, or plastics. Valves should be rubber or resin lined diaphragm, Saran lined with Teflon diaphragms, rubber sleeved pinch-type or plastic ball valves.
5. Glass rotameters should not be used in the feed system because ferric chloride will stain them.

Other Chemicals for Coagulation

Other aluminum and iron salts may be used as primary coagulants. Sodium aluminate ($NaAlO_2$) is an alternative source of aluminum ion which is usually a poor coagulant in soft wastewaters and a fair to good coagulant in hard waters. Parallel jar tests of sodium aluminate and alum will determine if sodium aluminate is an attractive alternative. Iron sulfates are an alternative source of iron ion. Ferric sulfate is available in granular form in bags, drums, and bulk, as $Fe_2(SO_4)_3 \cdot 3H_2O$ or $Fe_2(SO_4)_3 \cdot 2H_2O$. It is normally fed in dry form. The material is hygroscopic and must be stored in tight containers. It weighs 70–72 lb/ft^3. Ferrous sulfate ($FeSO_4 \cdot 7H_2O$), also referred to as copperas, is available in granular form in bags, barrels, and bulk. It weighs 63–66 lb/ft^3 and is normally fed dry.

Polyelectrolytes, or polymers, may be used in AWT plants as primary coagulants, as flocculation aids, as filter aids, or for sludge conditioning. The discussion at this point will be limited to the use of polymers as coagulants or flocculation aids. Their use for other purposes will be discussed later in connection with the appropriate subject.

In treatment of certain wastewaters, lime, alum or iron alone, or in combination, may produce a fine or light floc which settles very slowly. If so, there is probably a polymer which will produce a rapidly settling floc. There are certain general guidelines to follow in choosing polymers which may be useful, but the final selection is a matter of trial and error. At South Tahoe, Purifloc N-11, Calgon ST-270, and others, have been useful as flocculation aids in lime treatment at pH 11.0. At Orange County, several anionic coagulation–flocculation aids (Dow A-23, American Cynamide 835 A, and Calgon WT 2700) produced a larger, faster settling floc but did not measurably improve turbidity or COD removal efficiencies.

The usual dose for polymers used as an aid is about 0.10–0.25 mg/l. It is possible to overdose, and this may occur in the 1.0–2.0 mg/l range. Addition of the proper amount of polymer at the right point in treatment can improve removal of both turbidity and phosphorus. Jar tests are of some value for preliminary screening, but plant scale tests must be employed for final selection and for determining optimum dosage rates.

Some polymers sold as a dry powder require special procedures for preparation of water solutions. Specific instructions may be obtained from the supplier, but in general the following steps are necessary: thoroughly wet the polymer powder by means of a funnel-type aspirator, add warm water, and mix, usually for about an hour, with gentle, slow stirring until all of the polymer is in solution. Polymer feed solution strengths are usually in the range of 0.2–2.0 percent. Stronger solutions are often too viscous to feed. The solution can be fed by means of positive displacement metering pumps. Typical equipment layouts for polymer mixing and feeding are pictured in Chapter 4.

The addition of oxidizing agents improves chemical clarification of some wastewaters. During pilot studies in Orange County using alum to treat secondary effluent, it was found that the addition of 20 mg/l chlorine improved COD and turbidity removal. Other studies[11] have indicated that the use of ozone may substantially reduce the alum dosage required for turbidity and color removal from secondary effluent.

FLOCCULATION

The purpose of flocculation is to increase the collisions of coagulated solids so that they agglomerate to form settleable or filterable solids. Flocculation is accomplished by prolonged agitation of coagulated particles in order to promote an increase in size and/or density. The currently accepted theory states that the rate at which particles of different size collide when subjected to mixing is proportional to (1) the number of particles of each size, (2) the size of the particles, and (3) the local fluid velocity gradient in the suspension. This theory is described by the following equation:

$$N = \tfrac{1}{6} n_1 n_2 (d_1 + d_2)^3 \frac{dv}{dy}$$

where

N = the frequency of collision per unit volume of suspension between particles 1 and 2

n_1, n_2 = the numbers of 1 and 2 particles per unit volume of suspension.

d_1, d_2 = the respective diameters

$\dfrac{dv}{dy}$ = the local fluid velocity gradient in the suspension

In a continuously flowing or mixed tank, the local velocity gradient is not constant and so dv/dy must be approximated by the temporal mean velocity gradient, G, for the entire tank. G has the dimensions ft/sec/ft or sec^{-1}. Camp and Stein[12] defined the mean velocity gradient as follows:

$$G = \left(\frac{P}{\mu V} \right)^{1/2}$$

where

P = power input to the water, ft-lb/sec

μ = absolute viscosity of water in lb-sec/ft^2

V = volume of basin in ft^3

Camp[13] also presented information on the following relationships for G values in mechanically stirred flocculation basins:

$$G = \left(\frac{C_D A v_R}{2\zeta V}\right)^{1/2}$$

where

C_D = drag coefficient
A = area of paddles, ft^2
v = velocity of paddles relative to liquid, ft/sec
ζ = kinematic viscosity of fluid, ft^2/sec
V = volume of basin, ft^3

The value of the relative velocity v_R is difficult to determine and is usually assumed for design to be 50–75 percent of the paddle velocity. The coefficient of drag for rectangular plates moving face-on to the fluid is about 1.8.

Another parameter which has been used in flocculation basin design is Gt. Gt is dimensionless, since G is multiplied by the detention time t in the flocculation basin. G values for wastewater treatment have varied between 10 and 200 ft/sec/ft, with Gt ranging between 10,000 and 100,000.

Flocculation can be carried out in a separate basin or in an integral part of the clarifier structure. When carried out in a separate basin, the velocity in the conduits conveying the floc to the clarifier should be 0.5–1.0 ft/sec to prevent disruption of the floc. The velocity gradients necessary for flocculation can be induced by mechanical means, such as revolving paddles, or by air diffusion. The basic equipment is essentially the same as that used for many years for flocculation in water treatment plants. A typical paddle-type flocculator is shown in Figure 2-3.

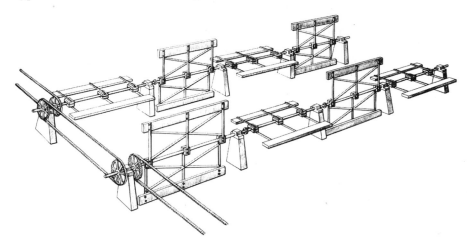

Fig. 2-3 Typical paddle-type flocculator. (*Courtesy Dorr-Oliver*)

The time required for flocculation varies depending upon wastewater characteristics. For example, lime coagulated activated sludge effluent at South Lake Tahoe is easily flocculated in 5 minutes in an air agitated basin (and good flocculation is obtained even without agitation). On the other hand, lime coagulated trickling effluent at Orange County is difficult to flocculate and requires 20–30 minutes of careful mixing to form a good settling floc.

Several flocculation basins in series will give better performance than one large basin of equal total volume. Multiple basins provide better mixing and enable mixing energy to be varied from one basin to another. The basin in Orange County is an example of the multiple basin concept and is discussed later in this chapter with the case studies.

SEDIMENTATION

Sedimentation in water treatment is often considered in terms of ideal settling. An ideal settling basin is divided into four zones, as illustrated in Figure 2-4: inlet, outlet, settling, and sludge zones.

Ideal settling theory results in the following equation for surface loading or overflow rate:

$$V_0 - \frac{Q}{A}$$

where

V_0 = settling velocity
Q = flow through the basin
A = surface area of the basin

Sedimentation basin loadings (Q/A) are often expressed as gpd/ft^2 (gallons per day per square foot). Thus, under ideal settling conditions, sedimen-

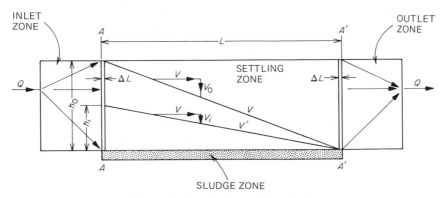

Fig. 2-4   Ideal sedimentation basin.

tation is independent of basin depth and detention time, and depends only on the flow rate, basin surface area and properties of the particle. Ideal settling assumes the following:

1. All particles settle discretely.
2. Particles which strike the bottom remain in the sludge zone.
3. Water and particulates are distributed uniformly over a vertical plane.
4. Incremental volumes of water move from inlet to outlet without changing shape.
5. Inlet and outlet conditions do not affect settling.

Sedimentation basins used in the chemical treatment of wastewaters will not perform in accordance with ideal settling theory for the following reasons:

1. Discrete particle settling is not obtained. Flocculation of particles occurs during settling.
2. Inlet and outlet conditions are not ideal and do affect settling.
3. Basin currents cause nonuniform flow and bottom scour.

Therefore, in the clarification of wastewater, removal is dependent on the basin depth as well as on flow, basin surface area, and properties of the particle. The performance of a sedimentation basin on a suspension of discrete particles can be calculated, but it is not possible to calculate sedimentation basin performance for a suspension of flocculating particles, such as a wastewater, because settling velocities change continually. Settling analyses, however, may be performed to predict sedimentation basin performance. Settling tests were described by Camp[14] in 1946 and have been detailed more recently by Zanoni and Bloomquist.[15]

Important sedimentation basin design considerations are inlet conditions, outlet conditions, and sludge removal. Basin design should strive to obtain ideal conditions, i.e., uniform distribution of flow at the inlet and outlet, and sludge removal conditions that minimize flow short circuiting and disruption. The following are typical design criteria used in chemical clarifier sedimentation basins:

Overflow rate: 500 to 2000 gpd/ft^2
Detention time: 1–4 hours
Depth: 7–15 ft
Velocity: 0.5–3 ft/minute through basin
Velocity, influent channel: 0.5–2.0 ft/sec
Weir rate: 5–35 gpm/ft (gallons per minute per foot)

Among the general types of clarifiers available for gravity sedimentation of settleable solids are the following:

1. Horizontal flow basins with external flocculation.

2. Horizontal flow basins with internal flocculation (usually circular).
3. Upflow basins with internal rapid mixing and flocculation (usually circular).
4. Solids contact units with recirculation of settled solids.

Several manufacturers offer equipment in the above categories with several variations on basin inlet and outlet systems, recirculation of the sludge, and control of the sludge blanket level available. Detailed literature describing each type of equipment is readily available from manufacturers and no attempt will be made here to review all available equipment. Rather, the experiences gained in field application of the general clarifier types available for advanced waste treatment processes will be examined. Only applications will be included which involve separation of floc resulting from chemical coagulation of sewage, since performance of conventional primary and secondary sewage clarifiers has been reviewed in detail in other publications (ASCE Manual No. 36).

Most field experience indicates that surface overflow rates recommended for application of specific equipment to removal of chemical floc in water treatment applications must be reduced for removal of floc from chemically coagulated sewage. Also, maintenance of sludge blankets in upflow basins has proven difficult in these applications.

Rose[16] reports that an overflow rate of 720 gpd/ft^2 should not be exceeded for the removal of alum floc from coagulated secondary effluent. Rose evaluated a circular clarifier with internal flocculation equipment normally used as a sludge blanket unit of the solids contact type in water treatment applications at overflow rates in excess of 1400 gpd/ft^2. However, Rose found that halving of the rate was required and that maintenance of a sludge blanket was not feasible. He reports the alum floc in sewage to be quite light and fragile and that polymers may be required as settling aids for operation even at 720 gpd/ft^2 to prevent massive overflow of floc. Rose emphasizes the importance of removing sludge at a rate adequate to prevent septicity of the sludge from occurring. He found that a sludge removal rate of 3–5 percent of the throughput was required to prevent septicity.

O'Farrell and co-workers[17] also report occasional difficulty with a solids-contact-type clarifier applied to lime coagulated secondary effluent. They report that the slurry pool in the clarifier was often unstable and expanded into a sludge blanket which would periodically overflow the clarifier. Septic solids occasionally forced shutdown of the clarifier. The chief difference between stable and unstable sludge blankets was reported to be the ratio of calcium carbonate to noncarbonate solids in sludge. Calcium carbonate concentrations in excess of 75 percent maintained a stable slurry pool. A sludge withdrawal rate of 0.5 percent of the total flow was adequate with a

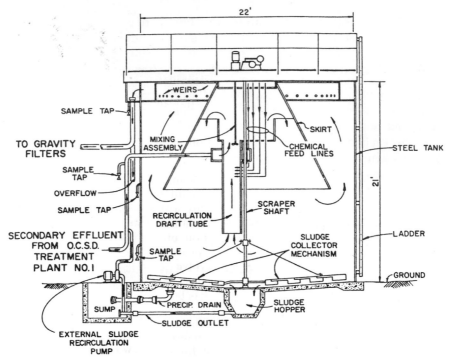

Fig. 2-5   Solids contact clarifier unit, Orange County, California.

OPERATING CHARACTERISTICS:

1. NOMINAL FLOW RATE	450 G.P.M.
2. MIXING SPEED (SINGLE SPEED)	350 R.P.M.
3. SLUDGE SCRAPER SPEED	8.3 R.P.M.
4. SLUDGE RECIRCULATION RATIO	6:1

stable slurry, and a sludge concentration of 10–12 percent solids was maintained. With a stable slurry, the clarifier produced effluent turbidities of 3 to 6.5 JTU (Jackson turbidity units) and a phosphorus concentration of about 0.5 mg/l as P at an overflow rate of 1500 gpd/ft^2 with a water temperature of 25°C. A decrease in wastewater temperature to 17°C caused the effluent turbidity to increase to 16 JTU and the phosphorus to 1.8 mg/l at the same overflow rate. The quality of the secondary effluent being coagulated and settled also worsened at the lower temperature.

Convery[18] recommends an overflow rate of 700 gpd/ft^2 for removal of alum floc from coagulated sewage. He reports that basin performance continued to improve as the rate was decreased to 300 gpd/ft^2 but that a rate of 700 gpd/ft^2 represented the best balance between basin performance and economic considerations. Weber and co-workers[19] also report successful operations at 700 gpd/ft^2 for separation of lime floc from coagulated raw sewage.

Data presented by Kalinske and Shell[20] also confirm that higher over-flow rates are permissible when using lime rather than alum. They suggest a design rate (solids contact unit) of 1800 gpd/ft^2 when using lime and 1200 gpd/ft^2 when using alum. Substantial polymer doses (0.25 mg/l) were used as settling aids in order to achieve these rates in a 25 gpm pilot unit. A rate of 1400 gpd/ft^2 was suggested for lime coagulated raw sewage. Plant scale experiences of others indicate that these overflow rates may be difficult to consistently maintain in a large scale clarifier.

A sludge-blanket-type clarifier illustrated in Figure 2-5 was used to treat secondary effluent in Orange County, California, at a rate of about 0.65 mgd (million gallons per day). A stable sludge blanket and consistently good effluent quality could not be maintained while operating this unit with alum as the coagulant.

Shallow Depth Sedimentation

There have been several attempts in the past to apply shallow depth sedimentation principles to operating basins. Most of these attempts involved wide, shallow trays, generally inserted within basins of conventional design. However, they met with only limited success because of two major problems: (1) unstable hydraulic conditions are encountered with very wide, shallow trays; and (2) the minimum tray spacing was limited by the vertical clearance required for mechanical sludge removal equipment. Both of these problems have been overcome by the development of tubes, relatively small in cross-sectional area, rather than wide, shallow trays. Flow through tubes with an area of a few inches offers theoretically optimum hydraulic conditions for sedimentation, and overcomes the hydraulic problems associated with tray settling basins.

Two basic configurations of settling tubes have been used. Essentially horizontal tube settlers have been used in combination with a filter system in small treatment plants. For larger treatment plants, or the expansion of existing plants, steeply inclined tube settlers have been utilized. The theoretical basis of the tube settling concept is discussed by McMichael[21] and Yao.[22] There is general agreement that the tube settler concept offers a theoretically sound basis for operating clarifiers at surface loading rates substantially (2 to 4 times) higher than in deep, conventional basins.

The two basic shallow depth settling systems now commercially available are illustrated in Fig. 2-6.[23,24] These configurations are (1) essentially horizontal or (2) steeply inclined. The operation of the essentially horizontal tube settlers is coordinated with that of the filter following the tube settler. The tubes essentially fill with sludge before any significant amount of floc escapes. Solids leaving the tubes are captured by the filter. Each time the filter backwashes, the settler is completely drained. The tubes are

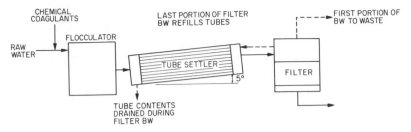

(A) ESSENTIALLY HORIZONTAL TUBE SETTLER

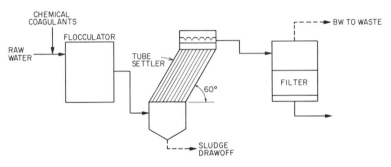

(B) STEEPLY INCLINED TUBE SETTLER

Fig. 2-6   Basic tube settler configurations, shown schematically. (*Courtesy Neptune Microfloc, Inc.*)

inclined only slightly in the direction of normal flow (5 degrees) to promote the drainage of sludge during the backwash cycle. The rapidly falling water surface scours the sludge deposits from the tubes and carries them to waste. The water drained from the tubes is replaced with the last portion of the filter backwash water, so that no additional water is lost in the tube draining procedure. This tube configuration is applicable primarily to small plants (1 mgd or less in capacity) and is often used in package plant systems.

Sediment in tubes inclined at angles in excess of 45 degrees does not accumulate but moves down the tube to eventually exit at the tube bottom. A flow pattern is established in which the settling solids are trapped in a downward-flowing stream of concentrated solids, as illustrated in Figure 2-7. The continuous sludge removal achieved in these steeply inclined tubes eliminates, in most cases, the need for drainage or backflushing of the tubes for sludge removal. The advantage of shallow settling depth coupled with that of continuous sludge removal extends the range of application of this principle to multi-mgd capacity installations.

Various manufacturers have developed alternative methods for incorpo-

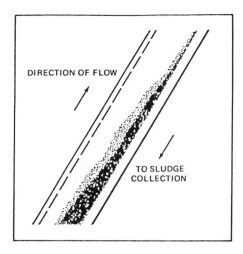

DIRECTION OF FLOW

TO SLUDGE
COLLECTION

Fig. 2-7   Flow of liquid and sludge in steeply inclined tubes.

rating steeply inclined tubes into a modular form which is economical to build and can be easily supported and installed in a sedimentation basin. One modular construction is shown in Figure 2-8 (Neptune Microfloc) in which the material of construction is normally PVC and ABS plastic. Extruded ABS channels are installed at 60 degrees inclination between thin sheets of PVC. By inclining the tube passageways rather than inclining the entire module, the rectangular module can be readily installed in either rectangular or circular basins. Alternating the direction of inclination of each row of the channels forming the tube passageways enables the module to become a beam which needs support only at its ends. The rectangular, tubular passageways are 2 in. × 2 in. square in cross section and normally are 24 in. long.

Another manufacturer, Permutit, has adopted a chevron shaped tube cross section, as shown in Figure 2-9. The manufacturer believes that the V-groove in the bottom of the tube improves the hydraulic characteristics and enhances the counterflow characteristics of the sludge.

A third manufacturer, Graver Water Conditioning Company, utilizes rectangular channels similar to that shown in Figure 2-8 but with all of the channels inclined in the same direction.

The Reynolds number for water at 70°F and a loading of 2000 gpd/ft^2 on the 2 in. × 1 in. square tube is about 43. Similar conditions for the chevron configuration shown in Figure 2–9 gives a Reynolds number of 19. Both of these values indicate that flow through the tubes is well within the laminar range, and it is doubtful whether the difference is significant for hydraulic conditions.

In these three settling tube systems, the flow passes upward through the

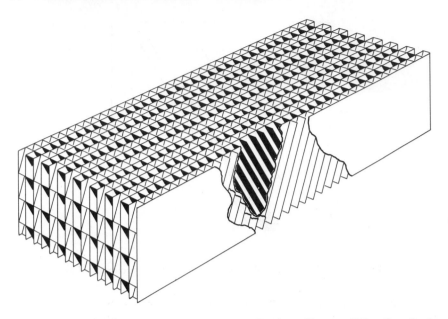

Fig. 2-8   Module of steeply inclined tubes. (*Courtesy Neptune Microfloc, Inc.*)

tubes. Another concept, recently marketed by the Parkson Corporation, is the Lamella separator, in which the flow is downward through a series of parallel plates, as illustrated in Figure 2-10. The sludge is collected at the bottom of the basin, with the sludge water flow being in the same direction rather than countercurrent. The clarified water is conveyed to the top of the basin by return tubes, as shown in Figure 2-10. The plates are typically 5 ft wide by 8 ft long, spaced 1.5 in. apart, and are inclined at 25–45 degrees. They are usually constructed of PVC.

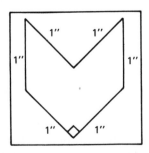

Fig. 2-9   Cross section of a chevron shaped tube. (*Courtesy Permutit*)

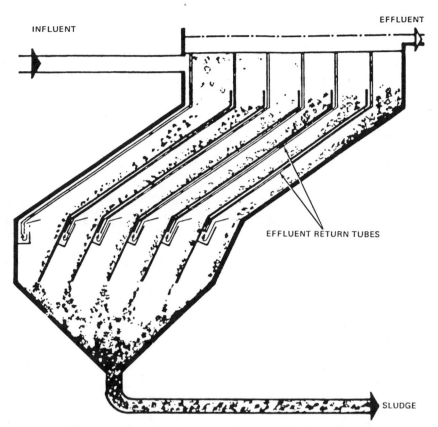

Fig. 2-10   Lamella separator. (*Courtesy Parkson Corp.*)

Steeply inclined tubes can be used in either upflow solids contact clarifiers or horizontal flow basins to improve performance and/or increase capacity of existing clarifiers. Of course, they can also be incorporated into the design of new facilities to reduce their size. Capacities of existing basins can usually be increased by 50–150 percent with similar or improved effluent quality. The overflow rate at which a basin equipped with tubes can be operated is dependent upon the design and type of sludge removal equipment, the character of the wastewater being treated, and the required effluent quality.

Figure 2-11 illustrates a typical cross section of a rectangular sedimentation basin design used for removal of chemical floc. The extensive effluent launder system is designed to insure adequate flow distribution through the tubes. With the exception of the added launders and the need for tube supports, the basin design would follow conventional practice.

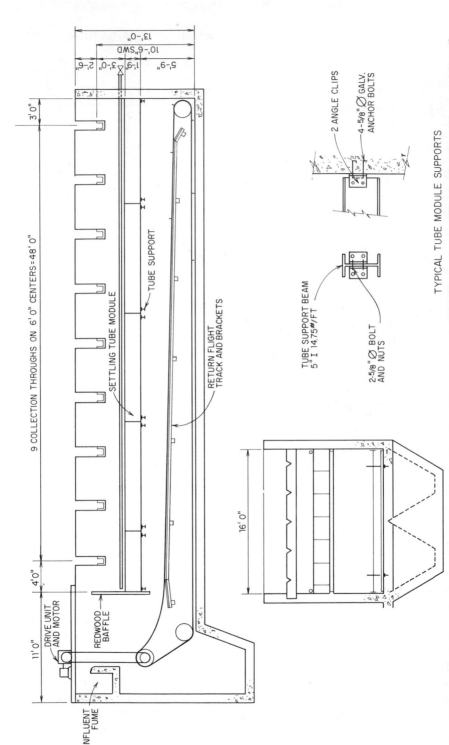

Fig. 2-11 Cross section of typical tube clarifier for clarification of chemically coagulated secondary effluent. (*Courtesy Neptune Microfloc, Inc.*)

Fig. 2-12   Tube modules being installed in rectangular clarifier. (*Courtesy Neptune Microfloc, Inc.*)

Fig. 2-13   Settling tube module being placed on support beam. (*Courtesy Neptune Microfloc, Inc.*)

Of course, the sludge removal system would have to be sized to handle the greater quantity of sludge than would occur in a conventional basin of the same size. Actual installation of settling tube modules is shown in Figures 2-12 and 2-13.

Tube Cleaning

In certain wastewaters, floc has a tendency to adhere to the upper edges of the tube openings. This is generally an infrequent occurrence and of no serious consequence other than detracting from the appearance of the installation. In some cases, the floc buildup eventually bridges the tube openings and results in a blanket of solids on top of the tubes which may reach 3–10 in. in depth unless some remedial action is taken. One method of removing this accumulation is to lower the water level of the basin occasionally to beneath the top of tubes. The floc particles are dislodged and fall to the bottom of the basin.

In some cases, it may not be possible to remove the basin from service and a mechanical cleaning system will be required. It is possible to use either a water or air wash system. A water wash system is illustrated in Figure 2-14; the general design guidelines are as follows:

1. Provide water distribution headers at intervals not exceeding 20 ft.

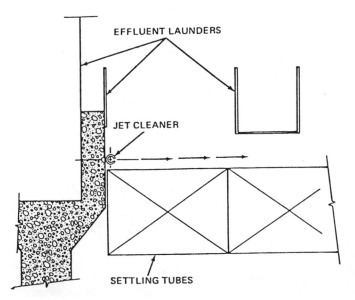

Fig. 2-14   Water wash settling tube cleaning system.

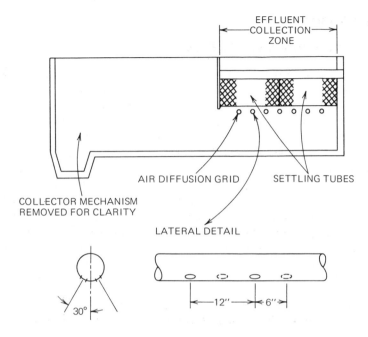

Fig. 2-15   Air wash settling tube cleaning system.

2. Provide nozzles at 1 ft centers.
3. Provide water supply at 60 psi and about 6 gpm per nozzle.
4. Provide valving so that cleaning system can be operated in sections in order to reduce water supply required.

An air wash system is shown in Figure 2-15 and general design guidelines include:

1. Provide air at a pressure of 6 psi and a rate of 1 cfm/ft^2 (cubic feet per minute per square foot) of tube surface area.
2. Provide air distribution laterals 1 ft below bottom of tubes and space at 1 ft centers.
3. Provide $\frac{1}{16}$ in. diameter orifices at 1 ft centers, 30 degrees from bottom centerline of lateral and staggered as shown in Figure 2-15.
4. Provide valving so that system can be operated in sections in order to reduce air supply required.

The cleaning system is normally operated for about 3 minutes with no flow through the basin during the cleaning period.

CASE EXAMPLES

South Tahoe Public Utility District

The South Tahoe AWT plant treats activated sludge effluent at a design average flow rate of 7.5 mgd. The chemical clarification system consists of separate rapid mix, flocculation, and sedimentation basins. The design criteria are summarized in Table 2-2 and the typical operating efficiency is shown in Table 2-3.

Duplicate equipment is provided for storing, feeding and slaking commercial quicklime and in-plant recalcined quicklime. Facilities are provided to feed up to 400 mg/l of lime as CaO to a maximum flow of 7.5 mgd.

Fresh, or makeup, lime is unloaded in bulk from delivery trucks using a pneumatic conveying system which has a rated capacity of 10 ton/hour and is discharged into a 35 ton capacity overhead steel storage bin. Recalcined lime is conveyed from the lime furnace to a second 35 ton bin by a separate pneumatic conveying system having a capacity of 0.75 ton/hour.

TABLE 2-2   Chemical Clarifier Design Criteria, South Tahoe Public Utility District. (*Design average flow, 7.5 mgd*)

Rapid Mixing
>1 basin with mechanical mixer
>Detention time: 30 sec

Flocculation
>1 basin with air mixing
>Detention time: 4.5 minutes

Sedimentation
>1 circular basin
>Diam: 100 ft
>Sidewater depth: 10 ft
>Surface overflow rate: 950 gpd/ft^2

Lime System
>Storage: Two 35 ton bins
>Feeders and slakers: 2 gravimetric feeders and paste-type
>   slakers @ 1500 lb/hour each

Polymer System
>Storage: Two 500 gal tanks
>Pumps: 4 @ 50 gph each

Alum System
>Storage: Two 5000 gal tanks
>Pumps: 2 @ 50 gph each

TABLE 2-3   Typical Efficiency of Chemical Clarifier, South Tahoe Public Utility District

Parameter	AVERAGE CONCENTRATIONS (mg/l)			Percentage removed
	Sec. Eff.	Recycle	Chem. Eff.	
BOD				
Soluble	15.5	9.3	5.1	54.5
Particulate	23.8	12.8	3.8	80.5
Total	39.3	22.1	8.9	73.5
COD				
Soluble	44.3	23.2	30.8	26.4
Particulate	44.6	29.5	3.8	91.0
Total	88.9	52.7	34.6	57.8
Total phosphorus[a]				
Soluble	10.33	0.53	0.03	99.7
Particulate	0.97	6.28	0.68	67.8
Total	11.30	6.81	0.71	93.7
Ortho phosphorus				
Soluble	8.79	0.35	0.04	99.4
Particulate	1.81	5.49	0.62	73.4
Total	10.60	5.84	0.66	93.4
Total SS	38	166	38	51.5
Organic SS[b]	33	29	6	89.2
Turbidity (JTU)	16	69	9	70.4
Color[c] (cobalt std.)	33	7	11	
Alkalinity, as $CaCO_3$	183	250	249	
Hardness, as $CaCO_3$	73	152	145	
pH	7.2	7.7	11	

[a]Acid digestion
[b]Pretreatment of sample to pH 2 with HCl
[c]Filtered at 0.45 microns

The piping to the lime bins is arranged so that either bin may be used to receive fresh or recalcined lime, if the need arises. Each overhead lime storage bin is located directly above a gravimetric belt-type lime feeder, which discharges to a paste-type (pug mill) lime slaker. Each unit has a capacity of 1500 lb of CaO per hour. Ordinarily, about 25 percent fresh makeup lime and 75 percent recalcined lime is fed to maintain the pH at 11.0 in the rapid mix basin. For the Tahoe wastewater, this requires a total of about 300 mg/l of quicklime. Also, about 0.25 mg/l of a polyelectrolyte (Calgon ST 2700 or Dow N-11) is usually added.

A 6 in. asbestos cement pipeline is used to convey the lime slurry by gravity flow over a distance of 300 ft to the rapid mix basin. This line is cleaned once each week by use of a high pressure hydraulic sewer cleaner.

A series of manholes on the lime feed line located about 50 ft apart provide easy access for this cleaning operation. The original feed line consisted of a 3 in. diameter rubber hose placed within the 6 in. asbestos cement carrier line. Two complete sets of hose with quick couplings located at each manhole were provided, which allowed each length of hose to be removed for cleaning as it was replaced with the clean duplicate length. This was a satisfactory arrangement, but less cleaning time is required using the hydrocleaning system for the larger line.

The rapid mix basin provides about one minute of mixing at design flow. A vertical shaft mechanical mixer is used. Lime deposits build up on the metal mixer paddles, which require frequent cleaning to avoid shaft vibration. Alum, polymer, or recycled settled lime solids may also be added in the rapid mix basin. Ordinarily no polymer is added in the rapid mix, but rather it is added between the flocculation and sedimentation basins.

Following rapid mixing, five minutes of flocculation is provided by air agitation, if desired. However, experience in operation at this plant indicates that better treatment is obtained without this additional mixing.

Lime feed rates are usually controlled by manual setting of the gravimetric feeders. This manual control has proven quite satisfactory. Automatic control is also available, but is seldom used. The automatic control system consists of a pH electrode in the rapid mix basin with a signal transmitted to an indicating-recording-control unit located near the lime feeders. This pH signal can be used to control the lime feed rate to maintain any desired pH in the rapid mix basin by making the appropriate setting on the controller. This control works satisfactorily except that the pH in the rapid mix basin is always just slightly above or below the set point and never exactly on it except for a very short time, because of time lag in the control and feed system. The operators prefer to use the manual control.

The sedimentation basin is a conventional center-inlet circular basin (Dorr-Oliver) 100 ft in diameter by 10 ft side water depth. Basin overflow rate at design flow is 950 gpd/ft^2, and the weir rate is 24,000 gpd/ft^2. The settled lime sludge withdrawn from the basin contains 0.5–2.0 percent solids. There are two sludge pumps. One is a variable speed drive centrifugal pump rated at 500 gpm. The other is a positive displacement pump (Moyno) with a capacity of 100 gpm. Both pumps discharge to a thickener, or, alternatively, to the primary clarifier. About 30 gpm of sludge is recycled to the rapid mix basin. The lime sludge suction line beneath the concrete bottom of the clarifier is glass lined pipe. The discharge line is cast iron pipe, which is cleaned about once a year by the use of a polyurethane pig.

A small dose of alum, in the range of 1–20 mg/l, is always added as a filter aid to the mixed media filter influent. In addition to this one regular point of application, alum may be added if the need arises at several other

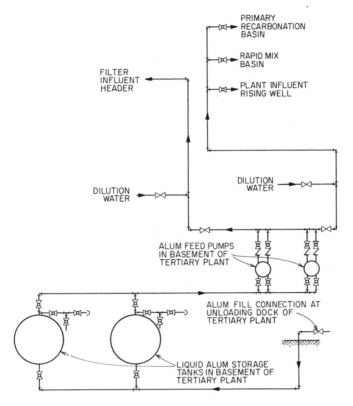

Fig. 2-16   Alum application system at South Tahoe.

points in the process, for example, in the primary influent, in the rapid mix basin, or in the first stage recarbonation basin. A schematic of the alum feed system at South Tahoe is shown in Figure 2-16.

Alum may be used as the primary coagulant in place of lime or in combination with lime, if desired. For example, if the recarbonation equipment is not operating for pH reduction following high lime treatment, then it is possible to feed about 200 mg/l of lime and 100 mg/l of alum and obtain the same treatment of the wastewater without the need for pH adjustment by recarbonation. This is done only for short periods of time, however, because the resulting chemical sludge is very difficult to thicken and dewater. There has been no need to add alum as a coagulant aid after first stage recarbonation, for the calcium carbonate floc settles quite rapidly without the use of alum or other aids. Addition of alum to the primary influent has not proved to be advantageous.

Liquid alum is stored in either or both of two 5000 gal capacity tanks.

One tank is rubber lined steel, and the other is made from fiberglass reinforced polyester resin. There are two positive displacement alum feed metering pumps, each with a capacity of 50.6 gph (25.3 gph for each of two feeder heads). Feed can be set over a 10:1 range by adjustment of the length of pump stroke. Dilution water is added to the alum feed pump discharge line to minimize line plugging problems, to reduce delivery time to the point of application, and to assist mixing with the water being treated.

For automatic alum feed control, two flow signals are provided. One measures the main plant influent flow and the other measures the filter effluent flow, which is always substantially greater than plant inflow because of the recycling of various process, cooling and scrubbing streams within the plant. Either flow signal can be selected for controlling the alum pumps. Ordinarily the dosage is set by the pump stroke adjustment, and the feed is paced to one or the other of the flow signals depending upon the point of alum application being served.

Orange County, California, Water District

This AWT plant, located in Fountain Valley, California, treats trickling filter effluent at a design average flow rate of 15 mgd. There are five major components of the chemical clarification system: rapid mixing, flocculation, settling, effluent pumping and chemical feeding. The chemical clarification unit is illustrated in Figures 2-17 and 2-18.

Lime is the primary coagulant used in this plant. The design criteria are summarized in Table 2-4. The following average removals were obtained during the first year of operation with an average lime dose of 550 mg/l as $Ca(OH)_2$ and an anionic polymer (Dow A-23) dose of 0.1 mg/l.

Constituent	Average influent concentration (mg/l)	Percentage removal
Turbidity	23	90–95
COD	95	35–40
Phosphate	18	96–99

The average daily effluent pH for the chemical clarifier demonstrated that excellent performance was maintained as long as the clarifier effluent pH was above 11.0. These data also demonstrate the effectiveness of the separate rapid mix, flocculation, sedimentation basin design concept to consistently remove more than 90 percent of the wastewater's turbidity. The chemical clarifier average effluent turbidity was 1.5 JTU.

Each of the two rapid basins is equipped with a mechanical mixer. The mixers are removable so that they can be cleaned without draining the

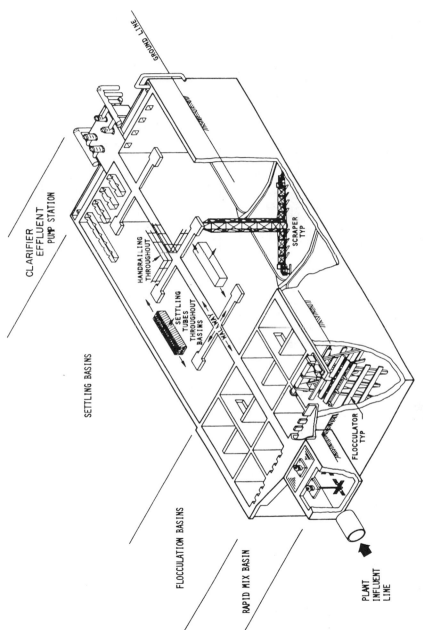

Fig. 2-17 Chemical clarifier, Orange County Water District.

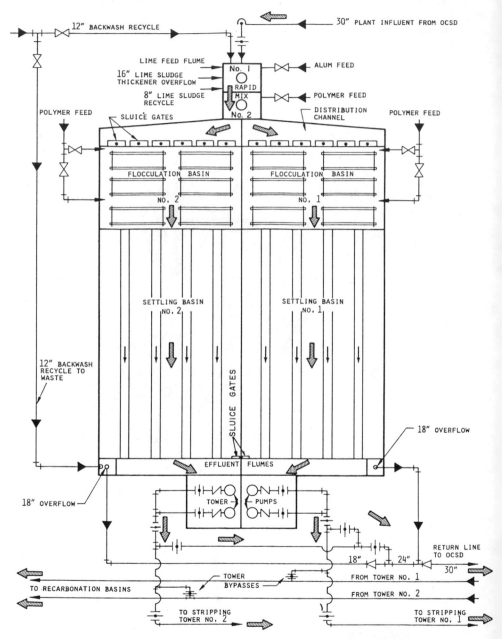

Fig. 2-18   Chemical clarifier schematic, Orange County Water District.

TABLE 2-4   Chemical Clarifier Design Criteria, Orange County Water District. (*Design Average Flow, 15 mgd*)

Rapid Mixing
  Number of basins: 2 in series
  Mechanical mixer in each basin
  Dimensions: length, 12 ft; width, 12 ft; depth, 12 ft
  Detention time: 2.4 min. total @ 15 mgd
  Chemical addition: First basin—lime, alum, recycled lime sludge
    Second basin—polymer

Flocculation
  Number of basins: 2, three compartments each
  Detention time: 10 minutes/compartment (30 minutes total) @ 15 mgd
  Chemical addition: Polymer, 1st and 3rd compartments
  Dimensions: length, 48 ft; width, 41 ft; depth, 11 ft
  Flocculator mechanism: Oscillating type with variable speed drive

Sedimentation
  Number of basins: 2 rectangular
  Dimensions: 120 ft long @ 40 ft wide, each
  Surface overflow rate: 1563 gpd/ft^2 @ 15 mgd
  Each basin equipped with settling tubes (supplied by Neptune Microfloc)

Clarifier Effluent Pump Station
  Number of pumps: 4
  Capacity: 3300 gpm @ 75 ft Total Dynamic Head; 3500 gpm @ 65 ft Total
    Dynamic Head
  Discharge: To ammonia stripping tower or to the Orange Co. Sanitation
    District plant or to the recarbonation basins

Lime System
  Storage: Two 35 ton bins
  Feeder and slakers: 2 gravimetric feeders and paste-type slakers @ 4000
    lb/hour each

Polymer System
  Number of mixing tanks: 2 (800 gal each)
  Number of feed pumps: 4 dual head
  Capacity: 0–25.3 gph each head

Alum System
  Number of storage tanks: 2 (4100 gal each)
  Number of feed pumps: 3 (2 double head and 1 single head)
  Capacity: 25.3 gph each head

basins. Secondary effluent flows into the first rapid mix basin where lime is added and mixed. Lime enters this basin in slurry form that flows by gravity from the slakers in an open channel. Liquid alum and recycled lime sludge can also be added to the first basin if it is determined that either or both will improve floc formation and settling.

Water flows from the first rapid mix basin to the second, where a polymer can be added if needed. This second basin also provides additional time for mixing so that all chemicals are thoroughly dispersed and dissolved before the water enters the flocculation basins.

Chemically treated secondary effluent flows from the second rapid mix basin to a distribution channel and then into two parallel flocculation basins. Each flocculation basin normally treats one-half the flow. One basin can be taken out of service by closing the sluice gates in the distribution channel. Each basin is divided into three bays or compartments, and each bay contains an oscillating paddle-type mixer. The speed of the oscillating paddles can be varied; to vary the G value from about 100 sec^{-1} in the first bay to about 20 sec^{-1} in the third bay.

Water flows from the bottom of the flocculation basin and enters near the bottom of two parallel sedimentation basins. Water flows horizontally into the basin and up through settling tubes, over notched weirs and out through collection troughs to the clarifier effluent pump station. Solids settle to the bottom and are removed continuously by three mechanical scrapers in each basin. The settled solids then flow by gravity to the sludge pump station. Each basin has an overflow line located in the effluent trough which prevents accidental overflow.

Recalcined lime and purchased or makeup lime are stored in two bins, each with a capacity of 35 tons. Normally, recalcined lime is stored in bin no. 1 and purchased lime in bin no. 2; however, lime from either source may be stored in either bin. The bins are equipped with vibrating devices at the bottom to prevent clogging of the discharge pipe.

Lime is withdrawn from the bottom of the bin(s) and fed to a paste-type slaker. Water is added to the slaked lime paste, which is then transferred in slurry form in an open flume to rapid mix basin no. 1. The lime feed rate is controlled by the rate of slaker operation. Usually both machines will be in operation, but, if necessary, either one alone can treat the entire plant flow. The feed rate can be adjusted manually or it can be paced to plant flow rate by means of an automatic pacing device.

There are four, dual head, polymer feed pumps in the plant, and one of these pumps supplies polymer to the chemical clarifier. The alum storage and feeding system serves both the chemical clarification and filtration treatment units. To date, good clarification has been accomplished using lime as the primary coagulant and a polymer as a coagulant aid. It is antici-

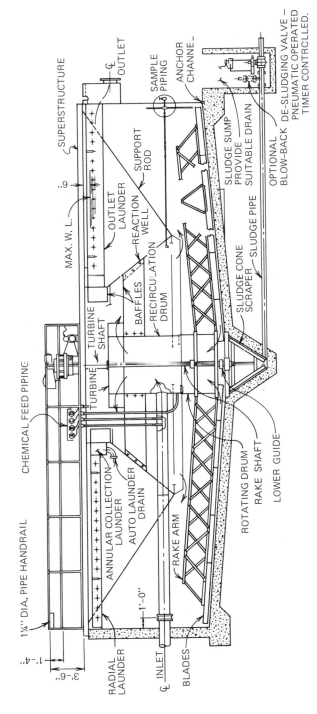

Fig. 2-19 Solids contact clarifier, Colorado Springs, Colorado.

pated that only under unusually difficult circumstances, if ever, will it be necessary to add alum to the rapid mix basin.

City of Colorado Springs, Colorado

The AWT plant at Colorado Springs treats activated sludge effluent for irrigation and industrial reuse at an average rate of 2.0 mgd. The chemical clarifier is a solids-contact-type unit manufactured by Eimco Corporation and is shown in Figure 2-19. It is 48 ft in diameter, with a 10 ft sidewater depth.

During the first year of operation, extreme problems were encountered in controlling the sludge blanket and the flow was reduced to 1.6 mgd or less. The unit's sludge recycling system was then modified by the manufacturer. Subsequently the unit has operated very satisfactorily at rates between 2.2 and 2.4 mgd.

Lime is fed at a rate of 300 to 350 mg/l as CaO to raise the pH to 11.5. Typical operating results achieved with this unit are summarized as follows:

Parameter	Influent; mg/l	Effluent, mg/l	Percentage Removal
BOD	17	4	76
COD	70	45	36
Turbidity	10	3	70
Total PO_4^{3-}	25	0.7	97
pH	7.4	11.5	—

Sludge withdrawn from the clarifier averages 7–8 percent dry solids. Lime is recalcined and reused; about 50 percent makeup lime is required.

REFERENCES

1. Anonymous, "State of the Art of Coagulation," *Journal American Water Works Assoc.*, February 1971, pp. 99–108.

2. Jordan, R. M., "Coagulation—Flocculation," *Water and Sewage Works*, January, February, March, 1971.

3. Weber, W. J., Jr., *Physiochemical Processes for Water Quality Control.* John Wiley & Sons, Inc., New York, 1972, pp. 61–91.

4. Merrill, D. T. and Jordan, R. M., "Lime Induced Reactions in Municipal Waste-waters," *Journal Water Pollution Control Federation*, December 1975, pp. 2783–2808.

5. Zoltek, J., Jr., "Identification of Orthophosphate Solids Formed by Lime Precipitation," *Journal Water Pollution Control Federation*, January 1976, pp. 179–182.

6. Schmid, L. A., and McKinney, R. E., "Phosphate Removal by a Lime—Biological Treatment Scheme," *Journal Water Pollution Control Federation*, 1969, p. 1259.

7. Mulbarger, M. C., et al., "Lime Clarification, Recovery, Reuse, and Sludge Dewatering Characteristics," *Journal Water Pollution Control Federation*, 41: 12, December 1969, p. 2070.

8. *Lime Handling, Application and Storage.* Bulletin 213, National Lime Association, May 1971, 73 pp.

9. Stumm, W., and Morgan, J. J., *Aquatic Chemistry.* Wiley–Interscience, New York, 1970, p. 521.

10. "The Use of Aluminum Sulfate for Phosphorus Reduction in Wastewaters." Allied Chemical Corporation, Industrial Chemicals Division, Morristown, N.Y., updated, 10 pp.

11. Trussell, R., et al., "Ozone Found Effective in Tertiary Pretreatment," *Bulletin CWPCA*, January 1976, pp. 75–83.

12. Camp, T. R., and Stein, P. C., "Velocity Gradients and Internal Work in Fluid Motion," *Journal Boston Society of Civil Engineers*, October 1943, pp. 219–237.

13. Camp, T. R., "Flocculation and Flocculation Basins," *Transactions American Society of Civil Engineers*, 1955, pp. 1–16.

14. Camp, T. R., "Sedimentation and the Design of Settling Tanks," *Transactions American Society of Civil Engineers*, 111, 1946, p. 895.

15. Zanoni, A. E. and Blomquist, M. W., "Column Settling Tests for Flocculant Suspensions," *Proc. American Society of Civil Engineers* 101: EE3, June 1975, pp. 309–318.

16. Rose, J. L., "Removal of Phosphorous by Alum." Paper presented at the Phosphorous Removal Conference sponsored by the Federal Water Pollution Control Administration, Chicago, Illinois, 1968.

17. O'Farrell, T. P.; Bishop, D. F.; and Bennett, S. M., "Advanced Waste Treatment at Washington, D.C." Paper presented at the Sixty-fifth Annual AIChE Meeting, Cleveland, Ohio, May 1969.

18. Convery, J. J., "Phosphorus Removal by Tertiary Treatment with Lime and Alum," Federal Water Pollution Control Administration Symposium on Nutrient Removal and Advanced Waste Treatment, Tampa, Florida, November 15, 1968.

19. Weber, W. J., Jr.; Hopkins, C. E.; and Bloom, R., Jr., "Physiochemical Treatment of Wastewater," *Journal Water Pollution Control Federation*, 1970, p. 83.

20. Kalinske, A. A., and Shell, G. L., "Phosphate Removal from Waste Effluent and Raw Wastes Using Chemical Treatment." Paper presented at the Phosphorus Removal Conference sponsored by the Federal Water Pollution Control Administration, Chicago, Illinois, 1968.

21. McMichael, F. C., "Sedimentation in Inclined Tubes and Its Application for the Design of High-Rate Sedimentation Devices." Paper presented at the International Union of Theoretical and Applied Mechanics Symposium on Flow of Fluid–Solid Mixtures; University of Cambridge, England, March 24–29, 1969.

22. Yao, K. M., "Theoretical Study of High Rate Sedimentation," *Journal Water Pollution Control Federation*, 1970, p. 218.

23. Hansen, S. P., and Culp, G. L., "Applying Shallow Depth Sedimentation Theory," *Journal American Water Works Assoc.*, 1967, p. 1134.

24. Culp, G. L.; Hansen, S. P.; and Richardson, G. H., "High Rate Sedimentation in Water Treatment Works," *Journal American Water Works Assoc.*, 1968, p. 681.

3

Recarbonation

PURPOSE

Recarbonation is a unit water treatment process which has been in use for many years in numerous municipal and industrial lime–soda softening plants throughout the world. More recently, with lime treatment of wastewaters, particularly with the massive doses of lime used in phosphorus removal, recarbonation has seen increasing use in wastewater treatment and water reclamation plants.

The addition of sufficient lime to wastewater raises the pH and converts bicarbonates and carbonates to hydroxides; this conversion is incidental to the main purposes of treatment, which are coagulation of nonsettleable matter and removal of phosphorus. Recarbonation is a term applied to the addition of carbon dioxide to a lime treated water. When carbon dioxide is added to high pH, lime treated water, the pH is lowered and the hydroxides are reconverted to carbonates and bicarbonates. Thus, the term recarbonation is very descriptive of the result of adding carbon dioxide to wastewater.

The basic purpose of recarbonation is the downward adjustment of the pH of the water. Properly done, this places the water in calcium carbonate equilibrium, and avoids problems of deposition of calcium scale which would occur without the reduction in pH accomplished by recarbonation. In water works practice, the carbon dioxide is added to the water ahead of the filters in order to avoid coating of the grains of filter media with calcium carbonate, which would eventually increase the grain size to the point that filter efficiency would be reduced. In water works it is also important to lower the pH of the lime treated water to the point of calcium carbonate stability to avoid deposition of calcium carbonate in pipelines. In advanced treatment of wastewaters, pH control by recarbonation has even greater significance than in water works practice, because of the effects of pH on treatment processes commonly found downstream.

There are several reasons for adjusting the pH of wastewater during treatment. Coagulation, flocculation, and clarification of wastewaters can be accomplished in a rather wide range of pH values, but particular coagulants or coagulant aids ordinarily yield optimum results within a rather narrow range of pH values. Lime usually gives good results in domestic wastewaters at pH values above 9.6 or 9.8. Alum is ordinarily good at pH values below 7.3 or above 8.6. The effective pH range for coagulants usually can be broadened by use of the correct polymer, activated silica, or other coagulant aids. Ammonia stripping is best at pH values above 10.8, because then most of the nitrogen is in the form of dissolved ammonia gas rather than as ammonium ions in true solution. Generally speaking, filtration through granular media is best in the pH range of 6.5–7.5, but the filterability of water depends upon many other factors, including the physical and chemical characteristics of the water, the coagulant used, and the chemicals employed as filter aids, if any. Activated granular carbon adsorption of organics from water generally takes place in the pH range of 5–9, with better adsorption at values below 7. Very often, if the pH is above 9, desorption will occur. That is, organics which have been adsorbed previously on the carbon will be released to the high pH water. For example, a column of granular activated carbon which has been operating at a pH of 7 for several days with good removal of color from the applied water may still have capacity for further color removal at a pH of 7. However, if water with a pH of 10 is passed through the column, the color of the effluent water from the column will undoubtedly be higher than that of the influent, because of desorption.

Massive lime treatment of wastewaters for phosphorus removal often raises the pH to values of about 11. Primary recarbonation is used to reduce the pH from 11 down to 9.3, which is near that of minimum solubility for calcium carbonate. In domestic wastewater, primary recarbonation to pH 9.3 results in the formation of a heavy, rapidly settling floc which is principally calcium carbonate, although some phosphorus is also removed from solution by adsorption on the floc. If sufficient reaction time, usually about 15 minutes in cold water, is allowed for the primary recarbonation reaction to go to completion, the calcium carbonate floc does not redissolve with subsequent further lowering of pH in secondary recarbonation. (However, there is a tendency for the magnesium salts to do so). If lime is to be reclaimed by recalcining and reused, this settled primary recarbonation floc is a rich source of calcium oxide, and may represent as much as one third of the total recoverable lime. The second stage of recarbonation to pH 7, is beneficial in several ways: it prepares the water for filtration; it lowers the pH to a value which increases the efficiency of carbon adsorption of organics, to an excellent range for effective disinfection by chlorine, and to a value suitable for discharge; and it stabilizes the water in respect to scale formation in pipelines. If the pH were not reduced to less than about 8.8 before application to the filters and carbon beds, extensive deposition of calcium carbonate would occur on the surface of the grains. This could reduce filter efficiency, and could also drastically reduce the adsorptive capacity of granular activated carbon for organics. It would produce rapid ash buildup in the carbon pores upon regeneration of the carbon, and would lead to early replacement of the carbon.

SINGLE STAGE VS. TWO STAGE RECARBONATION

It is possible to reduce the pH of a treated wastewater from 11 to 7 or to any other desired value in one stage of recarbonation. Single stage recarbonation eliminates the need for the intermediate settling basin which is used with two stage systems. However, by applying sufficient carbon dioxide in one step for the total pH reduction, little, if any, calcium is precipitated, with the bulk of calcium remaining in solution, thus increasing the calcium hardness of the finished water, and, in addition, causing the loss of a large quantity of calcium carbonate, which could otherwise be settled out, recalcined to lime, and reused. If lime is to be reclaimed or if calcium reduction in the effluent is desired, then two stage recarbonation is required. Otherwise, single stage recarbonation may be used with some savings in initial cost, and some reduction in the amount of lime sludge to be handled.

SOURCES OF CARBON DIOXIDE

In advanced wastewater treatment plants, the usual source of carbon dioxide for recarbonation probably will be the stack gas from either a lime recalcining furnace or a sludge incineration furnace. Other possible sources include the use of commercial liquid carbon dioxide; or the burning of natural gas, propane, butane, kerosene, fuel oil, or coke.

The stack gas from a sewage sludge incineration furnace which is fired with natural gas will contain about 16 percent carbon dioxide on a wet basis or about 10 percent on a dry basis. About 10 percent CO_2 on a dry basis is usually used for design purposes. The burning of 1000 cubic feet of natural gas produces about 115 lbs of CO_2.

The stack gas from a lime recalcining furnace contains not only the CO_2 produced by combustion of the fuel, but also the CO_2 driven off of the calcium carbonate sludge in the recalcining process. For design purposes, a value of 16 percent CO_2 in lime furnace stack gas is a conservative figure to use.

Kerosene and No. 2 fuel oil will yield about 20 lbs of CO_2 per pound of fuel. Coke will produce approximately 3 lbs of CO_2 per pound of coke burned. Commercial liquid CO_2 is 99.5 percent CO_2.

QUANTITIES OF CARBON DIOXIDE REQUIRED

In recarbonation, one molecule of CO_2 is required to convert one molecule of calcium hydroxide (caustic alkalinity) to calcium carbonate. In addition, it takes one molecule of CO_2 to convert one molecule of calcium carbonate to calcium bicarbonate. It follows then, that two molecules of CO_2 are required to convert one molecule of calcium hydroxide to calcium bicarbonate. These reactions are represented by the following equations:

$$Ca(OH)_2 + CO_2 \rightarrow CaCO_3 + H_2O$$
$$CaCO_3 + CO_2 + H_2O \rightarrow Ca(HCO_3)_2$$

Since all forms of alkalinity are expressed in terms of calcium carbonate (molecular weight = 100), the calculations are as follows:

1. Calcium hydroxide to calcium carbonate:
   (molecular weight of CO_2 = 44, and 1 mg/l = 8.33 lb/mg)
   CO_2 (in lb/mg)
   $$= \frac{44}{100} \times 8.33 \times (OH^- \text{ Alk. in mg/l as } CaCO_3)$$
   $$= 3.7 \times (OH^- \text{ Alk. in mg/l as } CaCO_3)$$
2. Calcium carbonate to calcium bicarbonate:
   CO_2 (lb/mg)
   $$= \frac{44}{100} \times 8.33 \times (CO_3^{2-} \text{ Alk. in mg/l as } CaCO_3)$$
   $$= 3.7 \times (CO_3^{2-} \text{ Alk. in mg/l as } CaCO_3)$$
3. Then, for calcium hydroxide to calcium bicarbonate:
   CO_2 (lb/mg)
   $$= 7.4 \times (OH^- \text{ Alk. in mg/l as } CaCO_3)$$

SAMPLE CALCULATIONS FOR THE AMOUNT OF CO2 REQUIRED

Assume that 400 mg/l of calcium oxide (CaO) have been added to a sample of a typical domestic wastewater and that the stirred and decanted liquor has the following characteristics:

$pH = 11.7$
Alkalinities (in mg/l as $CaCO_3$):
   $OH^- = 380$
   $CO_3^{2-} = 120$
   $HCO_3^- = 0$

After first recarbonation in the laboratory using bottled carbon dioxide, analysis of the lime treated wastewater shows the following:

$pH = 9.3$
Alkalinities (in mg/l as $CaCO_3$):
   $OH^- = 0$
   $CO_3^{2-} = 180$
   $HCO_3^- = 380$

Then, after secondary recarbonation in the laboratory, the water has the following analysis:

$pH = 8.3$
Alkalinities (in mg/l as $CaCO_3$):
   $OH^- = 0$
   $CO_3^{2-} = 0$
   $HCO_3^- = 750$

To change all caustic alkalinity and all carbonate alkalinity to bicarbonates then, the amount of CO_2 required is as calculated below:

$$7.4 \times 380 = 2812 \text{ lb/mg of } CO_2$$
$$3.7 \times 120 = \underline{444}$$
$$Total = 3256 \text{ lb/mg of } CO_2$$

For a 1 mgd flow, 3265 lb of CO_2 per day are required. If it is assumed that the CO_2 content of the stack gas to be used for recarbonation is 10 percent, then $(100/10 \times 3265 = 32,560$ lb of stack gas must be compressed in order to supply the necessary CO_2 to recarbonate 1 million gal of wastewater.

At standard conditions of 14.7 psia and 60° F (520° K), assume that the density of the stack gas is the same as for CO_2, or 0.116 lb/ft^3. Then,

$$\frac{32,560 \text{ lb}}{0.116 \text{ lb/ft}^3} = 280,600 \text{ ft}^3$$

If it is assumed that the stack gas is cooled in scrubbing to 110° F, then the temperature correction is

$$\frac{110 + 460}{60 + 460} \times 195 = 1.1 \times 195$$

With the same gas temperature, for a plant at 6300 ft above sea level (11.6 psia), the altitude correction is

$$\frac{14.7}{11.6} \times 215 = 1.26 \times 215 = 270 \text{ cfm/mgd}.$$

Since some CO_2 is not absorbed in the water but escapes at the water surface, it is customary to add about 20 percent to the theoretical requirements. If this is done then, at sea level 260 cfm/mgd of blower or compressor capacity is required, and at 6300 ft of altitude 325 cfm/mgd of plant capacity is needed.

It must be emphasized that these calculations are based on the following assumptions:

1. The water to be recarbonated has a $pH = 11.7$ with OH⁻ Alk. = 380 mg/l as $CaCO_3$, and with a CO_3^{2-} alkalinity of 120 mg/l as $CaCO_3$, all of which is to be converted to bicarbonates.
2. There is 10 percent CO_2 in the gas.
3. The flue gas temperature is 110° F.
4. An excess of 20 percent is added to the theoretical values to allow for absorption losses.

Conditions at each installation undoubtedly will differ from the assumptions used here, and calculations must be based on actual values rather than those given.

It should be noted that the quantities of CO_2 required for this type of wastewater treatment are many times greater than the amounts used in water softening plants. However, the observed actual CO_2 consumption at the South Tahoe Public Utility District Water Reclamation Plant corresponds quite closely to the figures obtained using the above method of computation.

NONPRESSURIZED CARBON DIOXIDE GENERATORS

If stack gas from a furnace operating at atmospheric pressure is to be used as a source of CO_2, the gas should be passed through a wet scrubber. Wet scrubbers provide contact between the gas and a flow of scrubbing water. Particulate matter is removed from the gas, and the gas is cooled. Wet scrubbers may be one of three general types: impingement, Venturi, and

Fig. 3-1 Water jet impingement-type stack gas scrubbers installed on sludge incineration and lime recalcining furnaces. (*Courtesy Cornell, Howland, Hayes, & Merryfield*)

surface area. Figure 3-1 shows two water jet impingement-type scrubbers manufactured by W. W. Sly Manufacturing Company, Cleveland, Ohio, as installed at the South Tahoe Public Utility District Plant. One serves a sludge incineration furnace (right), and the other a lime recalcining furnace. These scrubbers are very efficient in removing potential air pollutants from the exhaust gas, and provide some protection of the CO_2 compression equipment against plugging or scaling by particulates. The scrubbers cool the hot stack gas down to about 110° F.

When stack gas is used as the source of CO_2, the stack gas supply must exceed the maximum demands for CO_2. With this situation, control of the amount of CO_2 applied to the water is very simple. Air may be admitted through a valve into the suction line leading to the compressor as required to reduce the amount of CO_2 to that desired, or, as an alternative, part of the compressed gas may be bled off to the atmosphere through a valve in the compressor discharge line. As another method of control, compressed gas may be recirculated from the compressor discharge line back to the suction line through a bypass line and control valve. However, this method has the serious disadavantage of warming the gas by compression, and excessive recirculation can lead to compressor damage by overheating or increased corrosion at the elevated temperatures.

COMPRESSOR SELECTION

Even with thorough scrubbing, stack gas from sludge incineration furnaces or lime recalcining furnaces will contain sufficient particulate matter to cause plugging and seizure problems in some types of blowers and compressors, particularly those with limited clearance between moving metal parts. This problem is less severe with stack gas from atmospheric furnaces, which burn fuel primarily for production of CO_2.

Water sealed compressors similar to wet vacuum pumps are a good selection for handling dirty, corrosive gases. This type of compressor consists of a squirrel-cage-type rotor which revolves in a circular casing containing water. The rotor shaft is located off center toward the bottom of the casing. Figure 3-2 is a cross section through a water sealed compressor. As the rotor revolves, centrifugal force pushes the water out against the sides of the casing, leaving a doughnutlike hole of air in the center. Because the center drum of the rotor is located eccentrically, there is a large air space above it and very little below. Thus, as air is driven from the large space into the smaller one, it is compressed. Figure 3-3 shows an exterior view of a water sealed CO_2 compressor and disassembled parts. This is a simple, reliable piece of equipment with only one moving part. It has increased capacity when handling hot, saturated vapors, since the vapors are condensed by the cool liquid compressant. The water sealed compressor is a relatively quiet-running unit, free from pulsations and vibrations.

If the CO_2 distribution grids are submerged a minimum of 8 ft in water,

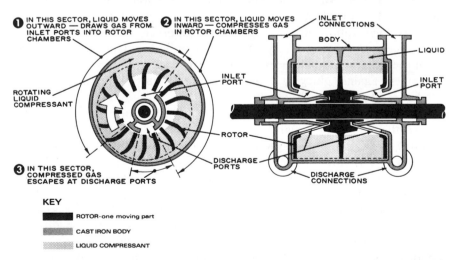

Fig. 3-2 **Section through water sealed carbon dioxide compressor.** (*Courtesy Nash Engineering Co.*)

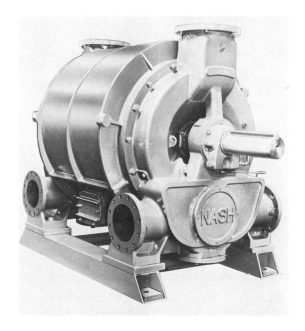

Fig. 3-3   Water sealed carbon dioxide compressor and dissassembled parts. (*Courtesy Nash Engineering Co.*)

as they usually are, the CO_2 compressor must deliver against a differential pressure across the machine of about 6–8 psi. The exact rating must be determined by calculation, taking into account not only the depth of submergence of the distribution piping, but also orifice losses and pipe friction losses. This is discussed in more detail later, because it is common to all types of CO_2 systems. The compressors may be of cast iron construction,

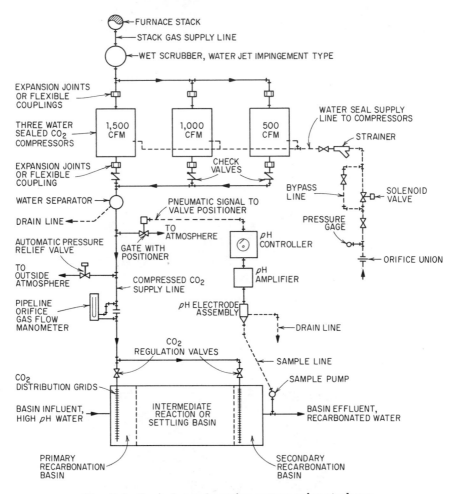

Fig. 3-4   Typical recarbonation system using stack gas.

or may be supplied with bronze rotor and cones at considerable extra cost. The following accessories are commonly required with water sealed compressor units: water separator with gage glass and bronze float valve; discharge check valve; expansion joints for inlet and outlet piping; water seal supply line with adjusting cock and orifice union, water line strainer, inlet water spray nozzles, and sealing water line solenoid valve. In addition, the discharge line is usually fitted with an automatic pressure relief valve and a bleed off valve, both of which should discharge to free atmosphere. It is not good practice to install shutoff or isolation valves on either the compressor suction or discharge lines because of the possibility of serious damage to compressor or pipelines in the event that the compressor is operated in error with either or both of such valves closed.

In selecting CO_2 compressor units to meet total capacity requirements, it is a good idea, except in very small installations, to provide at least three compressor units. By properly sizing these it is then possible to satisfy two needs–to secure a range in output and to provide standby service. If it is assumed that the total CO_2 capacity required is 1500 cfm, then units with individual capacities of 500, 1000, and 1500 cfm would represent a good choice. This combination gives a range of 500 to 1500 cfm to match plant needs, and it supplies standby with the largest unit out of service by using the two smaller units together. A typical recarbonation system using stack gas at atmospheric pressure is illustrated by Figure 3-4. As indicated in this figure, automatic pH control of the recarbonated effluent can be provided by continuously monitoring an effluent sample for pH to operate a pH controller, which in turn positions a bleed off valve in the CO_2 compressor discharge line to limit the amount of CO_2 to that necessary to maintain the desired pH.

PRESSURE GENERATORS AND UNDERWATER BURNERS

Generators designed specifically for the production of carbon dioxide for recarbonation are usually either pressure generators or submerged underwater burners. Most early installations were of the atmospheric type in which the fuel is burned at atmospheric pressure and the off gas is scrubbed and compressed. These systems are expensive to maintain because of the corrosive effects of the hot, moist combustion gases, and atmospheric generators have largely been replaced by pressure generators and underwater burners, except where waste stack gas is available from another source. Both types of pressure CO_2 generation equipment are now in commercial manufacture.

Pressure or forced draft generators produce CO_2 by burning natural gas, fuel oil, or other fuels in a pressure chamber. The fuel and excess air are first compressed and injected, and then burned at a pressure which is sufficiently high to allow discharge directly into the water to be recarbonated. The compressors handle only dry gas or dry air at ambient temperatures, and thus the corrosion problems involved in handling the hot, moist stack gases are avoided. One difficulty with this type of pressure generator is its limited capacity range, which may be 3 to 1, or at best 5 to 1. This low turndown ratio may necessitate the installation of two or more units in order to secure the required flexibility and process control. A wide range of sizes is commercially available in pressure CO_2 generators. This

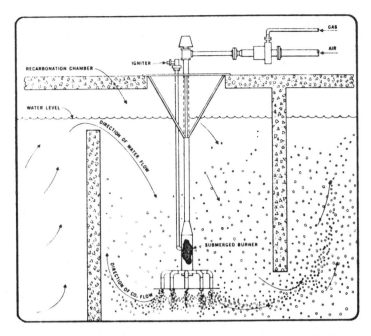

Fig. 3-5   Typical recarbonation system using submerged combustion burner. (*Courtesy Ozark Mahoning Co., Tulsa, Oklahoma*)

commercial equipment is well designed and reliable, and includes all auxiliaries and safety controls.

Submerged combustion of natural gas is another method for CO_2 generation (Figure 3-5). Air and natural gas are compressed and then burned under water at the point of application, that is, in the recarbonation basin. Automatic underwater electric ignition equipment is used to start combustion. Submerged combustion is a simple, efficient means of CO_2 generation which provides good control of recarbonation, and requires a minimum of maintenance. The turndown ratio of this type of burner is only about 2 to 1, so that it is necessary to provide enough burner assemblies to obtain the desired range of control in the amount of CO_2 applied.

LIQUID CARBON DIOXIDE

Commercial liquid CO_2 has found increasing use for recarbonation in water softening plants in the last few years, primarily because of its steadily decreasing cost. However, the price of liquid CO_2 depends greatly on the distance from the source of supply, and the first factor to be investigated is the cost of liquid CO_2 delivered to the plant under consideration. Current (1976) bulk liquid CO_2 prices delivered to points near a source of supply range from \$40 to \$155 per ton. Even in favorable locations, high cost is still the principal disadvantage of using liquid CO_2. Its advantages include flexibility, ease of control, high purity and efficiency, and the smaller piping required because of its high CO_2 content of 99.5 percent as compared to the 6–18 percent as obtained from other sources.

Liquid CO_2 may be delivered to customers in insulated tank trucks ranging from 10 to 20 tons in capacity. Rail car shipments of 30–40 tons are available to large volume users. Some manufacturers will lease tank cars so that they may be used for storage at the site, thus eliminating the need and expense of bulk storage tanks and auxiliaries at the plant. For small plants, liquid CO_2 is also available in 20–50 lb cylinders.

Bulk storage tanks may be purchased or leased. Capacities range from 1 to 100 tons, although the common sizes are 4–48 tons. Storage tanks must be insulated and equipped with Freon refrigeration and electric or steam vaporization equipment. The working pressure for storage tanks is 350 psi, and the ASME Code for Unfired Pressure Vessels requires hydrostatic testing to 525 psi, or 1.5 times the working pressure. The tanks may be insulated with pressed cork or polyurethane foam. The cooling and vaporizing systems are designed to maintain the liquid CO_2 at about $0°$ F and 300 psi. If temperature and pressure rise, the cooling system comes on, and on falling pressure, the vaporizer comes into service. In the event either of these systems fails, or in the event of fire or other accident the storage tanks are fitted with high and low pressure alarms, two safety pop valves, a manual bleeder relief valve, and a bursting disk.

Either liquid or gas feed systems may be used to apply the liquid CO_2 to the wastewater to be treated. In withdrawing CO_2 from the storage tank the pressure is reduced. This pressure reduction cools the CO_2, with the danger of dry ice formation if the expansion is too rapid. Consequently, it is common practice to reduce the pressure in two stages from the 300 psi tank pressure to the 20 psi pressure ordinarily required for feeding the CO_2. Vapor heaters may also be used just ahead of the pressure reducing valves.

For CO_2 gas feed, an orifice plate in the feed line with a simple manometer may be used to measure flow, and a manual valve installed downstream may be used to regulate or control the amount of CO_2 applied. Automatic control to a manual set point can be provided by using a differential pressure transmitter on the feed line orifice, and connecting it to an indicating controller which would operate the control valve. Optionally, the CO_2 feed could be made fully automatic by providing pH control. In this case, an electrode would be installed to measure the pH of the recarbonated water. This signal would be amplified and sent through a controller which would throttle the control valve on the feed line at low pH and open it at high, as set on the controller.

For solution feed of CO_2, equipment similar (except for materials of construction) to solution feed chlorinators may be used. Chlorinator capacity is reduced about 25 percent when feeding CO_2. Approximately 60 gallons of water are required to dissolve 1 lb of CO_2 at room temperature and atmospheric pressure. Absorption efficiency with solution feed of CO_2 approaches 100 percent.

CARBON DIOXIDE PIPING AND DIFFUSION SYSTEMS

Because in recarbonation systems the gas temperature is usually in the 70–100° F range, and pressure is about 6–8 psi, and because CO_2 pipe runs are usually less than 100 ft, it is convenient to use Table 3-1 to estimate the pipe size required. The pipe sizes obtained from Table 3-1 are, of course, an approximation. For greater accuracy, or for long lines, the following modification of the Darcy–Weisbach formula may be used for pressure loss in air piping:

$$\Delta p = \frac{f}{38,000} \frac{LTQ^2}{pD^5}$$

where

Δp = pressure drop in psi

$f = \dfrac{0.048\, D^{0.027}}{Q^{0.148}}$ (Note: usual values for f = 0.016–0.049)

L = pipe length in feet
T = absolute F temperature of the gas = °F + 460
Q = gas flow in cfm
P = absolute pressure of the gas in psi (or line pressure in psi + 14.7)
D = pipe diameter in inches

TABLE 3-1 Approximate Gas Carrying Capacity.

Pipe Diameter, In.	Capacity, cfm
1	45
2	250
3	685
4	1410
6	3870

The pressure loss in elbows and tees can be approximated by use of the following formula:

$$L = \frac{7.6\,D}{1 + \frac{3.6}{D}}$$

where

L = equivalent length of straight pipe in feet
D = pipe diameter in inches

The loss in globe valves is about

$$L = \frac{11.4\,D}{1 + \frac{3.6}{D}}$$

where L and D are the same as above.

Carbon dioxide absorption systems often consist of a grid of perforated pipe submerged in the wastewater. The recommended minimum depth of submergence is 8 ft. With lesser depths of submergence some undissolved CO_2 will escape at the water surface. Properly designed absorption systems will put into solution 85–100 percent of the applied CO_2. PVC pipe is an excellent material for the perforated CO_2 grid pipes. Current practice is to use 3/16 in. diameter orifices drilled in the bottom of the pipe at an angle of 30 degrees to the right of the vertical centerline, then 30 degrees to the left, alternating at a spacing of about 3 in. along the centerline of the pipe. Another arrangement is to point the orifices straight up at the top of the pipe and to direct a jet of water from a header down at the CO_2 orifice, in order to form fine bubbles of the gas, which dissolve more readily in the water. Since PVC does not corrode under acid conditions, the openings are not subject to plugging as they are in metal pipes. If 3/16 in. orifices are used, each opening is often rated at 1.1–1.65 cfm, which corresponds to head losses through the orifice of 3 and 8 in. of water column, respectively. This is sufficient loss through the orifice to insure good distribution of the CO_2 to each opening. CO_2 laterals must be laid with the same depth of submergence on each orifice. If the size of the pipe changes, then eccentric reducers should be used to keep the bottom of the pipe level (assuming that the holes are in the bottom of the pipe). Horizontal spacing between CO_2 diffusion laterals should be at least 1.5 ft. in order to get good absorption. To convey cool, dry CO_2, plain steel or cast-iron pipe may be used, but for hot, moist CO_2 gas, the use of stainless steel or other acid resistant metal is suggested. Special pipe is also required to convey liquid CO_2 in water; a 1.5 in. cotton fabric hose with openings of controlled size or porosity has been used successfully. Basin hydraulics must take into account raised water levels caused by CO_2 injection.

CARBON DIOXIDE REACTION BASINS OR INTERMEDIATE SETTLING BASINS

Contrary to many early reports in the literature, the recarbonation reaction is not instantaneous. Although about 90 percent of the applied CO_2 does dissolve in its very short upward journey from the distribution grid through 8 ft of water to the water surface, the time for complete reaction between the dissolved CO_2 and hydroxide and carbonate ions may be as great as 15 minutes in cold water. In the primary phase of two stage recarbonation, if the reaction is allowed to go to completion at a pH near 9.3, the calcium carbonate formed is not redissolved in the second phase of recarbonation to a low pH—say to pH 7.0. Magnesium salts do tend to redissolve under these conditions. In the recarbonation of domestic wastewaters, a dense, rapidly settling floc is formed following first stage recarbonation. This is a rich source of calcium carbonate from which lime (CaO) can be reclaimed and reused by recalcining at temperatures of about 1850° F. In this case, then, it is desirable to allow not only for reaction time (15 minutes) but for enough time to provide some separation of the calcium carbonate

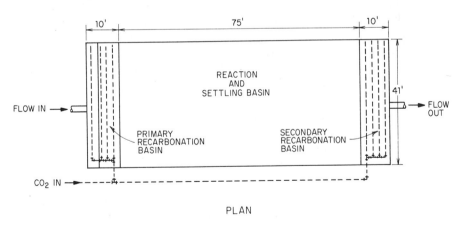

PLAN

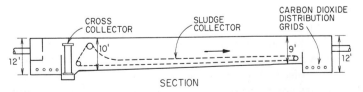

SECTION

Fig. 3-6 Plan and section showing two-stage recarbonation basin with intermediate settling for a 7.5 mgd plant.

by settling. This will require a settling basin with at least 30 minute detention at maximum flow rate, and a basin surface overflow rate of not more than 2400 $gal/ft^2/day$. This intermediate settling basin should be fitted with continuous mechanical sludge removal equipment. Figure 3-6 shows a two stage recarbonation system with intermediate reaction and settling for a design flow of 7.5 mgd.

Single stage recarbonation systems should be followed by 15 minutes of detention for completion of the chemical reactions, but no provisions for settling or sludge collection are required. The light, cloudy floc which may be formed at times with single stage recarbonation is removed quite readily by mixed media filtration with little effect on filter effluent turbidity, headloss, or length of filter run.

Recarbonated lime treated wastewater should not be applied directly to beds of granular activated carbon without filtration, because even at low pH (say 7.0) there is still sufficient deposition of calcium carbonate to cause serious problems in the carbon treatment, which easily can be avoided by prior filtration.

OPERATION AND CONTROL OF RECARBONATION

The operation and control of recarbonation systems is easy and simple. Automated control systems ordinarily use a single point of pH measurement following the last stage of recarbonation as the basis of control. In two stage recarbonation systems, the split of total CO_2 flow between the two stages of treatment is fairly constant once it is established for a given flow and the particular set of pH values desired, and control based on the final pH alone is satisfactory without readjustment of the valves supplying the first and second stage CO_2 supply headers.

An indirect but more sensitive control of recarbonation is provided by alkalinity measurements. Continuous reliable automatic monitoring and control equipment is available for either the pH or the alkalinity method, but the alkalinity measuring equipment is considerably more expensive than the pH equipment. Manual control is also quite satisfactory, based either on grab sampling and analysis or on observation of continuous automatic monitoring of pH or alkalinity of the recarbonated water. The CO_2 demands do not vary rapidly or widely, and manual control of dosage is better than might be expected.

SAFETY

Under certain conditions carbon dioxide can be dangerous, and there are safety precautions which must be observed. Prolonged exposure to concentrations of 5 percent or more CO_2 in air may cause unconsciousness and death. The maximum allowable daily exposure for a period of 8 hours is 0.5 percent CO_2 in air. Carbon dioxide is 1.5 times as dense as air, and therefore will tend to accumulate in low, confined areas. Filter-type gas masks are not useful in atmospheres containing excess CO_2, and self-contained breathing apparatus and hose masks must be used. Contact with liquid CO_2 by the skin can cause frostbite. Recarbonation basins must be located out of doors, and enclosed structures must not be built above them, because of the danger of excessive amounts of CO_2 accumulating within the structures. Before repairmen enter recarbonation basins, the CO_2 supply should be turned off and the space thoroughly ventilated. In the use of liquid CO_2 there are many other safety considerations too numerous and detailed to be covered completely here. Complete published information can be obtained from liquid CO_2 suppliers or from the Compressed Gas Association, Inc., 500 Fifth Avenue, New York, N.Y.

RECARBONATION AT WATER FACTORY 21

Two stage recarbonation and lime recovery and reuse are integral parts of the 15 mgd wastewater reclamation plant at the Orange County Water District. The two stage system used provides for maximum recovery of calcium and also reduces the total dissolved solids (TDS) content of the water by removing precipitated calcium carbonate. It is possible to reduce the TDS of the secondary effluent by 10–15 percent by proper operation and control of the recarbonation process.

Figure 3-7 is a schematic of the recarbonation system. Note that the normal source of CO_2 is from lime furnace stack gas, and that there is a connection for an auxiliary supply using commercial CO_2. Figure 3-8 is a plan of the recarbonation basins.

The design criteria are as follows:

Design Criteria

Number of basins	2 parallel basins (1st stage recarbonation, intermediate settling and 2nd stage recarbonation in each basin.)
1st stage recarbonation basin	Each basin divided into two compartments; length = 16 ft (each); width = 36 ft 9 in.; depth = 10 ft; water depth = 9 ft. Equipped with oscillating flocculators. Equipped with CO_2 distributors in each compartment, and water sprays for suppressing froth.

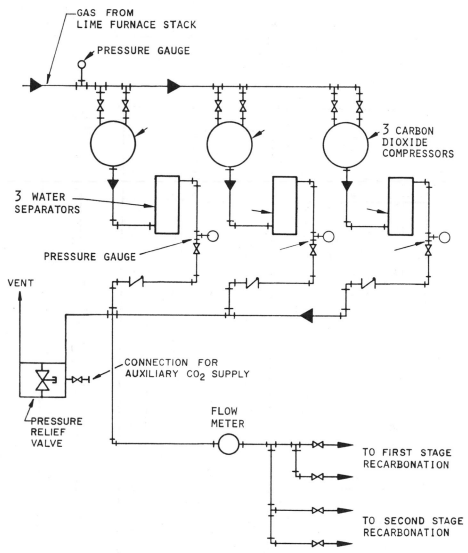

Fig. 3-7   Schematic of recarbonation system of the Orange County Water District.

Polymer addition

Intermediate settling   Length = 70 ft 9 in.; width = 73 ft 6 in.; sidewall depth = 12 ft 2 in.; sidewater depth = 10 ft 6 in.; overflow rate = 3000 gpd/ft^2 at 15 mgd; weir loading = 18.3 gpm/ft at 15 mgd.

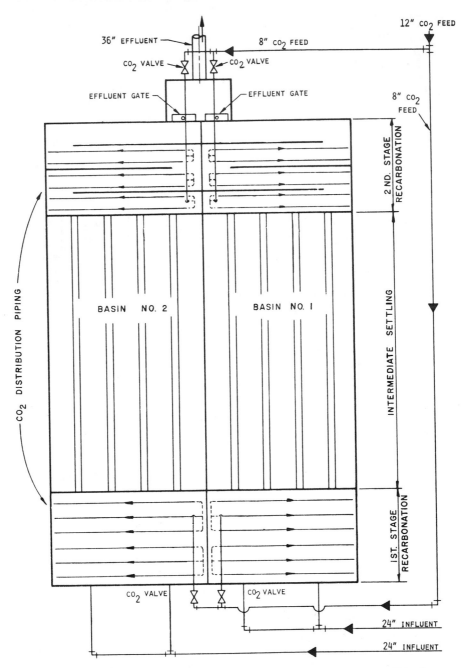

Fig. 3-8   Plan of recarbonation basins of Orange County Water District.

2nd stage recarbona- Each basin divided into four compartments;
tion   length = 7 ft 2 in. (each); width = 36 ft 9 in.;
depth = 12 ft; water depth = 10 ft. Three compartments equipped with CO_2 distributors and water sprays for suppressing froth.

Detention time at 15 Recarbonation, 1st stage = 15 minutes. Recar-
mgd   bonation, 2nd stage = 15 minutes. Intermediate settling = 40 minutes.

CO_2 compressors The total stack gas requirements for recarbonation at 12 percent CO_2 and 110° F at sea level are about 145 cfm/million gal, or about 2175 cfm for a flow of 15 mgd. There are three Nash CO_2 compressors, each with a capacity rating of 1600 cfm. Ordinarily, two units would be in service with the third unit on standby.

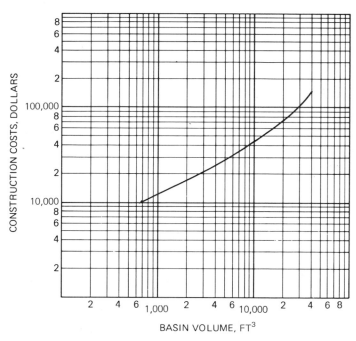

Fig. 3-9 Recarbonation basin capital costs—construction costs, 1975 (log–log plot).

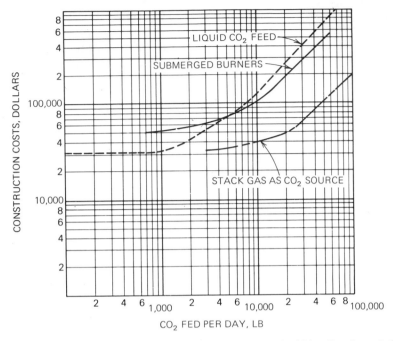

Fig. 3-10   Capital costs, 1975—CO₂ source and feed facilities (log–log plot).

First Stage Recarbonation

Flow enters the first stage recarbonation basin, where enough CO_2 is injected to lower the pH to about 10.2. As water flows through the basin the pH continues to drop and is below 10 when it enters the intermediate settling basin.

Intermediate Settling

Flow passes from the first recarbonation basin into the intermediate settling basin along the bottom of a common wall. The calcium carbonate particles formed in the first stage recarbonation basin settle by gravity. Each of the two intermediate settling basins is equipped with two circular mechanical sludge scrapers. Settled sludge is scraped to a central hopper and flows by gravity to the sludge pump station. Most of the settled sludge is pumped to the recalcining system, although a small portion can be recycled to the chemical clarifier rapid mix basin to aid coagulation and flocculation.

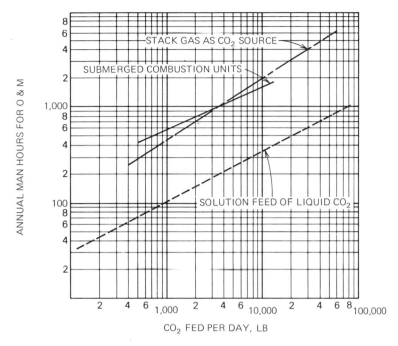

Fig. 3-11 CO_2 source and feed facilities—man-hour requirements for maintenance and operation (log–log plot).

Second Stage Recarbonation

Effluent from the intermediate settling basin flows over weirs and through launders by gravity to the second stage recarbonation basin. CO_2 is injected through perforated PVC pipe located along the bottom of this basin to lower the pH so that the water is in equilibrium with respect to calcium carbonate. The final recarbonated effluent then flows by gravity to the filter distribution box. Spray nozzles for foam suppression are located above the final recarbonated effluent collection box.

Process Control and Performance Evaluation

The total stack gas requirement at 12 percent CO_2 and 110° F at sea level is about 145 cfm per million gal, or about 2175 cfm for a flow of 15 mgd. Three compressors are provided, each with a capacity rating of 1600 cfm. Ordinarily, two units will be in service with the third unit on standby. Control of the amount of CO_2 applied is by trial and error based on observation of pH measurements through manual valve adjustments.

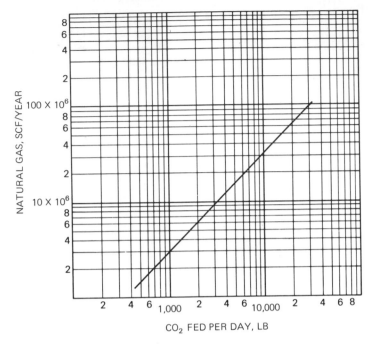

Fig. 3-12 Recarbonation—submerged combustion unit fuel requirements (log-log plot).

The first adjustment is to obtain the proper CO_2 flow for reduction of the pH in the second stage recarbonation basin accomplished by venting the excess CO_2 to the atmosphere through a muffled discharge line. The second adjustment is to split this flow between the two stages to obtain a pH of 10 at about the midpoint in the intermediate settling basin. Once the valves are set for a proper split, they require little or no further adjustment for changes in total CO_2 flow. At average wastewater temperatures, about 15 minutes is required for all bubbles of CO_2 gas to go into true solution and to react completely with hydroxide to form calcium carbonate. This is the reason for measuring pH at the midpoint of the settling basin rather than at the entrance.

A flowmeter measures total gas flow to the recarbonation basins. Pressure gages are provided on the common inlet header and each of the three individual compressor discharge lines. The compressors are designed to operate at differential pressure between inlet and discharge of not more than 8 psi. If the pressure exceeds this value, it is probable that one or more valves on the discharge lines are too far closed. In obtaining the split CO_2 flow between first stage and second stage basins, one set of valves should be fully open, with only the other set being throttled. It is never necessary to throttle both sets of valves in order to obtain the proper split.

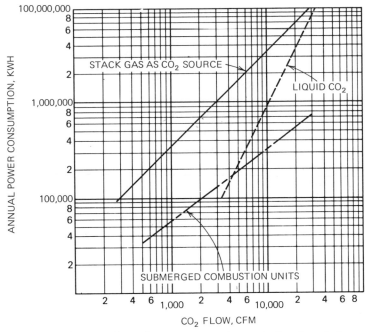

Fig. 3-13   **Recarbonation power requirements (log–log plot).**

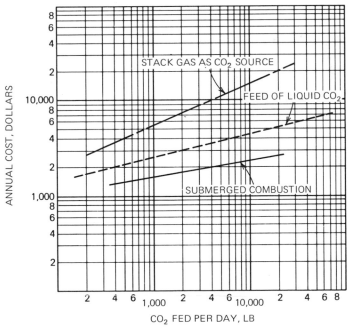

Fig. 3-14   **Recarbonation maintenance material costs, 1975 (log–log plot).**

Intermediate Settling Basin Loading

At design loading the surface overflow rate in the settling basin is 3000–3500 $gal/ft^2/day$ depending on the amount of recycle flow.

The two stage recarbonation process reduces the TDS by about 150 mg/l and the total hardness about 150 mg/l (as $CaCO_3$). The pH of the influent to the first stage basin and the effluent from the second stage basin are monitored and recorded continuously. In addition, composite samples of the second stage final effluent are tested daily for electrical conductivity, calcium, and alkalinity, and weekly for TDS.

Recarbonation Costs

To aid in the preparation of preliminary estimates, data are presented here for recarbonation costs including 1975 costs for construction and maintenance and operation using various methods of CO_2 generation (Figures 3-9–3-14).

REFERENCES

1. Anonymous, "Submarine Burners Make CO_2 for Softening Recarbonation," *Water Works Engineering*, 1963, p. 182.
2. Bulletin No. 7-W-83, "Carball CO_2 for Recarbonation," Walker Process Co. May, 1966.
3. Compressed Gas Association, *Handbook of Compressed Gases*. Van Nostrand Reinhold Co., New York, 1962.
4. Fair, M. F., and Geyer, J. C., *Water Supply and Waste Disposal*. John Wiley & Sons, New York, 1954.
5. Haney, P. D., and Hamann, C. L., "Recarbonation and Liquid Carbon Dioxide," *Journal American Water Works Assoc.*, 1969, p. 512.
6. Hoover, C. P., *Water Supply Treatment*, 8th ed. Bulletin 211, National Lime Association, Washington, D.C.
7. Pamphlet G-6, "Carbon Dioxide," 2nd ed. Compressed Gas Assoc., New York, 1962.
8. Pamphlet G-6, 1T, "Tentative Standard for Low Pressure Carbon Dioxide Systems at Consumer Sites." Compressed Gas Assoc., New York, 1966.
9. Ross, R. D., *Industrial Waste Disposal*. Van Nostrand Reinhold Co., New York, 1968.
10. Scott, L. H. "Development of Submerged Combustion for Recarbonation," *Journal American Water Works Assoc.*, 1940, p. 93.

4

Filtration

THE ROLE AND IMPORTANCE OF FILTRATION IN AWT

Filtration is the key process in the production of high quality effluents from wastewater. Efficient filtration can:

1. Remove particulate and colloidal matter not settleable after either biological or chemical flocculation or both.
2. Increase removal of suspended solids, turbidity, phosphorus, BOD, COD, heavy metals, asbestos, bacteria, virus, and other substances.
3. Improve the efficiency and reduce the cost of disinfection through removal of suspended matter and other interfering substances.
4. Assure continuous plant operation and consistent effluent quality.
5. Increase overall plant reliability by overcoming common irregularities in biological and chemical treatment.

6. Protect granular activated carbon against fouling, and increase carbon adsorption efficiencies by reducing the load of applied organics.
7. Increase aesthetic appreciation by producing an attractive, sparkling, high clarity effluent.
8. Be combined with biological denitrification by using the fine grains of the filter bed as an attached growth medium.

These filtration capabilities for wastewater were not demonstrated until the early 1960s with the development of the coarse to fine granular filter as embodied in the dual media filter and its successor, the mixed media filter. For almost 100 years engineers had attempted without success to clarify sewage using fine screens, slow and rapid sand filters, diatomaceous earth, and other types of strainers or surface-type filters. These efforts failed because of the tremendous volume of solids to be removed from wastewater and stored in the bed, and because of the fragile and sticky nature of these solids. With the use of any surface filter, solids accumulate rapidly at the top surface and form a seal. This produces high head loss followed by breakthrough of particulates as operating head increases.

The practical engineering solution to these problems came about by the development of in depth, coarse to fine filtration combined with the use of alum or polymers as filter aids. Only when good mixed media filter performance was demonstrated at pilot scale was it possible to construct the first full scale AWT plants.

Definition

The term *filtration* has different meanings or connotations to various people. Outside the water treatment profession, even in other technical disciplines, filtration is commonly thought of as a mechanical straining process. It may also have the same basic meaning in water practice as applied to the passage of water through a very thin layer of porous material deposited by flow on a support septum. This type of filter has a few rather specialized applications to water treatment, as described later. However, as most frequently used in water treatment parlance, filtration refers to the use of a relatively deep ($1\frac{1}{2}$–3 ft) granular bed to remove impurities from water. This general type of filter has a wide range of applications. In contrast to mechanical strainers, which remove only part of the coarse suspended solids, the newest types of filters used in water treatment remove all suspended solids, including virtually all colloidal particles. Over the years the meaning of filtration as used by the water treatment industry has changed as improved filters have been developed and as the nature of the physical and chemical processes involved in filtration have become better understood. It is important in discussions of filtration to exercise care in making known the specific details of the particular filter under consideration so that important distinctions in function or efficiency between filter types are understood and not overlooked.

Water filtration can be further defined as a physical-chemical process for separating suspended and colloidal impurities from water by passage through a bed of granular material. Water fills the pores of the filter medium, and the impurities are adsorbed on the surface of the grains or trapped in the openings. Filtration under controlled conditions in wastewater treatment plants is an indispensable unit process.

Filter Types

There are several ways to classify filters. They can be described according to the direction of flow through the bed, that is, downflow, upflow, biflow, radial flow, horizontal flow, fine to coarse, or coarse to fine. They may be classed according to the type of filter media used, such as sand, coal (or anthracite), coal–sand, multilayered, mixed media, or diatomaceous earth. Filters are also classed by flow rate. Slow sand filters operate at rates of 0.05–0.13 gpm/ft^2, rapid sand filters operate at rates of 1–2 gpm/ft^2, and high rate filters operate at rates of 3–15 gpm/ft^2. Another flow characteristic of filters is pressure or gravity flow. Gravity filter units are usually built with open top and constructed of concrete or steel, while pressure filters are ordinarily fabricated from steel in the form of a cylindrical tank. Available head for gravity flow usually is limited to about 8–12 ft, while it may be as high as 150 psi for pressure filters. Because pressure filters have a closed top, it is not possible to maintain routine visual observation of the condition of the filter media. Further, it is possible to disturb the media violently in a pressure filter by sudden changes in pressure. These two factors have in the past tended to limit municipal applications of pressure filters. The susceptibility to bed upset and the inability to see the surface of the media in pressure filters have been compensated for, to some extent at least, by the use of quick opening manholes and particularly by the recent development and application of recording turbidimeters for continuous monitoring of the filter effluent turbidity. The introduction of a 3 in. layer of coarse (1 mm) high density (specific gravity 4.2) garnet or ilmenite between the fine media and the gravel supporting bed has virtually eliminated the problem of gravel upsets, another of the concerns about the use of pressure filters.

Rapid or high rate filters operate at about 30 times the rate of slow filters, so they must be cleaned about 30 times as often. A common method for cleaning slow sand filters is to scrape a thin layer of media from the surface of the bed, discard it, or wash it and return it to the bed. Rapid filters are washed in place by reversing the flow through the media to expand and scour the media. This hydraulic cleaning process can be supplemented as necessary in various ways, such as water jet agitation of the expanded media, mechanical stirring by rakes, or injection of air into the bed before or during backwashing. With the scraping method used to clean slow sand filters, it was advantageous to collect as much of the foreign material as possible at the top surface of the bed. This favored the use of a fine-to-coarse filter medium. With modern hydraulic backwashing supplemented by water or air jet agitation, it is possible to thoroughly clean granular beds at all depths, making it feasible to tailor the graduation of the fine media (coarse to fine) for optimizing the filter cycle rather than the cleaning cycle of filter operation.

HOW WATER FILTERS ACT
Mechanisms

Several mechanisms are involved in particle removal by filtration. Some of the mechanisms are physical and others are chemical in nature. To fully explain the overall removal of impurities by filtration, the effects of both the physical and chemical actions occurring in a granular bed must be combined. Efficient filtration involves particle destabilization and particle transport similar to the mechanisms of coagulation. Good coagulants are also efficient filter aids. The processes of coagulation and filtration are inseparable and the interrelationships must be considered for best treatment results. One important advantage of filtration over coagulation in the removal of colloidal particles which are present in very dilute concentrations is the much greater number of opportunities for contact which are afforded by the granular bed as compared to the number afforded by stirring the water. The removal efficiency of a filter bed is independent of the applied concentration, while the time of flocculation is concentration dependent.

The removal of suspended particles in a filter consists of at least two steps: (1) the transport of suspended particles to the solid–liquid surface of a grain of filter media or to another floc particle previously retained in the bed, followed by (2) the attachment and adsorption of particles to this surface. For filter runs of practical length, filter aid to coat filter grains must be added continuously; precoated grains should not be used. When filter aids are used, the filtration process is so effective that conventional sand or other surface filters clog very rapidly. Effective filtration without excessive head loss can be accomplished best by use of a coarse to fine, in depth filter.

There are two basic approaches to obtaining optimum filterability of a water. One is to establish the dosage of primary coagulant for maximum filterability rather than for production of the most rapidly settling floc. A second approach is to add a dose of a second coagulant as a filter aid to the settled water as it enters the filter.

Adsorption of suspended particles to the surface of the filter grains is an important factor in filter performance. Physical factors affecting adsorption include both the nature of the filter and the suspension. Adsorption is a function of the filter grain size, flow particle size, and the adhesive characteristics and shearing of the floc. Chemical factors affecting adsorption include the chemical characteristics of the suspended particles, the aqueous suspension medium, and the filter medium. Two of the most important chemical characteristics are the electrochemical and van der Waals forces (molecular cohesive forces between particles).

For effective filtration, the objective of pretreatment should be to produce small, dense (rather than large and fluffy) floc so that the particles are small enough to penetrate the surface and into the bed. Removal of floc within a bed is accomplished primarily by contact of the floc particles with the surface of the grains or previously deposited floc, and adherence thereto. Contact is brought about principally by the convergence of stream lines at contractions in the pore channels between the grains. Of minor importance is the flocculation, sedimentation, and entrapment which occurs within the pores of the bed.

Filtration Efficiency

Filters are highly efficient in removing suspended and colloidal materials from water. Impurities affected by filtration included: turbidity, bacteria, algae, viruses, color, oxidized iron and manganese, radioactive particles, chemicals added in pretreatment, heavy metals and many other substances. Because filtration is both a physical and a chemical process, there are a large number of variables that influence filter efficiency. These variables exist both in the water applied to the filter, and in the filter itself. In view of the complexity of what is commonly considered to be a relatively simple process by plant designers and operators, it is surprising that the general level of filter performance is so high. Knowledge of the factors affecting filter efficiency increased quite rapidly during the 1960s. Use of this information in the design and control of filters will make possible remarkably better water quality with average or good filter operation, and will make it even more difficult to obtain unsatisfactory results with poor operation.

Filter efficiency is affected by the following properties of the applied water: temperature, filterability, and the size, nature, concentration, and adhesive qualities of suspended and colloidal particles. Cold water is notably more difficult to filter than warm water, but usually there is no control over water temperature. Filterability, which is related to the nature, size, and adhesive qualities of the suspended and colloidal impurities in the water, is the most important property. By recording filter effluent turbidities, appropriate adjustments can be made in chemical treatment of filter influent to obtain optimum filterability in the plant filters. Maximum filterability is much more important to production of a water of maximum clarity (minimum turbidity) than is maximum turbidity reduction prior to filtration.

Some properties of the filter bed which affect filtration efficiency are: the size and shape of the grains; the porosity of the bed (or the hydraulic radius of the pore space); the arrangement of grains, whether fine to coarse or coarse to fine; the depth of the bed; and the headloss through the bed. In general, filter efficiency increases with smaller grain size, lower porosity, and greater bed depth. Coarse to fine filters contain much more storage space for materials removed from the water and permit the practical use of much finer materials in the bottom of the bed than can be tolerated at the top of a fine to coarse filter. Because of the greater total surface area of the grains, smaller grain size, and lower porosity, the coarse to fine filter is more efficient than the fine to coarse filter. The much greater total grain surface area and the smaller grain size provided by mixed media as compared to dual media account for the greater resistance to breakthrough possessed by mixed media. Dual media, which should be considered merely an intermediate step in the development of mixed media, is less resistant to breakthrough than a rapid sand filter, while mixed media is more resistant than either. Hydraulic throughput rate also affects filter efficiency, although within the range of 2–8 gpm/ft^2 the rate is not nearly as significant as other variables are to effluent quality. In general, the lower the rate, the higher the efficiency. All other conditions being equal, a filter will produce a better effluent when operating at a rate of 1 gpm/ft^2 than when operating at 8 gpm/ft^2. However, it is also true that a given filter may operate entirely satisfactorily at 8 gpm/ft^2 on a properly prepared water, yet fail to produce a satisfactory effluent at 1 gpm/ft^2 when receiving an improperly pretreated water. With good filter design and operation, the optimum throughput rate is a matter of economics rather than a question of safety.

The efficiency of filters in bacterial removal varies with the applied loading of bacteria, but with proper pretreatment it should exceed 99 percent. Bacterial removal by filtration, however, should never be assumed to reach 100 percent. The water must be chlorinated for satisfactory disinfection. Coagulation, flocculation, and filtration will remove more than 98 percent of polio virus at filtration rates of 2–6 gpm/ft^2, but complete removal is dependent upon proper chlorination.

The turbidity of the effluent from a properly operating filter should be less than 0.2 JTU. With proper pretreatment, filtered water should be essentially free of color, iron, and manganese. Large microorganisms, including algae, diatoms, and amoebic cysts are readily removed from properly pretreated water by filtration.

The perfection of very sensitive, accurate, and reliable turbidimeters for the continuous monitoring of filter effluent quality has revolutionized the degree of control that can be exercised over filter performance. The turbidity of filter effluent is instantly and continuously determined, reported, and recorded. The signal obtained from the instrument can be used to sound alarms, adjust the coagulant dose, or, if necessary, shut down an improperly operating filter unit. This greatly increases the reliability with which filters of all types may be operated, and may broaden the range of safe application for pressure filters.

Filter Action in Denitrification

In wastewater treatment it is possible to combine nitrate removal with fine media filtration. The filter grains serve as a medium for attached growth of denitrifying organisms under anaerobic conditions. A supplemental source of carbon, such as methanol, is added in proper dosage to the filter influent water. Either mixed media beds or deep, coarse beds may be used for the combined filtration–denitrification processes. Provisions must be made to remove excess nitrogen gas and bacterial growths from the filter bed between normal backwashings (usually by filter "bumping") and to control slime growths in effluent piping. It is possible to obtain nitrate removals in excess of 98 percent using this process, which is often referred to as columnar denitrification. (See Chapter 7 for details).

The provision for attached growth of the denitrifying organisms is a distinct advantage over systems using a suspended growth reactor. The denitrifiers are slow growers, light in weight, and easily washed out of a suspended growth system. The Dravo Corporation and Neptune Microfloc Inc. have been active in developing filters and treatment techniques for combined downflow filtration and biological denitrification of nitrified wastewater.

BIOLOGICAL VS. CHEMICAL PRETREATMENT

The goal of filtration is the removal of particulate matter. Before selecting the filtration process to be used, it is necessary to determine accurately the total amount of filterable solids, the variability of their quantity, the fraction to be removed, and their characteristics.

The filtration characteristics of the solids found in a biological treatment plant effluent differ greatly from those of the floc resulting, for example, from chemical coagulation for phosphate removal. When no coagulants are used, the filterability of solids in a biological plant effluent is dependent upon the degree of flocculation achieved in the biological process. A trickling filter achieves only a poor degree of flocculation and efficient filtration of the effluent from a trickling filter plant will usually provide only about 50 percent or less removal of the suspended solids normally present. The activated sludge process is capable of much higher degrees of biological flocculation than is the trickling filter process. Culp and Hansen[20] found that up to 98 percent of the suspended solids found in an extended aeration plant effluent with 24 hour aeration of domestic sewage could be removed by filtration to produce turbidities as low as 0.3 JTU without the use of coagulants. These authors later reported[21] that pilot plant studies showed that the degree of biological flocculation achieved in an activated sludge plant was directly proportional to the aeration time and inversely proportional to the ratio of the amount of organic material added per day to the amount of suspended solids in the aeration chamber (load factor). Variation of mixed liquor suspended solids in the normal operating range of 1500–5000 mg/l did not significantly affect the filterability of the effluent at a given aeration time and load factor. For domestic wastes, aeration times of 10 hours or more were found to provide flocculation adequate to permit an efficient downstream filter to remove 90–98 percent of the effluent suspended solids. The flocculation provided by aeration times of 6–8 hours with domestic wastes enabled 70–85 percent suspended solids removal from the secondary effluent.

The strength of biological floc is much greater than that of floc resulting from chemical coagulation. Tchobanoglous and Eliassen[73] have also noted this basic difference. As a result, biological floc can be removed with coarser filters at higher filtration rates than can the weaker chemical flocs, which tend to shear and penetrate a filter more readily. Chemical floc strength can be controlled to some degree with the use of polymers as coagulant aids.

The types of filtration processes discussed in this book can be placed in two general categories: (1) in depth filtration devices—granular media filters graded coarse to fine in the direction of flow; and (2) surface filtration devices, including microscreens, diatomaceous earth filters, slow sand filters, rapid sand filters, and moving bed filters.

IN DEPTH FILTRATION—GENERAL THEORY

The previously mentioned sensitivity of a rapid sand filter to high suspended solids concentrations can be readily understood by examination of Figure 4-1A which is a cross section of a typical single medium filter, such as a rapid sand filter. During filter backwashing, the sand grades hydraulically, with the finest particles rising to the top of the bed. As a result, most of the material removed by the filter is removed at or very near the surface of the bed. Only a

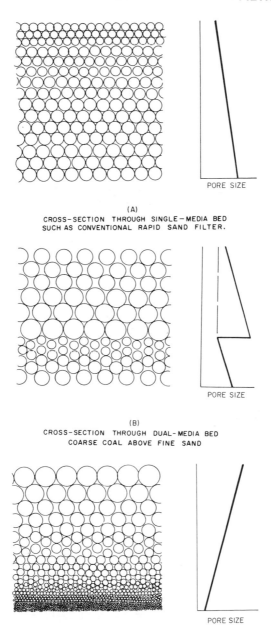

(A)
CROSS-SECTION THROUGH SINGLE-MEDIA BED
SUCH AS CONVENTIONAL RAPID SAND FILTER.

(B)
CROSS-SECTION THROUGH DUAL-MEDIA BED
COARSE COAL ABOVE FINE SAND

(C)
CROSS-SECTION THROUGH IDEAL FILTER
UNIFORMLY GRADED FROM COARSE TO FINE
FROM TOP TO BOTTOM

Fig. 4-1   Graphical representation of various media designs. *(Courtesy Neptune Microfloc, Inc.)*

small part of the total voids in the bed are used to store particulates, and head loss increases very rapidly. When the secondary effluent contains relatively high solids concentrations, a sand filter will blind at the surface in only a few minutes. Typically, 75–90 percent of the head loss occurs in the upper inch of rapid sand beds when they are filtering activated sludge plant effluent.

One approach to increasing the effective filter depth is the use of a dual media bed using a discrete layer of coarse coal above a layer of fine sand, as shown in Figure 4-1B. The work area is extended, although it still does not include the full depth of the bed, for there is some fine to coarse stratification within each of the layers, as shown by the graph depicting pore size. Effective size of the sand in a typical dual media filter is 0.4–1.0 mm; coal size is 0.8–2.0 mm; and the layers are 10 and 20 in. deep, respectively.

In designing a dual media bed, it is desirable to have the coal (specific gravity about 1.6) as coarse as is consistent with solids removal to prevent surface blinding, and to have the sand (specific gravity about 2.6) as fine as possible to provide maximum solids removal. However, if the sand is too fine in relation to the coal, it will actually rise above the top of the coal during the first backwash and remain there when the filter is returned to service. For example, if a 0.2 mm sand were placed below 1.0 mm coal, the materials would actually reverse during backwash, with the sand becoming the upper layer and the coal, the bottom. Although the sand has a higher specific gravity, its small diameter in this case would result in its rising above the coal at a given backwash rate. The only way to enable very fine silica sand to be used in the bottom filter layer would be to use finer coal, but this would defeat the purpose of the upper filter layer, since the fine coal would be susceptible to surface blinding. Experience has shown that it is not feasible to use silica sand smaller than about 0.4 mm, because smaller sand would require coal small enough to result in unacceptably high headloss at rates above 3 gpm/ft^2.

To overcome the above limitation and to achieve a filter which very closely approached an ideal one (Figure 4-1C), the mixed media concept was developed (patents held by Neptune Microfloc, Inc., Corvallis, Oregon). The problem of keeping a very fine medium at the bottom of the filter is overcome by using a third, very fine, very heavy material (garnet, specific gravity of about 4.2, or ilmenite, specific gravity of about 4.5) beneath the coal and sand. The garnet (or ilmenite), sand, and coal particles are sized so that intermixing of these materials occurs and no discrete interface exists between the three. This eliminates the stratification illustrated for the dual media filter in Figure 4-1B and results in a filter which very closely approximates the ideal of a uniform decrease in pore space with increasing filter depth, as shown in Figure 4-1C.

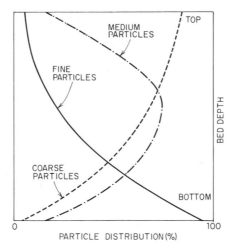

Fig. 4-2   Distribution of media in a properly designed mixed media filter *(Courtesy Neptune Microfloc, Inc.)*

Actually, the term *coarse to fine* refers more accurately to the pore space rather than to the media particles themselves. The illustrations in Figure 4-1 actually represent pore size rather than the size of individual grains. By selecting the proper size distribution of each of the three media, it is possible to construct a bed which has an increasing number of particles at each successively deeper level in the filter. A typical mixed media filter has a particle size gradation which decreases from about 2 mm at the top to about 0.15 mm at the bottom. The uniform decrease in pore space with filter depth allows the entire filter depth to be utilized for floc removal and storage. Figure 4-2 shows how particles of the different media are actually mixed throughout the bed. At all points in the bed there is some of each component, but the percentage of each changes with bed depth. There is steadily increasing efficiency of filtration in the direction of the flow.

There is no one mixed media design which will be optimum for all wastewater filtration problems. For example, removal of small quantities of high strength biological floc often found in activated sludge effluents may be satisfactorily achieved by a good dual media design. With a weaker floc strength or with an increase in applied solids loading, the benefits of the mixed, tri-media bed become more pronounced. Comparative data published by Conley[14] showed that a mixed media bed will consistently produce longer runs and lower effluent turbidities when removing alum floc at filter rates of 5 gpm/ft^2 than would a dual media bed. Conley and Hsiung[13] have presented techniques designed to optimize the media selection for any given filtration application. Their work clearly indicates the marked effects that the

TABLE 4-1   Illustrations of Varying Media Design for Various Types of Floc Removal. (*After Conley and Hsiung.*[13])

Type of Application	Garnet		Silica Sand		Coal	
	size	depth	size	depth	size	depth
Very heavy loading of fragile floc	− 40 + 80	8 in.	− 20 + 40	12 in.	− 10 + 20	22 in.
Moderate loading of very strong floc	− 20 + 40	3 in.	− 10 + 20	12 in.	− 10 + 16	15 in.
Moderate loading of fragile floc	− 40 + 80	3 in.	− 20 + 40	9 in.	− 10 + 20	8 in.

Size: − 40 + 80 = passing No. 40 and retained on No. 80 U.S. sieves.

quantity and quality of floc to be removed can have on media selection (Table 4-1). Removal of the poorly flocculated solids normally found in a trickling filter effluent can be improved by using smaller media than would be used for removal of activated sludge effluent suspended solids. Pilot tests of various media designs can be more than justified by improved plant performance in most cases.

Certainly, use of three media does not in itself ensure superior performance, as illustrated by the experiences of Oakley and Cripps.[49] Using the coal, sand, and garnet materials readily available to them resulted in a bed which did not have significant advantages over other filter types when filtering secondary effluent. Oakley[49a] reports the tri-media bed was made up of 8 in. of 0.7–0.8 mm garnet, 8 in. of 1.2–1.4 mm sand, and 8 in. of 1.4–2.4 mm coal. Our experience indicates that this bed was too shallow and too coarse and that a better media selection for the particular application would have been 3 in. of 0.4–0.8 mm garnet, 9 in. of 0.6–0.8 mm sand, and 24 in. of 1–2 mm coal.

One of the key factors in constructing a satisfactory mixed media bed is the careful control of the size distribution of each component medium. Rarely is the size distribution of commercially available materials adequate for construction of a good mixed media filter. The common problem is failure to remove excessive amounts of fine materials. These fines can be removed by placing a medium in the filter, backwashing it, draining the filter, and skimming the upper surface. The procedure is repeated until the field sieve analyes indicate that an adequate particle size distribution has been obtained. A second medium is added and the procedure repeated. The third medium is then added and the entire procedure repeated. Sometimes 20–30 percent of the materials may have to be skimmed and discarded to achieve the proper particle size distribution.

The remainder of this chapter will be devoted to filter system design considerations; these are largely independent of the exact media design selected, which must be optimized for each specific application.

PERFORMANCE OF IN DEPTH FILTERS

Plain Filtration

As previously discussed, the degree of solids removal in filtering of secondary effluents without the use of chemical coagulation is dependent upon the degree of biological flocculation achieved in the secondary plant. Plain filtration of primary effluent will provide very little suspended solids reduction because of the colloidal nature of most of the solids. Because of the high strength of biological floc, there will be little difference between the effluent quality produced by a well designed sand filter and by a mixed media filter. However, there will be a very significant difference in favor of the mixed media in length of filter runs, quantity of backwash water, and system reliability.

The advantage of in depth filtration is well illustrated by the data of Tchobanoglous,[72] which show that 90 percent of the head loss occurred in the top 2 in. of a rapid sand filter applied to activated sludge effluent (0.488 mm sand, 5 gpm/ft^2). As shown in Figure 4-3, about 90 percent of the head loss in a mixed media bed operating under similar conditions occurred in the upper 24 in. of the 30 in. bed, indicating excellent utilization of the bed depth for floc removal.[19]

Although a mixed media filter can tolerate higher suspended solids loadings than can the other filtration processes discussed, it still has an upper limit of applied suspended solids at which economically long runs can be maintained. With activated sludge effluent, suspended solids loadings of up to 120 mg/l, filter runs of 15–24 hr at 5 gpm/ft^2 have been maintained in operation to a terminal headloss of 15 ft of water. Suspended solids concentrations of 300 mg/l or more will lead to uneconomically short filter runs, even when a mixed media filter is used. Figure 4-4 illustrates the effects of influent solids on rate of headloss buildup. The media design shown in Figure 4-4 is somewhat finer in size than that now recommended for plain filtration of activated sludge solids by the media supplier but the general magnitude of headloss buildup can be determined from this curve. Should the secondary plant involved have a history of frequent, severe upsets resulting in secondary effluent suspended solids concentrations of 200–500 mg/l, an intermediate settling tank should be added between the secondary clarifier and the filter, with provision for chemical coagulation during upset periods.

Because of the variations in biological flocculation achieved in secondary processes, it is impossible to present accurate estimates of the efficiency of

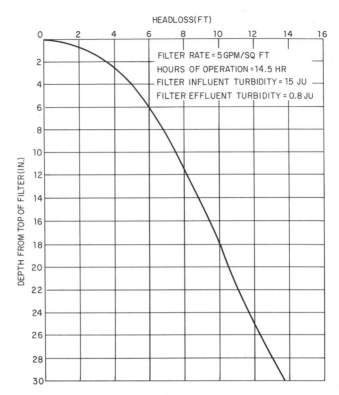

Fig. 4-3 Headloss distribution in mixed media filter applied to activated sludge effluent. *(Courtesy* **Water & Sewage Works).**

plain filtration without knowing the specific biological process loading parameters. The following filter effluent qualities are presented as general guides to the suspended solids concentrations which might be achieved in filtration of a secondary effluent of reasonable quality without chemical coagulation: high rate trickling filter, 10–20 mg/l; two stage trickling filter, 6–15 mg/l; contact stabilization, 6–15 mg/l; conventional activated sludge plant, 3–10 mg/l; activated sludge plant with load factor less than 0.15 mg/l, 1–5 mg/l. The degree of biological flocculation is the limiting factor in determining the solids removal. If effluent quality substantially better than that indicated by the general guides above is required, chemical coagulation must be used.

The effluent quality produced by plain filtration of secondary effluents is essentially independent of filter rate within the range of 5–15 gpm/ft^2; this is primarily due to the high strength of the biological floc. This statement is based on the results of many tests conducted by the authors which have

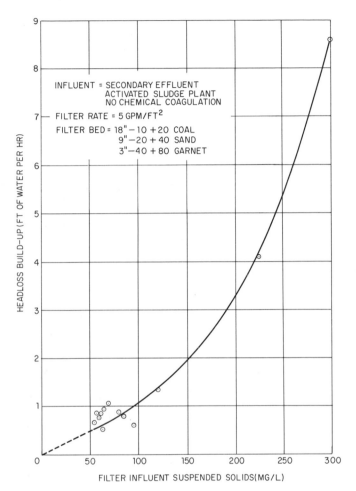

Fig. 4-4 Effect of filter influent (activated sludge effluent) suspended solids on headloss buildup for mixed media filter.

been confirmed by Tchobanoglous, who reports for his experiments on plain filtration that "for a given sand size, varying filter rate had little effect on the suspended solids removal characteristics of the filter bed."[72] It must be emphasized that this applies only to removal of strong biological floc and not to the weaker floc resulting from chemical coagulation. Tchobanoglous studied rates of 2–10 gpm/ft^2. The selection of design rate will depend primarily on the anticipated quality of filter influent. Should the operation of the secondary plant be extremely stable and rarely produce secondary effluent suspended solids in excess of 30 mg/l, a filter rate of 10–15 gpm/ft^2

would be satisfactory. Should the plant suffer from inconsistent operation resulting in frequent upsets and occurrences of 50–100 mg/l secondary effluent suspended solids concentrations, a loading rate of 5 gpm/ft^2 would be a better selection. Should the secondary effluent solids frequently exceed 100 mg/l even for short periods, consideration should be given to providing an intermediate settling basin between the secondary clarifier and the filter, with provisions for chemical coagulation during severe upsets, or to revising the secondary plant design or operation.

Filtration of Chemically Coagulated Effluents

The selection of design rate will be dependent upon the quality of filter influent and the required effluent quality. As noted earlier, the strength of the chemical floc is considerably less than for biological floc, with the result that the chemical floc shears more readily, causing filter breakthrough should the filter rate be excessive or the filter media too coarse. Mixed media filtration of properly coagulated secondary effluent will produce filter effluent essentially free of suspended solids as evidenced by turbidities of 0.1 JTU. The filter will also aid substantially in phosphorus removal by removing nonsettleable colloidal material which would contribute to the effluent phosphorus concentration. For example, proper chemical coagulation and settling of secondary effluent can produce a phosphate concentration of 0.5–2 mg/l when operating efficiently. Mixed media filtration of this effluent is capable of reducing the phosphorous concentration to less than 0.1 mg/l. O'Farrell and co-workers[52] report that the dual media filtration step provided 3–14 percent of the overall 90 percent or more phosphorus removal provided by the FWPCA pilot plant at Washington, D.C. At South Tahoe, the filtration step provides 8–12 percent of the overall 99 percent phosphorus removal.

The use of polymers as filtration aids to increase chemical floc strength is discussed in a subsequent section. Such use of polymers will permit higher filter rates and/or coarser filters, while reducing the risk of floc shearing.

Filtration rates of 5 gpm/ft^2 are routinely used for application of mixed media filters for chemical floc removal in potable water treatment. Rates higher than this, up to 10 gpm/ft^2, are practical in most tertiary applications where coagulation and efficient sedimentation precede the filtration step. If the pilot filter study required to optimize the filter rate and media design is not practical, selection of a filter rate of 5 gpm/ft^2 and sizing the filter hydraulics for twice this rate will enable the nominal capacity at 5 gpm/ft^2 to be exceeded should plant scale tests indicate higher filter rates are permissible.

The fact that mixed media beds can tolerate high solids loadings permits the application of coagulants directly to the filter influent without presettling. The coagulant doses required for phosphate removal are normally high enough that presettling is economically justified and should be used. However, if clarification only and not phosphate removal is the goal of co-agulation, coagulant doses up to 40–50 mg/l with secondary effluent sus-pended solids loadings of 25–35 mg/l can be handled with direct filtration at rates of 5 gpm/ft^2. If direct filtration of coagulated effluent is considered, a pilot study is mandatory. Conley and Hsiung[13] have reported experiences with some of the numerous water treatment plants using direct mixed media filtration of coagulated water. They report one case with 12 hour filter runs at 5 gpm/ft^2 with 32 mg/l of alum, 0.2 mg/l of polymer, and 25 units of tur-bidity applied directly to the filter without prior flocculation or settling. Filtered water turbidities were 0.2–0.4 JTU.

Use of Polymers as Filtration Aids

Polymers are high molecular weight, water soluble compounds which can be used as primary coagulants, settling aids, or filtration aids. They may be cationic, anionic, or nonionic in charge. Generally, the doses required for coagulation or as a settling aid in conjunction with another coagulant far exceed that needed as a filtration aid. Typical doses when used as a settling aid are 0.1–2.0 mg/l, while doses of less than 0.1 mg/l are often adequate to serve as a filtration aid. When used as a filtration aid, the polymer is added to increase the strength of the chemical floc and to control the depth of pene-tration of floc into the filter. The polymer is not added to improve coagula-tion but rather to strengthen the floc made up of previously coagulated material in order to minimize the prospects of shearing fragile floc and caus-ing filter breakthrough. For maximum effectiveness as a filtration aid, the polymer should be added directly to the filter influent and not in an upstream settling basin or flocculator. However, if polymers are used upstream as settling aids, it may not be necessary to add any additional polymer as fil-tration aid.

Figure 4-5 illustrates the effects of polymers as filter aids. The conditions represented by Figure 4-5A illustrate the results of a fragile floc shearing and then penetrating the filter, causing a premature termination of its run due to breakthrough of excessively high effluent turbidity. If the polymer dose is too high (Figure 4-5B), the floc is too strong to permit penetration into the filter, causing a rapid buildup of headloss in the upper portion of the filter and a premature termination by excessive headloss. The optimum polymer dose will permit the terminal headloss to be reached simultaneously with the first sign of increasing filter effluent turbidity (Figure 4-5C).

Many polymers are delivered in dry form. They are not easily dissolved, and special polymer mixing and feeding equipment is required. Typical equipment is shown in Figures 4-6 and 4-7. Many polymers are biodegradable

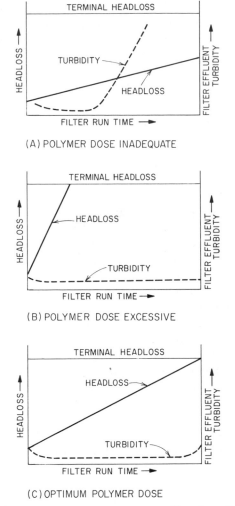

(A) POLYMER DOSE INADEQUATE

(B) POLYMER DOSE EXCESSIVE

(C) OPTIMUM POLYMER DOSE

Fig. 4-5   Effects of polymers as filtration aids.

and cannot be stored in dilute solution for more than a few days without suffering significant degradation and loss of strength.

In filtration of secondary effluents, the addition of chemicals to the filter influent does not materially improve suspended solids removal. If enough chemical is added to get good coagulation, the filter is not able to hold the material unless prior settling is provided.

For chemically coagulated secondary effluent and degritted raw waste-water, it is necessary to settle before filtration. The suspended solids con-

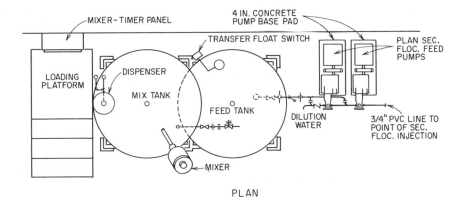

PLAN

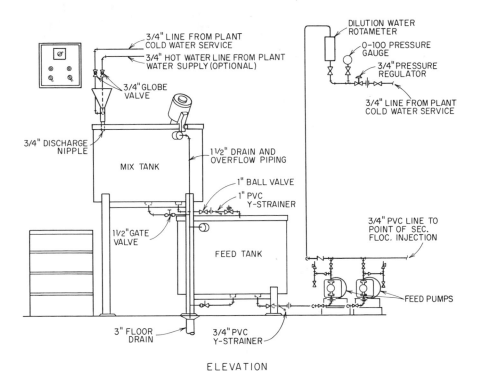

ELEVATION

Fig. 4-6 Polymer mixing and feed equipment. (Courtesy Neptune Microfloc, Inc.)

tent of these effluents is less than for secondary effluent that has not been chemically treated. The chemically coagulated effluents are more nearly like those from water treatment plants than like those from waste treatment plants. With the chemically treated effluents, filter breakthrough is always

Fig. 4-7 Typical polymer feed station. *(Courtesy Neptune Microfloc, Inc.)*

a potential problem, and this can be regulated by feeding polyelectrolyte just ahead of the filter. When alum is added to a water or to a wastewater, traces of polymer can be used to stop filter breakthrough. However, when alum is not used the polymers do not do any good.

Filter Aid Control by Interface Monitoring

This technique has been developed in conjunction with dual media filters. A sample of water is removed from the filter at a point near the interface between the coarse anthracite coal and the fine sand. The turbidity of this sample is continuously monitored and recorded, as is the filter effluent turbidity. Figure 4-8 shows the interface sampling system. The sample at the interface is obtained with a device which is a specially constructed well screen with slits small enough so that the coal and sand cannot pass through. The interface screen is placed 2 in. above the fine sand prior to anthracite placement. The turbidimeter is placed as close as possible to the filter. Samples flow by gravity to the turbidimeters. This prevents air bubbles, which frequently occur when turbidimeter samples are pumped, and also provides for the fastest possible response time. The interface turbidity sample is designed as a tool to aid the treatment plant operator in obtaining optimum performance from a dual media filter. The turbidity value obtained at the interface indicates where floc removal is occurring within the filter bed. For

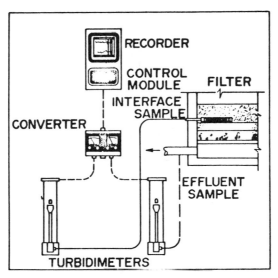

Fig. 4-8   Schematic diagram of an interface monitoring system. *(Courtesy Turbitrol Co.)*

instance, a high interface turbidity and a low filter effluent turbidity indicates the proper coagulant dosage is being used but that significant amounts of floc are penetrating through the anthracite layer and must be removed by the fine sand. If the sand is forced to carry too much of the load, either the headloss buildup becomes excessive or breakthrough of turbidity may occur into the effluent. When floc removal does not occur in the anthracite layer after the proper coagulant dose is applied, the floc strength is inadequate and should be increased by the feed of polymers. This method works with any material which can be successfully used as a filter aid or conditioner. The filter aid must be started at a known underdose and incrementally increased until the desired floc removal is obtained in the coal layer. If an overdose is applied, it is usually impossible to use the interface concept until the filter has been washed. It may also be desirable to adjust the filter aid dosage during the filter run, beginning with no aid or a small amount, and increasing the dosage as the run progresses.

DESIGN OF FILTER SYSTEMS

Selection of Type, Size, and Number of Units

The gravity and pressure filter structures commonly used in water treatment plants are readily adaptable to advanced wastewater treatment plant filtra-

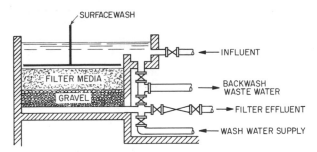

Fig. 4-9   Cross section through a typical gravity filter.

tion. Pressure filters are often advantageous in waste treatment applications for the following reasons:

1. In many wastewater applications, the applied solids loading is higher and more variable than in a water treatment application. Thus it is desirable to have higher heads (up to 20 ft) available than practical with gravity filter designs to provide maximum operating flexibility.
2. In advanced waste treatment processes, the filtration step is frequently followed by another unit process (carbon adsorption, ion exchange, etc.). The effluent from a pressure filter can be passed through a downstream process without having to pump the filter effluent, often eliminating a pumping step which would be required with a gravity filter.
3. All filter wash water must be treated in sewage applications. The ability to operate to higher headlosses with a pressure filter reduces the amount of wash water to be recycled.
4. Pressure filter systems are usually less costly in small and medium size plants.

A typical gravity filter section is shown in Figure 4-9. Gravity filter areas usually do not exceed 800–1000 ft² per filter unit. Two filter units may be combined in one structural cell.

Pressure filters may be either horizontal or vertical. The horizontal filters offer much larger filter areas per unit and would normally be used when plant capacities exceed 1–1.5 mgd. A typical horizontal pressure filter vessel is shown in Figure 4-10. Although an 8 ft diameter vessel is shown, 10 ft diameters are also commonly used with lengths up to 60 ft. The 10 ft diameter by 38 ft long horizontal pressure filters in use at the South Lake Tahoe Water Reclamation Plant are shown in Figure 4-11. A typical vertical filter is shown in Figure 4-12. Diameters up to 11 ft are commonly used, with working pressures up to 150 psig.

In medium and large plants, the minimum number of filter units is usually four. In small plants it may be two, or even one if adequate storage of filter

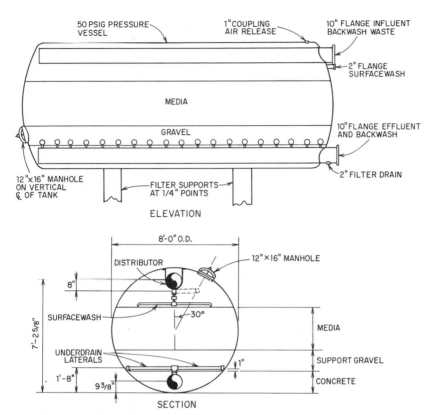

50 PSIG PRESSURE VESSEL

1" COUPLING AIR RELEASE

10" FLANGE INFLUENT BACKWASH WASTE

2" FLANGE SURFACEWASH

MEDIA

GRAVEL

10" FLANGE EFFLUENT AND BACKWASH

12" × 16" MANHOLE ON VERTICAL ℄ OF TANK

FILTER SUPPORTS AT 1/4" POINTS

2" FILTER DRAIN

ELEVATION

8'-0" O.D.

DISTRIBUTOR

12" × 16" MANHOLE

8"

SURFACEWASH

30°

7'-2 5/8"

MEDIA

UNDERDRAIN LATERALS

1"

SUPPORT GRAVEL

1'-8"

9 3/8"

CONCRETE

SECTION

Fig. 4-10   Typical pressure filter. *(Courtesy Neptune Microfloc, Inc.)*

Fig. 4-11   Pressure Filter installation at South Lake Tahoe water reclamation plant. *(Reprinted from* Water & Wastes Engineering, 6:4, Apr. 1969, p. 36. R. H. Donnelley Corp.)

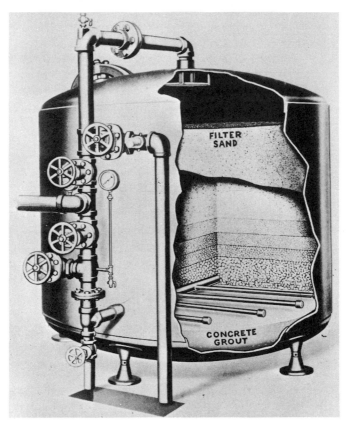

Fig. 4-12 A typical vertical pressure filter with concrete grout fill in the bottom head, pipe headers, and lateral underdrains, gravel supporting bed and filter sand. (*Courtesy Infilco Products*)

influent, effluent, or backwash water is provided for needs during filter shutdown for backwashing. The maximum size of a single filter unit is limited by the rate at which backwash water must be supplied, by difficulties in providing uniform distribution of backwash water over large areas, by reduction in filter plant capacity with one unit out of service for backwashing, and by structural considerations. The largest filter units in service are about 2100 ft² in area. Units larger than about 1000 ft² are usually built with a central gullet so that each half of the filter can be backwashed separately.

To filter a given quantity of water, the capital cost of piping, valves, controls, and filter structures is usually less for a minimum number of large filter units than for a greater number of small units.

Filter Layout

One popular arrangement is to place filters side by side in two rows on opposite sides of a central pipe gallery. One end of the rows of filters should be unobstructed to allow construction of additional units in future expansion. The best pipe gallery design includes a daylight entrance. This provides good lighting, ventilation, and drainage, and improves access for operation, maintenance, and repair.

Filters should be located as close as possible to the source of influent water, the backwash water supply, and the control room.

In the past (in water works practice) the overall depth of filters from water surface to underdrains was ordinarily 8–10 ft. The current trend is toward deeper filter boxes in order to reduce the possibility for air binding and to increase the available head for filtration. Filter effluent pipes and conduits must be water sealed to prevent the entry of air into the bottom of the filter when it is idle.

Filter Structures

The general practice is to house and roof filter pipe galleries. In the past, this was also the practice for filters and filter operating galleries. This was done to protect the filter controls and the water in the filters against freezing. Current practice is to locate the filter controls in a central control room and to provide perforated air lines around the periphery of each filter to control ice formation. This eliminates the need for housing the filters and filter gallery. If visual observation of filter washing without exposing the operator to the weather is considered essential, then it becomes a question of the relative costs and convenience of closed circuit television vs. housing the filters and filter gallery.

For gravity filter structures, either concrete or steel may be used. Concrete boxes may be either rectangular or circular. Steel tanks are usually circular. For small tanks and pressure filters, steel is usually preferred. For large filters, concrete is the material of choice.

There are a number of special structural considerations involved in reinforced concrete filter boxes which deserve the attention of designers. They must carry not only wind, dead, and live loads, but hydrostatic loads as well. To be watertight, walls should be not less than 9 in thick. Walls must often be designed to withstand water pressure from either of two directions, and to resist sliding and overturning. Slabs also may be subject to hydrostatic uplift as well as gravity loads. Shear and watertightness require special attention, particularly at keyways at the base of walls. Deflection at the top of cantilever walls and sliding at the base are common sources of trouble in water bearing concrete structures. The concrete should be impervious to minimize the deleterious effects of freezing and thawing, and ample cover should be provided for steel reinforcement. The difference in expansion between water cooled walls and floors and exposed slabs and roofs must be considered. The location and design of construction and expansion joints must be shown on the plans by the engineer and not left to the judgment of the field inspector or contractor.

Filter Underdrains

Filter underdrains have a twofold purpose. The most important is to provide uniform distribution of backwash water without disturbing or upsetting the filter media above. The other is to collect the filtered water uniformly over the area of the bed. Ideally, the filter underdrain system would also serve as a direct support for the fine media. Certain types of underdrains have been successfully applied for clean water service, but are not widely used in wastewater filtration because of greater danger of plugging and structural failure. The three most widely used underdrain systems (Wheeler, Leopold, and pipe lateral) all require an intermediate supporting bed of gravel, which should be topped with a layer of coarse garnet or ilmenite to prevent upsets of the gravel layer. Figures 4-13 and 4-14 illustrate two of these underdrain systems. All of the systems accomplish the uniform distribution of wash water by introducing a controlling loss of head, usually about 3–15 ft in the orifices of the underdrain system. The orifice loss must

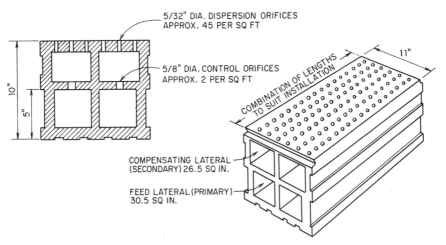

Fig. 4-13   Leopold filter bottom. (*Courtesy F. B. Leopold Co.*)

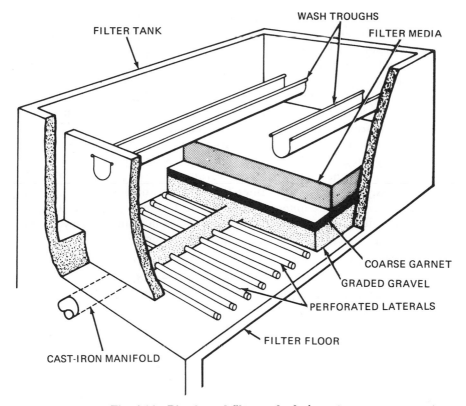

FILTER TANK

WASH TROUGHS

FILTER MEDIA

COARSE GARNET

GRADED GRAVEL

PERFORATED LATERALS

FILTER FLOOR

CAST-IRON MANIFOLD

Fig. 4-14   Pipe lateral filter underdrain system.

exceed a multiple of the sum of the minor (manifold and lateral) head losses in the underdrain to provide good backwash flow distribution.

The Wheeler bottom can be either a precast or poured in place false concrete bottom with a crawl space (22 in. high) between the structural and false bottoms. The crawl space and access manhole from the pipe gallery provide for very low headloss flow distribution, as well as means for removal of the concrete form. The bottom consists of a series of conical depressions at 1 ft centers each way. At the bottom of each conical depression is a porcelain thimble with a $\frac{3}{4}$ in. diameter orifice opening. Within each cone there are 14 porcelain spheres ranging in diameter from $1\frac{3}{8}$ to 3 in. and arranged so as to dissipate the velocity head from the orifice with minimum disturbance of the gravel above. Steel forms for the poured in place Wheeler bottom can be rented from either B-I-F, Providence, R.I., or Roberts Filter Co., Darby, Pa.

The Leopold bottom consists of vitrified clay blocks, each about 2 ft long by 11 in. wide by 10 in. high. The bottom half of each block contains two feeder channels (about 4 in. $\times$ 4 in.) with two $\frac{3}{8}$ in. diameter orifices per ft^2, which feed water into the two dispersion laterals (about $3\frac{1}{4} \times 4$ in.) above with a headloss of about 1.45 ft at a flow of 15 gpm/ft^2. In the top of the block there are approximately forty-five 5/32 in. diameter orifices per square foot for further distribution of the wash water and for dissipation of the velocity head from the lower orifices. These blocks and installation services are available from F. B. Leopold Co., Zelienople, Pa.

The perforated pipe lateral system uses a main header with several pipe laterals projecting from both sides. The laterals have perforations on the underside so that the velocity of the jets from the perforations during backwash is dissipated against the filter floor and in the surrounding gravel. Piping materials most commonly used are steel, transite, or PVC. Usually, orifice diameters are $\frac{1}{4}-\frac{1}{2}$ in., with spacings of 3–8 in. Fair and Geyer[28] give the following guides to pipe lateral underdrain design:

1. Ratio of area of orifice to area of bed served, .015–0.0001.
2. Ratio of area of lateral to area of orifices served, 2:1–4:1.
3. Ratio of area of main to area of laterals served, 1.5:1–3:1.
4. Diameter of orifices, $\frac{1}{4}-\frac{3}{4}$ in.
5. Spacing of orifices, 3–12 in. on centers.
6. Spacing of laterals, about the same as spacing of orifices.

Several underdrain systems have been developed with the goal of eliminating the need for the gravel support bed. A decision to use such underdrain systems should be based on the personal satisfactory experience of the designer in similar applications. In general currently available direct support systems have not yet proven to be as satisfactory as gravel support beds for fine media.

Filter Gravel

A graded gravel layer usually 14–18 in. deep is placed over the pipe lateral system to prevent the filter media from entering the lateral orifices and to aid in distribution of the backwash flow. A typical gravel design is shown in Table 4-2. A weakness of the gravel–pipe lateral system has been the tendency for the gravel to eventually intermix with the filter media. These gravel upsets are caused by localized high velocity during backwash, introduction of air into the backwash system, or use of excessive backwash flow rates. The gravel layer can be stabilized by using 3 in. of garnet or ilmenite as the top layer of the gravel bed. This coarse, very heavy material will not fluidize during backwash and provides excellent stabilization for the gravel. It also prevents the fine garnet or ilmenite used in a mixed, trimedia filter from mixing with the gravel support bed. The remaining major disadvantage of

TABLE 4-2   Typical Gravel Bed for Pipe Underdrain System.

Gravel Layer	Layer Thickness, in.	SQ. MESH SCREEN OPENING, in.	
		Passing	Retained
Bottom	—[a]	1	$\frac{3}{4}$
2	3	$\frac{3}{4}$	$\frac{1}{2}$
3	3	$\frac{1}{2}$	$\frac{1}{4}$
Top[b]	4	$\frac{1}{4}$	$\frac{1}{8}$

[a]Bottom layer should extend to a point 4 in. above the highest outlet of wash water.
[b]Plus coarse garnet.
Source: Culp/Culp, *New Concepts in Water Purification*, Van Nostrand Reinhold, 1974.

the gravel–pipe lateral system is the vertical space required for the gravel bed. Gravel layers are also used with several of the commercially available underdrain systems such as the Leopold bottoms (see Figure 4-15) and Wheeler bottoms. Gravel depths and gradations vary for these underdrain systems. For example, F. B. Leopold Co., Inc, Zelienople, Pa., recommends gradation listed in Table 4-3. The gravel to be used with Wheeler bottoms is shown in Table 4-4.

Gravel should consist of hard, rounded silica stones with an average specific gravity of not less than 2.5. Not more than 1 percent by weight of the material should have a specific gravity of 2.25 or less. Not more than

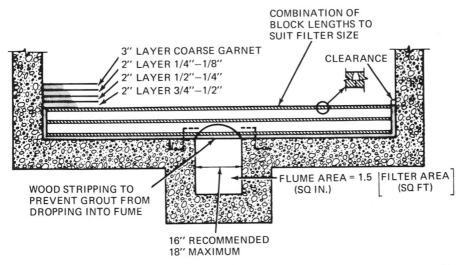

Fig. 4-15   Typical installation of Leopold filter bottoms. (*Courtesy F. B. Leopold Co.*)

TABLE 4-3 Gravel Size and Layer Thickness for Use with Clay Tile Bottoms.

Gravel Layer	Layer Thickness (in.)	Size Limit (in.)
Bottom	2	$\frac{3}{4} \times \frac{1}{2}$
Second	2	$\frac{1}{2} \times \frac{1}{4}$
Top[a]	2	$\frac{1}{4} \times \frac{1}{8}$

[a] Plus coarse garnet.
Source: Culp/Culp, *New Concepts in Water Purification*, Van Nostrand Reinhold, 1974.

2 percent by weight of the gravel should consist of thin, flat, or elongated pieces (pieces in which the largest dimension exceeds 5 times the smallest dimension). The gravel should be free from shale, mica, clay, sand, loam, and organic impurities of any kind. The porosity of gravel in any layer should not be less than 35 percent nor more than 45 percent. Gravel should be screened to proper size and uniformly graded within each layer. Not more than 8 percent by weight of any layer should be coarser or finer than the specified limit.

Coarse Garnet. It is recommended that a 3 in. layer of high density gravel (garnet or ilmenite) be used between the silica gravel bed and the fine media. This coarse, dense layer prevents disruption of the gravel. The specific gravity of the material should not be less than 4.2. The garnet or ilmenite particles in the bottom $1\frac{1}{2}$ in. layer should be $\frac{3}{16}$ in. by 10 mesh, and in the top $1\frac{1}{2}$ in. layer should be 1 mm in diameter. It is important to note than when the 3 in. coarse garnet layer is not used, another 3 in. layer of $\frac{3}{16}$ in. by 10 mesh silica gravel should be added as the top layer in Tables 4-2, 4-3, and 4-4. Otherwise there is apt to be migration of fine media down into the gravel supporting bed.

TABLE 4-4 Gravel Size and Layer Thickness for Use with Wheeler Bottoms.

Gravel Layer	Layer Thickness, in.	Size Limit, in.
Bottom	as required to cover the underdrain systems	$1 \times 1\frac{1}{4}$
Second	3	$\frac{5}{8} \times 1$
Third	3	$\frac{3}{8} \times \frac{5}{8}$
Top[a]	3	$\frac{3}{16} \times \frac{3}{8}$

[a] Plus coarse garnet.
Source: Culp/Culp, *New Concepts in Water Purification*, Van Nostrand Reinhold, 1974.

Gravel Placement. The filter tanks must be thoroughly cleaned before gravel is placed and kept clean throughout the placing operation. Gravel made dirty in any way should be removed and replaced with clean gravel. The bottom layer should be placed carefully by hand to avoid movement of the underdrain system and to assure free passage of water from the orifices. Each gravel layer should be completed before the next layer above is started. Workmen should not stand or walk directly on material less than $\frac{1}{2}$ in. in diameter, but rather should place boards to be used as walkways. If different layers of gravel are inadvertently mixed, the mixed gravel must be removed and replaced with new material. The top of each layer should be made perfectly level by matching to a water surface at the proper level in the filter box.

Mixed Media

For most wastewater applications, the use of mixed media (as developed and patented by the Neptune Microfloc, Inc.) will provide the optimum filtration efficiency.

There is no one mixed media design which will be optimum for all water filtration problems. Conley and Hsiung[13] have presented techniques designed to optimize the media selection for any given filtration application. Their work clearly indicates the marked effects that the quantity and quality of floc to be removed can have on media selection (Table 4-1). Pilot tests of various media designs can be more than justified by improved plant performance in most cases.

For production of high quality filtered water from chemically coagulated and settled wastewater, the following performance-type specification may be applicable:

The filter beds shall contain a 30 in. total depth of fine media made up of 18 in. of anthracite coal, specific gravity 1.5; 9 in. of silica sand, specific gravity 2.4; and 3 in. of garnet or ilmenite sand, minimum specific gravity 4.2. The relative size of the particles shall be such that hydraulic grading of the material during backwash will result in a filter bed with pore space graded progressively from coarse to fine in the direction of filtration (down). The total surface area of all the grains of fine media in a 30 in. deep bed shall not be less than 2650 square millimeters for each square millimeter of plan area. The number of grains of fine media per square millimeter of plan area shall not be less than 2750. Core samples of the completed backwashed filter bed shall show the number of grains per cubic millimeter given in the following table:

Depth in Bed	Number of Grains per Cubic Millimeter
Top 3 in.	0.59
3–9 in.	0.43
9–15 in.	0.41
15–18 in.	0.40
18–21 in.	0.76
21–24 in.	10.85
24–27 in.	9.85
27–30 in.	16.47

Source: Neptune Microfloc Inc., Corvallis, Ore.

When filtering coagulated and settled wastewater having a turbidity of 10–20 JTU at a rate of 5 gpm/ft^2, the filter shall produce an effluent having an average turbidity of less than 0.2 JTU and a maximum turbidity of 0.5 JTU even when the filter rate is suddenly changed from 2 to 5 gpm/ft^2. The initial head loss in a 30 in. deep clean filter shall not exceed 1.6 ft at a filter rate of 5 gpm/ft^2 and at a water temperature of 20°C. The time required to reach a total head loss of 10 ft (terminal headloss) under the above conditions shall not be less than 12 hours.

For filtration of secondary effluents to meet less rigorous effluent quality standards (say 10 mg/l SS) a coarser mixed media may be advisable. The mixed media can be made up of silica sand (specific gravity 2.65) and two grades of coal (specific gravities 1.35 and 1.55). Above a conventional dual media (coal–sand) bed (described below), a 6–8 in. layer of coarse (1–2 mm) low specific gravity (1.35) coal can be added. This modification will extend filter runs about 25 percent.

Dual Media

As compared to mixed media, the dual media (coal–sand) filter has less resistance to breakthrough because it is made up of coarser particles and has less total surface area of particles. Mixed media is capable of producing lower finished water turbidities than dual media, particularly from chemically treated wastewaters. For secondary effluents not subjected to chemical treatment it is very difficult to do very much about effluent quality by changing the filter design. The best that can be achieved is approximately a 20 percent variation in effluent quality from a dual media to a mixed media design. Large improvements in effluent quality must be obtained by better biological or chemical treatment upstream of the filters.

Typically, coal–sand filters consist of a coarse layer of coal about 18 in. deep above a fine layer of sand about 8 in. thick. Some mixing of coal and

TABLE 4-5   Typical Coal and Sand Distribution by Sieve Size in Dual Media Bed.

COAL DISTRIBUTION BY SIEVE SIZE	
US Sieve No.	Percent Passing Sieve
4	99–100
6	95–100
14	60–100
16	30–100
18	0–50
20	0–5
SAND DISTRIBUTION BY SIEVE SIZE	
US Sieve No.	Percent Passing Sieve
20	96–100
30	70–90
40	0–10
50	0–5

Source: Culp/Culp, *New Concepts in Water Purification*, Van Nostrand Reinhold, 1974.

sand at their interface is desirable to avoid the excessive accumulation of floc which occurs at this point in beds graded to produce well defined layers of sand and coal. Also, such intermixing reduces the void size in the lower coal layer, forcing it to remove floc which otherwise might pass through the coal layer. Typical gradations of sand and coal for use in dual media filters are given in Table 4-5.

Filter Backwashing

During the service cycle of filter operation particulate matter removed from the applied water accumulates on the surface of the grains of fine media and in the pore spaces between grains. With continued operation of a filter the materials removed from the water and stored within the bed reduce the porosity of the bed. This has two effects on filter operation: (1) it increases the head loss through the filter and (2) it increases the shearing stresses on the accumulated floc. Eventually the total hydraulic headloss may approach or equal the head necessary to provide the desired flow rate through the filter, or there may be a leakage or breakthrough of floc particles into the filter effluent. Just prior to either of these potential occurrences, the filter should be removed from service for cleaning. In the old slow sand filters the arrangement of sand particles in fine to coarse in the direction of filtration (down), so that most of the impurities removed from the water collect on the top surface of the bed and the bed can be cleaned by mechanically scraping the surface and removing about $\frac{1}{2}$ in. of sand and floc. In rapid sand filters, there is somewhat deeper penetration of particulates into the bed because of the coarser media and higher flow rates used. However, most of the materials are stored in the top 8 in. of the rapid sand filter bed. In dual media and mixed media beds, floc is stored throughout the bed depth to within a few inches of the bottom of the fine media. Rapid sand, dual media, and mixed media filters are cleaned by hydraulic backwashing (upflow). Thorough cleaning of the bed makes it advisable in the case of single media filters, and mandatory in the case of dual or mixed media filters, to use auxiliary scour or so called surface wash devices before or during the backwash cycle. Backwash flow rates of 15–20 gpm/ft^2 should be provided. A 20–50 percent expansion of the filter bed is usually adequate to suspend the bottom grains. The optimum rate of wash water application is a direct function of water temperature, as expansion of the bed varies inversely with viscosity of the wash water. Coarser beds may require higher backwash rates. The time required for complete washing varies from 3 to 15 minutes. Following the washing process, water should be filtered to waste until the turbidity drops to an acceptable value; this may require between 2 and 20 minutes, depending on pretreatment and type of filter. A recording turbidimeter for continuous monitoring of the effluent from each individual filter unit is of great value in controlling this operation at the start of a run as well as in predicting or detecting filter breakthrough at the end of a run.

Filters can be seriously damaged by slugs of air introduced during filter backwashing. The supporting gravel can be overturned and mixed with the fine media, which necessitates removal and replacement of all media for proper repair. Air can be unintentionally introduced to the bottom of the filter in a number of ways. If a vertical pump is used for the backwash supply, air may collect in the vertical pump column between backwashings. The air can be eliminated without any harm by starting the pump against a closed discharge valve and bleeding the air out from behind the valve through a pressure air release valve. The pressure air release valve must have sufficient capacity to discharge the accumulated air in a few seconds.

Also, air or dissolved oxygen, released from the water on standing and warming in the wash water supply piping, may accumulate at high points in the piping and be swept into the filter underdrains by the inrushing wash water. This can be avoided by placing a pressure air release valve at the high point in the line, and providing a $\frac{1}{2}$ in. pressure water connection to the wash water supply header to keep the line full of water and to expel the air.

The entry for wash water into the filter bottom must be designed to dissipate the velocity head of the wash water in such a manner that uniform distribution of wash water is obtained. Lack of attention to this important design factor has often led to difficult and expensive alterations and repairs to filters for correction.

Horizontal centrifugal pumps may be used to supply water for filter backwashing, provided they are located where positive suction head is available or are provided with an adequate priming system. Otherwise, a vertical pump unit may be used. The pumps should be sized to supply at least 15 gpm/ft^2 of filter area to be washed, and should be installed with an adequate pressure air release valve, a nonslam check valve, and a throttling valve in the discharge. Single wash water supply pumps are often installed, but consideration should be given to provisions for standby service in areas where power may be interrupted for several hours at a time, or where pump or motor repairs are not quickly available.

Wash water may also be supplied by gravity flow from a storage tank located above the top of the filter boxes. Wash water supply tanks usually have a minimum capacity equal to a 7 minute wash for 1 filter unit, but may be larger. The bottom of the tank must be high enough above the filter wash troughs to supply water at the rate required for backwashing as determined by a hydraulic analysis of the wash water system. This distance is usually at least 10 ft, but more often is 25 ft or greater. Wash water tanks should be equipped with an overflow line, and a vent for release and admission of air above the high water level. Wash water tanks are often filled by means of a small electric driven pump equipped with a high water level cutoff. Wash water tanks may be constructed of steel or concrete. Provisions should be made for shutdown of the tanks for maintenance and repair.

The wash water supply line must be equipped with provisions for accurately (within 5 percent either way) measuring and regulating the rate of wash water flow. The rate of flow indicator should be visible to the operator at the point at which the rate of flow is regulated.

The use of a high pressure (above 15 psi) source of filter backwash water through a pressure reducing valve is not advised. Numerous failures of systems using pressure reducing valves have so thoroughly upset and mixed the supporting gravel and fine media that these materials had to be completely removed from the filter and replaced with new media.

The backwash wastewater must be reprocessed. The rate of backwash flow, if returned directly to an upstream clarifier, is usually large enough in relation to the design flow through the clarifier to cause a hydraulic overload and upset of the clarifier. In this case, the backwash wastes should be collected in a storage tank and recycled at a controlled rate. The volume of backwash wastewater is typically 2–5 percent of the plant throughput, and plant components must be sized to handle this recycled flow. Provisions must be made to store the incoming flow during the backwash cycle or, if there are parallel units, to increase the rate on the other filters during the backwash cycle.

Wash Water Troughs

To equalize the head on the underdrainage system during backwashing of the filter, and thus to aid in uniform distribution of the wash water, a system of weirs and troughs is ordinarily used at the top of the filter to collect the backwash water after it emerges from the media and to conduct it to the wash water gullet or drain. The bottom of the trough should be above the top of the expanded media to prevent possible loss of media during backwashing. The clear horizontal distance between troughs is usually 5–7 ft. The cross section of the trough depends somewhat on the material of construction, but the semicircle bottom is a good design. Troughs may be made of concrete, fiberglass reinforced plastic, or other structurally adequate and corrosion resistant materials. The dimensions of a filter trough may be determined by use of the following equation:

$$Q = 2.49 \, bh^{3/2}$$

in which Q is rate of discharge in ft^3/second, b is width of trough in feet, and h is maximum water depth in trough in feet. Some freeboard should be allowed to prevent flooding of wash water troughs and uneven distribution of wash water.

The most popular material for construction of wash water troughs is reinforced concrete. Ordinarily, concrete troughs are precast on the job in an inverted position, and then placed in position after forms have been removed. Wash water troughs may also be factory fabricated from steel, fiberglass, or transite. Fiberglass reinforced plastic troughs may be purchased from F. B. Leopold Co., Inc., Zelienople, Pa.; Fischer & Porter, Warminster, Pa.; or other manufacturers. Transite wash troughs are available from Filtration Equipment Corp., Rochester, N.Y. and other suppliers.

If the filter troughs are to serve their purpose properly, after placement their weir edges must be honed to an absolutely smooth and perfectly level edge as determined by matching the finished edges of all troughs in a single filter to a still water surface at the desired overflow elevation.

Filter Agitators

Present practice in the U.S. leans heavily toward installation of the essential (but misnamed) surface wash on all new filters. "Auxiliary scour" or "filter agitation" better describe the function of this device, as it aids in cleaning much more than the filter surface. Rotary surface washers are the most common, but fixed jets are also used successfully. In the past, air scour and mechanical scour have not been used extensively in the U.S., but recent developments in this area may change this state of affairs in the future.

Adequate surface wash improves filter cleaning and prevents mudball formation and filter cracking. Rotary surface wash equipment consists of two arms on a fixed swivel supported from the wash water troughs about 2 in. above the surface of the unexpanded filter media. The arms are fitted with a series of nozzles, and revolve because of the water jet reaction. Water pressures of 40–100 psi are required for the operation of the rotary surface wash depending upon the diameter of the arms. The rate required is about 0.5 gpm/ft^2. Surface washers are usually started about one minute in advance of the normal backwash flow, and turned off a minute or so before the end of the backwash period. Rotary surface wash equipment can be purchased from Leopold. They can supply either the Palmer or Stuart rotary agitators. It is recommended that the nozzles be equipped with Stuart Flex-I-Jet rubber caps, which act to prevent entry of fine filter media and plugging of the nozzles.

A new concept in rotary agitator design has been developed by Roberts Filter Mfg. Co., Darby, Pa. (see Figure 4-16). This agitator utilizes the aspirator effect of the agitator water supply to introduce a controlled amount of air into the agitator arms. The air–water jet spray provides greatly improved scour and separation of accumulated floc. The agitators also include a rotary valve which cuts off the water supply to the end sprays during the instant that the arms are closest to the filter walls. This reduces erosion of the concrete at this point, which is sometimes a problem with conventional rotary surface washers.

Baylis has designed a fixed jet surface wash system consisting of a grid of distributing pipes extending to within a couple of inches of the top surface of the bed. Nozzles with five $\frac{1}{4}$ in. holes are spaced at about 24–30 in. centers each way. The required flow is about 2 gpm/ft^2 at the head of 20–60 ft.

The source of water used for the surface wash must be of high quality to ensure that plugging of the nozzles does not occur. A strainer with $\frac{3}{32}$ in. perforations in the surface wash line is a wise precaution. In pressure filter systems, an external indicator (available from Neptune Microfloc) should be used to show that the surface wash arm is actually rotating during backwash. A rotating agitator similar to the surface wash arms may be installed within the bed in very deep filters. These subsurface agitators have the nozzle discharge staggered 15 degrees up to 15 degrees down.

An alternative to mechanical surface wash employment is the use of air–water backwash techniques in which air is injected through the filter underdrain system to break up mudballs as it rises through the filter. These techniques have been used fairly extensively in Europe but lost favor in the United States about 30 years ago because of problems encountered with upsetting of beds and loss of media. Recent improvements in underdrain systems have led to renewed interest. Because of the tendency of the air

(a)

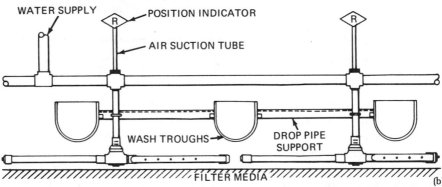

WATER SUPPLY   R   POSITION INDICATOR

AIR SUCTION TUBE

WASH TROUGHS   DROP PIPE SUPPORT

R

FILTER MEDIA
(b)

Fig. 4-16   The Roberts XL-500 rotary agitator. (a) Here the sweep arms are held stationary to illustrate the violent action of the sprays using an air–water mixture. At this point the control ports in the rotary joint are still closed as the end nozzle passes the wall of the filter. (b) Elevation Drawing of a typical installation showing relationship of the stator supports to wash troughs.

126

TABLE 4-6   Valve Positions During Various Treatment Operations.

| | VALVE POSITION | | Filtering to |
Valve	Filtering	Backwashing	waste
Influent	open	closed	open
Effluent	open	closed	closed
Filter to waste	closed	closed	open
Wash water supply	closed	open	closed
Wash water drain	closed	open	closed
Surface wash	closed	open	closed

to float coal out of a filter, the backwash water and the air cannot be applied at the same time with dual or tri-media beds utilizing coal. The air-water technique applied to these beds usually consists of the following steps: (1) stop the influent and lower the water level to within 2–6 in. of the filter surface; (2) apply air alone at a rate 2–5 cfm/ft^2 for 3–10 minutes; (3) apply a small amount of backwash water (about 2 gpm/ft^2) with the air continuing until the water is within 8–12 in. of the wash water trough (the air is then shut off); and (4) continue water backwash at 8–10 gpm/ft^2 until the filter is clean.

An important point to consider is that a mixed media bed will require 1–2 minutes of wash at 12–15 gpm/ft^2 at the end of the wash period to properly classify the media.

It has been the authors' experience that although lower backwash rates are used with the air–water techniques, the amount of backwash water required to achieve the *same degree of cleaning* as with surface wash in conjunction with backwash is not significantly different. The air–water techniques increase the complexity of the backwash cycle, lengthen the filter downtime for backwashing, increase the risk of media loss, and offer little, if any, advantage in the quantity of backwash water required to achieve a given degree of cleaning.

Valve positions for various cycles of filter operation are shown in Table 4–6.

Filter Rate of Flow Controllers

Ordinarily filters are equipped with a means of controlling the rate of flow through each bed. By controlling either the influent or the effluent flow it is possible to divide the flow among the filters, to limit the maximum flow through any unit, and to prevent sudden flow surges. Rapid changes in rate of flow through a filter are undesirable because they literally shake the floc particles down through the bed and into the effluent.

Filter rate of flow controllers consist of (1) a flow measuring device such as a Venturi tube or a Dall tube, (2) a throttling valve such as a rubber seated butterfly valve with hydraulic or pneumatic actuator, and (3) a valve stop or positioning device. Auxiliary or overriding control of the throttling valve may be provided to maintain a predetermined minimum water level above the filter media and a maximum water level in the receiving basin. The flow measuring device must be provided with the necessary straight pipe runs on either side if it is to provide the necessary accuracy. Controllers can be equipped to indicate or to record rate of flow through each filter, and to totalize flow.

Rate controllers may be set to operate at a constant fixed rate. This setting has several advantages from the standpoint of water quality; it means that flow through the plant must be nearly constant, or that filter units must be placed in or out of service to meet fluctuating demands. Another method of operation is to use a variable rate controller with a fixed maximum setting. This permits operation of all filters at the minimum possible rate all of the time, which also has advantages.

Declining Rate Filtration with Restricted Maximum Flow

As an alternative to the normal, constant-rate operation of filters, it is possible to operate them at a declining rate. Variable declining rate operation has the advantage of requiring less total available headloss and having the capability of producing an effluent of better quality. Each declining rate filter should be equipped with a device which limits the maximum flow rate through the filter unit so as to restrict the filter rate to the desired value, say 5–8 gpm/ft^2.

The reason that less available head is needed to operate a declining rate filter than is needed for a constant rate filter is that the flow rate decreases near the end of the filter run, with an accompanying decrease in headlosses through the unit. The lower flow rate at the end of a run, when the filter contains the most foreign material removed from the water, also may result in improved effluent quality.

Loss of Head

The loss of head through a filter provides valuable information about the condition of the bed and its proper operation. An increase in the initial loss of head for successive runs over a period of time may indicate clogging of the underdrains or gravel, the need for auxiliary scour, or unsufficient washing of the beds. The rate of headloss increase during a run yields considerable information concerning the efficiency of both pretreatment and filtration.

The determination of headloss through a filter is a very simple matter, since it involves only the measurement of the relative water levels on either side of the filter. The simplest form of headloss device for gravity filters is one made up of two transparent tubes installed side by side in the pipe gallery with a gage board between, graduated in feet. One tube is connected to the filter or filter influent line, the other to the effluent line. The headloss is the observed difference in water levels. More sophisticated methods for measuring the difference in water levels include float and differential pressure cell actuated indicating and recording devices. The headloss equipment may be used in connection with control systems for automatic backwashing of filters as one means of initiating the backwash cycle when the headloss reaches some preset maximum value.

Monitoring Filter Effluent Turbidity

Inexpensive, reliable turbidimeters are available to continuously monitor the quality of the filter effluent. Should the turbidity exceed a level suitable for a downstream process or for discharge, the signal from the turbidimeter can be used to initiate the backwash program or sound an alarm. Record-

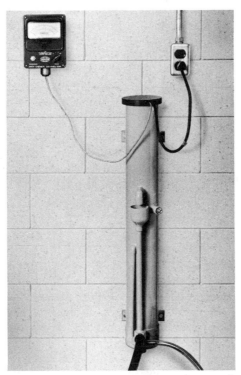

Fig. 4-17   A continuous flow turbidimeter capable of measuring turbidities of less than 0.1 JTU. *(Courtesy Hach Chemical Co.)*

ing of the turbidimeter output provides a continuous record of filter performance. The Hach Low Range Turbidimeter, shown in Figure 4-17, a relatively inexpensive unit, is well suited to this application. It is a continuous flow nephelometer. A strong beam of light passed through a water sample is scattered by particles of turbidity just as dust particles suspended in a room scatter sunlight streaming through a window. If the effluent is free of turbidity, no light is scattered and no light reaches the photoelectric light measuring cells, resulting in a zero reading. The presence of turbidity in the sample results in light being scattered and some of the light falls on the photocells. The most important advantage of a nephelometer is that a very intense beam of light can be passed through the sample, and this results in very high sensitivity, so that a trace amount of turbidity can be accurately measured. The output from the photocells is zero when the turbidity is zero and the output increases in proportion to the amount of turbidity read. The required sample stream volume is 0.25–0.50 gpm with a head of at least 6 in. for the Hach unit shown. The response time to detect a change in turbidity is 30 sec. The unit has ranges of 0–0.2 JTU, 0–1.0 JTU, 0–3.0 JTU, and 0–30 JTU. The approximate cost of a turbidimeter and recorder with alarm is $700. (1970 prices).

Control of Filter Operation

Filter runs are usually terminated when the headloss reaches a predetermined value, the filter effluent turbidity exceeds the desired maximum, or a certain amount of time passes. Each of these events is adaptable to instrumentation which can be used to signal the need for backwashing or to trigger a fully automated backwash system. Automatic control of filter backwashing may be provided by an automatic sequencing circuit (step switch) which is interlocked so that the necessary prerequisites for each step are completed prior to proceeding to the next step. At the receipt of a backwash start signal, the following events will occur in the sequence listed in this illustrative program:

Filter influent and effluent valves close; any chemical feed stops to the filter being backwashed; plant chemical feeds adjust to the new plant flow rate to maintain proper chemical feed to the filters still in service; the waste valve starts to open; when the waste valve reaches the fully open position and actuates a limit switch, the surface wash pump starts and the surface wash valves open; surface wash flow to waste continues for a period of time adjustable up to 10 minutes. At the end of the initial surface wash period, usually 1–2 minutes, the main backwash valve opens; backwash and surface wash both continue for a period of time, usually 6–7 minutes, adjustable up to 30 minutes; backwash flow rate is indicated on a controller and is controlled automatically to a manual set point; at the end of the combined wash periods, the surface wash valve closes and surface wash pump stops; backwash continues without surface wash for a period of usually 1–2 minutes, adjustable up to 30 minutes. At the completion of the backwash period, first the backwash valve closes, then the waste valve starts to close; when the latter valve has closed, the influent and filter to waste valves open, chemical feed to the clean filter is reestablished, and the bed filters to waste for a period of time, usually 3–7 minutes, adjustable to 30 minutes; the backwash delay timer resets and begins a new timing cycle adjustable up to 12 hours; the bed selector switch steps to the next filter; at the end of the filter to waste period, the filter to waste valve closes and the effluent valve opens to restore the cleaned bed to normal filter service.

Provision should be made for optional manual operation of all automatic features including automatic cycle initiation.

It may be desirable to alarm certain functions which affect filter operation on a conveniently located annunciator panel. These alarm functions include high effluent turbidity, high leadloss, low plant flow, low backwash flow rate, and excessive length of backwash (incomplete cycle).

Filter Piping, Valves, and Conduits

Cast iron pipe and fittings and coal tar enamel lined welded steel pipe and fittings are the most widely used materials for filter piping. The layout of filter piping must include consideration of ease of valve removal for repair and easy access for maintenance. Flexible pipe joints should be provided at all structure walls to prevent pipeline breaks due to differential settlement. The use of steel pipe can reduce flexible joint requirements. Color coding of the filter piping is a valuable operating aid. The filter piping is usually designed for the flows and velocities shown in Table 4-7.

The rubber seated butterfly valve with pneumatic cylinder or diaphragm operator has almost entirely replaced the hydraulically actuated and operated gate valves that were formerly used extensively as filter valves. The butterfly valve is smaller, lighter, easier to install, better for throttling services, and can be installed and operated in any position. The valves should be factory equipped with the desired valve stops, limit switches, and position indicators because field mounting of these devices is often unsatisfactory.

Each filter unit, except split beds, should have 6 valves for its proper operation: influent, effluent, wash water supply, wash water drain, surface wash, and filter to waste. The positions of these valves during the three cycles of filter operation were given in Table 4-6. The filter influent should enter the filter box in such a way that the velocity of the incoming water

TABLE 4-7   Filter Piping Design Flows and Velocities.

Description	Velocity, fps	Maximum flow, gpm/ft² of filter area
Influent	1–4	8–12
Effluent	3–6	8–12
Wash water supply	5–10	15–25
Backwash waste	3–8	15–25
Filter to waste	6–12	4–8

does not disrupt the surface of the fine media. This is often done by directing the influent stream against the gullet wall, thus dissipating the velocity head within the gullet. It can also be done by locating the influent pipe below the top of the filter troughs so that the water enters the filter through the troughs. A further precaution is to install an influent valve with throttling control for use in refilling the beds slowly, or to allow the final portion of backwash flow to fill the filter to normal level by closing the waste gate prior to stopping backwash flow. The filter to waste, effluent, and wash water supply lines usually are manifolded for common connection to the filter underdrain system.

In the design of pipe galleries, reinforced concrete flumes and box conduits and concrete encased concrete pipe may be used for wash water drains or other service when located adjacent to the pipe gallery floor, but should not be installed overhead because of difficulties with cracks and leaks. Invariably, pipe galleries with overhead concrete conduits are drippy, damp, unsightly places with a humid atmosphere which discourages good housekeeping by making it difficult to maintain. Rather, pipe galleries should be provided with positive drainage, good ventilation, plenty of light, and dehumidification equipment if required by the prevailing climate. Filter influent and effluent lines should be provided with sample taps.

FILTER PROBLEMS AND THEIR SOLUTIONS

Problems in filter operation and performance can arise either from poor design or poor operation. However, during recent years tremendous advances in engineering design of filters and filter controls and appurtenances have made water filtration an inherently stable, extremely efficient, and highly reliable unit treatment process. With proper design and good operation, all the problems of the past are easily solved.

Some potential filter problems are listed below:

1. Surface clogging and cracking.
2. Short runs due to rapid increases in headloss.

3. Short runs due to floc breakthrough and high effluent turbidity.
4. Variations in effluent quality with changes in applied water flow rate or quality.
5. Gravel displacement or mounding.
6. Mudballs.
7. Growth of filter grains, bed shrinkage, and media pulling away from sidewalls.
8. Leakage of fine media.
9. Loss of media.

Some solutions to these problems are suggested below:

1. Surface clogging and cracking are usually caused by rapid accumulation of solids on the top surface of the fine media. This is not a problem in dual or mixed media filters because of the greater porosity of their top surface as compared to sand. Also, when a filter aid is used with dual or mixed media, the dosage can be reduced as necessary (or eliminated), which allows particulates to penetrate deeper into the bed. In other words, regulation of the polymer dosage to the filter influent gives some control over the effective porosity of the filter so as to accommodate changes in incoming floc characteristics.

2. Short runs due to rapid increases in headloss are related to the problem just discussed, in that dual and mixed media beds collect particulates throughout the depth of the bed rather than mostly at the surface of the bed, as with a sand or other surface-type filter, and are much less susceptible to this problem. In filtration of chemically treated wastewater the flexibility provided by use of a polymer as a filter aid allows control of the rate of headloss buildup by changes in the amount of polymer used. As discussed previously, in filtration of secondary effluents, coarser beds of fine media should be used to lengthen filter runs.

3. Short filter runs due to floc breakthrough can be avoided by using mixed media. As mentioned earlier, this is one very important point of superiority of mixed media over dual media filters. It arises because of the much greater surface area of the grains in a mixed media filter as compared to dual media. The finest media in a mixed media bed is 40–80 mesh (10 percent of total bed), and 40–50 mesh (5 percent of total) in a dual media filter. The finest media (garnet) in a mixed media bed has the advantage not only of being finer but also in being located at the very bottom of the filter where the applied load is lightest and where it can serve its intended purpose as a polishing agent. Floc storage depths above the finest media in typical beds are as follows: coal–sand, 18 in.; mixed media, 27 in. A further advantage of mixed media in this regard lies in the greater total number of media

particles contained in an equal volume of bed. This tremendously increases the number of opportunities for contact between media and colloids in the water, thus greatly enhancing removal of these colloids.

4. Variation in effluent quality with changes in applied water are much less in mixed media beds, which are much less affected either by changes in flow or influent quality than are dual media. Again this relates to the greater total surface area of the fine media, the greater total number of fine media particles, and the smaller size of pore openings at the bottom of the filter bed.

5. Gravel displacement or mounding can be eliminated by use of the 3 in. layer of coarse garnet or ilmenite between fine media and gravel supporting bed, as previously recommended, and by limiting the total flow and head of water available for backwash.

6. Mudball formation can be eliminated by providing an adequate backwash flow rate (up to 20 gpm/ft^2), and a properly designed system for auxiliary scour (surface wash). The successful use of beds employing filter aids and in-depth filtration is dependent on provision and operation of a good system of auxiliary scour.

7. Growth of filter grains, bed shrinkage, and media pulling away from filter sidewalls are related problems. Again, the provision and use of adequate backwash facilities including surface wash are the keys. It is the compressibility of filter grains which are heavily coated with materials filtered, deposited, or adsorbed from the water which is the root of these difficulties. With the exception of calcium carbonate deposits, these problems can be avoided by proper backwashing. It should be mentioned that this growth of particles refers to a macroscopic increase in size, and not to the development of a microscopic film of polymer and other chemicals, which results from proper use of a filter aid and is beneficial in adsorption and retention of particulates for the period of a single operational cycle in the use of dual and mixed media beds, and which is not nearly thick enough to cause a problem by increasing the compressibility of the bed. In the case of calcium carbonate deposition on filter grains, an alum or polymer film on the grains may help to reduce the adherence of calcium carbonate and to facilitate its removal during backwashing, but in filtering lime treated waste waters it is important to adjust the pH of filter influent by addition of carbon dioxide or acid to a level at which calcium carbonate deposition is minimized.

8. Leakage of fine media can be minimized by using the coarse garnet or ilmenite layer between the fine media and gravel supporting bed, as recommended earlier, because this prevents the downward migration and escape of fines.

9. Loss of media during backwashing, particularly of coal, is one problem for which there is no complete solution. Losses can be reduced by increasing the distance between the top of the expanded bed during maximum backwash flows and the wash water troughs. It can also be helped by cutting off air wash or auxiliary scour 1 or 2 minutes before the end of the main backwash. Careful attention to design of air release facilities in the backwash piping will also help to minimize media loss.

Filter Design Checklist

1. Filter media sizing and selection should be based on pilot tests on the effluent involved. If this is not possible, data should be obtained from similar applications to determine the suitability of the media design.
2. For applications where direct granular filtration of secondary effluent is used (with no chemical coagulation and settling of chemical floc preceding filtration), an operating head of at least 15 ft of water should be provided.
3. Where chemical coagulation is used, provisions should be made for the addition of polyelectrolytes directly to the filter influent. Provisions should also be made for addition of supplemental dosages of coagulant to the filter influent.
4. The turbidity of each filter unit should be monitored continuously and recorded.
5. The flow and headloss through each filter should be monitored continuously and recorded.
6. Provisions should be made for the addition of disinfectant directly to the filter influent.
7. Pressure filters must be equipped with pressure and vacuum air release valves.
8. Provisions should be made to retain and retreat any filter effluent of unsatisfactory quality (i.e., provide a filter to waste cycle).
9. Provisions should be made for automatic initiation and completion of the filter backwash cycle. The filter controls and pipe galleries should be housed.
10. Filter piping should be color coded.
11. The filter system layout must enable easy removal of pumps and valves for maintenance.
12. The backwash rate selected must be based upon the specific filter media used and the wastewater temperature variations expected.
13. Filters should be backwashed with filter effluent in direct filtration applications. When coagulation and settling precede filtration,

chemical clarifier effluent may also be considered as a backwash source.

14. Filter backwash supply storage should have a volume at least adequate to complete two filter backwashes.
15. Adequate surface wash or air scour facilities must be provided.
16. There should be adequate backwash and surface wash pump capacity available, with the largest single pumps out of service.
17. Backwash supply lines must be equipped with air release valves.
18. Means should be provided to indicate the backwash flow rate continuously and to enable positive control of the filter backwash rate. A means should also be provided to limit filter backwash rate positively to a preset maximum value.
19. Backwash wastewaters should be collected in a storage tank and returned to the treatment process at a controlled rate unless direct return of these wastewaters will not adversely affect the treatment process.
20. The filter design must incorporate underdrains and backwash wastewater collection devices which insure uniform distribution of backwash water and filter influent.
21. The filter system should be equipped with an alarm system which will indicate major malfunctions.

OTHER TYPES OF IN DEPTH FILTERS

Deep, Coarse Beds

Deep, coarse, single media filters are finding increasing acceptance in wastewater treatment, particularly for attached growth nitrification and denitrification systems. Deep bed filtration systems use media from 1 to 6 mm in diameter in beds 6–10 ft deep and operate at rates of 1.5–15 gpm/ft^2. The media is silica sand chosen for its sphericity, and is closely sized to give as low a uniformity coefficient as possible. Air is used for backwashing and water is applied simultaneously to flush accumulated suspended material from the unit. Filter bottoms are nozzleless concrete blocks. Loadings obtained with these deep bed systems range from 1 to 10 lb/ft^3/cycle. In depth filtration is obtained because of the large size of the media particles used even with application of heavy loadings of relatively coarse solids. Only 10 percent bed expansion is obtained in backwash, which avoids classification of the media during this operation. There are many operating installations of deep, coarse bed filters, principally on industrial process water and wastewater.

Upflow Filters

Upflow filtration has an obvious theoretical advantage, because coarse to fine filtration can be achieved with a single medium such as sand with almost perfect gradation of both pore space and grain size from coarse to fine in the direction of filtration (upward). Since the bed is backwashed in the same direction but at higher flow rates, the desired relative positions of fine media are maintained or reestablished with each backwash. The inherent advantage of upflow filtration has long been recognized, and,

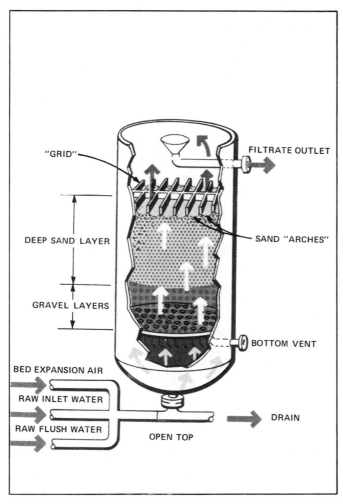

Fig. 4-18   Upflow filter with restraining grid.

under laboratory conditions, the anticipated high filtration efficiency has been well verified by several workers. The difficulty with upflow filtration comes when the headloss above a given level exceeds the weight of the bed above that level, at which time the bed lifts or partially fluidizes, allowing previously removed solids to escape in the effluent. In Russia, bed depths up to 6 ft are used in an attempt to minimize bed lifting. In the U.S., parallel plates or a metal grid are placed at the top of the fine media. The spacing of the plates or the size of the openings in the grid are such that the media grains arch across the open space to restrain the bed against expansion. These restraining bar systems have about 75 percent open area in the best designs developed to date. Figure 4-18 illustrates an upflow filter with a restraining grid system. Even with use of a restraining grid or a deep bed, there may be problems with excessive pressures or sudden variations in pressure which break the sand bridge or cause the bed to expand and lose its filter action. The frequency of breakthrough is rare, but the fact that it can occur at all, say with poor operation, has been sufficient to raise questions concerning public health implications and to limit the use of upflow filters for potable water applications. In other areas which are free of health considerations upflow filters have found wide application and have given excellent service. These areas include process water, wastewater treatment, deep well injection water, API separator effluent, cooling water, and other similar applications.

Biflow Filters

Biflow filters are an outgrowth of upflow filters, in that the divided flow [downflow from the top and upflow from the bottom (see Figure 4-19)] is an attempt to restrain the bottom upflow portion of the bed by placing above it a downflow filter. Biflow filters are used in Holland and Russia but not to any extent in the United States. Biflow filters permit filtration in two opposite directions at the same time. Essentially the top and bottom halves are completely independent filters equal in capacity, which results in some savings in structure and underdrains.

Unfortunately, the biflow filter has an inherent limitation which seems to preclude development of a unit which will produce an exceptionally high quality effluent. First consider a single media biflow bed. The finest materials are at the top of the upper downflow bed. This makes the top half of the bed a rapid sand or surface type filter, and the quality of water produced at best cannot exceed that from a rapid sand filter. The bottom half of this same filter is a coarse to fine filter, but unfortunately the finest material at the top outlet from the bed is somewhat coarser than the finest material which can be successfully used in a rapid sand filter. Obviously, the effluent from this bed will be of lesser quality than that from the rapid

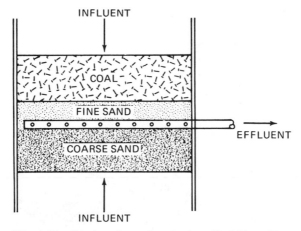

INFLUENT

COAL

FINE SAND

COARSE SAND

EFFLUENT

INFLUENT

Fig. 4-19   Section through a dual media biflow filter.

sand downflow filter above. This fact has been recognized by others and revealed by pilot tests. This leads to a consideration of the dual media biflow filter as illustrated in Figure 4-19. The idea here, and the advantage over the single media biflow bed, is that this arrangement places the fine sand closer to the mid-collector. It provides a dual media (coal–sand) downflow bed above a coarse to fine single media (sand) upflow bed. Again the limitation is the coarseness of the finest sand which can be used from a practical standpoint. If the sand is finer than that ordinarily used in a rapid sand filter, as it must be to build the best upflow filter in the bottom half of the bed, then the sand will be so fine that excessive amounts of sand will rise into the coal bed during backwashing. The gradation of a sand which will be suitable for the dual media bed in the upper half of the bed will be too coarse to provide the best possible filtration in the upflow bottom half of the bed. The quality of effluent from either half will not approach that from a mixed media filter. This problem is so basic that there does not appear to be an easy solution to this dilemma, but perhaps one will be found.

SURFACE FILTERS

Moving Bed Filter

The old "drifting sand filter" is being rejuvenated with modern mechanization to handle exceptionally heavy loads of applied solids in the form of the MBF (moving bed filter). The moving bed filter is being developed by Johns-Manville Products Corporation, see Figure 4-20. The MBF is a

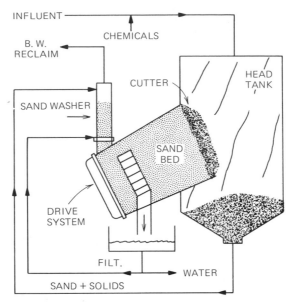

Fig. 4-20   Basic concept of the moving bed filter.

continuous sand filter in which the water moves countercurrent to the sand. The filter medium, sand, is driven through a cone in one direction while simultaneously passing the wastewater to be treated through the filter bed in the opposite direction. The solids are removed from the filter bed as rapidly as is required by their buildup. Movement of the filter bed is accomplished by means of a hydraulically actuated diaphragm. The diaphragm pushes the sand bed as a plug through the cone toward the inlet end. As the diaphragm relaxes, clean sand feeds by gravity into the void left in the bed in front of the diaphragm. The pushing cycle is then repeated. The sludge–sand mixture falls from the face down into the hopper bottom of the head tank, from which it is removed and passed through a washing tower. The cleaned sand is returned to the feed hopper of the MBF. Final effluent from the MBF is used for washing the sand, and to return the sand to the sand hopper. The removal and washing of the sand may be continuous or intermittent. Since the sand is constantly being removed, cleaned, and returned to the system, the filter unit does not have to be stopped for backwashing as do conventional units.

The filter medium usually used is 0.6–0.8 mm sand with a maximum sand feed rate of 12 in./hour. The MBF principle allows much higher solids loadings than permissible with a fixed sand bed. Preliminary results obtained while treating unsettled trickling filter effluent have been reported by Convery[18]. Following alum coagulation directly prior to filtration, influent turbidities were reduced from 22–41 JTU to 6.3–10.0 JTU, BOD from 40–64 mg/l to 8.8–10.0 mg/l, COD from 111–172 mg/l to 39–43 mg/l, and total phosphate from 30–40 mg/l to 1.5–2.5 mg/l. Polymer dosages of 0.5–1.0 mg/l were applied for the lower filter effluent turbidity values achieved. Plant scale experiences with the mixed media filters have shown that lower turbidities can be achieved in a similar direct coagulation application with substantially lower doses of polymer. It is felt that the coarseness of the media employed in the MBF must be compensated for with high polymer doses to strengthen the floc and prevent its penetration through the filter. The tests to date have clearly shown the MBF can tolerate much higher solids loadings than can a conventional rapid sand filter, a point of considerable merit in sewage filtration applications.

The manufacturer has released technical data (Johns-Manville Engineering Data MBF-2) on the MBF which indicates that the maximum unit capacity is 250,000 gpd/unit. The dimensions of the main unit are about 6 ft wide × 19 ft long × 21 ft 6 in. high, with the auxiliary units occupying a space of approximately the same area. The size and complexity of such a unit would appear to limit the applicatoin to plants of relatively small capacity. For example, the operation and maintenance of the 40 units required for a 10 mgd flow would appear to be impractical. The wastewater flow resulting from the sand cleaning operation is listed as 15 gpm for the 200 gpm influent flow of the 250,000 gpd unit, or $7\frac{1}{2}$ percent of the influent flow. The total motor horsepower is listed at 10 for the unit with a total

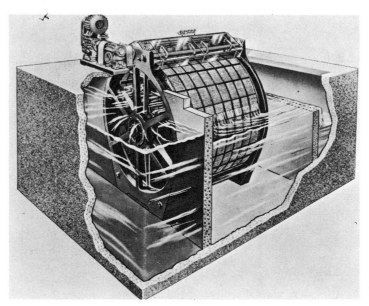

Fig. 4-21   Microscreen installed in concrete tank. *(Courtesy Crane-Cochrane Co.)*

unit weight of 78,000 lb. Capital and operating costs are not available at the time of this writing.

Microscreening

Microscreens are mechanical filters which consist of a horizontally mounted drum, the cylindrical surface of which is made up of a special metallic filter fabric, and which rotates slowly (peripheral speeds of up to 100 fpm) in a tank with two compartments, so that water enters the drum from one end and flows out through the filtering fabric. The drum is usually submerged to approximately two-thirds of its depth. The solids are retained on the inside of the rotating screen and are washed from the fabric by pumping strained effluent through a row of jets fitted on top of the machine. Figures 4-21, 4-22, and 4-23 illustrate typical microscreening equipment. The wash water containing the solids flushed from the screen is collected in a

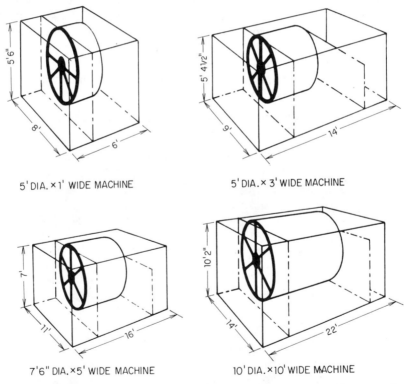

5' DIA. × 1' WIDE MACHINE   5' DIA. × 3' WIDE MACHINE

7'6" DIA. × 5' WIDE MACHINE   10' DIA. × 10' WIDE MACHINE

Fig. 4-22 Minimum internal dimensions of microscreen tanks. *(Courtesy Crane-Cochrane Co.)*

Fig. 4-23   5 × 3 ft microscreen in steel package unit. *(Courtesy Crane-Cochrane Co.)*

hopper or trough inside the drum for return to the secondary plant. The volume of wash water may vary from 3 to 5 percent and contain 700–1000 mg/l of suspended solids. The rate of flow through the microscreen is determined by the head applied (normally limited to about 6 in. or less) and by the concentration and nature of the suspended solids in the influent. Typical loading rates are in the range of 2.5–10 gpm/ft^2 of filtering fabric.

The fabrics are usually made of woven stainless steel of FSL quality (type 316, 18/8) containing 2–3 percent molybdenum.[24] Three aperture sizes are offered by one manufacturer: 60, 35, and 23 microns (μ).

Tertiary Filtration with Microscreens.   The operating results when applied to secondary effluents have been reported in several publications with somewhat mixed opinions. The finest fabric (23 μ) was found to remove 89 percent of the suspended solids from an activated sludge plant effluent in studies at Lebanon, Ohio.[6] The 35 μ screen removed 73 percent of the suspended solids. The percentage of backwash water averaged 5, with a range of 3–23 percent. The major operating problems noted at Lebanon were fouling of the screen with grease and the reduction of throughput during periods of high solids carryover from the secondary plant. In discussing the Lebanon test, Convery[18] reported that an increase in influent solids from 25 to 200 mg/l resulted in a decrease in throughput rate from 60 to 13 gpm. This sensitivity to variations in solids loading is cited frequently in the literature as a major weakness in the use of microscreening techniques for tertiary solids removal. In an attempt to overcome this problem, a system of automatic control has been devised by the manufacturers. The headloss across the screen is measured pneumatically and transmitted to a converter mechanism which sends a signal to increase the speed of the drum and backwashing pressure on rising headloss. However, no quantitative data defining the allowable variations in solids loading with this modification were available at the time of this writing.

Diaper[24] suggests the following guides on anticipated removals from secondary effluents:

| | PERCENTAGE ANTICIPATED REMOVAL | | |
Fabric aperture	Suspended solids	BOD	Flow, gpm/ft² of submerged area
23 microns	70–80	60–70	6.7
35 microns	50–60	40–50	10.0

Extensive tests at the Chicago Sanitary District were reported by Lynam and co-workers.[43] They report about 3 percent backwash water at 3.8 gpm/ft² hydraulic loading. Their data indicate that a final effluent suspended solids of 6–8 mg/l and a BOD of 3.5–5 mg/l could be expected when applying a good quality (20–25 mg/l suspended solids and 15-20 mg/l BOD) activated sludge effluent to a 23 μ microscreen. They also noted that the microscreen was more responsive to suspended solids loading than to hydraulic loading and that the maximum capacity of the microscreen was reached at the loading of 0.88 lb/ft²/day at 6.6 gpm/ft². Unfortunately, no solids capacity data were developed for lower hydraulic loading rates.

Diaper[24] reports that microscreens (35 μ aperture) have reduced the secondary effluent solids from 17 mg/l to 5.7 mg/l, on the average, for a trickling filter located at Berkshire, England. Tests at Brampton, Ontario, showed BOD and suspended solids reductions of 53.5 and 57 percent, respectively, in microscreening of an activated sludge effluent. Evans[27] reports that microscreening of the secondary effluent from the Luton, England, plant (activated sludge followed by trickling filter treatment) produced a final effluent with an average BOD of 6.6 mg/l and suspended solids of 6.3 mg/l. Dixon and Evans[25] reported application of a microscreen to a trickling filter plant effluent which contained 36–220 mg/l suspended solids. The 23 μ microscreen effluent contained 15–40 mg/l SS. The authors felt that a 20 mg/l BOD and 20 mg/l suspended solids requirement could generally be met by such a system. The somewhat varied results noted above are no doubt related to the variations in degree of biological flocculation achieved by the various secondary plants involved. The other references cited above indicate that microscreening of well flocculated biological solids, as found in some activated sludge plant effluents, can provide final effluent solids of less than 10 mg/l, in cases where the applied suspended solids are consistently less than 35 mg/l.

Clarification of Unsettled Trickling Filter Effluent. Truesdale and Birkbeck[76] have applied a 35 μ microscreen directly to the effluent from the second stage trickling filters at the two stage trickling plant at the Harpenden (England) sewage works. They reported that the performance of the microscreen was consistently superior to that of a parallel settling basin operating at an overflow rate of about 500 gpd/ft^2. The daily average concentrations of suspended solids in the effluents from the trickling filters, settling basin, and microscreen were 46.9, 22.3, and 12.8 mg/l, respectively. The maximum throughput was found to be 2.5–6 gpm/ft^2 of filter fabric. The suspended solids data showed that the concentration of suspended solids was greater than 20 mg/l only 6 percent of the time. They then attempted to apply the microscreen to the primary trickling filter effluent, but found that the microscreen throughput fell by nearly 70 percent in a few days because of heavy slime growths. Ultraviolet lamps and the use of hypochlorite for cleaning gave only temporary improvement. They concluded that the primary trickling filter effluent would not be amenable to microscreening. It has been reported by Diaper[23] that the promising results obtained on direct microscreening of secondary trickling filter effluents has led to the design of a permanent installation at Fleet, Hampshire, England, where the microscreens may take filter effluent directly. The results from this installation should enable the evaluation of the feasibility of supplementing or replacing two stage trickling filter plant secondary clarifiers with microscreens.

Clarification of Chemically Coagulated Sewage. Lynam and co-workers[43] report that microscreens cannot effect filtration of alum coagulated secondary effluent solids. This is not surprising, since the low strength of such chemical floc noted earlier leads to rapid shearing and penetration of the microscreen. This fact does limit the flexibility of a microscreening installation. Even if the microscreen presents a suitable solution to an immediate suspended solids and BOD removal problem, the microscreen could not be used for removal of chemical floc should such chemical treatment be required for phosphorus removal in the future.

Design of Microscreening Installation. There is no simple means, such as dividing total flow rate by suitable hydraulic filter loading rate, to determine the necessary filter area. The hydraulic capacity of a microscreen is governed by the rate of clogging of the fabric, the rotational speed of the drum, the area of submerged screen, and the available head. Boucher[7] derived a formula for the hydraulics of automatic strainers based on the concept of the filterability of fluids:

$$H = \frac{mQC_f e^{hIQ/s}}{A}$$

where

H = headloss across microscreen (inches)

Q = constant, total rate of flow through the unit (gallons per minute)

C = initial resistance of the clean filter fabric (feet) at a given temperature and standard flow conditions. (23 μ microscreen $C_f = 1.8$ ft; 35 μ microscreen $C_f = 1.0$ ft)

I = filterability index of influent measured on fabric in use (an expression of the volume of water obtained per unit headloss when passed at a standard rate through a unit area of standard filter)

S = speed of strainer expressed as number of square feet of effective fabric entering water in given time (square feet per minute)

A = Effective submerged area (square feet)

Constants: $m = 0.0267$

The filterability index can be determined in a number of ways. The most accurate method actually measures the headloss by means of a mercury manometer across a section of filter being tested using the feed in question with the flow controlled by a constant head box. A portable instrument (Figure 4-24) is also available.

Bodien and Stenburg[6] determined the filterability index for the activated sludge effluent used in their studies at Lebanon. Their data showed the following relationships: $I = 0.494c$, where c is equal to the concentration of influent suspended solids in mg/l. They determined an average value of $I = 17$ for the 23 μ screen and calculated the capacity of a 5×1 ft screen with a maximum headloss of 6 in. and a maximum peripheral drum velocity of 508 fpm for $S = 58$ gpm or 0.0835 mgd. The following table summarizes the microscreen sizes available from Glenfield and Kennedy:

Drum sizes, ft		Motors, bhp		Approx. ranges of	Recommended maximum flow for tertiary sewage applications, mgd	
diam.	width	drive	wash pump	capacity, mgd	23μ	35μ
5	1	$\frac{1}{2}$	1	0.05 to 0.5	0.075	0.11
5	3	$\frac{3}{4}$	3	0.3 to 1.5	0.20	0.30
7$\frac{1}{2}$	5	2	5	0.8 to 4	0.70	1.00
10	10	4	7$\frac{1}{2}$	3 to 10	2.00	3.00

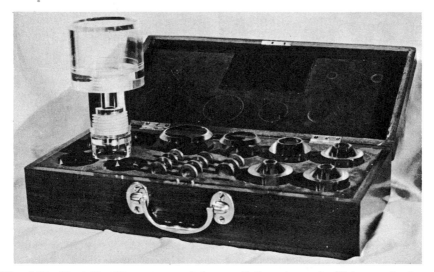

Fig. 4-24 Portable filtrameter used for predicting size and efficiency of micro-screens. *(Courtesy Crane-Cochrane Co.)*

Another manufacturer (Zurn Industries) offers the following units:

Drum sizes, ft		Screen area, ft^2
diam.	width	
4	2	24
4	4	48
6	4	72
6	6	108
6	8	144
10	10	315

Slow Sand Filters

Slow sand filters are usually 12–30 in. deep, with a layer of sand composing about one-half the depth. The sand rests on a layer of coarser material, which in turn rests on an underdrain system. These filters are constructed without a means of backwashing. When such filters are used, effluent is applied at rates of about 3 gal/ft^2/hr. The filter is left in service until the headloss reaches the point that the applied effluent rises to the top of the filter wall. The filter is then drained and allowed to dry partially, and the surface layer of sludge is manually removed. Sand must be added to make up for the losses which occur during the cleaning operation. Such filters obviously require very large land areas and considerable maintenance. For these reasons, their use has been primarily restricted to polishing the effluents from small package plants. However, a study of such installations for the New England Interstate Water Pollution Control Commission[48] led members to conclude that slow sand filters were ineffective because of the rapid clogging of the filters.

Truesdale and Birkbeck[76] report on slow sand filter installations at two works in England. They reported that the frequency of cleaning filters varied from once to twice a month with no improvement in performance observed after cleaning. When in service, the filters were found to remove about 60 percent of the suspended solids and 40 percent of the BOD at hydraulic loading rates of 1.5–2.5 $gal/ft^2/hour$ with applied suspended solids (trickling filter plant effluent) of 20–90 mg/l. Cold winter weather caused the filter to be removed from service for 4–6 weeks during the winter.

Another British publication[4] estimated the cost of slow sand filtration at three times the cost of rapid sand filters and twice the cost of microscreens. The advantage of simplicity of construction is more than offset by the disadvantages of at best moderate performance, high maintenance, large space requirements, and high cost. Slow sand filtration is not a competitive tertiary treatment process.

Rapid Sand Filtration

Surface filtration results when effluent is passed downward through a filter composed of a single type of granular medium which is backwashed by reversing the flow through the filter. During filter backwashing, the medium grades hydraulically, with the finest particles rising to the top of the bed. As a result, most of the material removed by the filter is removed at or very near the surface of the bed. Only a small part of the total voids in the bed are used to store particulates, and the headloss increases very rapidly at high filter rates or high solids loadings. When the secondary effluent contains relatively high solids concentrations, a sand filter will blind at the surface in only a few minutes.

Two extensive discussions of rapid sand filtration of secondary effluents have been prepared by Lynam and co-workers[43] and Tchobanoglous.[72]

Lynam reported on the Chicago Sanitary District tests which involved filtration of secondary effluent from an activated sludge plant through an 11 in. deep bed of sand with an effective size of 0.58 mm and a uniformity coefficient of 1.62. The sand was contained in a continuous backwashing filter of the Hardinge design. Flow passes through the influent ports and down through the media and porous plate underdrain. A constant head is maintained above the filter by setting a filtrate overflow weir elevation. At a present headloss, the backwashing mechanism is activated. During the cleaning cycle, filtrate is pumped from the effluent channel up through one filter compartment at a time. The other filter compartments remain in service. The cleaning mechanism travels the length of the filter, cleaning each filter compartment. The effluent from the operating compartments is used for backwash supply, eliminating the need for separate backwash storage facilities. Investigators examined the rapid sand filter with and without chemical coagulation.

Without prior chemical coagulation, suspended solids reductions through the filters were on the order of 70 percent at rates of 2–6 gpm/ft^2, with BOD reductions on the order of 80 percent. The maximum operating headloss was only 11 in., which is much lower than most rapid sand filter designs permit. The high floc strength of the biological floc permitted these relatively high removals in such a coarse, shallow filter. The significance of floc strength is well illustrated by the fact that tests conducted with alum coagulation and settling of the effluent prior to filtration showed a slight deterioration compared to filtration without pretreatment by alum coagulation. Lynam and his co-authors felt that the poorer performance was due to insufficient coagulation, resulting in floc too fine to be removed. Filter performance was improved by the use of an anionic polymer in conjunction with the alum treatment. The authors of this book feel that the improved performance was primarily due to an increase in floc strength rather than to an increase in floc size. A filter as coarse (0.58 mm effective size) and shallow as that used at Chicago will be very prone to shearing and breakthrough of fragile chemical floc. The data certainly illustrate the point that a coarse filter medium design is better suited for removal of strong biological floc than for removal of fragile chemical floc. The Chicago investigators concluded that the rapid sand filter produced a given degree of treatment more economically than the microscreening device operated in parallel in their study.

Tchobanoglous[72,74] evaluated rapid sand filtration of activated sludge effluent without prior chemical coagulation in a rapid sand filter. As would be expected with a single medium filter, 75–90 percent of the headloss occurred in the upper inch of the rapid sand beds, confirming that such filters are indeed surface filtration devices. At a hydraulic loading rate of 5 gpm/ft^2, a decrease in sand grain size from 1.1 to 0.5 mm increased the suspended solids removal from 10 to 40 percent. The activated sludge process was a high rate process which provided oly a limited degree of biological flocculation. The effect of sand size was much more pronounced than was the effect of the rate of filtration for the biological floc being filtered. The lower limit on sand size is determined, of course, by the initial headloss and the rate of headloss buildup with the single medium filter.

The limitations of the single medium rapid sand filter result from its behavior as a surface filtration device. The advantages of in-depth filtration achieved by dual or tri-media filters can be achieved at lower capital and operating costs than with conventional rapid sand filters. Therefore, if a granular filter is contemplated for use, the filter design selected should provide in-depth filtration rather than merely the surface filtration provided by a rapid sand filter design. Thus, detailed design considerations for granular filters are presented in the sections on in depth filtration. The conventional rapid sand filter medium offers no advantage over the in-depth filter media designs now available, while suffering several severe disadvantages, and is not recommended for use in advanced waste treatment plants.

Diatomaceous Earth Filtration

Several investigators have studied diatomaceous earth (D-E) filtration of secondary effluents. D-E is fed at a controlled rate to the secondary effluent, which is then passed through a precoated filter septum. Shatto[63a] found D-E filtration to produce a final effluent with "no detectable BOD and only a trace of suspended solids." However, the ability of a D-E filter to produce an excellent quality effluent is accompanied by extremely high cost. A major problem associated with D-E filtration of secondary effluents is the inability of the D-E filter to tolerate significant variations in suspended solids concentrations. Eliassen and Bennet[26] confirmed the excellent quality obtainable and also the erratic operation experienced with the varying amounts of suspended solids normally encountered in secondary effluents. They felt that new designs would have to be developed with automated backwash and body feed equipment to cope with these load variations. The unfavorable economics when applied to sewage filtration have, to the author's knowledge, discouraged any attempts by manufacturers to do this development work.

Filtration Costs

Capital costs and the costs for operation and maintenance of wastewater filtration vary widely from place to place. For purposes of preliminary cost estimates, see Chapter 11.

FILTRATION CASE HISTORIES

South Lake Tahoe

The key to the reliable performance and continuous operation of the advanced wastewater treatment plant at South Lake Tahoe is the successful development of the mixed media filter. It is a unit process which is vital to the proper functioning of the plant as a whole. The beds will accept heavy shock loads of suspended solids from upsets in biological or chemical pretreatment without interruption in service or deterioration in effluent quality. They remove all suspended solids, and produce excellent water clarity (turbidity = 0.01–1.0 JTU). They remove significant amounts of colloidal and dissolved phosphorus from the wastewater. They protect the granular carbon treatment which follows from the prolonged interruptions in service and from the serious loss of efficiency which would occur in the absence of the filters because of poor applied water quality.

For the design flow of 7.5 mgd, there are 3 pairs of pressure beds in series, each 10 ft in diameter by 38 ft long. The design filter rate is 5 gpm/ft^2, but rates as high as 8 gpm/ft^2 have been employed at full treatment efficiency. The backwash rate is 15 gpm/ft^2. Each pair of beds is washed as a unit in series. The surface wash consists of four 7 ft diameter rotary filter agitators per bed. Each bed consists of 3 ft of mixed media (as supplied by Neptune Microfloc), supported on 3 in. of coarse garnet and 2 ft 4 in. of graded gravel. The underdrains are perforated plastic pipe. The influent rate of flow controller consists of a Dall flow tube and a rubber seated butterfly valve. Loss of head across each bed is continuously measured and recorded. Turbidity of separation bed effluent is continuously measured by a Hach CR Turbidimeter to tenths of a Jackson unit and recorded. All filter operations are fully automatic. Backwash is initiated by time clock, high headloss, high turbidity, or manually. The beds are backwashed, filtered to waste, and restored on line automatically by a program timer. The filters are backwashed with filter influent water by means of a pump. There is a pressure booster pump for surface wash supply. Waste wash water discharges into an 80,000 gallon steel tank (which holds the water from two backwashes). The water from this tank is returned to the treatment process slowly over a period of about 2 hours. One pair of these beds has been in service for 11 years, and the other two pairs have been in use for 8 years at this writing (1976).

All except one end of each bed is installed out of doors. An allowance was made in the design for the formation of 4 in. of ice inside the steel filter shell. This ice then insulates the tank against further freezing under conditions of normal water flow through the bed.

The performance of these beds and the control system has been excellent. The length of filter runs varies from 4 hours, under very bad conditions to about 60 hours under good. They have been used with alum as the primary coagulant in pretreatment, and with Calgon ST 270 or Purifloc N-11 (0.1–0.8 mg/l) as a filter aid. They have also been used with lime as the primary coagulant in pretreatment and either alum (1–20 mg/l) or ST-270 or N-11 (0.1–0.10 mg/l) as a filter aid applied directly to the filter influent. Normally the beds are backwashed when the headloss through each bed is

TABLE 4-8   Typical Removals by Tahoe Mixed Media Filters.

Substance	Typical Concentrations (mg/l) Influent	Effluent	Range (% removal)
Phosphorus, total	0.65	0.05	70–95
Phosphorus, dissolved	0.45	0.05	65–90
Phosphorus, particulate	0.20	0.00	100
COD	23	15	20–45
BOD	9	5	40–70
SS	15	0	100
Turbidity, JU	7.0	0.3	60–95

about 8 ft (16 ft total). However, they have been backwashed successfully after headlosses through the two beds totaled as much as 40 ft. This high headloss would be excessive for continuous operation. Table 4-8 indicates typical removals of several materials by the separation beds.

Orange County Water District, Water Factory 21

The Orange County plant takes up to 15 mgd of secondary effluent and treats it for groundwater injection after blending with desalted seawater. The injected water serves as a barrier to the intrusion of seawater into the aquifer. Water from the recharged aquifer is withdrawn for agricultural, municipal, industrial, and domestic purposes.

Treatment in the wastewater reclamation plant consists of:

1. Chemical clarification
2. Ammonia stripping
3. Recarbonation
4. Mixed media filtration
5. Carbon adsorption
6. Chlorination

In this treatment scheme it is the purpose of the filtration system to provide the maximum water clarity achievable. This is important for several reasons: (1) the discharge requirements specify that the effluent turbidity shall not exceed 1 JTU; (2) filtration ahead of carbon adsorption reduces the load of organics, suspended solids, and colloidal matter reaching the carbon, thus increasing carbon efficiency and reducing carbon fouling; and (3) the effectiveness of disinfection by chlorination is significantly improved by removing particulate matter which might otherwise trap virus and bacteria, preventing the necessary contact with the chlorine.

The recarbonation system effluent passes through four open, gravity, downflow filter beds which operate in parallel. Figure 4-25 illustrates the

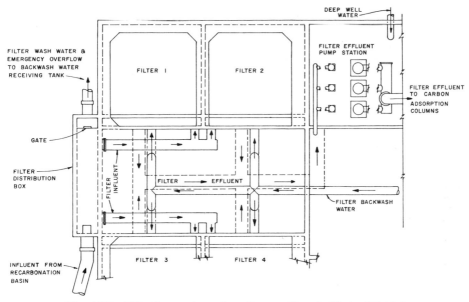

Fig. 4-25   Filtration system plan, Orange County Water District.

layout of the filter system. Each filter is a concrete box 22 ft wide and 24 ft long. Each bed has a capacity of 3.75 mgd at a throughput rate of 5 gpm/ft^2 of surface area. The filters are of the mixed media type, which provide coarse to fine filtration in the direction of flow.

The filter media (30 in. deep) is made up of coarse coal, silica sand, and garnet sand; the supporting media is composed of garnet gravel, silica gravel, and an underdrain, as shown below:

Filter Media	Depth, in.	Specific Gravity	Grain Size Range, mm
Anthracite coal	16.5	1.6	0.84–2.00
Silica sand	9	2.6	0.42–0.48
Garnet sand	4.5	4.0	0.18–0.42
Garnet gravel	1.5	4.0	1
Garnet gravel	1.5	4.0	2.00–4.76
Silica gravel	2	2.6	3.18–6.36
Silica gravel	2	2.6	6.36–12.72
Silica gravel	2	2.6	12.72–19.08
Leopold blocks	—	—	

As a filter operates in removing suspended matter from the wastewater, the headloss or pressure drop increases. Eventually the headloss reaches

TABLE 4-9 Reported Efficiencies for Direct Filtration of Trickling Filter Plant Effluents[1]

Location	Ref.	MEDIA Type	Size² mm	Depth in.	Filter Rate, gpm/ft² (U.S.)	N³	SUSPENDED SOLIDS (mg/l) Influent Range	Avg.	Effluent Range	Avg.	BOD₅, mg/l Influent Range	Avg.	Effluent Range	Avg.	Percentage Removal
Luton, England (1949) Lab study		sand	0.85–2.0	21	2.0	5	34–77	53	1–20	6					
		coal	1.0–2.0	21	2.0	7	41–67	51	4–13	7					
		sand	0.5–0.85	21	2.0	2	40–59	50	1–2	2					
		sand	0.5–1.0	21	2.0	2	49	49	0	0					
Pilot study		sand	0.85–1.7	24	2.33	3m	20–37	28	2–5	3	24–40	30	8–15	11	
		sand	0.85–1.7	24	2.66	1m	20	20	3	3	31	31	15	15	
		sand	0.85–1.7	24	2.83	5m	15–25	19	2–5	3	17–28	21	6–9	7	
		coal	0.85–2	18	1.7	2m	20–37	29	3–5	4	27–40	34	10–14	12	
			2–5	10	2.8	7m	15–28	21	2–5	3	17–28	23	6–13	8	
Finham, England Pilot study (1949)		sand	1.0–2.0	24	1.6	m	32		5		26		10		
		sand	1.0–2.0	24	1.9–2.4	m	40		6		31		10		
		sand	1.0–2.0	24	2.9–3.2	m	35–36		7–8		22–25		10–13		
		sand	1.0–2.0	24	3.2–3.8	m	34–35		7–9		27–38		11–14		
				24	1.4–1.9	m	43		10		32		14		
					2.0–2.3	m	38		5		27		10		
		coal	1.0–2.0	3	2.4–2.8	m	32		11		35		14		
			2–5		2.9–3.2	m	34		8		28		13		
					3.2–3.7	m	34–35		7–8		30–37		14		
Luton, England Full scale (1967)	27	sand	0.85–1.7	36	2.5	12m	7–18	13	1–7		6–15	9	2–8	4	
		sand	0.85–1.7	36	3.4	3m	28–35		9–10		9–10		3–4		
Pilot scale upflow		sand	0.85–1.7	60	5	3m	13		8		5		3		
		sand	0.85–1.7	60	3.4	3m	29–35		5		9–10		3		
					4.8	3m	13		6		5		3		
Derby, England Pilot study (1970)		sand	1.2–2.3	24	2.0	4	25–35	29	7–14	10					34–45
					3.0	2	29–31	30	10–13	12					11–39
					4.0	5	27–31	29	12–14	13					
					8.0	2	24–29	27	16–18	17					
		sand	1.2–1.7	24	2.0	4	26–35	29	7–15	10					45
					3.0	2	29–31	30	9–13	11					15–47
					4.0	5	24–32	28	11–14	13					

Study	Media	Size	Depth	Rate	N								
				8.0	2	23-29	26	16-18	17				33
Triple media "a"	coal	1.4-2.3	8	2.0	2	25-35	29	6-12	8				
	sand	1.2-1.4	8	4.0	4	27-28	27	11-13	12				
	garnet	0.7-0.85	8	8.0	3	23-29	26	15	15				
Upflow sand "a"	sand	0.7-2.3	24	2.0	2	28-37	33	9-15	12				
				4.0	3	27-30	27	13-14	13				
				8.0	2	23-29	26	16-17	17				
Triple media "b"	coal	1.4-2.3	8	3.0	2	29-31	30	7-11	9				38
	sand	0.85-1.0	8	4.0	2	29-32	31	10	10				38
	garnet	0.7-0.85	8										
Upflow sand "b"	sand	0.85-2.3	24	3.0	2	29-31	30	10-14	12				
				4.0	2	28-32	30	11-14	13				
Derby, England Pilot study	sand	1.2-2.3	24	3.0	2	22-25	24	9-10	10				31
				4.0	5	20-24	22	7-10	9				50-64
				6.0	2	19-26	23	10-11	11				35
After trickling filter improvement	sand	1.2-1.7	24	3.0	2	22-25	24	9-10	10				54
				4.0	4	21-24	23	8-10	9				53-65
Triple media "b"	triple (see above)			3.0	1	22	22	8	8				
				4.0	5	20-24	22	6-9	8				
				6.0	2	19-26	23	9	9				
Upflow sand "b"	sand (see above)			3.0	2	22-25	24	8-10	9				69
				4.0	4	21-24	23	8-11	10				59-71
Upflow sand	sand	1.2-2.3	36	6.0	4	19-26	23	9-11	10				43-67
Ames, Iowa Pilot study (1965)	sand	0.55 ES / 2.36 UC	24	2.0	15	11-49	20	1-15	6	38-115	56	13-49	24
				4.0	15	10-58	19	1-24	6	29-130	53	15-65	23
				6.0	15	8-60	18	2-27	6	25-132	50	13-74	24
Pilot study (1973) unpublished	dual media coal	0.9 ES	12	1.7	12	21-75	33	1-8	4	39-85	57	6-19	13
	sand	0.4 ES	12										

[1]Blank spaces in table are due to data missing or not presented in manner needed for table, e.g., for averaging. All mg/l values rounded to nearest 1 mg/l.

[2]Range in size given, British practice, or ES (effective size) and UC (uniformity coefficient), U.S. practice.

[3]N = number of values reported in the range and average presented. N generally represents individual filter runs unless followed by the letter m which indicates the number of average monthly values presented; m without numeral means average of several months data (unspecified duration).

the point (about 10 ft) that the filter must be backwashed to remove the accumulated solids. A filter is backwashed at a rate of about 15 gpm/ft^2 (7920 gpm) for a period of 5–10 minutes. In addition to the backwashing action achieved by reversing the flow through the filter, added cleaning is achieved by the use of rotary surface wash devices. These devices are an essential part of the backwash system. The surface washers are usually operated for 1–2 minutes before the main backwash flow is started, and are turned off 1–2 minutes before the main flow is shut off.

When a filter is returned to service after a backwash cycle, the filter effluent turbidity may be higher than desired. Thus, following the backwash operation, the filter is operated in the filter to waste cycle for about 15–20 minutes as determined by the filter effluent turbidity. It is desirable to filter to waste until the turbidity of the effluent is less than 0.2 JTU before returning the clean filter to service.

The flow rate of filter backwash wastewater would cause a significant hydraulic surge if recycled directly to the plant; therefore, these flows are collected in a backwash water receiving basin. This basin has a capacity of about 160,000 gal and serves several purposes. Its main function is to store flow surges of wastewater from filter backwash, filter to waste, carbon column reverse flow, carbon column to waste, and carbon wash tanks. It also receives emergency overflows from the filters, recarbonation basins, chlorination basin, and blending reservoir. By means of either or both of two pumps, water in the basin can be (1) returned slowly and at a fairly uniform rate to the head of the plant for reprocessing and salvage (the normal mode of operation), (2) wasted to the return line to the secondary treatment plant through the overflow chamber of the chemical clarifier, or (3) wasted to a sewer. The receiving basin also has a gravity emergency overflow to the sewer.

Numerous indicators, recorders, and controls are installed to aid in filter operation. The filter monitoring and control system features are summarized below:

1. The common filter level in the influent distribution box is measured, indicated and controlled.
2. Each of the filter flows is measured, transmitted, recorded, and controlled.
3. Any filter flow rate can be set and then controlled automatically.
4. The headloss for each filter is measured, transmitted, and recorded on the filter flow recorder–controller.
5. The filter backwash flow is measured, transmitted, indicated, and totalized. Special provisions are included for control of backwash flow.
6. The water level in the backwash water receiving basin is measured, transmitted, and indicated.

TABLE 4-10 Reported Efficiencies for Direct Filtration of Activated Sludge Plant Effluents[1]

Location	Ref.	MEDIA Type	MEDIA Size[2] mm	MEDIA Depth in.	Filter Rate, gpm/ft[2] (U.S.)	N[3]	SS Influent Range	SS Influent Avg.	SS Effluent Range	SS Effluent Avg.	BOD Influent Range	BOD Influent Avg.	BOD Effluent Range	BOD Effluent Avg.	Percentage Removal
West Hertfordshire, England (1968)		gravel	40–50	6	2.16		10–89	44	1–2	2		58		3.9	
		gravel	8–12	10	4.0		9–70	37	2–7	3.7		53		4.6	
Pilot scale "boby" upflow filter		gravel	2–3	10	5.0		12–128	55	1–17	7.1		42		5.6	
		sand	1–2	60	6.0		5–97	37	2–22	9.9		35		4.7	
Letchworth, England (1968)		gravel	20–30	4	3–4	81	10–26	22	1–12	5.5					
		gravel	10–15	4	4–6	66	10–28	16	2–15	6.7					
		gravel	2–3	4	6–8	65	7–24	14	6–14	8.8					
Pilot scale "boby" upflow filter		sand	1–2	60											
Los Angeles, Calif. (1961) Preliminary tests		sand	0.95 ES / 1.6 UC	11	2	5	19–34	27	7–21	15	6–15	10	2–8	4	
Philomath, Ore. (1967) (Extended aeration AS)	20	mixed media		30	5		30–2180	59	1–20	4.6	17–36	26	1–4	2.5	
Peoria, Ill. (1964) (High rate AS)		sand			1.1			35		8		45		17	
Cleveland, Ohio		dual media / coal	1.78 ES / 1.63 UC	60	8	1	20		5		19		14		
					16	1	27		8		9		5		
		sand	0.95 ES / 1.41 UC	24	24	2	22–23		9–11		9–10		4–6		
					32	1	29		14		9		7		
Declining rate filters (Avg. filter rate presented)		dual media / coal	4.0 ES / 1.5 UC	60	8	1	13		4		7		5		
					16	1	13		4		7		5		
		sand	2.0 ES / 1.32 UC	24	24	1	13		6		7		5		

[1]Blank spaces in table are due to data missing or not presented in manner needed for table, e.g., for averaging. All mg/l values rounded to nearest 1 mg/l.

[2]Range in size given, British practice, or ES (effective size) and UC (uniformity coefficient), U.S. practice.

[3]N = number of values reported in the range and average presented. N generally represents individual filter runs unless followed by the letter m which indicates the number of average monthly values presented; m without numeral means average of several months data (unspecified duration).

157

TABLE 4-11 **Reported Efficiencies for Filtration of Secondary Effluents After Chemical Treatment**

Location	Ref.	MEDIA Type	Size,[2] mm	Depth, in.	Filter, Rate, gpm/ft[2] (U.S.)	N[3]
ACTIVATED SLUDGE PLANTS						
Lebanon, Ohio		dual				
300 mg/l lime		media			2	
		coal	0.75 ES	18		
			1.5 UC			
		sand	0.46 ES	6		
			1.4 UC			
Lake Tahoe		mixed media		36	5	
400 mg/l lime; alum or poly for		two filters in series				
filter aid						
Hanover Park, Chicago, Ill.	43	sand	0.58 ES	11	0.3–1.9	
38 and 60 mg/l alum & settling			1.62 UC			
Nassau County, N.Y.		dual				
200 mg/l alum & settling; 0.5 mg/l		media				
annionic polymer		coal	0.9 mm	30		
		sand	0.35 mm	6		
TRICKLING FILTER PLANTS						
Ames, Iowa		dual				
		media				
1973-tertiary		coal	0.9 ES	12	2	75
200 mg/l alum & settling		sand	0.4 ES	12		
Richardson, Texas, 1971–72		mixed media		30	2–5	
Alum treatment before secondary						
settling						

[1]Blank spaces in table are due to data missing or not presented in manner needed for this table, e.g., for averaging. All mg/l values rounded to nearest 1 mg/l.
[2]Range in size given, British practice, or ES (effective size) and UC (uniformity coefficient), U.S. practice.
[3]N = number of values reported in the range and average presented. N generally represents individual filter runs unless followed by the letter m which indicates the number of average monthly values presented; m without numeral means average of several months data (unspecified duration).

7. The level in the filtered water pump sump is measured, transmitted, indicated, and controlled.
8. Filter backwash can be initiated manually or automatically. Automatic backwash can be initiated by high headloss or high turbidity. The automatic backwash cycle is completely interlocked so that only one filter can be backwashed at a time.
9. All filter valves close in the event of a power failure. If the level in the blending and storage tank falls below the intermediate level, backwashing cannot be started, but if a backwash is in progress, it will be completed in the normal manner.

TABLE 4-11 *(Continued)*

SUSPENDED SOLIDS (mg/l) OR TURBIDITY (JTU)				BOD$_5$, mg/l				
Influent		Effluent		Influent		Effluent		Percentage
Range	Avg.	Range	Avg.	Range	Avg.	Range	Avg.	Removal
13–36	16	0.07–0.14 (JTU)						
	15		0 0.3 (JTU)		9		5	
		1–28	3			1–9	3	89
			0.4 (JTU)					
3–28	10.3 2.4 (JTU)	0–4	1.4 0.7 (JTU)	3–28	10.8	2–22	5.3	
	7		1–2		42 (COD)		21 (COD)	

Operating data indicate that when an alum dose of 5–10 mg/l is combined with the addition of 0.05–0.1 mg/l polymer as a filter aid, approximately 12–24 hour filter runs are typical with effluent turbidities of about 0.2 JTU.

SUMMARY OF OTHER RESULTS

In March, 1974, Gary A. Rice and John L. Cleasby of Iowa State University tabulated reported efficiencies from a number of operating plants for (1) direct filtration of trickling filter plant effluents, (2) direct filtration of activated sludge plant effluents, and (3) filtration of secondary effluents after chemical treatment. These results are shown in Tables 4-9, 4-10, and 4-11.

REFERENCES

1. Adin, A., and Rebhum, M., "High-Rate Contact Flocculation-Filtration With Cationic Polyelectrolytes," *Journal American Water Works Assoc.*, February, 1974, p. 109.

2. AWWA Committee Report, "State of the Art of Water Filtration," *Journal American Water Works Assoc.*, October, 1972, p. 662.

3. AWWA, "Diatomite Filters for Municipal Use," Task Group Report, *Journal American Water Works Assoc.*, 1965, p. 157.

4. Anonymous, *Renovation and Reuse of Wastewaters in Britain*. Water Pollution Research Laboratory, London, 1967.

5. Bernade, M. A., and Johnson, Barry, "*Schistosoma Cercariae* Removal by Sand Filtration," *Journal American Water Works Assoc.*, 1971, p. 449.

6. Bodien, D. G., and Stenburg, R. L., "Microscreening Effectively Polishes Activated Sludge Plant Effluent," *Water and Wastes Engineering*, September, 1966, p. 74.

7. Boucher, P. L., "A New Measure of the Filterability of Fluids With Application to Water Engineering," *ICE Journal*, 1947, p. 415.

8. Boyd, R. H., and Ghosh, M. M., "An Investigation of the Influences of Some Physicochemical Variables on Pourous-Media Filtration," *Journal American Water Works Assoc.*, February, 1974, p. 94.

9. Brown, T. S.; Malina, J. F., Jr., and Moore, B. D., "Virus Removal by Diatomaceous-Earth Filtration—Part I," *Journal American Water Works Assoc.*, February, 1974, p. 98.

10. Camp. T. R., "Theory of Water Filtration," *Proceedings ASCE*, 1964, p. 1.

11. Cleasby, J. L., "Filter Rate Control Without Rate Controllers, *Journal American Water Works Assoc.*, 1969, p. 181.

12. Cleasby, J. L.; Stangel, E. W.; and Rice, G. A., "Developments In Backwashing of Granular Filters," *Journal Sanitation Engineering Division, ASCE*, October, 1975, p. 713.

13. Conley, W. R., Jr., and Hsiung, K., "Design and Application of Multi-media Filters," *Journal American Water Works Assoc.*, 1969, p. 97.

14. Conley, W. R., Jr., "Integration of the Clarification Process," *Journal American Water Works Assoc.*, 1965, p. 1333.

15. Conley, W. R., Jr., and Pitman, R. W., "Innovations in Water Clarification," *Journal American Water Works Assoc.*, 1960, p. 1319.

16. Conley, W. R., Jr., "Experiences With Anthracite Sand Filters," *Journal American Water Works Assoc.*, 1961, p. 1473.

17. Conley, W. R., "Operations Guide to High-Rate Filtration," *Journal American Water Works Assoc.*, March, 1972, p. 203.

18. Convery, J. J., "Solids Removal Processes." Paper presented at the FWPCA Symposium on Nutrient Removal and Advanced Waste Treatment, Tampa, Florida, November, 1968.

19. Culp, G. L., "Secondary Plant Effluent Polishing," *Water and Sewage Works*, April, 1968, p. 145.

20. Culp, G. L., and Hansen, S. P., "Extended Aeration Effluent Polishing by Mixed Media Filtration," *Water and Sewage Works*, February, 1967, p. 46.

21. Culp, G. L., "How to Reclaim Wastewater for Reuse," *American City*, June, 1967, p. 96.

22. Culp, R. L., "Filtration," Chapter 8, *Water Treatment Design Manual*. AWWA, Denver, 1969.

23. Diaper, E. W. J., "Tertiary Treatment by Microstraining," *Water and Sewage Works*, June, 1969, p. 202.

24. Diaper, E. W. J., "Microstraining and Ozonation of Sewage Effluents." Paper presented at the Forty-first Annual Conference of the Water Pollution Control Federation, Chicago, Illinois, Sept., 1968.

25. Dixon, R. M., and Evans, G. R., "Experiences With Microstraining on Trickling Filter Effluents in Texas." Paper presented at the Forty-eighth Texas Water and Sewage Works Associations Short School, College Station, Texas, March, 1966.

26. Eliassen, K., and Bennett, G. E., "Renovation of Domestic and Industrial Waste Water." Paper presented at the International Conference on Water for Peace, Washington, D.C., May, 1967.

27. Evans, S. C., "Ten Years Operation and Development at Luton Sewage Treatment Works," *Water and Sewage Works*, 1957, p. 214.

28. Fair, G. M., and Geyer, J. C., *Elements of Water Supply and Wastewater Disposal*. John Wiley & Sons, New York, 1958.

29. Geise, G. D.; Pitman, R. W.; and Wells, G. W., "The Use of Filter Conditioners in Water Treatment," *Journal American Water Works Assoc.*, 1967, p. 1303.

30. Hamann, Carl L., and McKinney, Ross E., "Upflow Filtration Process," *Journal American Water Works Assoc.*, 1968, p. 1023.

31. Haney, B. J., and Steimle, S. E., "Potable-Water Supply by Means of Upflow

Filtration (l'Eau Claire Process)," *Journal American Water Works Assoc.*, February, 1974, p. 117.

32. Harmeson, R. H., et al., "Coarse Media Filtration for Artificial Recharge," *Journal American Water Works Assoc.*, 1968, p. 1396.

33. Harris, W. Leslie, "High-Rate Filter Efficiency," *Journal American Water Works Assoc.*, 1970, p. 515.

34. Hudson, H. E., Jr., "Functional Design of Rapid Sand Filters," *Journal Sanitation Engineering Division, ASCE*, 1963, p. 17.

35. Hudson, H. E., Jr., "Physical Aspects of Filtration," *Journal American Water Works Assoc.*, 1969, p. 33.

36. Hudson, H. E., Jr., "High Quality Water Production and Viral Disease," *Journal American Water Works Assoc.*, 1962, p. 1265.

37. Hutchinson, W., and Foley, P. D., "Operational and Experimental Results of Direct Filtration," *Journal American Water Works Assoc.*, February, 1974, p. 79.

38. Ives, K. J., "Progress in Filtration," *Journal American Water Works Assoc.*, 1964, pp. 1–25.

39. Jung, H., and Savage, E. S., "Deep-Bed Filtration," *Jouirnal American Water Works Assoc.*, Feb., 1974, p. 73.

40. Kirchman, W. B., and Jones, W. H., "High-Rise Filtration," *Journal American Water Works Assoc.*, March, 1972, p. 157.

41. Kreissl, J. F.; Robeck, G. G.; and Sommerville, G. A., "Use of Pilot Filters to Predict Optimum Chemical Feeds," *Journal American Water Works Assoc.*, 1968, p. 299.

42. Laughlin, J. E., and Duvall, T. E., "Simultaneous Plant Scale Tests of Mixed Media and Rapid Sand Filters," *Journal American Water Works Assoc.*, 1968, p. 1015.

43. Lynam, B.; Ettelt, G.; and McAloon, T., "Tertiary Treatment at Metro Chicago By Means of Rapid Sand Filtration and Microstrainers," *Journal Water Pollution Control Federation*, 1969, p. 247.

44. Minz, D. M., "Modern Theory of Filtration," Special Subject No. 10, International Water Supply Assoc., London, 1966.

45. Mohanka, S. S., "Multilayer Filtration," *Journal American Water Works Assoc.*, 1969, p. 504.

46. Morey, E. F., "High-Rate Filtration-Media Concepts," *Public Works*, May, 1974, p. 80.

47. Moulton, F.; Bedc, C.; and Guthrie, J. L., "Mixed-Media Filters Clean 50 MGD Without Clarifiers," *Chemical Processing*, March, 1968.

48. New England Interstate Water Pollution Control Commission, *A Study of Small, Complete Mixing, Extended Aeration Activated Sludge Plants in Massachusetts.* 1961.

49. Oakley, H. R., and Cripps, T., "British Practice in the Tertiary Treatment of Sewage," *Journal Water Pollution Control Federation*, 1969, p. 36.

49a. Oakley, H. R., personal communication (Jan. 7, 1969).

50. O'Conner, J. T., and Baliga, K. Y., "Control of Bacterial Growths in Rapid Sand Filters," *Journal Sanitation Engineering Division, ASCE*, December, 1970, p. 1377.

51. Oeben, R. W.; Haines, H. P.; and Ives, K. J., "Comparison of Normal and Reverse-Graded Filtration," *Journal American Water Works Association*, 1968, p. 429.

52. O'Farrell, T. P.; Bishop, D. F.; and Bennett, S. M., "Advanced Waste Treatment at Washington, D. C." Paper presented at the Sixty-fifth Annual AICHE meeting, Cleveland, Ohio, May, 1969.

53. O'Melia, C. R., and Cripps, D. K., "Some Chemical Aspects of Rapid Sand Filtration," *Journal American Water Works Assoc.*, 1964, p. 1326.

54. O'Melia, C. R., and Stumm, Werner, "Theory of Water Filtration," *Journal American Water Works Assoc.*, 1967, p. 1393.

55. Pitman, R. W., and Wells, G. W., "Activated Silica as a Filter Conditioner," 1968, p. 1667.

56. Rice, A. H., "High-Rate Filtration," *Journal American Water Works Assoc.*, April, 1974, p. 258.

57. Rice, A. H., and Conley, W. R., "The Microfloc Process in Water Treatment," *Tappi*, 1961, p. 167A.

58. Riddick, T. M., "Filter Sand," *Journal American Water Works Assoc.*, 1940, p. 121.

59. Robeck, G. G.; Clarke, N. A.; and Dostal, K. A., "Effectiveness of Water Treatment Processes in Virus Removal," *Journal American Water Works Assoc.*, 1962, p. 1275.

60. Robeck, G. G.; Dostal, K. A.; and Woodward, R. L., "Studies of Modifications in Water Filtration," *Journal American Water Works Assoc.*, 1964, p. 198.

61. Sedore, J. K., "We Converted to High Rate Filtration," *American City*, April, 1968, p. 00.

62. Segall, B. A., and Okun, D. A., "Effect of Filtration Rate on Filtrate Quality," 1966, p. 468.

63. Seth, A. K., et al., "Nematode Removal by Rapid Sand Filtration," *Journal American Water Works Assoc.*, 1968, p. 962.

63a. Shatto, J., "Aerobic Digestion and Diatomite Filter," *Public Works*, Dec., 1960, p. 82.

64. Shea, T. G.; Gates, W. E.; and Argaman, Y. A., "Experimental Evaluation of Operating Variables in Contact Flocculation," *Journal American Water Works Assoc.*, 1971, p. 41.

65. Shull, K. E., "Experiences With Multi-Bed Filters," *Journal American Water Works Assoc.*, 1965, p. 314.

66. Smit, P., "Upflow Filter," *Journal American Water Works Assoc.*, 1963, p. 804.

67. Smith, R., "Cost of Conventional and Advanced Treatment of Wastewaters," *Journal Water Pollution Control Federation*, 1968, p. 1546.

68. Sprink, S. M., and Monscitz, J. T., "Design and Operation of a 200 MGD Direct-Filtration Facility," *Journal American Water Works Assoc.*, February, 1974, p. 127.

69. Sweeney, G. E., and Prendville, P. W., "Direct Filtration: An Economic Answer to a City's Water Needs," *Journal American Water Works Assoc.*, February, 1974, p. 65.

70. Syrotynski, S., "Microscopic Water Quality and Filtration Efficiency," *Journal American Water Works Assoc.*, 1971, p. 237.

71. Talley, D. G.; Johnson, J. A.; and Pilzer, J. E., "Continuous Turbidity Monitoring," *Journal American Water Works Assoc.*, March, 1972, p. 184.

72. Tchobanoglous, G., "Filtration Techniques in Tertiary Treatment." Paper presented at the Fortieth Annual Conference of the California Water Pollution Control Association, April 26, 1968.

73. Tchobanoglous, G., and Eliassen, R., "Filtration of Treated Sewage Effluent," *Journal Sanitation Engineering Division, ASCE*, 1970, p. 243.

74. Tchobanoglous, G., and Bennett, G., "Progress Report Water Reclamation Study Program." FWPCA Demonstration Grant, WPD 21-05, October, 1967.

75. Tredgett, R. G., "Direct-Filtration Studies for Metropolitan Toronto," *Journal American Water Works Assoc.*, February, 1974, p. 103.

76. Truesdale, G. A., and Birkbeck, A. E., "Tertiary Treatment Processes for Sewage Works Effluents." Paper presented to the Scotish Centre of the Institution of Public Health Engineers, March 28, 1966.

77. Tuepker, J. L., and Buescher, C. A., Jr., "Operation and Maintenance of Rapid Sand and Mixed-Media Filters in a Lime Softening Plant," *Journal American Water Works Assoc.*, 1968, p. 1377.

78. Walker, J. D., "High Energy Flocculation and Air-Water Backwashing,' *Journal American Water Works Assoc.*, 1968, p. 321.

79. Walton, Graham, *Effectiveness of Water Treatment Processes*. U.S. Public Health Service, Publication No. 898., 1962.

80. Weber, W. J., Jr., *Physicochemical Processes for Water Quality Control*. Wiley–Interscience, New York, 1972.

5

Activated carbon adsorption and regeneration

INTRODUCTION

Conventional wastewater treatment processes, such as activated sludge and trickling filters, remove many organic materials by biological oxidation. These conventional processes may remove nearly all of those organics measured by the biochemical oxygen demand (BOD) test, but are not as effective in removing the so called refractory organic materials measured by the chemical oxygen demand (COD) test. Even well treated secondary effluents may contain 50–120 mg/l of organics. These materials include tannins, lignins, ethers, proteinaceous substances, and other color and odor producing organics, as well as MBAS (methylene blue active substances), herbicides, and pesticides such as DDT.

There are a rather limited number of unit processes capable of removing refractory organic materials from wastewater. Available processes include reverse osmosis, freezing, chemical oxidation, distillation, and adsorption on granular or powdered activated carbon. The purpose of this chapter is to review the major operating parameters, relative effectiveness, and costs of both granular and powdered activated carbon treatment systems.

Activated carbon is the best developed and one of the most efficient processes available for the removal of most organic and some inorganic materials from wastewater. Activated carbon was first used in water and wastewater treatment for the removal of organic material. Concentrations of organics in water in the microgram per liter (μg/l) range can impart taste, odor, and color; some organic compounds, such as herbicides and pesticides, are toxic. Little is known about the non-bio-degradable or more stable organics in wastewater. It has been suggested that prolonged ingestion of organics originating in municipal wastewater may have long term effects on human health.[1]

There is also evidence that activated carbon is effective in the removal of some inorganic materials, including some potentially toxic trace metals. Therefore, activated carbon may have application in the treatment of wastewaters for the removal of some inorganic as well as organic contaminants.

A second edition of the U. S. Environmental Protection Agency manual on the design of granular activated carbon systems has been published.[2] This EPA manual contains detailed information on the design and costs of granular carbon systems and there is no attempt to duplicate the EPA data in this chapter.

ADSORPTION BY ACTIVATED CARBON

Activated carbon removes some organic and inorganic material from water by the process of adsorption or the attraction and accumulation of one substance on the surface of another. In general, pore structure and high surface area are the most important characteristics of activated carbon in adsorption of material from water and the chemical nature of the chemical nature of the carbon surface is less important.

There are several factors which affect adsorption by activated carbon, including: (1) characteristics of the activated carbon; (2) characteristics and concentration of the material to be adsorbed; (3) characteristics of the wastewater, such as pH and suspended solids content; and (4) the contacting system and its mode of operation.

Most of the surface area available for adsorption by carbon is located in the pores within the carbon particles created during the activation process. The major contribution to surface area is located in pores of molecular dimensions. For example, the surface area of several activated carbons manufactured for wastewater treatment is about 1000 m^2/g, with a mean particle diameter of about 1.6 mm and density of 1.4 g/cm^3. Assuming spherical particles, only about 0.0003 percent of the total surface area is furnished by the outside surface of the carbon particle. Similarly, for powdered activated carbon (less than no. 300 sieve size) only about 0.014 percent of the total surface area is on the outside surface of the particle.

The organic material in municipal wastewater is a heterogeneous mixture of a multitude of compounds which are largely unknown. There have been a few attempts to identify the organics in raw or treated municipal wastewater, and the nature of the residual organics which are not removed by treatment with activated carbon. As of August 1976, research was being conducted by Stanford University to identify organic materials in carbon treated wastewater produced at the Orange County Water District treatment plant in Fountain Valley, California.

Bunch and co-workers[3] analyzed samples of secondary effluent from five treatment plants in the United States and found the organic content to be approximately as shown in Table 5-1. The components listed in Table 5-1 are given as a percentage of total COD. The organic acids that are ether extractable are the largest identifed constituent. Note, however, that 65 percent of the organic content of secondary effluent remains unidentified.

Data on the specific analyses of an English effluent are shown in Table 5-2.[4] The largest single constituent of the particulate fraction was protein

TABLE 5-1   Organic Matter in Secondary Effluent. (*After Bunch et al.*[3])

Organic Matter	Percentage of Total COD
Ether extractables	10
Organic acids	65
Neutral	27
Others	8
Proteins	10
Carbohydrates and polysaccharides (No simple sugars)	5
Tannins and lignins	5
Anionic detergents	10
Unidentified	65

Average COD, 100 mg/l; BOD, 20 mg/l.

TABLE 5-2   Composition of an English Secondary Effluent (Trickling Filter). (*After Helfgott et al.*[4])

Constituent	Effluent, mg/l	
	Particulate	Soluble
Fat—acid	0.12	0
Fat—ester	0.12	0
Protein	2.74	0.25
Amino acid	0	0.06
Carbohydrates	1.39	0.24
Soluble acids	0.13	1.63
MBAS	0.05	1.40
Amino sugars	0.38	0
Muramic acid	0.05	0

(2.74 mg/l). The soluble fraction contained relatively little protein (0.25 mg/l) and even less amino acids (0.06 mg/l).

Trickling filter effluent from the Haifa, Israel municipal wastewater treatment plant was analyzed for total organic content and the results reported as percent of total COD.[5] The total COD was about 180 mg/l and the organics were classified as: 40 to 50 percent humic substances (humic, fulvic, and hymathomelanic acids); 8.3 percent ether extractables; 13.9 percent anionic detergents; 11.5 percent carbohydrates; 22.5 percent proteins; and 1.7 percent tannins. A later and more extensive study including activated sludge effluents was conducted[6] and the results are summarized in Tables 5-3 and 5-4. These data show that the distribution of the main organic fractions is similar for the various biological treatment units studied.

The three broad classes of organics in municipal wastewater—fats, carbohydrates, and proteins—are usually considered removable by sedimentation and biological treatment, protein being somewhat less readily removed than fats and carbohydrates. Other potentially harmful materials such as herbicides, pesticides, and trace metals might also be present in municipal wastewater.

Studies on the removal of pesticides from water supplies indicated that adsorption on activated carbon is effective for a variety of chlorinated hydrocarbon, organic phosphorus, and carbamate compounds.[7] Weber and Gould[8] investigated the adsorption by activated carbon of several organic pesticides from aqueous solution and found carbon to have a fairly high capacity for all the types studied. Studies of the adsorption on activated carbon of several carboxylic acids and three herbicides[9] showed

TABLE 5-3   Distribution of Organic Groupings in Secondary Effluents—Mean Values. (*After Manka et al.*[6])

Organic Groupings and Fractions	Municipal Wastewater; High Rate Trickling Filter	Municipal Wastewater; Stabilization Pond	Domestic Wastewater; Extended Aeration Activated Sludge
	PERCENTAGE OF TOTAL COD		
Proteins	21.6	21.1	23.1
Carbohydrates	5.9	7.8	4.6
Tannins and lignins	1.3	2.1	1.0
Anionic detergents	16.6	12.2	16.0
Ether extractables	13.4	11.9	16.3
Fulvic acid	25.4	26.6	24.0
Humic acid	12.5	14.7	6.1
Hymathomelanic acid	7.7	6.7	4.8

that there was a marked increase in the removal of all solutes from water on lowering the pH below 7.0. In general, adsorption of typical organic pollutants from water is increased with decreasing pH.

Some of the materials which have been adsorbed from water by activated carbon and identified include phenyl ether, orthonitrochlorobenzene, naphthalene, styrene, xylenols, phenols, DDT, aldrin, alkylbenezene-sulfonate, and a wide variety of aliphatic and aromatic hydrocarbon materials.[10] Eleven individual odorous pollutants were adsorbed by activated

TABLE 5-4   Molecular Weight Distribution of Humic Substances from Secondary Effluents. (*After Manka, et al.*[6])

Molecular Weight Range	Fulvic Acid	Humic Acid	Hymathomelanic Acid
	PERCENTAGE OF HUMIC COMPOUND PRESENT		
<500	27.5	17.9	4.5
500–1000	7.8	6.2	12.2
1000–5000	35.7	29.4	48.0
5000–10,000	15.3	7.8	28.0
10,000–50,000	9.4	36.7	7.5
>50,000	4.3	2.0	0

carbon and identified in the Kanawha River at Nitro, West Virginia.[11] Recent isotherm tests showed good removals from distilled water of aldrin, dieldrin, endrin, DDD, DDE, DDT, toxaphene, and Arochlor 1242 and 1254.[12] The first seven are pesticides, and the last two are mixtures of PCB containing 42 percent and 54 percent chlorine by weight, respectively.

There are organic materials in secondary effluent that resist removal by activated carbon. In one study of municipal wastewaters, 77 organic compounds were detected in the primary effluent and 38 in the secondary effluent, and apparently some additional compounds were being produced during secondary treatment.[13] It was found in another study[4] that soluble organics are produced in biological treatment that are more refractory to further treatment than are the organics in sewage. It was hypothesized by these same investigators that the residual organics not removed by activated carbon are intermediate breakdown products of protein, and that these are most likely the protein that originates from the cell walls of microorganisms present in biological treatment processes.

A recent study found that chlorine containing, stable, organic constituents are present after chlorination of effluents from domestic sanitary sewage treatment plants.[14] Over 50 chlorine containing constituents were separated from chlorinated secondary effluents; 17 of these chlorine containing, organic compounds were tentatively identified and quantified at the 0.5 to 4.3 μg/l level. In addition to the 17 chlorine containing compounds that were identified, 32 stable organic constituents were identified and 23 of these were quantified at 2–190 μg/l levels in the effluents domestic sanitary primary sewage treatment plants.

Some inorganic materials are also removed by activated carbon. For example, chlorine is believed to react on the carbon surface as follows:

$$C + 2Cl_2 + 2H_2O \longrightarrow CO_2 + 4HCl$$

The use of activated carbon is believed to have considerable potential in industrial waste treatment for removing the last portion (1–2 mg/l) of metal following electrodeposition or cementation, such as in the processing of electroplating wastes.[15]

The potential of activated carbon to adsorb a variety of inorganic compounds was reported by Sigworth and Smith,[16] as shown in Table 5-5. This information is based on a literature review by the authors, laboratory research by Westvāco Carbon Department, and a two year research study conducted by the Colorado School of Mines Research Foundation.

A pilot plant treatment study on trickling filter effluent achieved good removals of silver, cadmium, chromium, and selenium on 14 $\times$ 40 mesh granular activated carbon.[17] The data from this study are summarized

TABLE 5-5 Potential for Removal of Inorganic Material by Activated Carbon. (*After Sigworth and Smith*[16])

Constituents	Potential for Removal by Carbon
Metals of High Sorption Potential	
Antimony	Highly sorbable in some solutions
Arsenic	Good in higher oxidation states
Bismuth	Very good
Chromium	Good, easily reduced
Tin	Proven very high
Metals of Good Sorption Potential	
Silver	Reduced on carbon surface
Mercury	CH_3HgCl sorbs easily Metal filtered out
Cobalt	Trace quantities readily sorbed, possibly as complex ions
Zirconium	Good at low pH
Elements of Fair-to-Good Sorption Potential	
Lead	Good
Nickel	Fair
Titanium	Good
Vanadium	Variable
Iron	Fe^{3+} good, Fe^{2+} poor, but may oxidize
Elements of Low or Unknown Sorption Potential	
Copper	Slight, possibly good if complexed
Cadmium	Slight
Zinc	Slight
Beryllium	Unknown
Barium	Very low
Selenium	Slight
Molybdenum	Slight at pH 6–8, good as complex ion
Manganese	Not likely, except as MnO_4
Tungsten	Slight
Miscellaneous Inorganic Water Constituents	
Phosphorus	
P, free element	Not likely to exist in reduced form in water

TABLE 5-5 (Continued)

Constituents	Potential for Removal by Carbon
PO_4^{3-} phosphate	Not sorbed but carbon may induce precipitation of $Ca^3(PO_4)_2$ or $FePO_4$
Free halogens	
F_2 fluorine	Will not exist in water
Cl_2 chlorine	Sorbed well and reduced
Br_2 bromine	Sorbed strongly and reduced
I_2 iodine	Sorbed very strongly, stable
Halides	
F^- fluoride	May sorb under special condition
Cl^-, Br^-, I^-	Not appreciably sorbed

as follows:

Trace Metal	Percentage Removal
Ag (Ag^+)	85.5
Cd (Cd^{2+})	92.3
Cr (Cr_2O^{2-})	94.1
Se (SeO_3^{3-})	33.7

Concentrations of these metals in the trickling filter effluent varied from about 13 to 55 $\mu g/l$.

A study of the behavior of inorganic ions on charcoal was reported by several Japanese investigators.[18] They presented data on observed adsorption, and indicated that of the 50 inorganic ions studied, most anionic species readily attach on charcoal. The major exceptions were the alkali and alkaline earth metal ions.

It is evident that activated carbon is capable of removing a wide variety of organic and inorganic materials from wastewater; knowledge of the mechanism of removal, however, particularly for inorganic material, is somewhat speculative.

MANUFACTURE OF ACTIVATED CARBON

Activated carbons can be made from a variety of carbonaceous materials including wood, coal, peat, lignin, nutshells, bagasse (sugar cane pulp), sawdust, lignite, bone, and petroleum residues. In the past, carbons used in industrial applications have been produced most frequently from wood, peat, and wastes of vegetable origin, such as fruit stones, nutshells, and sawdust. The present tendency in the manufacture of granular carbons for wastewater treatment applications in the U.S., is toward use of various kinds of natural coal and coke, which are relatively inexpensive and readily available. Of interest is the potential utilization of different wastes as raw materials; for example: waste lignin, sulfite liquors, processing wastes from petroleum and lubricating oils, and carbonaceous solid waste.

General descriptions of activated carbon manufacturing processes are available from manufacturers and in several books.[19,20,21] However, the production of activated carbon is a highly competitive industry, and specific details are not available. Among the major U.S. producers are:

Calgon—"Filtrasorb" granular carbons
Westvāco—"Nuchar" granular carbons; "Aqua Nuchar" powdered carbons
Husky Industries—"Husky" powdered carbon
ICI United States—"Hydrodarco" powdered and granular carbons
Barneby Cheney—granular and powdered carbons
Union Carbide—"Columbia" granular carbons
Witco Chemical—"Whitco" granular carbons

Activated carbons can be classed in two groups: powdered and granular. In wastewater applications, powdered carbons are predominantly (60–75 percent) smaller than 325 mesh, while granular carbons are typically larger than 40 mesh (0.42 mm). Powdered activated carbon is usually produced by activating lump material, or chips of wood charcoal, or lumps of paste prepared from sawdust, and subsequently grinding the activated product. Depending on the method of grinding the shape of the particles can differ to a certain extent, and this may markedly influence the properties the product displays in industrial application; for example, the hydrodynamic resistance to a liquid decolorized by filtration through a layer of active carbon. Active carbon ground in ball mills has oval particles, and that ground in hammer mills has elongated particles.

Granular carbons may be formed by either crushing or pressing. Crushed activated carbon is prepared by activating a lump material, which is then crushed and classified to desired particle size. Pressed activated carbon is formed prior to activation. The appropriate starting material is prepared in a plastic mass, then extruded from a die and cut into pieces of uniform length. These uniform cylindrical shapes are then activated. The necessary hardness is acquired in the activation process. The physical characteristics of hardness, a very important factor, are directly related to the nature of the starting material for crushed activated carbons. For activated carbons formed from pulverized materials, which are then bound together, the nature of the binding agent is a major factor in determining the hardness of the finished product.

Activated carbon is produced by a process consisting of raw material dehydration and carbonization followed by activation. The starting material is dehydrated and carbonized by slowly heating in the absence of air, sometimes using a dehydrating agent such as zinc chloride or phosphoric acid.

Excess water, including structural water, must be driven from the organic material. Carbonization converts this organic material to primary carbon, which is a mixture of ash (inert inorganics), tars, amorphous carbon, and crystalling carbon (elementary graphitic crystallites). Noncarbon elements (H_2 and O_2) are removed in gaseous form, and the freed elementary carbon atoms are grouped into oxidized crystallographic formations. During carbonization, some decomposition products or tars will be deposited in the pores, but will be removed in the activation step.

Activation is essentially a two phase process requiring burnoff of amorphous decomposition products (tars), plus enlargement of pores in the carbonized material. Burnoff frees the pore openings, increasing the number of pores, and activation enlarges these pore openings. Activated carbon can be manufactured by two different procedures: physical activation and chemical activation. Although both processes are widely used, physically activated carbons are utilized in wastewater treatment while chemically activated carbons are utilized elsewhere—as in the recovery of solvents.

Particle size is generally considered to affect adsorption rate, but not adsorptive capacity. As shown previously in this chapter, the external surface constitutes a small percentage of the total surface area of an activated carbon particle. Since adsorption capacity is related to surface area, a given weight of carbon gains little capacity upon being crushed to smaller size.

CHARACTERISTICS OF CARBON USED IN WASTEWATER TREATMENT

There are several characteristics of activated carbon of importance in evaluating its suitability to wastewater treatment, including: surface area, apparent density, bulk density, effective size, pore volume, sieve analysis, abrasion number, ash percentage, iodine number, molasses number, and pore size distribution. Typical properties of four U.S. commercial granular activated carbons are given in the EPA carbon manual[2] and are shown in Table 5-6; similar information for powdered activated carbons is shown in Table 5-7.

TABLE 5-6 Properties of Several Commercially Available Granular Activated Carbons.

Physical Properties	ICI America Hydrodarco 3000	Calgon Filtrasorb 300 (8×30)	Westvaco Nuchar WV-L (8×30)	Witco 517 (12×30)
Physical Properties				
Surface area, m²/g (BET)	600–650	950–1050	1000	1050
Apparent density, g/cm³	0.43	0.48	0.48	0.48
Density, backwashed and drained, lb/ft³	22	26	26	30
Real density, g/cm³	2.0	2.1	2.1	2.1
Particle density, g/cm³	1.4–1.5	1.3–1.4	1.4	0.92
Effective size, mm	0.8–0.9	0.8–0.9	0.85–1.05	0.89
Uniformity coefficient	1.7	≤1.9	≤1.8	1.44
Pore volume, cm³/g	0.95	0.85	0.85	0.60
Mean particle diameter, mm	1.6	1.5–1.7	1.5–1.7	1.2
Specifications				
Sieve size, U.S. std. series[a]				
Larger than No. 8, max. percentage	8	8	8	c
Larger than No. 12, max. percentage	c	c	c	5
Smaller than No. 30, max. percentage	5	5	5	5
Smaller than No. 40, max. percentage	c	c	c	c
Iodine No.	650	900	950	1000
Abrasion No., minimum	b	70	70	85
Ash, percentage	b	8	7.5	0.5
Moisture as packed, Max. percentage	b	2	2	1

[a]Other sizes of carbon are available on request from the manufacturers.
[b]No available data from the manufacturer.
[c]Not applicable to this size carbon.

TABLE 5-7   Properties of Some Commercially Available Powdered Activated Carbons.

Property	ICI America Hydrodarco C	Pittsburgh activated carbon type RC
Particle Size, percentage less than No. 325	65	65–75
Tamped density, g/ml	0.50	0.48
Surface area, m^2/g	550	1100–1300
pH	10.5	—
Molasses No.	95	300
Water solubles, %	5.5	—

EVALUATION OF ACTIVATED CARBON

The evaluation of alternative activated carbons for a particular application is at present a difficult task. Evaluation of activated carbon includes (1) determination of adsorption rate and capacity and (2) determination of attrition losses that will occur during carbon handling and regeneration. Procedures for these determinations are not well developed.

Mattson and Kennedy investigated adsorption capacity, rate of adsorption and resistance to attrition.[22] It was concluded that additional standards for technical evaluation of activated carbons are needed for the determination of adsorption rate and capacity as well as standard tests for determining attrition losses.

The ability of granular carbon to withstand handling and slurry transfer is very important in wastewater treatment because attrition losses create carbon fines and can increase headlosses during flow through the beds. Presently the best carbons for treatment of wastewater appear to be those made from select grades of coal. These carbons are hard and dense and can be conveyed in water slurry with no appreciable deterioration. There is no precise method for predicting the performance of carbons founded on their basic properties or those of the adsorbing molecules. In wastewater, the substances to be removed, such as color, odor, COD, MBAS, and other refractory organics, are always a composite of ingredients of unknown identity. Further, the use of granular carbon is fairly new for treatment of wastewater, the potential market is great, and the competition among carbon manufacturers is likely to be keen. Therefore, it is not unreasonable to expect not only a future reduction in prices for granular carbon, but also a possible improvement in the kind and quality of carbons commercially available.

The approximate adsorptive capacity of a carbon can be measured by determining the adsorption isotherm experimentally in the system under consideration. Simpler capacity tests, such as the Iodine Number or the Molasses Number, also may be appropriate measures of adsorptive capacity.

Isotherm Tests

The adsorption isotherm is the relationship, at a given temperature, between the amount of a substance adsorbed and its concentration in the surrounding solution. If a color adsorption isotherm is taken as an example, the adsorption isotherm would consist of a curve plotted with residual color in the water as the abscissa, and the color adsorbed per gram of carbon as the ordinate. A reading taken at any point on the isotherm gives the amount of color adsorbed per unit weight of carbon, which is the carbon adsorptive capacity at a particular color concentration and water temperature. In very dilute solutions, such as wastewater, a logarithmic isotherm plotting usually gives a straight line. In this connection, a useful formula is the Freundlich equation, which relates the amount of impurity in the solution to that adsorbed as follows:

$$x/m = KC^{1/n}$$

where

x = amount of color adsorbed
m = weight of carbon
x/m = amount of color adsorbed per unit weight of carbon
k and n are constants
C = unadsorbed concentration of color left in solution

In logarithmic form, this equation is

$$\log x/m = \log k + 1/n \log C$$

in which $1/n$ represents the slope of the straight line isotherm.

Data for plotting isotherms are obtained by treating fixed volumes of the liquid to be tested with a series of known weights of carbon. The carbon–liquid mixture is agitated for a fixed time at constant temperature. After the carbon has been removed by filtration, the residual organic content of the solution is determined. From these measurements, all of the values necessary to plot an isotherm may be calculated.

The isotherm tests may be performed at room temperature unless the anticipated plant operation will be at a significantly different one. To determine the needed contact time for the isotherms, a preliminary experiment should be run in which fixed volumes of the wastewater are contacted with a fixed weight of carbon for 1, 2, 3, and 4 hour periods. A contact time sufficiently long to ensure a reasonable equilibrium should be chosen from these data for the isotherms.

The pH of the wastewater being tested will affect the carbon efficiency. Care should be taken to ensure that the pH of the test sample is representative of anticipated plant scale conditions.

In order to minimize the variation of carbon particle size in the isotherm test, the granular carbon should be pulverized so that 95 percent passes a 325 mesh screen. If suitable mechanical equipment, such as a ball mill, is not available to pulverize the carbon, a mortar and pestle can be used, although this is a tedious approach. The manufacturer of the carbon may also be able to supply pulverized material.

For the initial isotherm test, carbon dosages of 50, 100, 150, 200, and 300 mg/l of wastewater may be used. If the highest dosage is not adequate to effect the desired degree of treatment, higher dosages should be tried until that degree is achieved. If one of the intermediate dosages is sufficient to effect complete or satisfactory removal, one or two lower dosages should be tried. In order to obtain a satisfactory isotherm, as wide a range of organic removal should be obtained as is practical.

Isotherm Procedure:
1. Pulverize a representative sample of the granular carbon (a 10–20 g sample is usually adequate) so that 95 percent will pass through a 325 mesh screen. Oven dry the pulverized sample for 3 hours at 150°C.
2. Obtain a representative sample of the wastewater to be tested. Suspended matter should be removed by filtration.
3. Transfer four different weights of the oven dried pulverized carbon to the test containers. Stoppered flasks or pressure bottles are satisfactory containers.
4. To one container, add 100 ml of wastewater from a delivery burette or graduate cylinder, and clamp the container on a mechanical shaker. The samples must be constantly agitated during the isotherm test and a mechanical shaker is desirable. A Burrell Wrist Action Shaker, Catalog No. 75-775, is satisfactory. Agitate the mixture for the chosen contact time. The bottles may be filled and placed on the shaker at 10 or 15 minute intervals to give the analyst sufficient time to filter each sample immediately after the contact time has elapsed. The same volume of wastewater should be added to a container without carbon and subjected to the same procedure in order to obtain a blank reading.
5. After the chosen contact time has elapsed, filter the contents of the

flask through either a laboratory pressure filter fitted with an asbestos disk or through a Buchner funnel containing a filter paper inserted in a filter flask connected to a vacuum. The blank should be filtered in the same manner as the other samples. It is desirable to discard the first and last portions of the filtrate and save only the middle portion for analysis.

6. Determine the organic content of the filtrate.

7. Tabulate the data as shown in Table 5-8. The residual solution COD concentration c is obtained directly from the filtrate analysis. The amount adsorbed on the carbon, x, is obtained by subtracting the value of c from that of c_0, the influent concentration. Dividing x by m, the weight of carbon used in the test, gives the amount adsorbed per unit weight of carbon.

8. On log paper plot c on the horizontal axis against x/m on the vertical axis and draw the best straight line through the points, as illustrated in Figure 5-1.

From the isotherm, it is apparent whether or not the desired degree of purification can be attained with the particular activated carbon tested. If a vertical line is erected from the point on the horizontal scale corresponding to the influent concentration (c_0) and the isotherm is extrapolated to intersect that line, the x/m value at the point of intersection can be read from the vertical scale.

This value, $(x/m)c_0$, represents the amount of COD adsorbed per unit weight of carbon when that carbon is in equilibrium with the influent concentration. Since this should eventually be attained during column treatment, it represents the ultimate capacity of the carbon. However, experience has shown that it is difficult to predict the carbon dosage accurately from isotherm tests for plant scale carbon columns. Pilot carbon column tests conducted over several weeks are the only accurate means of determining the required carbon dosage.

TABLE 5-8   Tabulation of Isotherm Data.

m Weight of carbon; mg/100 ml solution	c Residual COD, mg/l	x COD adsorbed, mg	x/m COD adsorbed per unit weight
0	40	—	—
5	27	1.3	0.26
10	19	2.1	0.21
15	14	2.6	0.17
20	10	3.0	0.15

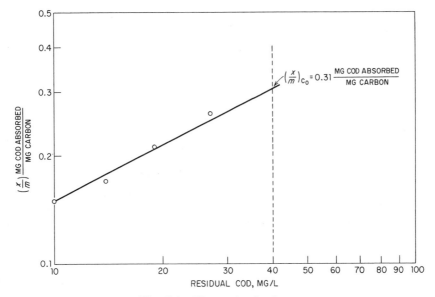

Fig. 5-1   Illustrative isotherm.

The chief value of the isotherm test lies in comparison of various types of carbon and the ability to determine the quality of effluent achievable following carbon adsorption. From an isotherm test it can be determined whether or not a particular purification can be effected. The isotherm will also show the approximate capacity of the carbon for the application, and provide a rough estimate of the carbon dosage required. Isotherm tests also afford a convenient means of studying the effects of pH and temperature on adsorption. Isotherms put a large amount of data into concise form for ready evaluation and interpretation. Isotherms obtained under identical conditions using the same test solutions for two test carbons can be quickly and conveniently compared to reveal the relative merits of the carbons.

Figures 5-2 and 5-3 are presented to illustrate the interpretaton of ad-

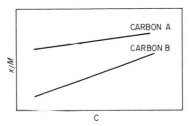

Fig. 5-2   Adsorption isotherm, carbons A and B.

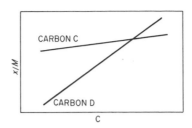

Fig. 5-3   Adsorption isotherm, carbons C and D.

sorption isotherms. In Figure 5-2 the isotherm for carbon A is at a high level and has only a slight slope. This means that adsorption is large over the entire range of concentrations studied. The fact that the isotherm for carbon B in Figure 5-2 is at a lower level indicates proportionally less adsorption, although adsorption improves at higher concentrations over that at low concentrations. An isotherm having a steep slope indicates that adsorption is good at high concentrations but much less at low concentration. In general, the steeper the slope of its isotherm the greater the efficiency of a carbon in column operation. In Figure 5-3, carbon D is better suited to countercurrent column operation than carbon C. It has a higher capacity at the influent concentration, or more reserve capacity. Carbon C would be better than carbon D for batch treatment.

As mentioned previously, the Iodine Number and the Molasses Number also give indications of the adsorptive capacity of a carbon. The Iodine Number is the milligrams of iodine adsorbed from a 0.02 N solution at equilibrium under specified conditions. The Molasses Number is an index of the adsorptive capacity of the carbon for color bodies in a standard molasses solution as compared to a standard carbon. The procedures for determining the Iodine and Molasses Numbers are given in the EPA Manual[2] and are therefore not repeated here. These tests are generally used for screening purposes. The Iodine Number gives a general indication of the efficiency of the carbon in adsorbing small molecules, and the Molasses Number the same for large molecules.

Plant scale use of granular activated carbons involves a dynamic system; therefore, the rates of adsorption are as important as the equilibrium adsorption properties. The rate determinations can best be made by dynamic pilot tests. The general range of the variables to be investigated in granular systems are flow rates of 2–10 gpm/ft^2 and bed depths of 1–30 ft. The diameter of the carbon columns can be scaled down to 2–8 in. by appropriate reductions in total flow, and still simulate full scale plant conditions. If the laboratory cannot accommodate full height columns, the same total bed depth can be provided by several shorter columns operated in series.

Fig. 5-4 Upflow pilot carbon column. *(Courtesy Cornell, Howland, Hayes & Merryfield)*

Figure 5-4 illustrates a pilot column arrangement used at the South Tahoe Water Reclamation Plant.

Pilot Carbon Column Tests

Pilot plant carbon column tests may be performed for the express purpose of obtaining design data for full scale plant construction, rather than for research purposes.

What information from pilot plant studies is needed for plant design? It is assumed that one or two carbons have been selected which are effective in treating the particular wastewater, since this can be done in the laboratory by adsorption isotherms and Iodine Numbers, as already discussed. It is not necessary to run pilot column tests to reach this point. Pilot column tests make it possible to accomplish the following things:

1. Compare the performance of two or more carbons under the same dynamic flow conditions.
2. Determine the minimum contact time required to produce the desired quality of carbon column effluent, which is the most important of all design factors.
3. Check the manufacturer's data for headloss at various flow rates through different bed depths.
4. Check the backwash flow rate necessary to expand the carbon bed for cleaning purposes.
5. Establish the carbon dosage required, which will determine the necessary capacity of the carbon regeneration furnaces and auxiliaries.
6. If the overall process or plant flow sheet has not been firmly established, check the effects of various methods of pretreatment (influent water quality) upon carbon column performance, carbon dosage, and overall plant costs.
7. Determine many of the practical advantages and disadvantages that cannot be reached by reading the experience of others, regarding the use of upflow or downflow carbon columns or the particle size of carbon to be used.

Pilot column tests are also described in the EPA Manual.[2]

ALTERNATIVE ADSORPTION SYSTEMS

The remainder of this chapter will present information on alternative carbon adsorption systems for the treatment of wastewaters, with the discussion divided into two major categories—granular carbon systems and powdered carbon systems. The design and costs of granular carbon systems are covered in detail in the EPA Design Manual[2]; therefore, operating results are reviewed in this chapter and design considerations for powdered carbon systems emphasized.

Granular Carbon Systems

Granular activated carbon treatment has been used for many years in several industrial processes such as sugar refining, and much of the technology for municipal wastewater treatment is based on this industrial experience. Also, during the past 10 years in the United States, a considerable amount of research and development work has been directed toward the use of granular activated carbon for municipal wastewater treatment. Several pilot scale studies have been conducted and one full scale (7.5 mgd) treatment system at South Lake Tahoe, California, has utilized granular activated carbon for the treatment of municipal wastewater for about 8 years. Another full scale (15.0 mgd) treatment system recently began operation in Orange County, California, and also uses granular carbon to treat municipal wastewater.

There are currently two approaches for the use of granular carbon in wastewater treatment: (1) tertiary treatment or AWT and (2) physical-chemical treatment (PCT). Figures 5-5 and 5-6 summarize typical tertiary and PCT processes. Activated carbon in a tertiary treatment (AWT) sequence follows conventional primary and biological secondary treatment. Tertiary treatment processes involving carbon range from treatment of the secondary effluent with only activated carbon to systems with chemical clarification, nutrient removal, filtration, carbon adsorption, and disinfection. PCT involves treatment of raw wastewater in a primary clarifier with chemicals prior to carbon adsorption. Filtration and disinfection may also be included in PCT, but biological processes are not used.

The two most important features of a granular carbon treatment system are the type of contacting system and the contact time. These parameters and other design criteria currently being used in both PCT and tertiary systems are summarized below:

Contractor type	upflow or downflow pressure or gravity
Contact time (empty bed basis)	10–50 minutes
Hydraulic loading	2–10 gpm/ft
Backwash rate	15–20 gpm/sq ft
$\dfrac{\text{COD removed (weight)}}{\text{Activated carbon (weight)}}$	0.3–0.8
Carbon requirement	
PCT plant	500–1800 lb carbon per 10^6 gal
Tertiary plant	200–500 lb carbon per 10^6 gal

There are two full scale PCT plants just beginning operation, but none with any appreciable operating history. Culp and Shuckrow reviewed the data from several PCT pilot studies.[23] These data were updated by Suhr and Culp in 1974[24] and are summarized in Table 5-9.

There are five full scale AWT plants currently operating which treat municipal wastewater: South Lake Tahoe, California; Colorado Springs, Colorado; Piscataway, Maryland; Orange County, California; and Windhoek, South West Africa. The most reliable information on the performance of granular activated carbon treatment systems will develop from the operating experience of these and other large scale treatment plants. The available data from these full scale plants and some of the pertinent pilot data are presented in the following sections of this chapter.

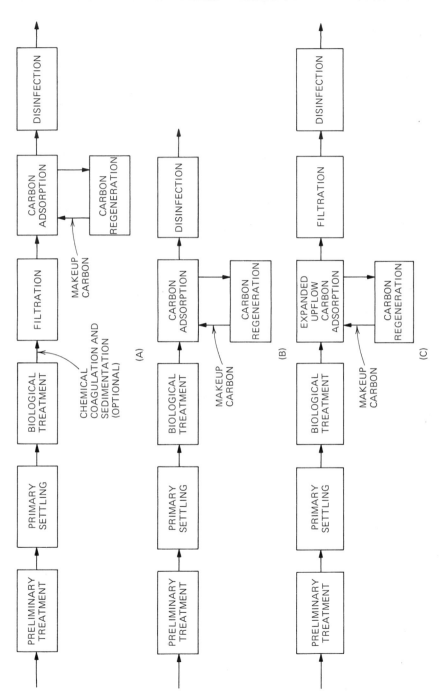

Fig. 5-5 Typical treatment schemes using carbon adsorption as a tertiary step.[2]

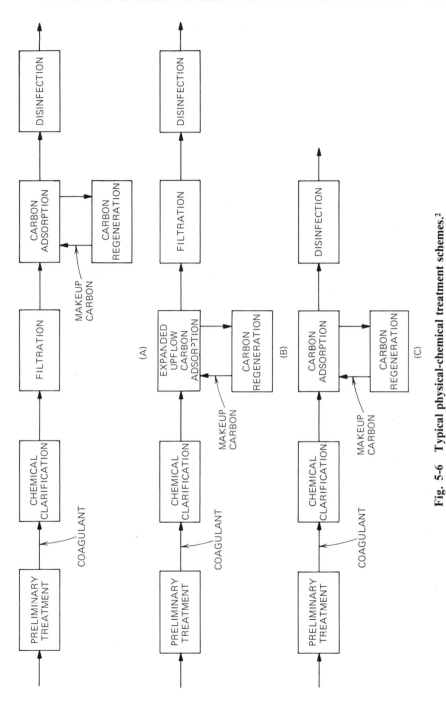

Fig. 5-6 Typical physical-chemical treatment schemes.[2]

TABLE 5-9   Summary of Efficiency of Physical-Chemical Treatment of Raw Wastewaters. (After Suhr and Culp[24])

Test Site	INFLUENT CHARACTERISTICS, mg/l				EFFLUENT CHARACTERISTICS, mg/l			
	BOD	COD	SS	P	BOD	COD	SS	P
Rocky River, Ohio	118	235	107	—	8	44	7	—
Ewing Lawrence, N.J.	100	—	—	10	2–3	—	—	1
Z-M Pilot Plant at New Rochelle, N.Y.	—	220	—	7.6	—	2–14	—	0.2
Cleveland, Ohio	235	523	207	5.4	24	50	13	0.5
Blue Plains, D.C.	129	307	—	8.4	6.2	15.5	—	0.13
Salt Lake City, Utah	100	200	100	6	10–20	12–30	5	0.3–0.1
Albany, N.Y.	104	276	130	—	10–20	10–40	—	—

Pomona, California.   A 0.3 mgd plant treated activated sludge effluent continuously from June 1965 through July 1969.[25,26] The activated sludge effluent was pumped directly to four downflow carbon contactors operated in series. The contactors were 6 ft in diameter and provided a total superficial contact time of 40 minutes at a flow rate of 200 gpm and hydraulic loading of 7 gpm/ft.^{2}

The average quality of the carbon adsorption influent and effluent is shown in Table 5-10. A steady state condition was achieved, after the carbon had been thermally regenerated several times, at a loading of 0.45–0.50 lb COD/lb carbon.

TABLE 5-10   Pomona Carbon Sorption Pilot Plant Average Water Quality Characteristics, June 1965 to July 1969. (After English et al.[26])

Parameter	Influent	Effluent
Suspended Solids, mg/l	9	0.6
COD, mg/l	43	10
Dissolved COD, mg/l	30	8
Total Organic Carbon, mg/l	12	3
Nitrate as N, mg/l	8.1	6.6
Turbidity, JTU	8.2	1.2
Color, Platinum–Cobalt	28	3
Odor, Threshold Odor Number	12	1
CCE, Carbon Chloroform Extract, mg/l	—	0.026
BOD, mg/l	3	1

Orange County, California.   A tertiary treatment pilot plant was operated here from May 1970 to June 1971.[27,28] This plant treated chemically clarified trickling filter effluent in two 20 in. diam. downflow carbon contactors in series which provided a superficial contact time of about 30 min at a flow rate 6.7 gpm and a hydraulic loading of 3.1 gpm/ft^2. The contactors each contained 360 lb of 8 × 30 mesh granular activated carbon. The typical quality of the pilot plant influent and effluent is shown in Table 5-11.

Los Angeles, California.   The City of Los Angeles Department of Water and Power operated a pilot plant from January 1968 through November 1970.[29] Activated sludge effluent from the Hyperion treatment plant was pumped directly to two carbon contactors. The contactors were operated in three different modes:

January 1968 to January 1970—Two stage downflow.
February 1970—Two stage upflow-downflow.
April 1970 to November 1970—Single stage downflow.

The contactors were 24 in. in diameter and each contained 13.5 ft of 12 × 40 mesh granular activated carbon. The pilot studies included investigation of contact time, rate of filtration, adsorptive capacity, and plant configuration. Some of the pilot study conclusions were:

Contact time: Contact times of 12, 25 and 50 min were investigated. The use of a 50 minute contact time produced the best quality water with the longest comparative regeneration cycle. A general decreasing organic removal was correlated with a decreasing contact time.

Rate of Filtration: Various rates of filtration over the range of 2–8 gpm/ft^2 were tested. The straining and headloss results for the various rates indicated that 8 × 30 mesh carbon should be used, that two stage provided better straining than single stage, and that the depth of carbon, 13.5 ft used in each bed, may be more important for straining than the rate of filtration.

Adsorptive Capacity: Three carbons were tested in the pilot plant for adsorptive capacity: Filtrasorb 400 (12 × 40 mesh), Filtrasorb 300 (8 × 30 mesh), and Nuchar WV-L (8 × 30 mesh). Under the test conditions, the carbon dosage requirements for Filtrasorbs 400 and 300 were approximately 380 lb carbon per million gallons of water treated, while the value for Nuchar WV-L was 590 lb carbon per million gallons of water treated. (System changes may have been responsible for the greater dosage with Nuchar WV-L).

TABLE 5-11   Orange County Pilot Plant, Typical Water Quality of Pilot Waste Water Reclamation Plant Influent and Effluent.

Constituent	CONCENTRATION, mg/l EXCEPT ODOR AND COLOR	
	Influent	Effluent
Calcium	70–110	50–80
Magnesium	20–45	1–3
Sodium	240–260	240–260
Potassium	20–35	20–35
Bicarbonate	200–450	150–170
Sulfate	270–350	270–350
Chloride	300–350	300–350
Phosphate	20–25	<1
Nitrogen		
Organic	5–15	<1
Ammonia	15–30	<2
Nitrite	<1	<1
Nitrate	<1	<1
Total dissolved solids	1200–1400	1000–1100
Total hardness	255–460	130–210
Suspended solids	30–80	<1
BOD	30–80	<2
COD	50–200	10–30
MBAS	3–4	0.1
Threshold odor no.	50–100	<2
Color, units	30–50	<2
Arsenic	0.00–0.01	0.01
Barium	<0.02	<0.02
Cadmium	0.10–0.13	0.000–0.005
Chromium (hexavalent)	0.10–0.20	0.00–0.04
Copper	0.09–0.39	0.02–0.30
Iron	–	0.05–0.25
Lead	0.00–0.05	0.00–0.04
Manganese	–	0.05–0.10
Mercury	<0.001–0.003	<0.001–0.003
Selenium	0.00–0.09	0.000–0.003
Silver	0.00–0.01	0.00–0.01
Zinc	0.07–2.08	0.02–0.07

Plant Configuration: Of the three types of plant configuration investigated (two stage downflow, two stage upflow-downflow, and single stage downflow), two stage downflow was believed to have several advantages. The two stage downflow carbon efficiency was better than the single stage, although the difference may have been the carbon and not the system. Also, two stage was more reliable for operational consistency than either of the other systems, especially the two stage upflow-downflow.

The average quality of the pilot plant influent and effluent is shown in Table 5-12. These data indicate that about 64 percent of the COD in the activated sludge effluent was removed by activated carbon treatment. It should be noted that the Hyperion effluent is extremely high quality secondary effluent, with suspended solids = 5.4 mg/l, BOD = 5.7 mg/l, and COD = 29.9 mg/l. In addition to organic removals, the activated carbon also effected good removals of several trace constituents: aluminum, boron, cadmium, chromium, iron, lead, lithium, selenium, silver, titanium, and strontium-90.

South Lake Tahoe, California.   This 7.5 mgd AWT plant treats activated sludge effluent by chemical clarification, mixed media filtration, and granular activated carbon adsorption. The carbon contactors operate in the upflow packed bed mode and provide 17 minutes contact time at average flow.

Results of carbon treatment at South Lake Tahoe have been reported.[2,30] During a 32 month period between April 1968 and December 1970, the activated carbon adsorption columns treated approximately 3600 million gallons of chemically clarified and filtered activated sludge effluent. The monthly average for the period was 113 million gallons; the lowest monthly flow was 80 million gallons and the highest 154 million gallons. The average detention or superficial contact time was 19.2 minutes, which provided an average loading rate of 4.7 gpm/ft.^{2}

The average COD concentration in the carbon influent was 20.3 mg/l, which resulted in approximately 612,000 lb of COD being applied to the carbon during the period. The average COD concentration in the carbon column effluent was 10.0 mg/l. Approximately 311,000 pounds of COD were removed by the activated carbon, which resulted in a removal efficiency of 50.8 percent.

The average MBAS concentration in the carbon influent was approximately 0.6 mg/l, which resulted in 18,800 lb of MBAS being applied to the carbon. The average carbon column effluent MBAS concentration was slightly more than 0.1 mg/l, which provided a removal efficiency of 77 percent.

TABLE 5-12   Los Angeles Activated Carbon Pilot Plant Average Water Quality. (All concentrations in mg/l except as noted.)

Constituent	Secondary Effluent 1968–1969	Activated Carbon Effluent 1968–1970
Specific conductance (μmho/cm)	1450	1440
Total dissolved solids	897	839
Suspended solids	5.4	2.4
Biochemical oxygen demand	5.7	2.4
Turbidity (units)	1.5	0.8
Chemical oxygen demand (COD)	29.9	10.7
Dissolved COD	24.6	8.1
Total organic carbon	—	—
Dissolved organic carbon	—	—
Color (color units)	30	<5
Nitrate as N	3.5	3.0
Nitrite as N	0.4	0.5
Ammonia as N	7.4	7.1
TKN and N	10.0	9.3
Dissolved oxygen	2.0	<0.1
Carbon chloroform extract	1.130	0.027
Detergents—MBAS	0.15	<0.02
Total hardness as $CaCO_3$	229	229
Alkalinity as $CaCO_3$	211	215
pH (pH scale)	—	7.5
Odor (no units)	I-M	0
Temperature (°C)	—	22.5
Chloride	207	207
Cyanide	.003	<0.001
Fluoride	1.0	1.0
Phenols	<0.001	—
Phosphate as P	2.9	2.9
Silica	24	24
Sulfate	207	214
Arsenic	0.02	0.01
Aluminum	0.14	0.09
Antimony		<0.1
Barium	0.18	0.14
Bismuth		<0.01
Boron	0.91	0.03
Cadmium	0.01	0.004
Calcium	55	55
Chromium—Total	0.08	0.046
Chromium—Hexavalent	<0.01	<0.01

TABLE 5-12 (Continued)

Constituent	Secondary Effluent 1968–1969	Activated Carbon Effluent 1968–1970
Copper	0.21	0.21
Gallium		<0.01
Iron	0.02	<0.01
Lead	0.013	<0.01
Lithium	0.14	0.075
Magnesium	22	22
Manganese	<0.05	<0.04
Mercury	<0.0001	<0.0001
Molybdenum	0.019	0.017
Nickel		0.09
Potassium	13	13
Selenium	0.018	0.005
Silver	0.004	0.002
Sodium	219	219
Strontium	0.42	0.40
Tin		<0.01
Titanium	0.01	0.006
Tungsten		<0.2
Vanadium	0.03	0.03
Zinc	0.10	0.12
Zirconium	0.006	0.004
Radium-226 (pCi/l)	0.17	0.22
Strontium	9.0	3.8
Gross beta (pCi/l)	22.3	19.5
Gross alpha (pCi/l)	2.3	1.7

The carbon dosage rate was about 207 lb regenerated carbon/million gal carbon column flow. For the same period, an average of 0.8 lb of COD were applied per pound of regenerated carbon, and 0.38 lb of COD were removed per pound of carbon. An average of 0.027 lb of MBAS were applied per pound of regenerated carbon, and 0.021 lb of MBAS were removed per pound of carbon.

Orange County, California.   This 15 mgd AWT plant began operation in 1975 and initial results were reported by Argo.[31] Trickling filter effluent is chemically clarified, filtered in gravity mixed media filters, and pumped upflow through activated carbon columns that provide 30 minutes contact time. Initial operating results showed that activated carbon adsorption removed 75–85 percent of the COD, 90–99 percent of the MBAS, and

TABLE 5-13   Colorado Springs AWT Plant March–July 1971 Operation. (Average flow: 1.15 mgd.)

	Trickling Filter Effluent (mg/l)	CARBON COLUMNS			Total Plant, Percentage Removal
		Influent (mg/l)	Effluent (mg/l)	Percentage Removal	
COD	315	139	39	72	88
TOC	83	43	13	70	84
BOD	129	57	24	58	81
MBAS	4.85	2.9	0.1	96	98
Total SS	62	15	3	79	95
Turbidity (JTU)	52	62	6	0	89
Color (units)	173	39	18	54	90
Total P	10.8	0.7	0.9	—	92
NH_3–N	34.6	23.9	26.3	—	24
Total N	42.5	29.5	28	—	34
pH	7.25	6.9	6.9	—	—

65–75 percent of the total organic carbon (TOC). Average carbon column effluent concentrations for COD, MBAS, and TOC were 7.9, 0.05, and 4.4 mg/l, respectively.

Colorado Springs, Colorado.   This 3 mgd AWT plant began operation in December 1970. Secondary effluent (trickling filters later changed to activated sludge) is treated by chemical clarification, dual media filtration, and carbon adsorption in downflow pressure contactors that provide 30 minutes contact time. The results from five months' operation in 1971 are summarized in Table 5-13.

Pitscataway, Maryland.   This AWT plant receives a constant 5 mgd flow of municipal wastewater treated by the activated sludge process. This facility recently began operations and performance data are not yet available. The expected effluent characteristics have been reported[32] and are shown below:

> *Total organic carbon:* < 2.0 mg/l
> *Biochemical oxygen demand:* < 5.0 mg/l
> *Suspended solids:* < 2.0 mg/l
> *Turbidity (as JTU):* < 1.0
> *Phosphorus (as PO_4):* < 1.0 mg/l

The plant has three parallel two stage granular activated carbon contactors that provide 37 minutes contact time. The contactors are 15 ft in diameter and are the pressure downflow type.

Windhoek, South West Africa. This 1.3 mgd AWT plant receives highly treated secondary effluent and processes this wastewater for subsequent use in the city's municipal water system. It has been operating successfully since October 1968. The pilot studies and results of the full scale plant operation have been reported.[33,34,35] The BOD and MBAS concentrations in the carbon effluent are less than 1 mg/l. At the present time there are no facilities for the on-site regeneration of carbon.

Powdered Carbon Systems

Powdered carbon systems may also be classed as either tertiary or AWT and PCT. In addition, systems in which powdered carbon is fed directly to the mixed liquor of the activated sludge process are being studied and are discussed in this chapter. Powdered carbon systems are less sensitive to upstream treatment than are granular carbon systems, and, when applied to high BOD wastewaters, also permit ready control of hydrogen sulfide generation.

Powdered carbons are applied to wastewaters, mixed, and then removed by settling, rather than being used in packed beds. The small size of powdered carbon particles would create excessive headloss in a packed bed. Powdered carbon contact systems may be single stage or multistage. The multistage systems attempt to utilize more of the carbon capacity than can be utilized in a single stage. Usually, in wastewater applications no more than two stages are used. In two stage systems the fresh and/or regenerated carbon is added to the second stage, settled, and then pumped to the first stage. Some additional adsorption capacity may be used by contacting the partially spent carbon with the most concentrated wastewater, before the carbon is removed for regeneration or disposal.

Two major investigations in the use of powdered carbon for the treatment of municipal wastewaters have been conducted. One study was conducted by Battelle Northwest[36,37] and the other was conducted by the Eimco Division of the Envirotech Corporation.[38-41]

Battelle Northwest Study. The Battelle Northwest study developed the process shown schematically in Figure 5-7. The process involves contacting raw sewage with powdered activated carbon to effect removal of dissolved organic matter. An inorganic coagulant, alum, is then used to aid in subsequent clarification. Addition of polyelectrolyte is followed by a short flocculation period. Solids are separated from the liquid stream by

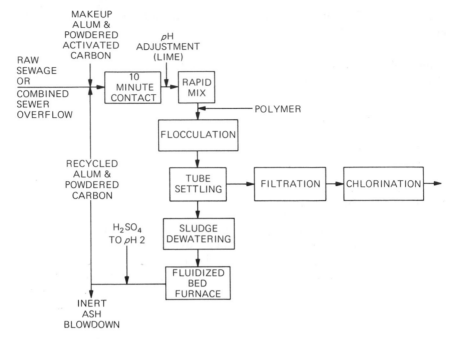

Fig. 5-7   Battelle powdered carbon process.

gravity settling, and the effluent is then disinfected and discharged, or it can be filtered prior to disinfection. Carbon sludge from the treatment process is thermally regenerated by a fluidized bed process. Alum is recovered by acidifying the regenerated carbon–aluminum oxide mixture to pH 2 with sulfuric acid. This reclaimed alum is then reused in the treatment process. A pH adjustment, accomplished with a lime slurry, is required to raise the pH to 6.5–7.0 for aluminum hydroxide precipitation when reclaimed alum is recycled.

The process was first evaluated in a nine month laboratory study.[37] Aqua Nuchar A* was selected as the carbon from 15 different commercial carbons. The Nuchar carbon appeared to offer the best benefit/cost ratio for this particular application. From several sets of experiments conducted to examine the adsorption characteristics of the powdered carbon, it appeared that adsorption of COD was essentially complete at a carbon concentration of 800–1000 mg/l. Increasing the carbon concentration beyond this point had little or no effect on the residual COD. Additional tests indicated that TOC removal was virtually completed at a carbon dose of 500–600 mg/l, and a 10 minute detention time proved adequate at this carbon dose.

Alum and Magnifloc 985 N were used as coagulants throughout the initial laboratory investigations, with a carbon dose of 1000 mg/l. The alum [$AL_2SO_4)_3 \cdot 18H_2O$] dose was varied in a series of jar tests. Adequate turbidity removal was achieved at an alum dose of 150 mg/l. However, objectionable quantities of powdered carbon remained in the effluent and good flocculation was not achieved until the alum dose had been increased to 350 mg/l. Additional jar tests at this alum concentration and at poly-electrolyte concentrations of 5–10 mg/l indicated that COD removals of over 90 percent could be obtained consistently. It was observed that the coagulation was impaired in some instances if alum and carbon were added simultaneously. A carbon contact time of at least 5 minutes prior to alum addition was required in order to insure consistently good floc formation.

As part of the laboratory study, an investigation was initiated to study the possibilities of aluminum recovery from the regenerated carbon—aluminum oxide mixture. Aluminum hydroxide can be heated to form aluminum oxide as follows:

$$Al(OH)_3 \xrightarrow{500°C} \gamma\text{-}Al_2O_3 \xrightarrow{1000°C} \alpha\text{-}Al_2O_3$$

The γ-oxide readily dissolves in H_2SO_4 to reform Al ions while the α-oxide is insoluble at reasonable acid levels. Since carbon regeneration takes place below 1000°C, the majority of the aluminum should be recoverable. Alum recovery of 86 to 100 percent was achieved in the laboratory.

A similar line of investigation was pursued for a $Fe_2(SO_4)_3$ coagulant. However, a red powder, believed to be Fe_2O_3 (which should form at 200°C), formed in the thermal regeneration step. This iron oxide was very acid resistant and a concentrated sulfuric acid solution was required to effect good dissolution in a reasonable length of time. It was concluded that the recovery of $Fe_2(SO_4)_3$ in this manner was not feasible. Similarly, bentonite clays were rejected as coagulants because of the collapse of the lattice struc-ture upon heating and subsequent buildup of colloidal ashes.

The laboratory studies indicated that it was not necessary to separate the reclaimed alum from the regenerated powdered carbon prior to reuse. It was found that the recycled carbon could, in fact, remove substances which otherwise interfered with alum coagulation if the wastewater-carbon–alum mixture was maintained at an acid pH for the first few min-utes of contact. Thus, as shown in the process schematic, the pH depres-sion which results from the recycling of the acidified alum–carbon stream is not neutralized until after the first 10 minutes of contact. The labora-tory study indicated that organic removals substantially greater than 90 percent could be achieved while producing effluent turbidities less than 1

JTU. The laboratory study also indicated that excellent recovery of the adsorptive capacity of the carbon could be achieved.

Based upon the favorable results of the laboratory study, a 100,000 gpd mobile treatment plant was constructed. This pilot plant was operated in Albany, New York from April to June, 1972, and from June to October, 1973. The pilot plant was composed of two major systems: a liquid treatment system and a carbon regeneration facility. The liquid treatment system was contained, almost entirely, in a forty foot mobile trailer van. It was designed for a nominal capacity of 100,000 gpd. Carbon, alum, and polyelectrolyte were added in a pipe reactor, providing rapid mixing of the chemicals, which precedes flocculation and separation in a tube settler. Clarified effluent was chlorinated and released, with the option of routing through a gravity filter prior to chlorination. Sludge was dewatered in a centrifuge. Carbon was regenerated in the fluidized inert sand bed unit illustrated later in this chapter in Figure 5-17.

The pilot study confirmed that proper control of pH within the system was critical. A pH of 4 or less in the first few minutes of carbon contact was found necessary to prevent excessive carryover of carbon particles from the downstream clarifier. The pH was adjusted to near neutral with lime prior to flocculation. The tube clarifier was found to perform well, producing effluent turbidities less than 2 JTU at overflow rates as high as 2880 gpd/ft.^{2} Filter runs averaged 10 hours at a loading rate of 4.4 gpm/ft.^{2} Polymer filter aid doses of 2 mg/l were normally used.

The carbon sludge was found to dewater readily in a 6 in. solid bowl centrifuge. The dewatered sludge ranged from 20 to 35 percent solids at 70 percent recovery with no conditioning polymer. Use of polymers increased the solids recovery to 95 percent.

The pilot system was operated both on storm flows from the combined sewer and on dry weather, municipal wastewater flows. Excellent degrees of wastewater purification were achieved in both cases. During the dry weather conditions, average plant effluent BOD, COD, and suspended solids concentrations for the 1971 studies were 17.8, 35, and 7.7 mg/l, respectively. This represents removals of 82.3 percent BOD, 87.3 percent COD, and 94 percent suspended solids.

Plant operational data for the 1972 studies were comparable to those observed in the 1971 portion of the program. During the 1972 operations the average effluent turbidity, suspended solids, COD, and BOD concentrations were 0.67 JTU, 3.1 mg/l, 39 mg/l, and 17 mg/l, respectively. This represents average removals of 98.1 percent suspended solids, 82.6 percent COD and 81.3 percent BOD.

These results were achieved at total plant detention times which averaged slightly less than 90 minutes. Recovery of 91 percent of the powdered carbon was achieved.

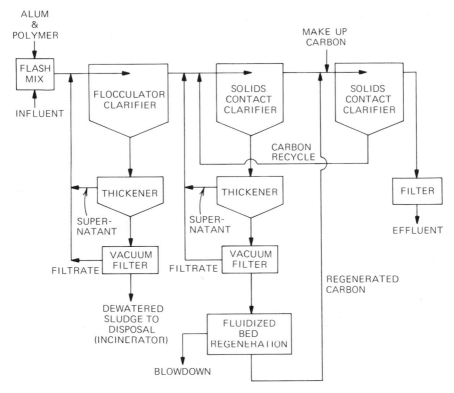

Fig. 5-8   Eimco powdered carbon process.

Data from the pilot studies indicate that at times there was a significant nonadsorbable organic component present in the Albany wastewater. When present, this nonadsorbable fraction represented BOD and COD residuals of 10–20 mg/l and 20–50 mg/l, respectively, which could not be removed even at carbon doses as high as 1000 mg/l.

Eimco Study.   Eimco constructed a 100 gpm pilot plant in Salt Lake City, Utah for evaluation of powdered activated carbon treatment of raw sewage. The plant is shown schematically in Figure 5-8. It was operated for 16 months to evaluate lime, alum, and ferric iron coagulation and single and two stage countercurrent carbon treatment. Screened and comminuted raw wastewater was obtained from the main Salt Lake City pump station discharge line. The desired flow was pumped to the chemical treatment unit, a 12 ft diameter solids contact clarifier provided with a surface skimmer. Chemicals were added to achieve coagulation-precipitation and aid flocculation and clarification. The settled solids were removed and collected for gravity thickening and vacuum dewatering tests.

The chemically treated effluent then flowed by gravity to the carbon contactors which could be operated either single stage (parallel) or two stage countercurrent (series). The carbon contactors were 10 ft diameter solids contact units. Powdered activated carbon was fed and maintained as a concentrated slurry. Spent carbon was periodically withdrawn to control slurry concentration. The spent carbon removed was gravity thickened in a 5 ft diameter unit and then dewatered on a 3 ft by 3 ft vacuum filter.

The effluent from carbon treatment was filtered through a 3.5 ft diameter granular media filter. The filter bed consisted of 1.5 ft of 1–1.5 mm coal over 1.0 ft of 0.6–0.8 mm sand. The backwash water, containing spent carbon, was collected and recycled back to the carbon contactors. The final effluent was collected in a clear well and used for backwashing the filter and plant water.

It was found that single stage carbon contact in a slurry contactor actually provided the equivalent of 2 to 4 contacts due to the biological action occurring in the slurry contactor. Hydrogen sulfide problems could be controlled by maintaining the solids detention in the clarifier to 3 days or less. Based on these results, single stage carbon contact appeared adequate. The influent soluble COD of 80–100 mg/l was reduced to 15 mg/l with 300 mg/l of powdered carbon and to 30 mg/l with 75 mg/l of carbon.

Thickening and dewatering of spent carbon were effective accomplished. Spent carbon was concentrated from 25–50 g/l in the carbon contactor blowdown to 70–100 g/l at a gravity thickener solids loading averaging 10 lb/day/ft.^{2} The thickened material was readily dewatered to 78 percent moisture at vacuum filtration rates of 6–9 lb/hr/ft.^{2} About 0.2 percent cationic polyelectrolyte by weight was required for conditioning to obtain about 90 percent solids recovery across the vacuum filter and produce a readily dischargeable filter cake.

Based on the pilot studies, the process shown schematically in Figure 5-8 was recommended by the Eimco investigators. It was also concluded that sizing criteria for powdered carbon systems should be developed from pilot tests because it is impossible to model practically, or predict theoretically, the performance of powdered carbon systems.[42]

Biophysical Process.   Another method of gaining benefits from the use of powdered carbon is the addition of carbon directly to the mixed-liquor in an activated sludge plant aeration basin, referred to as the PACT process.[43] The benefits which are attributed to this system also are:

Improved BOD and COD removal by adsorption and improved settling even at lower than optimum temperatures, lower MLVSS (mixed liquor volatile suspended solids) and/or at higher than design flow rates.

Adsorption of color and toxic agents that cannot be removed by merely expanding a plant.

Reduction of aerator and effluent foam by adsorption of detergents.

More uniform plant operation and plant effluent quality during periods of widely varying organic and hydraulic loads.

Improved solids settling (lower sludge volume index, increased sludge solids, and lower effluent solids).

Increased aerobic digester capacity through foam reduction.

The mechanisms which account for these benefits are postulated to be as follows:

Adsorption on the extensive surface area of the carbon.

Biological adsorption and degradation. The carbon settles in the sludge with pollutants adsorbed, and remains in the system rather than escaping in the effluent. The longer the sludge ages, the greater the chance for bio-oxidation of poorly degraded organics.

Continuous regeneration of the carbon by biological action. While the carbon and organisms are adsorbing organic pollutants, the organisms continuously degrade the pollutants, thereby freeing surface areas again for adsorption of more pollutants.

Improved solids settling. Improved settling in the secondary clarifier leads to lower suspended solids and BOD in the effluent. The settling rate of some powdered carbons plus biosolids is greater than that for biosolids alone.

There is little capital investment, with the chief cost being the carbon itself. As this process approach is new, there are still some uncertainties associated with optimizing the operation. The following procedure is recommended by the carbon manufacturer:

During the first week, add sufficient carbon to increase the MLVSS by 25 percent (carbon is measured as a volatile solid). Maintain that level in the second week by adding only enough carbon to make up the amount lost in sludge wasting. This process is repeated over successive two week periods until desired effluent quality is achieved; 50 percent, 75 percent, or even 100 percent of the original MLVSS may be needed.

The goal of the PACT process may be to improve organic removal or to improve sludge settling characteristics.

DuPont evaluated the PACT process for several months in a 75 gpm pilot

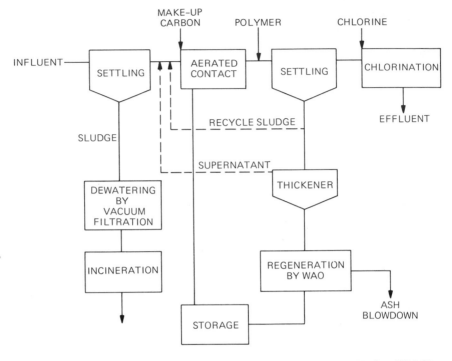

Fig. 5-9   Biophysical powdered carbon process with wet air oxidation (WAO).

facility and is installing a 30 mgd PACT system to treat industrial wastewaters at its own Deepwater, New Jersey plant. The carbon will be regenerated in a multiple hearth furnace. Zimpro, Inc. has also evaluated the PACT process at the Rothschild, Wisconsin sewage treatment plant, using wet air oxidation to regenerate the carbon. A flow diagram for a biophysical process using wet air oxidation for carbon regeneration is shown in Figure 5-9.

Jet Propulsion Laboratory (JPL) Process.   The JPL process is a two stage countercurrent adsorption system using powdered activated carbon. A flow diagram is shown in Figure 5-10. Fresh activated carbon is mixed with wastewater in the secondary mixing basin and settled, and then the entire mixture of settled sewage solids and activated carbon is transferred to the primary mixing basin. Settled solids and carbon are removed from the primary settling basin, dewatered, and transferred to a pyrolysis reactor. The reactor produces activated carbon and a burnable gas. The activated carbon is then recycled to the secondary mixing basin.

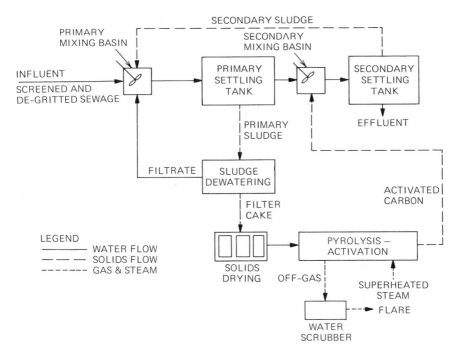

Fig. 5-10   JPL activated carbon process.

Activated carbon is intended to serve two functions: (1) as an adsorbent of organics and other pollutants, and (2) as a settling aid in both the primary and secondary sedimentation basins. It is also believed that the carbon acts as a filtration aid and prevents compression of sewage solids during dewatering.

A trailer mounted pilot plant was constructed by JPL in Pasadena and operated at Orange County Sanitation District Plant No. 1 in Fountain Valley, California beginning in February 1974. Typical operating conditions for the pilot study are shown in Table 5-14. Results reported by Humphrey and co-workers[44] are summarized in Table 5-15. These results were achieved with carbon from the pyrolysis reactor at dosages from 300–600 mg/l.

A sample of activated carbon was dry screened into the following size fractions:

Above 100 mesh
100–200 mesh
200–300 mesh
below 325 mesh

TABLE 5-14   JPL Pilot Plant Operated at County Sanitation Districts of Orange County Plant No. 1.

Capacity	7 gpm
Primary mixing	0.25 hp mixer @ 1725 rpm detention time = 28 min.
Primary settling	overflow rate = 335 gpd/ft^2 detention time = 86 min.
Secondary mixing	0.33 hp mixer @ 1725 rpm detention time = 28 min.
Secondary settling	equipped with 1 Microfloc settling tube module, 8 ft long overflow rate = 335 gpd/ft^2 detention time = 86 min.
Solids handling system	No equipment for continuous sludge removal; therefore, system shut down for batch removal of sludge from primary and secondary settling basins

Dewatering	1. Rotary vacuum filter, or
	2. Netzsh plate filter:
	    11 plates, 14 × 14 in.
	    total filter area = 32.7 ft^2

Pyrolysis reactor	8 in. ID stainless steel tube equipped with an external gas fired jacket bed temperature = 1800° F

Samples of degritted raw wastewater were treated with 600 mg/l of carbon from these four size fractions and analyzed for COD at various times from 0 to 30 minutes after addition of the carbon. The results of this test showed that, in the size ranges studied, COD removal was not related to particle size. It was concluded that carbon sizes which are difficult to settle are not necessary for adsorption in the treatment process.

A 1 mgd plant utilizing the JPL process began operation in June 1976 at the Orange County Sanitation Districts Plant No. 2. Data from this plant were not available at the time of this writing.

GRANULAR CARBON REGENERATION

Wastewater Treatment Plant Size for Economic Regeneration of Granular Carbon

The decision for a particular project as to whether to regenerate and reuse granular carbon or to use granular or powdered carbon on a once through,

TABLE 5-15   JPL Pilot Plant Results. (After Humphrey, et al.[44])

(Pyrolysis reactor carbon dosages 300–600 mg/l; filtration with a "Blue Lake swimming pool filter.")

| | CONCENTRATION | | | |
Parameter	Influent	Secondary effluent	Secondary effluent (filtered)	Percentage removal
Total suspended solids	404	117	1	99
Volatile suspended solids	298	101	0	100
Grease	44	7	2	95
Biochemical oxygen demand	182	14	6	97
Cadmium	0.30		0.026	84
Chromium	0.94		0.36	40
Copper	0.74		0.017	97
Lead	0.21		0.05	73
Nickel	0.32		0.13	63
Silver	0.016		0.002	88
Zinc	1.01		0.07	93

throwaway basis is based primarily on economics. If the cost of carbon regeneration exceeds the cost of new carbon, then there is little incentive to install regeneration facilities, but, on the other hand, if the total costs of regeneration per pound of carbon are less than the cost of new carbon, regeneration should be considered. The costs for carbon regeneration vary with the size of plant and the type of regeneration equipment to be used. At the time of this writing (1976) most installations employ multiple hearth furnaces as the means for regeneration. For plants requiring less than 200 lb/day of carbon, regeneration probably is not economical because of the capital cost of minimum size equipment of this type. For carbon needs in excess of about 1500 lb/day, granular carbon regeneration is probably warranted. For carbon usages between 200 and 1500 lb/day which would include tertiary plants with capacities between 0.80 and 6.0 mgd and PCT plants with capacities between 0.17 and 0.80 mgd, the cost of on site regeneration must be compared to the costs of central regeneration and to the costs of alternative treatment methods for removing refractory organics.

Recently, an all electric furnace has been developed and placed into use in a few installations, which may make it possible to regenerate granular carbon economically at dosages less than 200 lb/day.

In any event, the use of granular activated carbon usually involves the regeneration and reuse of the carbon. Indeed, the amenability of carbon to regeneration and reuse, and the fact that there are no liquid or solid by-products from its use requiring disposal, are among activated carbon's principal process advantages for removing organics from wastewater.

There are four general methods for reactivating granular carbon: solvent wash, acid or caustic wash, steam reactivation, and thermal regeneration. With the use of a solvent which will dissolve the adsorbed material, the adsorbate is passed through the carbon bed in the direction opposite that of the service cycle until it is removed. The bed is then drained or purged of the solvent, and the regenerated carbon is ready to go back on stream. If an acid or caustic is more effective than a solvent in dissolving a particular adsorbate, it may be used in a manner identical to that described for solvent washing. Adsorbates with low boiling temperatures can sometimes be removed by steam. Approximately 3–5 lb of steam per pound of adsorbate is passed through the carbon opposite the normal flow, and is either vented to the atmosphere or condensed and recovered. These first three methods do not appear to have application as the primary methods for regeneration of activated carbon which has been used for the treatment of wastewater. The fourth method, that of thermal regeneration, is universally used for this purpose at the present time.

Carbon Regeneration Systems

Juhola and Tupper[45] have shown that thermal regeneration of granular carbon consists of three basic steps: (1) drying; (2) baking (or pyrolysis) of adsorbates; and (3) activating, by oxidation of the carbon residues from decomposed adsorbates. Drying may be accomplished at 212°F, baking between 212°F and 1500°F, and activating at carbon temperatures above 1500°F.

The total regeneration process requires about 30 minutes. The first 15 minutes is a drying period during which the water retained in the carbon is evaporated; this is followed by a 5 minute period during which the adsorbed material is pyrolyzed and the volatile portions thereof are driven off, and finally by a 10 minute period during which the adsorbed material is oxidized and the granular carbon reactivated.

A typical plant sequence for thermal regeneration of granular carbon consists of transporting the spent carbon from a contactor in water slurry to a dewatering drain bin. After dewatering, the carbon is fed to a furnace with a controlled oxygen atmosphere at 1500–1700°F, where the adsorbed organics and other impurities are volatilized or oxidized. The hot regenerated carbon is quenched in water. Except for use in expanded upflow contactors, the cooled, regenerated carbon is washed to remove fines, and then is hydraulically transported to a carbon contactor or storage. The furnace off gases may be passed through an afterburner to complete oxidation of exhaust volatiles, and then scrubbed. Scrubber water is usually returned to the liquid treatment process.

The pounds of carbon that must be regenerated per day can be estimated by multiplying the anticipated carbon dosage in pounds per million gallons by the flow rate in mgd. An allowance of 10–40 percent downtime should be made in arriving at the furnace size. Multiple hearth furnaces used for regenerating carbon should provide a hearth area of about one square foot per 40 lb of carbon to be regenerated per day. The regeneration furnace should provide for control of hearth temperatures, furnace feed rate, rabble arm speed, and steam addition. The hot gases leaving the top hearth of the regeneration furnace contain fine carbon particulates visible as smoke and odorous materials. Since the wet spent carbon enters the top of the furnace near the exit point for the exhaust gases, some of the more volatile adsorbate is removed from the carbon and carried into the atmosphere without being completely oxidized. This smoke and the carbon particulates might present air pollution problems if left uncontrolled. Therefore, it is important that air pollution control equipment be included in the design of the carbon regeneration furnace. Systems are available which include an afterburner for removal of smoke and odors, and a wet scrubber or bag filter for removal of particulates. These are designed as integral parts of the furnace installation. Stringent air emission standards can be met by use of a variable throat, Venturi-type scrubber with 20 in. pressure drop. The fuel requirement for the carbon furnace, exclusive of fuel for the afterburner, can be estimated at about 4300 Btu/lb of carbon regenerated. This includes fuel for production of auxiliary steam at the rate of 1 lb steam/lb carbon.

Regeneration Equipment

Electric Furnace.   An all electric furnace using an infrared heat source is currently under development by Shirco Co., Dallas, Texas (see Figure 5-11). Recent developments in infrared lamps, coupled with the advent of silicon controlled rectifiers, semiconductor controls, and ceramic reflector materials, have provided an economic means for applying and controlling radiant energy. By properly applying this technology, it may be possible to approach operating costs of natural gas with an all electric infrared furnace utilizing smaller gas scrubber systems, less air, and lower capital investments than are used in other furnaces presently on the market. The shortage of natural gas and oil supplies is a factor enhancing the appeal of this approach. Also, areas remote from natural gas or other petroleum fuel sources may find electric infrared regeneration an attractive alternative. Carbon is conveyed to the belt conveyor, which discharges the carbon into the machine onto a

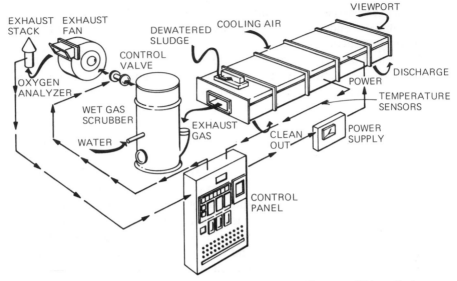

Fig 5-11   Infrared carbon regeneration system. *(Courtesy Shirco Co.)*

conveyor belt. The high temperature belt conveyor carries the carbon through a drying zone and then into a regeneration zone. In the regeneration zone, mounted just above the belt, is a battery of infrared lamps which maintains the regeneration temperature. The belt then discharges the regeneration carbon into a hopper at the end of the machine. The lamps and end seals are cooled by drawing outside air through the cooling air ducts. The exhaust air is then discharged through a wet gas scrubber or other air pollution control equipment.

Because the heat is transferred by radiation rather than by conduction or convection, the air is not heated. The other potential advantages of this system appear to be:

1. The capital cost is less than for multiple hearth furnaces.
2. Operation and maintenance costs are low.
3. The temperature can be brought from ambient to 1600–1800° F within one hour.
4. There is no explosion danger.

Although there is no doubt that natural gas or oil provides a cheaper source of heat than electricity, the other savings associated with the infrared system may offset the higher unit fuel costs. Based on data available at this time, it appears the total operating costs will be comparable to a multiple hearth system and capital costs will be only 60–75 percent of a multiple hearth system. The infrared approach also scales down much better than other systems and may be particularly attractive for small plants. A 50 lb/hour unit is in operation for carbon regeneration in an industrial application in Baton Rouge, La.

Multiple Hearth Furnace. At South Tahoe, a commercial size B-S-P Corporation multiple hearth furnace has been in operation since 1965 (over 10½ years at this writing) in the successful regeneration of spent granular carbon used in full scale wastewater reclamation. This installation is typical for the processing of wastewater, and will be used as an example to illustrate some of the principles involved in design and operation of carbon regeneration facilities. Nichols Engineering Company also makes similar multiple hearth furnaces.

The design capacity of the Tahoe plant is 7.5 mgd. At the present time, the carbon dosage averages about 250 lb/million gal. The theoretical required furnace capacity is, then, $7.5 \times 250 = 1875$ lb/day. This computation does not allow for furnace downtime or other contingencies. Also, the smallest commercial furnace available is one with a 54 in. diameter, which is rated at 6000 lb/day; at 7.5 mgd, this furnace would handle carbon dosages up to 800 lb/million gal, which gives a comfortable margin of safety. The furnace has satisfactorily regenerated carbon at a feed rate of 4800 lb of dry carbon per day at temperatures on the fired hearths of 1650–1700° F.

Determining When to Initiate Carbon Regeneration. In starting up a new plant, this may be done initially by observation of effluent COD concentrations. If the value approaches predetermined maximum, say 10–20 mg/l of COD, for a period of one or two days, then it may be time to start a regeneration campaign. In upflow contactors, it is usual to withdraw 5–10 percent of the carbon from the bottom of each contactor during each regeneration cycle. In downflow contactors, the entire contents of the lead contactor are removed for regeneration and, if there are two beds in series, the second bed is placed in the lead position by changing valve positions accordingly. As plant operating experience is gained, there may be other ways to determine the need for regeneration. It can be done on the basis of pounds of COD adsorbed per pound of carbon, or the required carbon dosage in terms of total flows processed by the plant. The carbon dosage required depends upon the effluent quality standards and the pretreatment provided. Typically, tertiary effluents would require 200–400 lb/million gal, secondary effluents 400–600 lb/million gal, and physical chemical effluents 1500–1800 lb/million gal. The above figures are a rough guide only, and may vary widely from place to place, even with the same process, because of differences in water quality and in plant operation, and in accordance with other variables. Once the dosage is known, then a combination of average COD concentration in the applied water and total flow through the bed will establish the need for transfer of carbon.

Control of Carbon Regeneration.   Quality control of regenerated carbon is provided by measuring the A.D. (apparent density) of the spent and regenerated carbon, and, later, by laboratory tests to measure the capacity of the regenerated carbon to adsorb a standard iodine solution, thus determining the Iodine Number of the carbon. The A.D. of virgin carbon is about 0.48 g/cm^3. As carbon becomes saturated with adsorbed organics, the A.D. may increase to 0.50 or 0.52, or more. As the carbon is regenerated, organics are removed, and the carbon loses weight. If properly regenerated, the A.D. will return to 0.48. The A.D. is easily and rapidly determined by weighing a known volume of carbon.

The A.D. of the regenerated carbon can be controlled by several means. To increase the A.D. increase the furnace hearth temperatures, decrease the spent carbon feed rate, increase the use of auxiliary steam, or decrease the rabble arm speed. To decrease the A.D. do the reverse for one or more of the above four items. It is best to change not more than one process variable per half hour. Typically, there will be about 5–10 percent loss of carbon during each regeneration cycle. The Iodine Number of virgin carbon is about 935, of a spent carbon about 580, and of regenerated carbon 800–900. The ash content of carbon may be used to detect any buildup of calcium or other undesirable foreign material. The ash content of virgin carbon is about 5.2 percent.

With tertiary pretreatment the maximum buildup of ash in the regenerated carbon should not exceed about 6.5 percent, but with lower degrees of pretreatment it may be greater. Carbon may lose some adsorptive capacity as measured by the Iodine Number during the first regeneration, but probably not in subsequent regeneration cycles.

Regeneration Procedures at Tahoe

When the carbon column effluent reaches the minimum effluent standards, spent carbon is removed from the columns. These standards are as follows: COD, 15 mg/l (or less than 50 percent of applied COD); color, 5 units; and MBAS, 0.5 mg/l. Before the advent of biodegradable detergents, ABS was the governing criterion, but now the soft detergents are broken down in pretreatment to the extent that COD controls. Experience in operating the Tahoe plant has shown that the COD breakthrough occurs when the carbon dosage is about 215 lb/million gal, but that the carbon is much easier and cheaper to regenerate if withdrawn earlier, at the time the dosage is about 250 lb/million gal, so this has become the basic signal for beginning carbon withdrawal. More highly saturated carbon takes longer to regenerate.

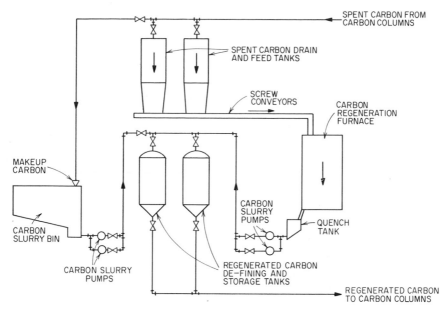

Fig. 5-12 Carbon regeneraton system. *(Courtesy Clair A. Hill & Assocs.)*

The carbon columns are pressurized to withdraw the spent carbon. Carbon is withdrawn while the column is in service. Upward flow is increased to about 10 gpm/ft² in order to expand the carbon bed slightly and to assist in uniform withdrawal of the spent carbon, that is, to prevent "ratholing" of the carbon. About 11 percent (5200 lb) of the contents of one column are removed at one time and replaced with regenerated carbon.

A flow chart of the regeneration system is shown in Figure 5-12. Spent carbon taken in slurry form from the columns passes either of two drain and feed tanks. These tanks are located at an elevation above the top of the furnace. Each dewatering bin will hold a slug of about 200 ft³ (5200 lb dry weight) of carbon, or a total of 400 ft³ (10,400 lb dry weight) in the two tanks. When carbon slurry is deposited in the spent carbon bin the water must be drained off through screens in the tank sides at the bottom to about 50 percent moisture before it can be fed to the carbon regeneration furnace. This takes about 15 minutes.

The partially dewatered carbon is fed from the drain bins to the furnace by means of a screw conveyor. This conveyor is constructed entirely of 304 stainless steel to resist corrosion. The screw conveyor is equipped with a variable speed drive so that the rate of carbon feed to the furnace can be controlled accurately. The furnace has a turndown ratio of 6 to 1, so that it can operate at rates between 1000 and 6000 lb of dry carbon per day. This cor-

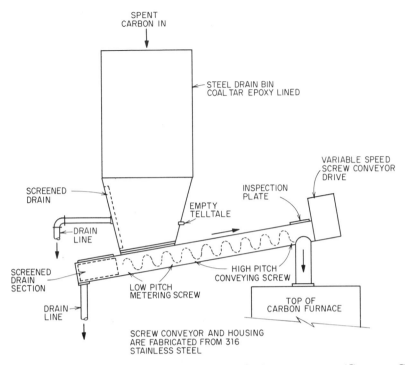

Fig. 5-13 Spent carbon drain bin and furnace feed arrangement. *(Courtesy Cornell, Howland, Hayes & Merryfield)*

responds to 33.3–200 ft³/day, or 0.023–0.138 cfm. The 9 in. diameter screw has a 3 in. pitch at the lower (drain bin) end, and a 9 in. pitch at the upper (furnace) end (Figure 5-13). The low pitch, metering section of the screw under the bin runs full of carbon for accurate feed control, and the higher pitch, conveying section runs only partially full to reduce the total driving torque required. At 100 percent efficiency (no slippage) the 9 in. pitch section would deliver 0.25 ft³/rev, and the 3 in. pitch section 0.083 ft³/rev. At 100 percent efficiency, the screw speed for the 3 in. pitch section should be 0.3–1.8 rpm in order to get the desired rate of carbon delivery to the furnace. The screw is driven through a Winsmith Reducer (952.6 to 1 reduction) by a Reeves Drive which can be varied from 216 to 2160 rpm. After reduction, the speed range for the drive is 0.2–2.0 rpm. To deliver 6200 lb/day then, the 9 in. portion of the screw must have an efficiency of at least 30 percent, and the 3 in. pitch section an efficiency of 90 percent (10 percent slippage).

Near the bottom of the drain and feed bin is a low level alarm (Bindicator) which indicates when the spent carbon level reaches the minimum level adequate to maintain a carbon seal in the conveyor and to prevent gases produced in the furnace from escaping into the building.

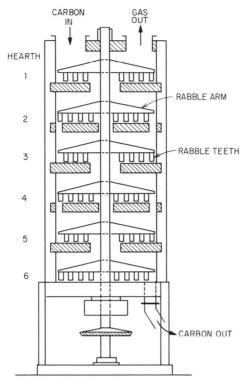

CARBON
IN

GAS
OUT

HEARTH

1

2

3

4

5

6

RABBLE ARM

RABBLE TEETH

CARBON OUT

Fig. 5-14   Cross sectional view of multiple hearth furnace.

The spent carbon is discharged into the top of the carbon regeneration furnace through a 10 in. inlet port. Figure 5-14 is a cross sectional view of the B-S-P Corporation multiple hearth furnace. This is a 54 in. diameter, 6 hearth, gas fired furnace with propane standby. The exhaust gases pass through an afterburner and a wet scrubber to eliminate air pollution. The carbon is moved downward through the six hearths (numbered 1 through 6 from top to bottom) by stainless steel rabble arms.

There are two burners on both hearths 4 and 6. The temperatures on these two hearths are independently controlled by an automatic temperature controller which maintains the temperature within 10°F of the desired temperature. A proportional flowmeter is provided in each of the air–gas mixture lines to the burners to ensure a constant percentage of excess oxygen. In addition, steam can be added to hearths 4 and 6.

An ultraviolet scanner is provided on each of the burners. In case of a flameout, the furnace is automatically shut down. Other safety features which shut down the furnace are high or low gas pressure; high combustion air pressure; low scrubber water pressure; high stack gas temperature; and draft fan or shaft cooling air fan not operating. In addition, the shaft cooling air fan is on a standby power circuit to avoid possible damage to the furnace rabble arms in the event of an electrical outage.

Temperatures on the various hearths are about as follows: No. 1, 800° F; No. 2, 1000° F; No. 3, 1300° F; No. 4, 1680° F; No. 5, 1600° F; and No. 6, 1680° F. In the furnace the carbon is thermally regenerated in the presence of steam. The heat vaporizes and drives off as gas the impurities which are adsorbed on the carbon, and restores the carbon essentially to pristine activity. The addition of steam on hearths 4 and 6 gives a more uniform distribution of temperatures throughout the furnace. It also reduces the apparent density and increases the Iodine Number of the regenerated carbon. About 1 lb of steam per pound of dry carbon is used.

The gases produced from the carbon regeneration leave the top of the furnace and enter an afterburner. When needed, the afterburner is operated at 1200° F to burn volatile and noxious gases. Under most conditions the afterburner is not needed, and wet scrubbing of the stack gas is sufficient. The wet scrubber removes carbon dust and odorous substances, and the stack gas poses no air pollution problems.

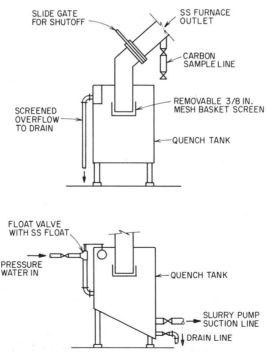

Fig. 5-15   Carbon quench tank. *(Courtesy Cornell, Howland, Hayes & Merryfield)*

Fig. 5-16   Carbon furnace discharge into quench tank with connection to slurry pumps in pit below floor. *(Courtesy Cornell, Howland, Hayes & Merryfield)*

The regenerated carbon is discharged from the bottom of the furnace into a small quench tank. Figures 5-15 and 5-16 are a drawing and a photograph, respectively, of a quench tank arrangement. Note that the quench tank has two separate pump suction lines from it, each with several water jets strategically placed to keep the carbon moving in the quench tank and connecting pipelines. Considerable agitation of the carbon slurry is necessary at this point to prevent buildup of carbon in the quench tank and plugging of the pump suction lines. From the quench tank the carbon slurry is pumped to the regenerated carbon storage tanks or wash tanks. Two diaphragm slurry pumps manufactured by Dorr-Oliver, Inc., each having a capacity of 3–20 gpm, are used to pump the carbon. The pumps discharge through a 1 in. line to the storage or wash tanks. The process of washing out the fines from the regenerated carbon is the same as that described previously for new carbon when filling carbon columns for the first time. Following the washing, the wash tank is pressurized and the regenerated carbon is forced in water slurry through a 2 in. line to the top of the carbon columns. Makeup carbon is added to the system to replace the carbon lost in the regeneration process by dumping bags of new carbon into a concrete slurry bin. The makeup carbon is pumped from this bin by two diaphragm slurry pumps to the wash tanks, where it is washed prior to being placed in the carbon columns. The carbon bag dump is equipped with a hinged cover and dust collector.

Operation of the furnace is controlled routinely by the apparent density of the carbon, supplemented at times by tests of Iodine Number, adsorption isotherms, or other laboratory tests to check the activity of the regenerated carbon. The carbon is regenerated to an apparent density of 0.48. If the ap-

Fig. 5-17   Carbon regeneration system. *(Reprinted from* Journal of the American Water Works Assoc., *60:1, Jan. 1968. Copyright 1968 by AWWA Inc., 2 Park Ave., New York, 10016)*

parent density is greater than 0.49, the furnace feed rate is too high or the temperature too low. If the apparent density is less than 0.48, the furnace temperature is too high or the feed rate too low, and chances are that some carbon is being burned.

Progressive decreases in the Iodine Numbers of regenerated carbon indicate that the regenerating temperature is too low and that adsorbates are being carbonized and left in the original pore structure. The temperature of the carbon itself during regeneration must be at least 1500° F to avoid this deterioration, and the indicated correction is to raise furnace temperatures. Temperatures as high as 1750° F or 1800° F may be required, and full restoration of the Iodine Number may not be obtained until after several successive regenerations. This might be avoided by using the special equipment required to measure the actual carbon temperatures during regeneration to be sure that they are not less than 1500° F. The procedures for performing the regeneration control tests have been discussed earlier in this chapter.

Figure 5-17 is a photograph of the complete carbon regeneration system at South Tahoe.

Results of Regeneration

The accumulation of full scale plant data on regenerated carbon is an extremely slow process because of the long period of time between regeneration cycles. Eight to ten slugs of carbon must be withdrawn from each carbon column per cycle, and the period between slugs is from four to six weeks depending upon plant flows. After $10\frac{1}{2}$ years of operation at Tahoe, the oldest carbon in the plant is in the eleventh cycle of reuse. To date, the carbon has been regenerated to original apparent density and nearly virginal Iodine Number. No appreciable ash buildup has been detected. In accelerated laboratory tests, granular carbon saturated with organics from wastewater has been regenerated through 20 cycles of use.

Table 5-16 summarizes the results of plant scale regeneration at Tahoe

TABLE 5-16   South Tahoe Water Reclamation Plant—Plant Scale Regeneration Through Three Cycles of Reuse.

Carbon Property	Virginal Carbon	Spent Carbon	Carbon After 1st Regeneration (1965)	Carbon After 2nd Regeneration (1967)	Carbon After 3rd Regeneration (1969)
Apparent density	0.48	0.52–0.59	0.47–0.50	0.47–0.50	0.48–0.49
Iodine number	935	550–843	900–969	782–933	743–875
Ash (%)	5.0	4.5	4.7	4.8–5.9	6.6–8.2

TABLE 5-17 Carbon Regeneration Data, South Tahoe Public Utility District.

Date	Regeneration Cycle	Carbon	Carbon Feed (dry lb/hr)	Btu (per lb)	Apparent Density	Iodine No.	Ash (%)	FURNACE HEARTH TEMPERATURES, °F					
								1	2	3	4	5	6
11-18-69	First	Spent	217	3,200	0.520	674	5.5	635	805	1115	1620	1500	1640
		Regen.			0.479	818	6.3						1640
11-20-69	First	Spent	158	4,450	0.531	658	6.4	595	780	1120	1630	1505	1640
		Regen.			0.485	788	6.7						1640
1-12-70	First	Spent	202	3,315	0.560	626	6.2	720	860	1110	1645	1500	1670
		Regen.			0.482	894	6.7						1670
2-16-70	Second	Spent	176	3,700	0.577	539	6.4	765	935	1200	1695	1550	1700
		Regen.			0.486	860	7.2						
2-20-70	Second	Spent	185	2,890	0.571	546	6.1	570	835	1120	1685	1550	1685
		Regen.			0.483	830	6.7						
2-21-70	Second	Spent	162	3,530	0.570	607	5.7	650	865	1230	1675	1540	1675
		Regen.			0.479	874	6.3						
1- 3-70	Third	Spent	99	5,070	0.553	599	7.8	730	970	1310	1635	1490	1635
		Regen.			0.501	789	7.3						
1- 7-70	Third	Spent	177	4,150	0.515	676	7.6	805	1000	1320	1630	1460	1650
		Regen.			0.488	823	7.3						
2-12-70	Third	Spent	119	5,230	0.580	524	6.5	835	1045	1330	1695	1525	1690
		Regen.			0.488	816	7.7						
2-14-70	Fourth	Spent	116	5,070	0.580	528	6.4	855	1100	1420	1695	1540	1695
		Regen.			0.487	820	7.1						

through the third cycle of reuse. Table 5-17 presents more detailed information for each cycle of regeneration, including the regeneration of fourth cycle carbon.

Typically the furnace was operated at 1650° F on the fourth and sixth hearths. The burners on hearth 6 were set to deliver 2 percent excess oxygen and the burners on hearth 4 were set to deliver 4 percent. The carbon feed rate is varied, of course, to suit spent carbon properties, but averages about 240 lb/hour. The moisture in spent carbon fed to the furnace varies from 40 to 44 percent. The total steam added to the furnace averages about 1 lb per pound of dry carbon. When steam is added, temperatures on hearths 1, 2, and 3 rise by about 200° F, thus giving a more uniform distribution of temperatures through the furnace. At the same temperature and feed rate, steam addition markedly reduces apparent density (0.50–0.49) and increases the Iodine Number (880–910). The heat input to the furnace varies from 2500 to 5000 Btu/lb of carbon.

Carbon losses and labor are the two major items of cost in carbon regeneration. The cost of makeup carbon may be as much as 50 percent of the total cost of carbon treatment. Minimizing carbon losses is essential to economic carbon regeneration. Measuring short term carbon losses precisely is difficult. The most accurate method of determination is the measurement over a long period of time of the quantities of makeup carbon required on a weight basis to maintain the proper level in the carbon columns. Pinpointing the location where carbon losses occur is not easy. Juhola and Tupper[45] determined the mechanical attrition losses in an externally heated rotating tube regenerator to be about 0.2 percent by operating the system without heat. Observations at Tahoe also indicate that mechanical attrition losses are low. In the Tahoe plant carbon losses have varied from 2 to 12 percent/cycle. The average loss is about 5 percent/cycle. The carbon loss varies with the feed rate to the furnace, losses increasing at increased feed rates and highest temperatures. In some sugar refineries carbon is precooled before quenching in an attempt to reduce carbon losses. It is claimed to be effective, but no definitive data are available. Burning of carbon in the furnace during operation, and particularly during startup and shutdown is a major factor. Continuous furnace operation will avoid the startup and shutdown losses, of course, and these losses can be minimized by operating as long as possible during a single regeneration campaign.

The expanded 7½ mgd plant at Tahoe completed its first full year of carbon operation in March, 1969. During the previous year of operation, the plant inflow and output was 673 million gal. A total of 85 tons of carbon were regenerated, and the average carbon dose was 228 lb/million gal. The actual volume of water passed through the carbon was 1104 million gal. This is greater than the plant flow due to recycling of various waste process streams

TABLE 5-18   Results of Carbon Treatment at Tahoe March 1, 1968 to February 28, 1969.

	Influent (mg/l)	Effluent (mg/l)	Applied (lb)	Removed (lb)	Pounds Removed per Pound Applied
COD	20.7	11.4	190,500	85,500	0.45
MBAS	0.46	0.13	4,010	2,790	0.70

within the plant. Table 5-18 shows the effect of carbon treatment on the wastewater, based on analyses of 24 hour composite samples and duplicate COD tests.

POWDERED CARBON REGENERATION

Powdered carbon is used by many water purification plants. It is also widely used in industry to remove objectionable organic constituents from liquids. The residue is a wet, organic loaded, spent carbon waste. Historically, these wastes have been disposed of by discharge to streams, incineration, or burial. Powdered carbon has potential applications to wastewater treatment which could be greatly increased if practical applications of powdered carbon re-activation can be successfully demonstrated. In the past, problems with large quantities of carbon dust have discouraged such attempts, and caused abandonment of some full scale facilities. However, recent developments appear to have overcome the dust problem. There are at least six potential systems that have appeared recently which show promise. These are the wet air oxidation process developed by Zimpro, Inc.; the transport reactor system developed by FMC Corporation; the fluidized bed process developed by Battelle Institute; a multiple hearth furnace developed by Nichols Engineering and Research Company; a reactor developed by Westvāco Corporation; and a pyrolisis unit developed by Jet Propulsion Laboratories.

The Zimpro wet air oxidation process is said to effectively restore the adsorptive properties of powdered activated carbon without dewatering of the spent carbon slurry. Carbon losses are estimated at about 5 percent per regeneration cycle. Zimpro has investigated the feasibility of applying their wet air oxidation process to regeneration of powdered activated carbon used in wastewater treatment. The spent carbon slurry at approximately 6–8 percent solids is pumped along with air through a heat exchanger to a reactor designed to operate at a preselected temperature and pressure. Under these

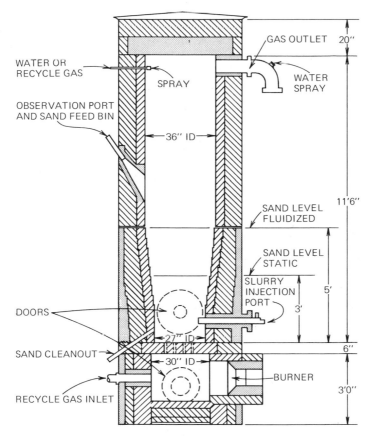

Fig. 5-18 Fluidized bed regeneration unit for powdered activated carbon.[36]

conditions the sorbed organics are selectively burned off the carbon. The regenerated carbon is returned through the heat exchanger, giving up its heat to the incoming slurry, and is conveyed to a storage tank. Makeup carbon is added to the system to maintain the carbon dose rate at the required value.

In order to evaluate this regeneration method, a pilot plant was set up in which primary effluent from the Village of Rothschild, Wisconsin sewage treatment plant was processed through a two stage countercurrent carbon sorption system. The liquid treating phase reduced the COD from an average of 233 mg/l to 34 mg/l, with an average carbon loading of 0.394 g COD/g carbon. The carbon was used through 23 cycles. The percentage makeup and changes in carbon composition through these 23 cycles are shown on following page:

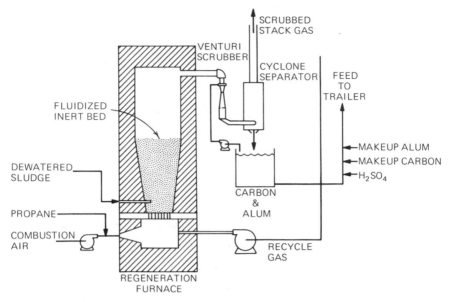

Fig. 5-19 Regeneration system schematic flowchart of the Battelle Northwest powdered activated carbon pilot plant.[36]

	Virgin Carbon	After 23 Recycles
Carbon	80	60
Ash	5.4	14.0
Na$_2$O and K$_2$O	0.86	0.14
CaO and MgO	1.24	3.74
Al$_2$O$_3$ and Fe$_2$O$_3$	0.35	3.52
P$_2$O$_5$	0.07	2.52
SO$_3$	1.68	0.75
SiO$_2$	0.49	2.21

The activated carbon loss per cycle was 4.7 percent.

Tests are also underway using this approach on the mixture of carbon and activated sludge from the DuPont PACT process.

Based on preliminary cost estimates, there is no clear cut economic difference between the wet air oxidation system and the fluidized bed furnace system described below.

The fluidized bed furnace (FBF) is the regeneration system used by both Battelle Northwest and Eimco in their work. The unit used in the 100,000 gpd Battelle pilot plant is shown in Figure 5-18. This unit consisted of three main sections: a firebox housing the burner, 30 in. ID by 20 in. high; a bed section

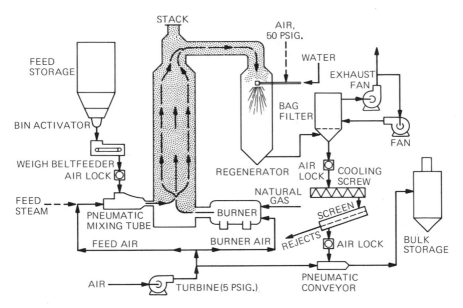

Fig. 5-20   Schematic diagram of the Westvaco carbon regeneration system.[43]

containing inert sand, 27 in. ID bottom, 36 in. ID top by 60 in. high; and a freeboard 36 in. ID by 72 in. high. A schematic diagram of the overall carbon regeneration system is shown in Figure 5-19. The pilot furnace used in the Battelle study was built by Nichols Engineering. The conclusions from several months of operation of this regeneration facility were:

Powdered activated carbon can be successfully regenerated in a fluidized bed furnace.

Satisfactory regeneration can be achieved at a temperature of 1250° F with a stack gas oxygen concentration of less than 0.5 percent.

After 6–7 regenerations, the regenerated carbon is as effective as virgin carbon in removing organic matter from raw sewage.

Average carbon losses per regeneration cycle were 9.7 percent.

Hearth plugging problems during pilot plant operations resulted from corrosion of the recycle gas system. Such corrosion problems can easily be precluded in design of a full scale system.

Inert material buildup averaged 2.9 percent per cycle during the pilot plant operations.

Sand carryover from the fluidized bed furnace is believed to represent the most significant fraction of this buildup.

Stack gases from the regeneration furnaces should not present significant air pollution problems.

These promising pilot plant results indicate that the FBF may offer a practical means of regenerating powdered carbon used in wastewater treatment. Its potential can better be evaluated when a full scale installation is placed in operation.

Westvāco Corporation has developed and commercialized a patented method for powdered carbon regeneration (U.S. Patent 3,647,716). A 20,000 lb/day unit was placed in operation at Covington, Virginia in early 1971 to regenerate spent carbon from corn syrup refineries. A schematic of the system is shown in Figure 5-20.

The carbon is dispersed and suspended by a metered oxidizing air stream and pneumatically conveyed to the top of a vertical Venturi-type section inside the furnace. Steam may also be utilized as the oxidizing stream on carbons with low organic loading. High velocity, high temperature flue gas enters just below the Venturi section and mixes intensively with the carbon–water–air stream above the Venturi, resulting in instantaneous heat transfer and optimum gas–solid contacting.

The reactor processing steps are drying, volatilization of organics, burning of volatiles, and steam activation of residual carbon. These steps occur almost simultaneously in the reactor above the Venturi section in the space of a few seconds. The steam selectively activates any carbon residual left in the carbon micropore structure. Overall reactor temperature for these steps is 1750–1850° F, depending on spent carbon loading.

In the 10 ton/day unit at Covington, no more than 1 lb of carbon is suspended in the furnace at any instant, allowing almost immediate response to control changes.

Suspended particles at 1800° F exit the reactor via a horizontal refractory lined duct to a downflow evaporative cooler in which the gas stream is cooled to 450° F with a water spray. Regenerated carbon is then collected in a three compartment, glass cloth, bag filter. The carbon is continuously discharged by air locks, conveyed by a water cooled screw conveyor, and screened for any foreign material. A pneumatic conveyor then delivers the regenerated carbon to bulk storage tanks. Carbon losses of 5–25 percent have been reported.

The Westvāco transport system for powdered carbon regeneration has been piloted in conjunction with a pilot study of the DuPont PACT process. In this application, a mixture (approximately 50:50) of waste activated sludge and powdered carbon are supplied to the furnace. About 80 percent carbon recovery has been achieved with simultaneous burning of the biological sludges. The recovered carbon is acid washed to remove inert materials. An investigation of landfill of the dewatered carbon–activated sludge mixture indicated that color and other organics would be leached from the sludge.

The Jet Propulsion Laboratory pyrolysis unit not only reactivates powdered carbon, but also produces new powdered carbon from the carbon containing, organic solids in wastewater.

At this time it appears that the only remaining step in proving the practical applicability of powdered carbon regeneration to wastewater treatment is the construction and operation of a full scale plant. Such operation would answer questions concerning dust control, costs, regeneration efficiency, and carbon losses that can be determined accurately in no other way.

POWDERED VS. GRANULAR CARBON SYSTEMS
Advantages

Lower capital cost. No columnar contactors and associated piping and valving are required.

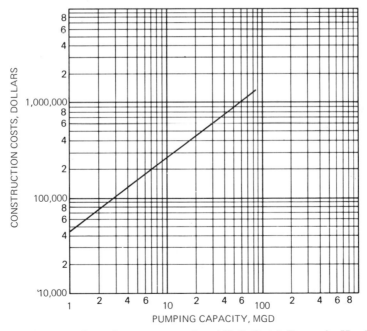

Fig. 5-21 Carbon adsorption pump station (40 ft Total Dynamic Head) construction costs.

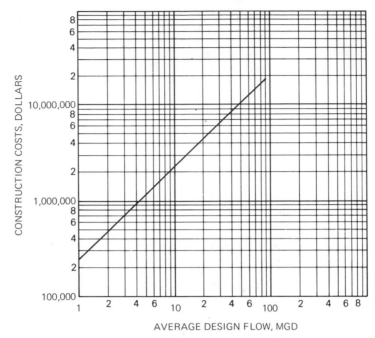

Fig. 5-22   Granular carbon contactor system (30 minutes contact at design flow) construction costs.

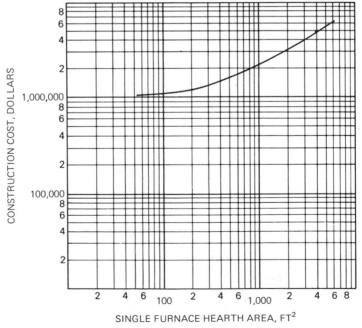

Fig. 5-23   Multiple hearth incineration construction costs.

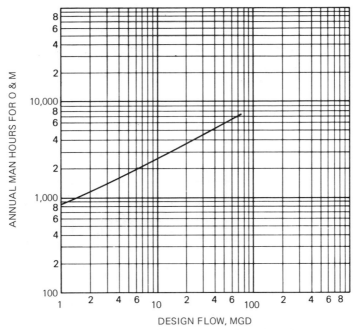

Fig. 5-24   Granular carbon adsorption and pumping (30 minutes contact, 40 ft Total Dynamic Head) man hour requirements.

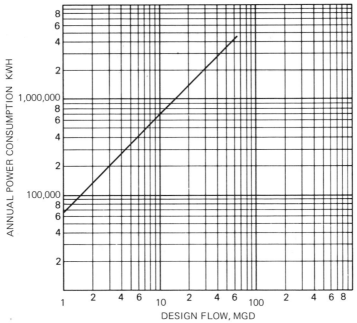

Fig. 5-25   Granular carbon adsorption and pumping (30 minutes contact, 40 ft Total Dynamic Head) power requirements.

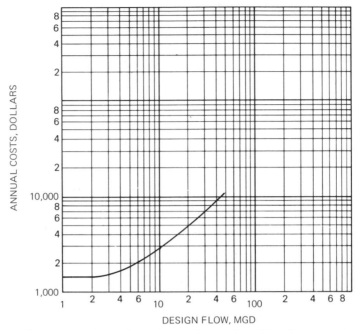

Fig. 5-26   Granular carbon adsorption and pumping (30 minutes contact, 40 ft Total Dynamic Head) maintenance materials costs.

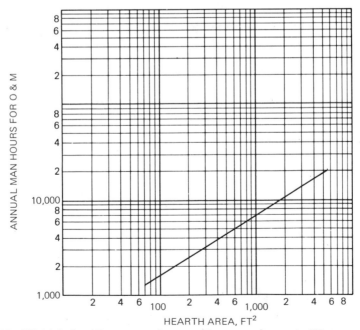

Fig. 5-27   Multiple hearth regeneration man hour requirements (70 percent operation time, 30 percent downtime).

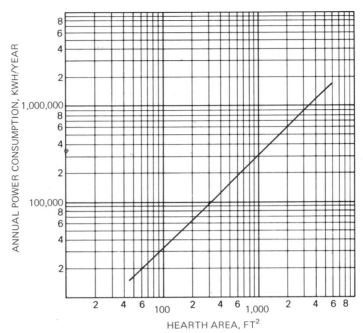

Fig. 5-28   Multiple hearth regeneration power requirements (70 percent operation time, 30 percent downtime).

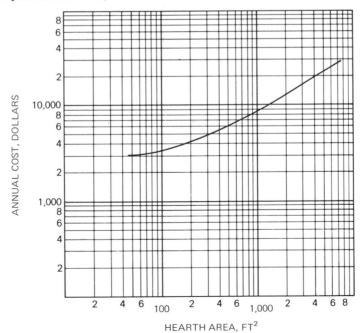

Fig. 5-29   Multiple hearth regeneration maintenance materials costs (70 percent operation time, 30 percent downtime).

TABLE 5-19   Granular Carbon Regeneration Costs. (Annual Costs, $1,000.)

	1,000 lbs Carbon regenerated per year				
	547.5	2,737	5,474	13,685	27,370
Amortized capital (7 percent, 20 years)	151	184	236	341	511
Labor ($9/hour)	8.5	36	61	108	171
Power ($0.02/kwh)	0.9	4.0	6.5	16	20
Fuel ($1.50/10⁶ Btu)	15	27	55	136	271
Maintenance materials	2	6	8	12	16
Makeup carbon ($0.50/lb, 8 percent loss/cycle)	22	109	219	547	1,096
Total	199.4	366	585	1,160	2,085
Cost/lb regenerated carbon	36.4¢	13.4	10.7	8.5	7.6

TABLE 5-20   Assumed Composition of Raw Sewage and Primary Effluent. (Values in mg/l unless indicated.)

	Raw Sewage	Primary Effluent
Solids, total	700	
Dissolved, total	500	
Fixed	300	
Volatile	200	
Suspended, total	200	80
Fixed	50	
Volatile	150	
Settleable solids, ml/l	10	
BOD_5–20°C	200	130
TOC	200	
COD	500	
Total N	40	
Organic	15	32
Free Ammonia	25	
Nitrites	0	
Nitrates	0	
Total P	10	
Organic	3	
Inorganic	7	
Chloride[a]	50	
Alkalinity[a] as $CaCO_3$	100	
Grease	100	

[a]Increase by unit in carriage water.

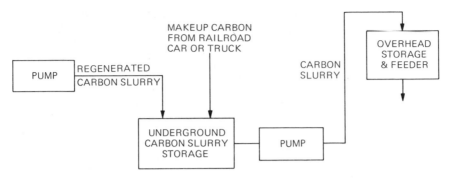

Fig. 5-30 Powdered activated carbon feed system (5–50 mgd).

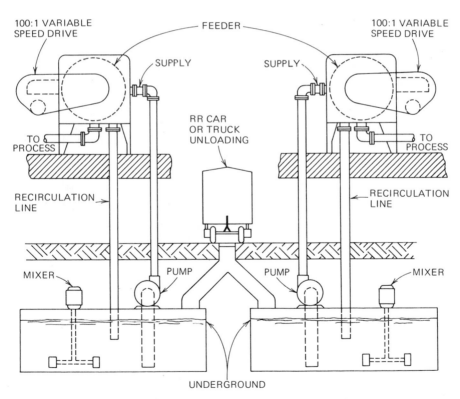

Fig. 5-31 Powdered carbon storage and feeding.

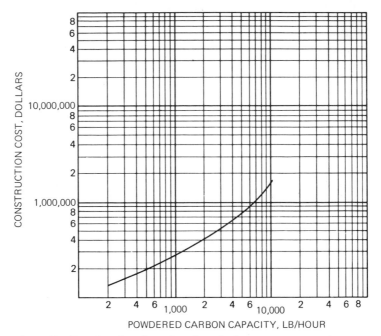

Fig. 5-32 Powdered activated carbon feed system construction costs.

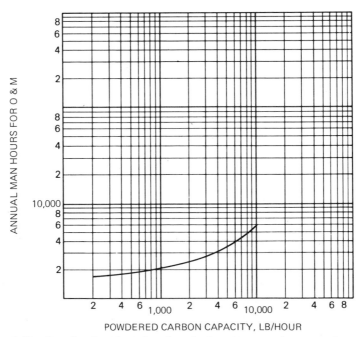

Fig. 5-33 Powdered activated carbon feed system man hour requirements.

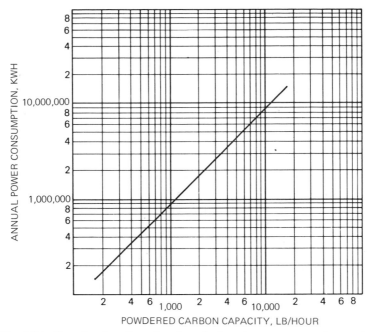

Fig. 5-34   Powdered activated carbon feed system power requirements.

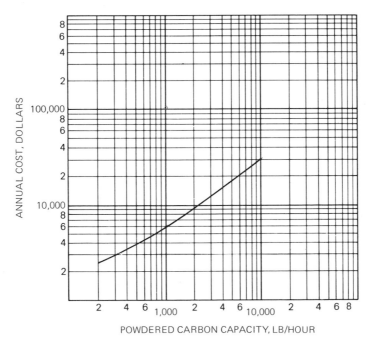

Fig. 5-35   Powdered activated carbon feed system maintenance materials costs.

TABLE 5-21   Battelle Process System Design Parameters.

Treatment System	
Carbon contact	
Time at pH 4, min.	10
Time at pH 7, min.	5
Flocculation	
Velocity gradient, fps/ft	75
Time, min.	10
Tube settler loading rate, gpd/ft^2	2880
Filter	
Length of filter run, hr	12
Loading rate, gpm/ft^2	5
Chlorine contact time, peak dry weather flow, min.	30
Chemical storage capacity, hr	12
Sludge storage, days	1
Carbon dose, mg/l	600
Alum dose, mg/l	200
Polyelectrolyte dose, mg/l	2.0
Lime dose, mg/l	150
Sulfuric acid, lb/lb carbon	0.5
Sludge dewatering polyelectrolyte dose, lb/ton dry solids	1
Regeneration System	
Combustion chamber temperature, °F	2000
Bed temperature, °F	1500
Fluidizing gas velocity, ft/sec	1.3
Maximum bed diameter, ft	22
Carbon recovery, %	91
Alum recovery, %	91
Blowdown, %	5
Sludge quantity, lb/million gal	7380
Settler underflow concentrations, % solids	4.5
Sludge carbon content, % on dry basis	57.3
Sludge inerts content, % on dry basis	17.2
Dewatered sludge solids content, %	22
Dewatered sludge flow, lb/hr/mgd (wet)	1419
Carbon feed rate, $/hr/mgd	
(100% operation of FBF)	179
(70% operation of FBF)	255

Ability to control hydrogen sulfide problems. PCT granular carbon systems have experienced great difficulty in controlling hydrogen sulfide gas production. Powdered carbon systems have successfully overcome this problem.

Requires minimal pretreatment.

TABLE 5-22   Capital Costs, Battelle Northwest Process, 10 mgd.

Chemical Feed	
Alum	34,000
Carbon	430,000
Lime	200,000
Polymer (wastewater)	86,000
Polymer (sludge)	86,500
Sulfuric acid	66,000
Carbon contact	45,000
Rapid mix	52,000
Flocculation	45,000
Sedimentation tank	160,000
Tube settling modules	80,000
Chemical sludge pumps	180,000
Centrifuge	350,000
Fluidized bed furnace	4,100,000
Chlorine contact tank	200,000
Chlorination equipment	38,000
Filtration	1,050,000
Subtotal	$7,202,500
Yardwork	1,008,350
Total Construction Cost	$8,210,850
Engineering, fiscal, legal	985,000
Interest during construction	821,000
Total Capital Cost	$10,016,850

Disadvantages

Lack of experience with full scale regeneration systems.

Uncertain operating costs.

Greater handling problems due to dust associated with powdered carbon.

Does not appear cost competitive for tertiary treatment.

Costs

Cost data for granular carbon systems have been presented in the EPA Technology Transfer Manual on carbon adsorption.[2] Chapter 11 of this text also presents preliminary cost curves on granular carbon adsorption and regeneration. Figures 5-21 through 5-29 present cost data on components of granular carbon systems as developed by Culp/Wesner/Culp under EPA contract 68-03-2186. Costs are based on fourth quarter 1975

TABLE 5-23   Battelle Northwest Process, Operation and Maintenance Costs, 10 mgd.

	Annual Labor, Hours	Annual Power Consumption, kwh	Annual Fuel Consumption scf, Nat'l Gas	Annual Cost of Maintenance Mat'ls, Dollars
Chemical Feed				
Alum[a]	250	2,700	—	200
Carbon	2,500	1,800,000	—	10,000
Lime[b]	2,800	27,000	—	1,300
Polymer (wastewater)[c]	790	7,000	—	490
Polymer (sludge)[c]	800	7,100	—	500
Sulfuric acid	500	1,500	—	420
Carbon contact	170	11,000	—	1,000
Rapid mix	800	190,000	—	2,800
Flocculation	170	11,000	—	1,000
Sedimentation	860	—	—	1,000
Chemical sludge pumps	170	36,000	—	2,800
Centrifuge	15,000	390,000	—	32,000
Fluidized bed furnace	3,500	10,500,000	168×10^6	9,000
Chlorination	890	—	—	2,200
Filtration	4,000	140,000	—	6,800
Total	33,200	13,123,300	168×10^6	71,510

[a]Liquid alum.
[b]Quicklime, pumped feeder.
[c]Dry polymer.
SCF = standard cubic foot.

cost levels. Regeneration fuel requirements (based on data furnished by a carbon manufacturer during a recent design project) are as follows:

	Btu per lb Granular Carbon Reactivated
Furnace gas	3000
Steam	1250
Afterburner	2400
Total	6650

Typical multiple hearth regeneration loading rates for granular carbon are 40 lb carbon/ft^2/day. Projected granular carbon regeneration costs

TABLE 5-24   Battelle Northwest Process, Total Costs, 10 mgd.

Total Annual Costs	Dollars
Amortized capital,	
10,016,850 × 0.09439	945,490
Labor,	
33,200 hours @ $9/hour	298,800
Power,	
13,123,300 kwh @ $0.02/kwh	262,466
Fuel,	
168 × 10^6 scf @ $1.50/tcf	252,000
Maintenance materials	71,510
Chemicals	
Alum, 420 tons/year @ $70/ton	29,400
Carbon, 1,270 tons/year @ $650/ton	825,500
Lime, 2,280 tons/year @ $37/ton	84,360
Polymer (wastewater), 60,440 lb/year @ $0.30/lb	18,132
Polymer (sludge), 51,830 lb/year @ $2/lb	103,660
Sulfuric Acid, 4,560 tons/year @ $57.30/ton	261,288
Chlorine, 46 tons/year @ $220/ton	10,120
Total	$3,075,568

$$\text{Cost/million gal @ capacity} = \frac{\$3,162,726}{365 \times 10} = \$867/\text{million gal}$$

scf = standard cubic foot
tcf = thousand cubic feet

are shown in Table 5-19. Virgin carbon costs were about $0.50/lb in the fourth quarter of 1975.

No full scale powdered carbon treatment systems were operating in 1976. The following cost estimates for the various powdered carbon systems described earlier were prepared by Culp/Wesner/Culp under EPA contract 68-01-0183 utilizing the best available data and the following criteria:

1. Construction costs are based on fourth quarter 1975 levels.

2. Construction costs include equipment and installation and the following allowances:
    25 percent contractor overhead and profit
    15 percent for electrical system
    15 percent contingency
    15 percent for miscellaneous items

TABLE 5-25   Design Parameters for Eimco System.

Chemical Treatment	
Coagulant	Alum
Coagulant dose, mg/l	125
Polyelectrolyte dose, mg/l	0.25
Flash mix time, min.	1
Clarifier	
Type	flocculator-clarifier
Hydraulic loading, gpm/ft^2	
(peak)	0.8
Gravity thickener	
Solids loading, lb/day/ft^2	10
Underflow solids, %	5
Vacuum filter	
Feed solids, %	5
Yield, lb/hr/ft^2	2.8
Cake moisture content, %	75
Lime dose, % by weight	40
Carbon Treatment	
Carbon contractor	internal solids recycle
Peak hydraulic loading, gpm/ft^2	0.8
Carbon dose, mg/l	300
Carbon slurry concentration, g/l	10
Underflow concentration, %	3
Gravity thickener	
Solids loading, lb/day/ft^2	20
Underflow solids, %	12
Vacuum filter	
Feed solids, %	12
Polyelectrolyte dose, lb/ton/day	
solids	10
Yield, lb/hr/ft^2	8
Cake solids, %	27
Granular Media Filter	
Type	tri-media
Average hydraulic loading gpm/ft^2	5
Average Backwash Recycle, % of	
filtrate	3
Fluidized Bed Furnace	
Solids loading, lb/hr/ft^2	3
Freeboard velocity, ft/sec	1.2
Firebox temperature, °F	2000
Operating temperature, °F	1250
Primary Sludge Incineration	
Multiple hearth loading rate,	
lb/hr/ft^3 (wet basis)	7

TABLE 5-26   Capital Costs, Eimco Process, 10 mgd.

Rapid mix	37,000
Floc-clarifier	700,000
Solids contact clarifiers	1,800,000
Filter	1,050,000
Primary sludge	
Thickener	160,000
Vacuum filter	400,000
Incinerator	1,900,000
Carbon sludge	
Thickener	160,000
Vacuum filter	300,000
FBF	2,300,000
Chemical feed	
Wastewater	
Alum	70,000
Poly	18,000
Carbon	280,000
Primary sludge	
Lime	160,000
Carbon sludge	
Poly	100,000
Chlorine contact	200,000
Chlorine feed	38,000
Sludge pumping	
Primary sludge	60,000
Carbon sludge	52,000
Subtotal	9,785,000
Yardwork	1,370,000
Total Construction Cost	11,155,000
Engineering, fiscal, legal	1,339,000
Interest during construction	1,116,000
Total Capital Cost	$13,610,000

3. Construction costs do not include engineering, legal, fiscal, and administrative fees.

4. The assumed quality of raw sewage and primary effluent is shown in Table 5-20.

5. The powdered carbon storage and feeding system is shown in Figures 5-30 and 5-31. Figures 5-32 through 5-35 summarize cost data for powdered carbon feed systems.

TABLE 5-27   Eimco Process, O & M Costs, 10 mgd.

	Annual Labor, Hours	Annual Power Consumption, kwh	Annual Fuel Consumption, SCF Nat'l Gas	Annual Cost of Maintenance Mat'ls, Dollars
Rapid mix	600	250,000	—	900
Floc-clarif.	2,100	54,000	—	3,900
Solids contact				
clarifiers	5,600	420,000	—	18,000
Filter	4,000	140,000	—	6,800
Primary sludge				
Thickener	420	9,000	—	650
Vacuum filter	5,000	340,000	—	35,000
Incinerator	6,000	420,000	32×10^6	10,000
Carbon sludge				
Thickener	420	9,000	—	650
Vacuum filter	3,900	230,000	—	27,000
FBF	2,500	4,900,000	63×10^6	5,000
Chemical feed				
Wastewater				
Alum	650	3,000	—	470
Poly	400	2,900	—	150
Carbon	2,000	930,000	—	6,100
Primary sludge				
Lime feed	2,000	15,000	—	900
Carbon sludge				
Poly feed	450	7,000	—	550
Sludge pumping				
Primary sludge	200	50,000	—	3,700
Carbon sludge	160	26,000	—	2,200
Total	36,400	7,805,900	95×10^6	$121,970

The Battelle system costs are based on the process shown in Figure 5-7 and the design parameters given in Table 5-21. The capital costs and detailed operation and maintenance data for a 10 mgd plant are summarized in Tables 5-22 through 5-24.

The Eimco system costs are based on the process shown in Figure 5-8 and the design parameters given in Table 5-25. Capital costs and detailed operation and maintenance data for a 10 mgd plant are summarized in Tables 5-26 through 5-28.

Similar design and cost data for a biophysical system, as shown in Figure 5-9, are given in Tables 5-29 through 5-32.

TABLE 5-28   Eimco Process, Total Costs, 10 mgd.

Total Annual Costs[a]	Dollars
Amortized capital @ 7%, 20 years	
13,610,000 × 0.09439	1,284,648
Labor	
36,500 hr @ $9/hour	327,600
Power	
7,805,900 kwh @ $0.02/kwh	156,118
Fuel	
95 × 10^6 scf @ $1.50/tcf	142,500
Maintenance materials	121,970
Chemicals	
Powdered carbon—makeup	
690 tons/year @ $650/ton	448,500
Alum	
1,900 tons/year @ $70/ton	133,000
Polymer—wastewater	
7,610 lb/year @ $0.30/lb	2,300
Polymer—carbon sludge	
70,080 lb/year @ $2.00/lb	140,160
Chlorine	
46 tons/year @ $220/ton	10,120
Lime—primary sludge	
1,532 tons/year @ $37/ton	56,700
Total	$2,823,616

$$\text{Cost/million gal @ Capacity} = \frac{\$2,823,616}{365 \times 10} = \$774/\text{million gal}$$

[a]Excluding effluent filtration and carbon feeding system.

Because powdered carbon processes are still in the developmental stage, the cost estimates given in the tables are based on assumptions, some of which may prove to be either overly optimistic or pessimistic. Thus, it is desirable to evaluate the potential impact of some key assumptions on relative costs.

The Eimco process is based on two stages of carbon contact preceded by chemical coagulation and sedimentation. Although carbon requirements would increase, capital and O & M costs would decrease by eliminating the second stage contactors. The maximum potential gain for cost savings of about 6¢/1000 gal by use of a single stage system results

TABLE 5-29   Design Parameters for Biophysical Process with Wet Air Oxidation

Primary Sedimentation	
Surface loading rate, gpd/ft^2	800
Detention time, hr	2.5
Solids removal efficiency, %	65
Sludge moisture, %	95
Sludge specific gravity	1.03
Activated Sludge	
Air rate, scf m/mgd	1,275
Recycle, %	50
Mixed liquor solids, mg/l	
Volatile	4,000
Carbon	8,000
Total	13,000
Growth yield coefficient, lb vs/ lb BOD	0.5
Return sludge solids, mg/l	
VSS	12,000
Carbon	24,000
Total	39,000
Detention time, hr	4.5
Sludge age, days	12.5
Secondary Sedimentation	
Overflow rate, gpd/ft^2	400
Polymer dose, mg/l	5
Flow to thickener, gal/mgd	5,000
Gravity Thickener	
Loading rate, lb/ft^2/day	10
Thickened sludge, % solids	8
Wet Oxidation System	
Temperature, °F	450
Pressure, psi	700
Blowdown	
Volume, gal/mgd	100
Solids, lb/mgd	166
Ash content of solids, %	75
Carbon Losses, mg/l	
Blowdown	5
Oxidation	7
Effluent	5
Carbon Dose mg/l	
Makeup carbon	17
Regenerated carbon	103

TABLE 5-29 (*Continued*)

Primary Sludge Dewatering	
Vacuum filtration, $lb/ft^2/hr$	6
Polymer, lb/ton	1
Primary Sludge Disposal	
Multiple hearth incinerator, $lb/ft^2/hr$	7

scfm = standard cubic feet/minute.
Vs = volatile suspended solids.

from using the same carbon dosage as in the two stage system—realizing that some of this potential gain would probably be offset by increased carbon costs. Eimco's work indicates that carbon dosages as low as 100 mg/l may be practical (producing an effluent quality of 5 mg/l COD, 5 mg/l SS, and 0.3 mg/l phosphorus) under some conditions. This lower dosage

TABLE 5-30   Capital Costs, Biophysical Process, 10 mgd.

Primary sedimentation tanks	440,000
Primary sludge pumping	140,000
Primary sludge thickener	90,000
Vacuum filtration	200,000
Incineration	1,500,000
Chemical feed systems	
Carbon	170,000
Polymer, wastewater	200,000
Polymer, sludge	61,000
Aeration basins	650,000
Aerators	800,000
Secondary sedimentation basins	800,000
Return sludge pumping stations	230,000
Gravity thickener	130,000
Wet air oxidation	1,400,000
Chlorine contact basin	200,000
Chlorine feed equipment	38,000
Subtotal	7,049,000
Yardwork	987,000
Total Construction Cost	8,036,000
Engineering, fiscal, legal	964,000
Interest during construction	804,000
Total Cost	$9,804,000

TABLE 5-31   Biophysical Process, O & M Costs, 10 mgd.

	Annual Labor, Hours	Annual Power Consumption, kwh	Annual Fuel Consumption, scf Nat'l Gas	Annual Cost of Maintenance Mat'ls, Dollars
Primary sedimentation	1,800	3,300	—	2,700
Primary sludge pumping	170	3,000	—	3,000
Primary thickening	420	—	—	240
Vacuum filtration	2,700	140,000	—	18,000
Incineration	3,600	120,000	15×10^6	5,200
Chemical feed systems				
Carbon	1,700	350,000	—	3,600
Polymer, wastewater	800	11,000	—	900
Polymer, sludge	640	4,500	—	410
Aeration basins	—	—	—	—
Aerators	5,600	3,600,000	—	11,000
Secondary sedimentation	2,600	6,600	—	4,700
Return sludge pumping	1,100	78,000	—	1,300
Gravity Thickening	440	—	—	350
Wet air oxidation	3,500	1,000,000	0.09×10^6	14,000
Chlorine contact basin	—	—	—	—
Chlorine feed equipment	890	—	—	2,200
Filtration	—	—	—	—
Total	25,960	5,316,400	15.09×10^6	67,600

TABLE 5-32   Biophysical Process, Total Costs, 10 mgd.

Total Annual Costs	Dollars
Amortized capital @7%, 20 years	
$9,804,000 \times 0.09439$	925,400
Labor	
25,960 hr @ \$9/hr	233,648
Power	
5,316,400 kwh @ \$0.02/kwh	106,328
Fuel	
15.09×10^6 scf @ \$1.50/tcf	22,635
Maintenance materials	67,600
Chemicals	
Carbon	
260 tons/year @ \$650/ton	169,000
Polymer	
152,205 lb/year @ \$0.30/lb	45,660
Chlorine	
46 tons/year @ \$220/ton	10,120
Total	1,580,383

$$\text{Cost/million gal @ capacity} = \frac{1,580,383}{365 \times 10} = \$433/\text{million gal}$$

would result in an overall cost reduction of about 20¢/1000 gal for the two stage system.

A dominant factor in determining the cost of the basic Battelle process is the large carbon (600 mg/l) and alum (200 mg/l) dosages used in the original Battelle pilot work. These in turn affect the cost of sludge handling and regeneration facilities. The costs using the same alum dosage (125 mg/l) used in the Eimco process and a carbon dosage of 200 mg/l were calculated. The impact on costs is dramatic—reducing the costs at 10 mgd from 87¢/1000 gal to 46¢/1000 gal. The Battelle data do indicate that the need for effluent filtration is marginal. If effluent filtration were eliminated, a savings of 5¢/1000 gal treated would result at 10 mgd.

The criteria used for the design of the biophysical process are reported to reliably provide nitrification and also represent the basis on which most of the available data on this process have been collected. If the biophysical process provides a degree of stability of nitrification comparable to two stage activated sludge, it may offer an economic advantage for plants of 5 mgd capacity or larger. It does not appear to offer an economic advantage over single stage nitrification activated sludge plants.

REFERENCES

1. Ongerth, H. J., et al. "Public Health Aspects of Organics in Water," *Journal American Water Works Assoc.*, 65, July 1973, pp. 495–498.

2. *Process Design Manual For Carbon Adsorption.* Environmental Protection Agency, Office of Technology Transfer, October 1973.

3. Bunch, R. L., et al. "Organic Materials in Secondary Effluents," *Journal Water Pollution Control Federation*, 33, February 1961, pp. 122–126.

4. Helfgott, T., et al. "Analytic and Process Classification of Effluents," *Proc. ASCE*, SA 3, June 1970, pp. 779–803.

5. Rubhun, M. and Manka, J., "Classification of Organics in Secondary Effluents," *Environmental Science and Technology*, 5, 1971, pp. 606–609.

6. Manka, J.; Rubhun, M.; Mandlebaum, A.; and Bortinger, A., "Characterization of Organics in Secondary Effluents," *Environmental Science and Technology*, November 1974, p. 1017.

7. Schwartz, H. G., Jr., "Adsorption of Selected Pesticides on Activated Carbon and Mineral Surfaces," *Environmental Science and Technology*, 1, April 1967, pp. 332–337.

8. Weber, W. J., Jr., and Gould, J. P. "Organic Pesticides in the Environment," in *Advances in Chemistry* Vol. 60, ed. by R. G. Gould, ACS, Washington, D.C., 1966, p. 280.

9. Ward, T. M. and Getzen, F. W., "Influence of pH on the Adsorption of Aromatic Acids on Activated Carbon," *Environmental Science and Technology*, 4, January 1970, pp. 64–67.

10. Hoak, R. D. "Recovery and Identification of Organics in Water," in *Advances in Water Pollution Research*, Vol. 1, Macmillan, New York, 1964, pp. 163–180.

11. Rosen, A. A., et al. "Relationship of River Water Odor to Specific Organic Contaminants," *Journal Water Pollution Control Federation*, 35, June 1963, pp. 777–782.

12. Hager, D. G., and Rizzo, J., "Removal of Toxic Organics From Wastewater by Adsorption With Granular Activated Carbon." Paper presented at EPA Technology Transfer Session on Treatment of Toxic Chemicals, Atlanta, Georgia, April 1974.

13. Rosen, A. A., et al. "The Determination of Stable Organic Compounds in Waste Effluents at Microgram per Liter Levels by Automatic High Resolution Ion Exchange Chromatography, *Water Research*, 6, 1972, p. 1029.

14. Jolley, R. L., *Chlorination Effects on Organic Constituents in Effluents From Domestic Sanitary Sewage Treatment Plants*. Oak Ridge National Lab., Publ. No. 565, ORNL-TM-4290, October 1973.

15. Dean, D. G., et al. "Removing Heavy Metals from Waste Water," *Environmental Science and Technology*, 6, June 1972, pp. 518–522.

16. Sigworth, E. A. and Smith, S. B., "Adsorption of Inorganic Compounds by Activated Carbon," *Journal American Water Works Assoc.*, June 1972, pp. 386–391.

17. Linstedt, K. D., et al. "Trace Element Removals in Advanced Wastewater Treatment Processes," *J. Water Pollution Control Federation*, 43, July 1971, pp. 1507–1513.

18. Akatsu, E., et al., "Radiochemical Study of Adsorption Behavior of Inorganic Ions on Zirconium Phosphate, Silica Gel and Charcoal," *Journal Nuclear Science and Technology*, 2, 1965, p. 141.

19. Mantell, C. I., *Carbon and Graphite Handbook*. Wiley–Interscience, New York, 1968.

20. Hassler, J. W., *Purification With Activated Carbon*. Chemical Publishing Company, New York, 1974.

21. Smisek, M. and Cerny, S., *Activated Carbon, Manufacture, Properties and Applications*. Elsevier Publishing Co., New York, 1970, p. 41.

22. Mattson, J. S. and Kennedy, F. W., "Evaluation Criteria for Granular Activated Carbons," *Journal Water Pollution Control Federation*, 43, November 1971, pp. 2210–2217.

23. Culp, G. L., and Shuckrow, A. J., "Physical-Chemical Techniques for Treatment of Raw Wastewater," *Public Works*, July 1972.

24. Suhr, L. G. and Culp, G. L., "State of The Art—Activated Carbon Treatment of Wastewater," *Water and Sewage Works*, Reference Number, 1974, p. R-104.

25. Parkhurst, J. D., et al., "Pomona Activated Carbon Pilot Plant," *Journal Water Pollution Control Federation*, 39:10, Part Z, October 1967, pp. R 70-R 81.

26. English, J. N., et al., "Removal of Organics from Wastewater by Activated Carbon." Paper presented at the Advanced Waste Treatment Seminar, San Francisco, October 28 and 29, 1970.

27. Wesner, G. M. and Argo, D. G., *Pilot Wastewater Reclamation Study May 1970–June 1971.*" Orange County Water District, July 1973.

28. Wesner, G. M. and Culp, R. L. "Wastewater Reclamation and Seawater Desalination," *Journal Water Pollution Control Federation*, 44:10, October 1972, pp. 1932–1939.

29. *Los Angeles Water Recycling Project.* Los Angeles Department of Water and Power, September 1972.

30. *Advanced Waste Water Treatment as Practiced at South Tahoe.* Project 17010 ELQ(WPRD 52-01-67), August 1971.

31. Argo, D. G., "Wastewater Reclamation Plant Helps Manufacture Fresh Water," *Water and Sewage Works*, Reference Number, 1976, p. R-160.

32. McDonnell, G. M., "Water Reuse Through Advanced Wastewater Treatment," *Water and Sewage Works*, Reference Number, 1973, pp. R-60–R-64.

33. Stander, G. J., and Van Vuuren, L. R. J., "The Reclamation of Potable Water from Wastewater," *Journal Water Pollution Control Federation*, 41, March 1969, pp. 355–367.

34. Stander, G. J., and Funke, J. W., "Direct Cycle Water Reuse Provides Drinking Water Supply in South Africa," *Water and Wastes Engineering*, May 1969, pp. 66–67.

35. Van Vuuren, L. R. J., et al., "The Full-Scale Reclamation of Purified Sewage Effluent for the Augmentation of the Domestic Supplies of the City of Windhoek." Paper presented at the 5th International Water Pollution Research Conference, July–August, 1970.

36. Shuckrow, A. J.; Dawson, G. W.; and Bonne, W. F., *Powdered Activated Carbon Treatment of Combined and Municipal Sewage.* Environmental Protection Agency Contract Report #14-12-519, 1972.

37. Shuckrow, A. J.; Dawson, G. W.; and Olesen, D. E., "Treatment of Raw and Combined Sewage," *Water and Sewage Works*, April 1971, p. 104.

38. Burns, D. E., and Shell, G. C., "Physical-Chemical Treatment of a Municipal Wastewater Using Powdered Activated Carbon." Paper presented at the 44th Annual Water Pollution Control Federation Conference, San Francisco, California, October 1971.

39. Burns, D. E., and Shell, G. L., *Physical-Chemical Treatment of a Municipal Wastewater Using Powdered Activated Carbon*. Sanitary Engineering Research and Development, Eimco Process Machinery Division, Envirotech Corporation, Final Report, FWQA Contract No. 14-12-585, 1971.

40. Shell, G. L., et al., "Regeneration of Activated Carbon," in *Applications of New Concepts of Physical-Chemical Wastewater Treatment*, Pergamon Press, Elmsford, New York, 1972, pp. 167–198.

41. Shell, G. L., and Burns, D. E., "Powdered Activated Carbon Application, Regeneration and Reuse in Wastewater Treatment Systems." Envirotech, Corporation (unpublished).

42. Wallace, R. N. and Burns, D. E., "Factors Affecting Powdered Carbon Treatment of a Municipal Wastewater," *Journal Water Pollution Control Federation*, March 1976, pp. 511–519.

43. *Powdered Hydrodarco Activated Carbons Improve Activated Sludge Treatment*. ICI America, PC-4, October 1972.

44. Humphrey, M. F., et al., "Carbon Wastewater Treatment Process," *Journal ASME*, 74-ENAs-46, July 1974.

45. Juhola, A. J., and Tupper, F., *Laboratory Investigation of The Regeneration of Spent Granular Activated Carbon*. FWPCA Report N. TWRC-7, February 1969.

6

Disinfection

BIOLOGICAL CONTAMINANTS OF CONCERN

It was not until the establishment of the germ theory of disease by Pasteur in the mid-1880s that water as a carrier of disease producing organisms really could be understood, although the epidemiological relation between water and disease had been suggested as early as 1854. In that year London experienced the "Broad Street Well" cholera epidemic, and Dr. John Snow conducted his now famous epidemiological study. He concluded the well had become contaminated by a visitor who arrived in the vicinity with the disease. Cholera was one of the first diseases to be recognized as capable of being waterborne.

Sanitary engineering and preventive medical practices have combined to reach a point where waterbone disease outbreaks of epidemic proportions have largely been eliminated. However, the potential for disease transmission through the water route has not been eliminated. With a few exceptions, the disease organisms of epidemic history are still present in today's sewage. Cooper[1] has presented an excellent review on the infectious agents of greatest concern—those whose origin is in the intestinal discharge of man or animals. These can be broadly classified as bacterial, parasitic, and viral.

Bacteria

Bacteria of the genus *Salmonella* contain a wide variety of species pathogenic for man and animals. These bacterial agents can be transmitted from man to man by means of contaminated water. Typhoid fever caused by *Salmonella typhosa* is the most severe enteric fever form of salmonellosis, and man is the only host. At the turn of this century, typhoid fever death rates of more than 50 per 100,000 persons were not uncommon in cities of the United States; at present, however, death due to this disease is practically nonexistent. Over the last seventy years, typhoid related morbidity has also shown significant change; for example, the U.S. morbidity for typhoid fever in 1930 was 12.6 per 100,000, whereas in 1971 there were fewer than 0.2 isolations of *S. typhosa* per 100,000. Although water is no longer a significant vehicle for the transmission of this disease, there was an isolated, recent outbreak of apparently waterborne typhoid fever among farm workers in Dade County, Florida in which 213 cases occurred; this event underscores the continued potential for typhoid transmission.

Acute gastroenteritis is the circumstance under which other *Salmonella* are most commonly encountered. Water, while an important potential source, contributes only a little more than 2 percent of the total cases.[1] Although the incidence of waterborne salmonellosis is very low, there are a number of occurrences in which large numbers of persons were infected by this route. One of the best documented is the outbreak in Riverside, California in 1965, in which 18,000 people were infected. At least 70 were hospitalized, and 3 deaths occurred. The supply was a groundwater supply that had not been chlorinated; initiation of chlorination stopped the outbreak. Just why the outbreak occurred when it did is not known. The water supply was clearly incriminated, but the exact source of contamination could not be determined. Of interest is the fact that before the outbreak no coliform organisms were detected during the routine surveillance of the water system. During the investigation, *S. typhimurium* was isolated from the water. A surprising result of the quantitative analysis of a composite water sample from all parts of the city was that *Salmonella* were estimated to be 10 times as numerous as *E. coli*. Thus, the potential for waterborne transmission remains a real one.

Bacteria of the genus *Shigella* produce bacillary dysentery. The disease spreads rapidly under conditions of overcrowding and improper sanitation. The mode of transmission is primarily from person to person and through contaminated food. However, transmission by water certainly should not be overlooked. There have been three reports of waterborne outbreaks of shigellosis in small communities, involving small, unchlorinated wells which were contaminated with sewage.[1] *Shigella sonnei* was isolated in all instances.

Cholera is a serious disease of both historical and current interest. It is similar in many respects to typhoid fever but more rapid in onset, more virulent, and more often fatal. Death rates of 25–85 percent of observed cases are commonly reported. The etiologic agent of this disease is the small, gram negative, motile, non–spore forming rod, *Vibrio cholerae*. Man is the only known natural host, and since a prolonged carrier state is uncommon, the disease must be maintained by an unbroken chain of mild inapparent infections. Infected water is the frequent mode of transmission. The first case of cholera to occur in the U.S.A. since 1911 has been associated with a well contaminated by the leachate from a septic tank system.[1]

Mycobacterium tuberculosis has been isolated from sewage, particularly where an institution treating tuberculosis patients is involved or where industries such as dairies and slaughterhouses handling tubercular animals may be expected to discharge their wastes. Human cases associated with wastewater usually involve swimmers in highly contaminated water who breathe water into their lungs.

There are many reports of waterborne gastroenteritis of unknown etiology in which bacterial infections are suspected. It is important, therefore, to mention certain gram negative bacteria normally considered nonpathogenic as potential sources for these disturbances. Both enteropathogenic *Escherichia coli* and certain strains of pseudomonas are often mentioned in relation to gastroenteric disturbances in the newborn.[2] The publicized gastrointestinal disease outbreak that occurred at Crater Lake National Park has been shown to be due to waterborne, enterotoxigenic *E. coli*.[1]

Parasites

There are myriad diseases of protozoan and metazoan etiology which are transmitted from man to environment to man in fecal material; many of these may be associated with water.

In the United States, perhaps the most serious parasitic disease to be associated with wastewater is amoebic dysentery. *Entamoeba histolytica*, a protozoan, can infect the human colon, causing erosion of the superficial mucous membrane. The amoeba has the ability to form heavy walled cysts, and transmission occurs when mature cysts are excreted with feces into water or food. The cysts are most important because they are resistant to environmental forces. The disease is worldwide and normally occurs as inapparent infections. The incidence of infection in the United States is not well established because of the frequent lack of clinical manifestations; it is most probably of considerable magnitude.[1] Waterborne outbreaks of disease caused by *E. histolytica* are well documented. The most infamous one in this country occurred in 1933 in Chicago. The cause proved to be plumbing inadequacies which allowed fecal pollution of the drinking water of the Congress and Ambassador Hotels, resulting in many cases of amoebic dysentery and some 23 deaths.

Giardiasis is an intestinal disease caused by protozoa, particularly *Giardia lamblia*. These protozoans grow in the human small intestine and frequently cause gastroenteritis. A large, water associated outbreak occurred among skiers in Aspen, Colorado in 1969.[1]

Other parasites which may be present in municipal wastewater include the eggs of *Ascaris lumbricoides* (giant roundworm), whose host and reservoir is man, and various cestodes (tapeworms), in particular *Taenia* spp. Cestode eggs are transmitted to an intermediate host (swine or cattle), and man would be infected by eating the meat of an infected animal. Cestode infestation is of greatest importance where sewage sludge is spread on agricultural areas.

Viruses

Many types of human virus are discharged in the feces of infected persons and are disseminated in the water environment from sewage and other sources of fecal pollution. Improved laboratory technology for recovery of viruses from water has resulted in a growing number of reports describing widespread occurrence of viruses in sewage effluents and surface waters.

Once virus enters the water environment there are a number of potential modes of transmission to man; the chief of these are ingestion and bodily contact, including such activities as bathing, water contact sports, contact with crop or landscape irrigation waters, drinking from untreated or improperly treated water sources, and ingestion of shellfish. The potential for viral contamination has increased greatly with the growth in the volume of domestic wastes being discharged to natural watercourses.

Viruses in the stool of infected persons commonly range from 1,000 to 100,000 infective units per gram of feces. It has been calculated that the average enteric virus density in domestic sewage is probably about 500 virus units per 100 ml, and in polluted surface water not more than one virus unit per 100 ml.[3] These quantities are very small in comparison to the quantities of coliforms. The coliform–virus ratio is about 92,000:1 in sewage and about 50,000:1 in polluted surface water.[4] The human enteric viruses and the diseases associated with them are as follows:[1]

Virus Subgroup	No. of types	Disease
Poliovirus	3	paralytic poliomyelitis, aseptic meningitis
Coxsackie virus		
Group A	26	herpangina, aseptic meningitis, paraly-
Group B	6	sis pleurodynia, acute infantile myo-carditis
Echovirus	34	aseptic meningitis, rash and fever, diarrheal disease, respiratory illnesses
Infectious hepatitis	—	infectious hepatitis
Reovirus	3	fever, respiratory infections, diarrhea
Adenovirus	32	respiratory and eye infections

It is only with the virus(es) of infectious hepatitis that proven, documented outbreaks of waterborne virus disease have occurred. The December, 1955 epidemic in Delhi, India was the largest scale epidemic of hepatitis related to direct contamination of a water supply by sewage. Ten outbreaks have been documented in the United States, most of which involved gross contamination of small supplies.[5] Fourteen cases of infectious hepatitis among thirty individuals who were swimming in a North Carolina recreational lake of questionable sanitary quality have been reported,[1] indicating the potential of contracting this disease while swimming in sewage polluted water. A number of outbreaks of infectious hepatitis due to the ingestion of raw or steamed shellfish taken from sewage polluted waters have been reported. Although no outbreaks of shellfish associated disease have been attributed to other members of the enteric virus group, several outbreaks of shellfish associated "gastroenteritis" have been reported.

Poliovirus has frequently been mentioned as being possibly waterborne. Several epidemics of poliomyelitis attributable to water have been cited, but evidence has not been sufficiently complete to carry conviction. In the late 1930s it was discovered that the virus of poliomyelitis could be found in the feces not only of persons with paralytic disease but also of asymptomatic carriers, and that the virus could be excreted in the feces of such carriers for several weeks. This observation led to the successful attempts to isolate poliovirus from sewage and from raw surface waters, and the hypothesis that waterborne transmission of poliomyelitis could occur. Epidemiological seaches for episodes of waterborne poliomyelitis continued throughout much of the 1940s and the 1950s, and yielded eight outbreaks in which contaminated drinking water was suspect, one of which occurred in the United States.

In 1974, five children showed symptoms of a disease with similar clinical characteristics. The disease was positively diagnosed as caused by Coxsackie virus type A16.[1] In this instance, the infections were acquired while swimming in lake water that had relatively high fecal coliform counts and from which the specific virus was successfully isolated. This is one of the first instances in which a type A Coxsackie virus has been shown to be transmitted to bathers.

No other enteric viruses have been specifically implicated as causative agents in documented outbreaks of waterborne viral disease. However, it should be pointed out that for many outbreaks of gastroenteritis and diarrhea, no agent can be specifically incriminated. The cause of such outbreaks is often listed as viral in nature. For the years 1946 through 1960, 142 epidemics of waterborne gastroenteritis and diarrhea involving 18,790 cases were reported in the United States.[6] In terms of magnitude, gastroenteritis and diarrhea are the most important diseases transmitted by water.

Indicator Organisms

None of the microbes discussed above are readily isolated from water, nor are they readily enumerated. Thus, control analyses related to control of these pathogens must rely upon indicator microorganisms, those that can be readily detected and counted. The commonly used indicator organism is the coliform group of bacteria, a relatively large group of bacteria which are commonly present in the gut of warm blooded animals. These microorganisms are not usually considered to be pathogenic, and are used as indicators of the presence of fecal material. Since the direct measurement of enteric pathogens is difficult, health authorities rely on the more easily determined coliform indices to alert them to the presence of fecal material and, therefore, the potential presence of pathogenic microbes of enteric origin.

Unfortunately, some coliforms are ubiquitous and are not necessarily associated with sewage or similar materials. These are commonly found in soil and, during times of heavy runoff, are flushed into local waters. A test has been developed which differentiates fecal coliform, based upon the fact that coliforms of fecal origin will grow at elevated temperatures.

The measurement of fecal coliforms is a fairly reliable index of the presence of waste contaminated with potentially dangerous fecal material. Reductions in fecal coliform by a treatment step are generally in proportion to the reduction of any bacterial pathogen which might be present. Unfortunately, this relationship does not necessarily hold for viruses, particularly when chlorination is involved, as many viruses are more resistant than bacteria to such disinfection.

Dose Response Considerations

The dose of microorganisms required to cause disease (i.e., dose response) in a given individual is a most difficult factor to determine. To be conservative, it could be assumed that one infectious agent (bacteria, virus, etc.) is enough to cause disease; however, as might be expected, there is a great deal of variation both in the various agents and in the susceptible hosts. Dose response data for human cases of typhoid fever for a human volunteer study in which each subject was given one dose of a given number of *Salmonella typhosa* have been published.[1] Doses of 10^5 viable organisms produced disease in 28 percent of the volunteers, while 10^9 viable organisms produced a 95 percent response. These data emphasize the point that it often takes more than one infectious unit to produce disease.

In the case of viruses, the consensus is that any dose capable of causing a tissue culture infection should be considered capable of causing infection in man. In this regard, it has been shown that as little as one tissue culture infective dose ($TCID_{50}$) of polio is infective for susceptible infants.[1] It has been estimated that a $TCID_{50}$ of poliovirus may be equivalent to between 30 and 100 viral particles.

The relationship between indicator organisms and the dose of pathogenic microbes is most conservative. When exposed to various treatments and procedures, the coliforms have survival capabilities similar to if not slightly greater than those of bacterial pathogens. Since there are more coliforms than pathogens in sewage, their reduction to low numbers would be indicative of the reduction of pathogenic bacteria to proportionately lower numbers. The relationship between coliform levels in wastewater and the number of typhoid bacteria present as a function of typhoid morbidity in the community have been estimated.[7] At a reported rate of *Salmonella* of 12.0 per 100,000 persons in the United States, one would expect about 27.5 *Salmonella* per one million coliforms found in sewage, while the number of *S. typhosa* would be far less than one per million coliforms.

Recently, the number of fecal coliforms has been related to isolations of *Salmonella* in surface waters[8] with the following results:

Fecal Coliform per 100 ml	Total Samples Examined	Percent Positive for Salmonella
1–200	29	27.6
201–2000	27	85.2
>2000	54	98.1

A marked increase in *Salmonella* occurred when fecal coliform concentrations exceeded 200 per 100 ml.

Recreational water quality has been integrated with dose response data and probability of contact with disease organisms (*Salmonella*) via water to construct a curve (Figure 6-1) relating numbers of organisms per volume of water with risk of contracting disease or infection.[9] It was assumed in making the risk analysis that a swimmer imbibes 10 ml of water; thus, when relating this curve to drinking water the risk would be increased by one or two orders of magnitude. Extrapolating this curve to estimate the risk of drinking a liter of water that contains one coliform per 100 ml (approximate drinking water standard), one finds a risk of one in 10,000,000 of contracting salmonellosis when a liter of water is consumed. A similar associ-

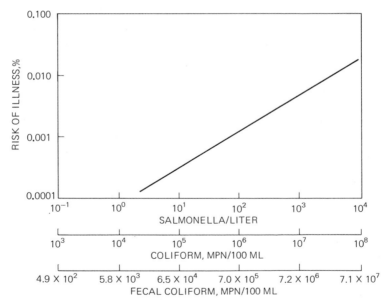

Fig. 6-1   Relationship between disease risk and *Salmonella*, coliforms, and fecal coliforms.[9]

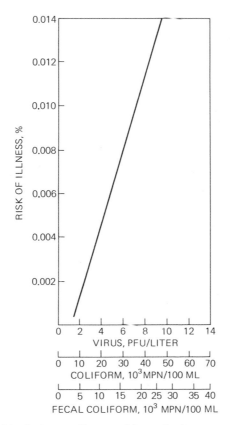

Fig. 6-2 Relationship between disease risk and viruses, coliforms, and fecal coliforms.[9]

ation was developed for virus infection (Figure 6-2), but this relationship is based upon virus dose response data which are recognized as uncertain.

ALTERNATIVE MEANS OF DISINFECTION

Wastewater disinfection can be accomplished by several means. Specific disinfection processes of current significance include one or a combination of the following: (1) physical treatment, by means of storage, ultrasonic waves, or the application of heat or other physical agents; (2) irradiation, with gamma radiation or ultraviolet light; (3) the addition of metal ions, such as copper and silver; (4) the addition of alkalis and acids; (5) the addition of surface active chemicals, such as quaternary ammonium compounds; and (6) the addition of oxidants, such as chlorine, chlorine-ammonia, chlorine dioxide, bromine, iodine, ozone, and others. Fair and co-workers[9a] have summarized criteria for appraising the suitability of various potential disinfectants as follows:

1. Ability of the disinfectant to destroy the kinds and numbers of organisms present within the contact time available; the range of water temperatures encountered; and the anticipated fluctuations in composition, concentration, and condition of the water being treated.
2. Ready and dependable availability of the disinfectant at reasonable cost and in a form that can be conveniently, safely, and accurately applied.
3. Ability of the disinfectant, in concentrations employed, to accomplish the desired objectives without rendering the water toxic or objectionable, aesthetically or otherwise, for its intended purpose.
4. Ability of the disinfectant to persist in residual concentrations as a safeguard against recontamination.
5. Adaptability of practical, replicable, quick, and accurate assay techniques for determining disinfectant concentration to use in operational control of the treatment process and as measures of disinfecting efficiency.

COAGULATION, SETTLING, FILTRATION, AND ADSORPTION PROCESSES

Although effluent chlorination is the process that is mostly commonly used to further reduce the pathogen population of secondary effluents, some other AWT processes also provide significant reductions, and also make the water more susceptible to action of disinfectants. The disinfecting capabilities of these other processes will be discussed before the disinfecting processes proper.

The chemical coagulation process used in conjunction with sedimentation (and also filtration in some cases) has been noted by several researchers[5, 10-19] as providing high degrees of removal of viruses. For example, alum coagulation was found to remove 95–99 percent of Coxsackie virus and ferric chloride coagulation was found to remove 92–94 percent of the same virus.[10] With both alum and ferric chloride, good virus removal was contingent upon good floc formation and the absence of interfering substances. The effectiveness of removal was not temperature dependent with either coagulant. It has been determined that viruses were not inactivated by alum coagulation but could be partially recovered from the sludge. The presence of organic material was shown to decrease the amount of virus removed by alum or ferric chloride coagulation.

In a 20 month study of the removal of poliovirus Type 1 from water, Robeck, Clarke, and Dostal observed that the poliovirus organism is removed by flocculation and filtration with about the same efficiency as coliforms.[12] They noted that if a low but well mixed dose of alum was fed just ahead of the filters operated at 6 or 2 gpm/ft^2, more than 98 percent of the viruses were removed by 16 in. of coarse coal on top of 8 in. of sand. If the alum dose was increased and conventional flocculators and settling were used, the removal was increased to over 99 percent. Floc breakthrough of the filter, sufficient to cause a turbidity of less than 0.5 units, was usually accompanied by virus breakthrough. Addition of a polyelectrolyte to the filter influent decreased virus in the effluent by increasing floc strength and floc retention in the filter.

The report of an AWWA committee on viruses in water noted that viruses, because of their small size, become enmeshed more easily in a protective coating of turbidity contributing matter than would bacteria.[5] For most effective disinfection they concluded, turbidities should be kept below 1 JTU; indeed, they felt it would be best to keep the turbidity as low as 0.1 JTU. With turbidities as low as 0.1–1, they concluded that a chlorine feed in water plants need be only enough to have a 1 mg/l free chlorine residual after 30 minutes contact time.

While the bulk of studies on virus removal using alum coagulation have been laboratory studies, Walton reported a study of coliform bacteria removal in over 80 full scale water treatment plants.[14] He found that the removal of coliform bacteria by coagulation, sedimentation, and filtration averaged about 98 percent for the several plants studied. The average coliform densities in filtered by unchlorinated waters ranged from 2.9 to 200 per 100 ml. Also, large scale pilot studies conducted in Dallas, Texas[11] demonstrated that virus removals from secondary effluents by alum coagulation–sedimentation and coagulation–sedimentation–filtration processes are essentially the same as described in the water treatment literature using smaller scale processes. Removals of bacterial virus as high as 99.845 percent for coagulation–sedimentation and 99.985 percent for coagulation–sedimentation–filtration processes were observed using alum coagulation. Tests at a 300 gpm pilot plant in Orange County, California found no virus in 64 samples of alum coagulated, filtered and chlorinated secondary (trickling filter) effluent.

Lime coagulation has been demonstrated to be capable of effectively removing and inactivating viruses at high pH values. The mechanism of inactivation under alkaline conditions is probably caused by denaturation of the protein coat and by disruption of the virus. In some cases complete loss of structural integrity of the virus may occur under high pH conditions. The pH reached in lime coagulation is a critical factor in determining the degree of virus inactivation. Figure 6-3 shows the marked difference in

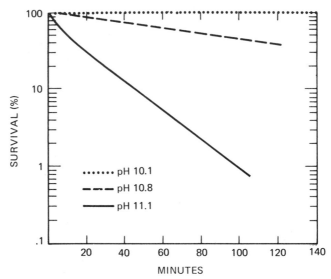

Fig. 6-3   Inactivation of poliovirus 1 by high pH at 25°C in lime treated [500 mg/l Ca(OH)$_2$], sand filtered secondary effluents.[10]

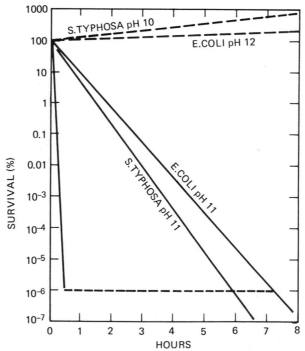

Fig. 6-4   Survival of *E. coli* at pH 11 and 12 and of *S. typhosa* at pH 10, 11, and 12 at 25°C.[10]

virus inactivation as the pH is increased from 10.1 to 10.8 and then to 11.1.[10] Figure 6-4 illustrates the effects of pH on *Escherichia coli* and *Salmonella typhosa*.

The pilot plant at Dallas, Texas also found that only a few gram positive rods could survive high pH lime treatment (pH of 11.2–11.3 with contact times of 1.56–2.40 hours). The virucidal effect of the pH treatment is reflected by the fact that no viable poliovirus were recovered from the sludges resulting from the high pH treatment.[11]

Activated carbon has the ability to adsorb, although not deactivate, virus from wastewater—with maximum adsorption at near neutral pH values.[38] However, the significance of this ability in carbon treatment of wastewaters has yet to be demonstrated in the field.

CHLORINATION

The destruction of pathogens by chlorination is dependent upon water temperature, pH, time of contact, degree of mixing, turbidity, presence of interfering substances, and concentration of chlorine available. Chloramines (which are formed if ammonia is present) are much less effective disinfectants than free chlorine. For example, in one study, at normal pH values, approximately 40 times more chloramine than free chlorine was required to produce a near 100 percent kill of *E. coli* in the same time period.[21] For *S. typhosa* this ratio was about 25:1. To obtain a near 100 percent kill with the same amounts of residual chloramine as with free chlorine required approximately 100 times the contact period for chloramine.

When chlorine is dissolved in water at temperatures between 49°F and 212°F it reacts to form hypochlorous and hydrochloric acids:

$$Cl_2 + H_2O \longrightarrow HOCl + HCl$$

This reaction is essentially complete within a very few seconds. The hypochlorous acid ionizes or dissociates practically instantaneously into hydrogen and hypochlorite ions:

$$HOCl \longrightarrow H^+ + OCl^-$$

These reactions represent the basis for use of chlorine in most sanitary applications.

Hypochlorite salts also ionize in water and yield hypochlorite ion, which establishes equilibrium with hydrogen ions:

$$Ca(OCl)_2 + 2H_2O \longrightarrow 2HOCl + Ca(OH)_2$$
$$NaOCl + H_2O \longrightarrow HOCl + NaOH$$

Generally, the HOCl form has been considered to be a far more effective disinfectant than OCl⁻. Several investigators have reported that HOCl is 70–80 times as bactericidal as OCl^-, and that increasing the pH reduces germicidal efficiency because most of the chlorine exists as the less microbicidal OCl^- at the higher pH levels. However, one recent study reported that one virus (poliovirus 1) was more rapidly inactivated at pH levels (pH 10) where the chlorine is in the form of OCl^- rather than $HOCl$.[39]

The reactions of chlorine with ammonia in solution are of great significance in wastewater treatment. In the presence of ammonia, a three chloramines are obtained:

$$NH_3 + HOCl \longrightarrow NH_2Cl + H_2O \text{ (monochloramine)}$$
$$NH_3 + 2HOCl \longrightarrow NHCl_2 + 2H_2O \text{ (dichloramine)}$$
$$NH_3 + 3HOCl \longrightarrow NCl_3 + 3H_2O \text{ (trichloramine)}$$

These chloramines differ in many properties from $HOCl$ and OCl^- salts. They exist in various proportions depending on the relative rates of formation of monochloramine and dichloramine, which change with the relative concentrations of chlorine and ammonia, as well as with pH and temperature. Above pH about 9, monochloramines exist almost exclusively; at pH about 6.5, monochloramines and dichloramines coexist in approximately equal amounts; below pH 6.5 dichloramines predominate; and trichloramines exist below pH about 4.5.

The point where all ammonia is converted to trichloramine or oxidized to free nitrogen is referred to as the breakpoint. Chlorination below this level is combined available residual chlorination; that above this level is free available residual chlorination.

Just as free available chlorine forms differ in germicidal capacity, so do inorganic and organic chloramine combined available chlorine forms. Inorganic chloramines have a substantially lower oxidation–reduction potential and less germicidal capacity than does $HOCl$; most organic chloramines have little or no germicidal capacity and are of concern primarily for the chlorine they consume that otherwise would be available for disinfection, and in the measurement and interpretation of residual chlorine.

Figure 6-5 illustrates the relationship between chlorine concentration and the contact time required for 99 percent destruction of *E. coli* for three different forms of chlorine. Figure 6-6 illustrates the relative resistance of three viruses and *E. coli*. While the polio and Coxsackie viruses are considerably more resistant than *E. coli* of $HOCl$, the adenovirus tested is apparently more sensitive. One-tenth ppm of chlorine as $HOCl$ destroyed 99 percent of *E. coli* in about 99 seconds. The same quantity of the adenovirus tested was destroyed in about one-tenth of that time by the same amount of $HOCl$, but at this $HOCl$ concentration the same amount of poliovirus required about 15 minutes more, and the Coxsackie virus required over 40 minutes.

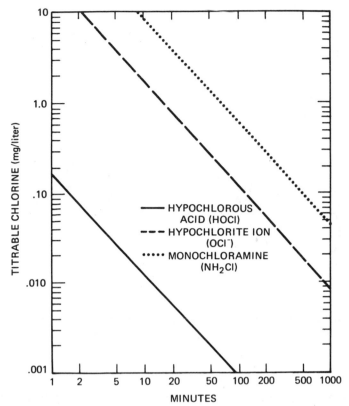

Fig. 6-5 **Relationship between concentration and time for 99 percent destruction of *E. coli* by three forms of chlorine at 2–6°C.**

Other researchers,[40] conclude that with the exception of adenovirus 3, from 3 to 100 times more chlorine is required to kill the viruses than to kill *E. coli*, *Aerobacter aerogenes*, *Elbertella typhosa*, or *Shigella dysenteriae*.

Both *p*H and temperature have a marked effect on the rate of virus kill by chlorine. Table 6-1 summarizes data on virus inactivation rates at varying *p*H at 10°C in filtered secondary sewage effluent. The constants in Table 6-1 were determined from experimental results using model f₂ virus which was seeded in filtered secondary sewage effluent buffered to the desired *p*H value. Chlorine solution was flash mixed at a dose of 30 mg/l and the time of persistence of any free chlorine as well as virus survival was determined.[41]

Since most sewage effluents are near natural *p*H, the 30 mg/l chlorine dose would result in less than 80 percent viral inactivation. However, by merely lowering the *p*H to 6.0, viral kill could be increased from 80 to 99.93

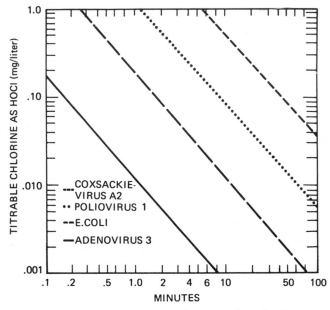

Fig. 6-6   Relationship between concentration and time for 99 percent destruction of *E. coli* and three viruses by hypochlorous acid (HOCl) at 0–6°C.

TABLE 6-1   Viral Inactivation Rate Constants at Varying pH of Sewage at 10°C. (After Kruse[41])

| Sewage pH | Flash Mixing of 30 mg/l Chlorine | | Viral Kill, % |
	Viral Inactivation Rate Constant (min⁻¹)	Free Chlorine Duration (sec)	
5.0	120.0	110.0	>99.9999
5.5	72.0	24.0	>99.9999
6.0	44.0	10.0	99.93
6.5	26.0	5.6	92.0
7.0	16.0	5.4	76.0
7.5	9.5	5.0	55.0
8.0	5.6	5.0	40.0
8.5	3.5	4.9	25.0[a]
9.0	2.0	4.8	15.0[a]
9.5	1.2	4.6	10.0[a]

[a]Essentially control survival.

percent. The results dramatically show how the kill of virus may be enhanced, especially when hypochlorites are fed for sewage disinfection by proper pH control. For four logs (99.99 percent) of viral inactivation, the flash mixing of the following dosages were required: 40 mg/1 of chlorine solution (pH 6.3), 50 mg/1 of hypochlorite solution (pH 7.4) and 25 mg/1 hypochlorite in acidified sewage (pH 5.0).

Other studies[42] show that decreasing the pH from 7.0 to 6.0 reduced the required virus inactivation time by about 50 percent and that a rise in pH from 7.0 to 8.8 or 9.0 increased the inactivation period about six times. In the presence of low free chlorine residuals, the virus inactivation rate is markedly affected by variations of temperature and pH. In the destruction of viruses by chlorine, Clarke's work suggests that the temperature coefficient for a 10°C change is in the range of 2–3, indicating that the inactivation time must be increased 2–3 times when the temperature is lowered 10°C. Data also indicate that the chlorine concentration coefficient lies in the range of 0.7–0.9. This means that the inactivation time is reduced a little less than half when the free chlorine concentration is doubled. To increase virus kill, therefore, there is some advantage in increasing the contact time instead of raising the chlorine content.

The effects of chlorine on a mixture of cysts consisting predominantly of *Entamoeba histolytica* and *Entamoeba coli* has been reported.[43] The cysticidal efficiency was related to the residual concentration of halogen measured chemically after 10 minutes of contact time. It was concluded that the chemical species of free and combined halogens which prevail in low pH waters were superior cysticides when compared to forms which predominate in waters of high pH. The most rapid acting cysticide was found to be the hypochlorus acid (HOCl) form of free available chlorine.

The AWWA Committee report on virus in water[5] concluded that "in the prechlorination of raw water, any enteric virus so far studied would be destroyed by a free chlorine residual of about 1.0 ppm, provided this concentration could be maintained for about 30 minutes and that the virus was not embedded in particulate material. In postchlorination practices where relatively low chlorine residuals are usually maintained, and in water of about 20°C and pH values not more than 8.0–8.5, a free chlorine residual of 0.2–0.3 ppm would probably destroy in 30 minutes most viruses so far examined."

The Committee also states: "there is no doubt that water can be treated so that it is always free from infectious microorganisms—it will be biologically safe. Adequate treatment means clarification (coagulation, sedimentation, and filtration) followed by effective disinfection. Effective disinfection can be carried out only on water free from suspended material." Turbidities should be less than 1 JTU, or preferably less than 0.2 JTU.

Alternative Forms of Chlorine Available

For purposes of disinfection of municipal supplies, chlorine is primarily used in two forms: as a gaseous element or as a solid or liquid, chlorine containing (hypochlorite) compound. Gaseous chlorine is generally considered the least costly form of chlorine that can be used in large facilities. Hypochlorite forms have been used primarily in small systems (fewer than 5000 persons) or in large systems where safety concerns related to handling the gaseous form outweigh economic concerns.

Gaseous Chlorine.   In the gaseous state, chlorine is greenish yellow in color and about 2.48 times as heavy as air; the liquid is amber colored and about 1.44 times as heavy as water. Unconfined liquid chlorine rapidly vaporizes to gas; 1 volume of liquid yields about 450 volumes of gas.

Chlorine is only slightly soluble in water, its maximum solubility being approximately 1 percent at 49.2°F. Chlorine confined in a container may exist as a gas, as a liquid, or as a mixture of both. Thus, any consideration of liquid chlorine includes that of gaseous chlorine. Vapor pressure of chlorine in a container is a function of temperature, and is independent of the contained volume of chlorine; therefore, gage pressure is not an indication of container contents.

Chlorine gas is a respiratory irritant with concentrations in air of 3–5 ppm by volume being readily detectable. Chlorine can cause varying degrees of irritation of the skin, mucous membranes, and the respiratory system, depending on the concentration and duration of exposure. In extreme cases, death can occur from suffocation. The severely irritating effect of the gas makes it unlikely that any person will remain in a chlorine contaminated atmosphere unless he is unconscious or trapped. Liquid chlorine may cause skin and eye burns upon contact with these tissues; when unconfined in a container, it rapidly vaporizes to gas producing the effects given above.

Chlorine is shipped in cylinders, tank cars, tank trucks, and barges as a liquefied gas under pressure. Containers commonly employed include 100 and 150 lb cylinders; 1 ton containers; 16, 30, 55, 85, and 90 ton tank cars; 15–16 ton tank trucks; and barges varying in capacity from 55 to 1110 tons. Regulations limit the maximum filling of chlorine cylinders to 150 lb. Most cylinders are designed to withstand a test pressure of 480 psig. Ton containers, loaded to about 2000 lb of chlorine and having a gross weight as high as 3700 lb, are designed to withstand a test pressure of 500 psig. Single unit chlorine tank cars have 4 in. corkboard or self extinguishing foam insulation and are designed to withstand pressures of either 300 or 500 psig. Chlorination tank barges have either 4 or 6 tanks, each with a capacity of from 85 to 185 tons.

Hypochlorites.   Calcium hypochlorite, a dry bleach, has been used widely since the late 1920s. Present day commercial high test calcium hypochlorite products contain at least 70 percent available chlorine and have from less than 3 percent to about 5 percent lime.

Calcium hypochlorite is readily soluble in water, varying from about 21.5 g/100 ml at 0°C to 23.4 g/100 ml at 40°C. Tablet forms dissolve more slowly than the granular materials, and provide a fairly steady source of available chlorine over an 18–24 hour period.

Granular forms usually are shipped in 35 lb or 100 lb drums, cartons containing 3¾ lb resealable cans, or cases containing nine 5 lb resealable cans. Tablet forms are shipped in 35 lb and 100 lb drums, and in cases containing twelve 3¾ lb or 4 lb resealable plastic containers.

Commercial sodium hypochlorite usually contains 12–15 percent available chlorine at the time of manufacture, and is available only in liquid form. It is marketed in carboys and rubber lined drums of up to 50 gal volume, and in trucks. All NaOCl solutions are unstable to some degree and deteriorate more rapidly than calcium hypochlorite (see Table 6-2). The effect can be minimized by care in the manufacturing processes and by controlling the alkalinity of the solution. Greatest stability is attained with a pH close to 11.0, and with the absence of heavy metal cations. Storage temperatures should not exceed about 85°F; above that level, the rate of decomposition rapidly increases. While storage in a cool, darkened area very greatly limits the deterioration rate, most large manufacturers recommend a maximum shelf life of 60–90 days. All hypochlorite solutions are corrosive to some degree and affect the skin, eyes, and other body tissues that they contact.

Chlorine Dioxide.   Until the present time, at least, the use of chlorine dioxide has been limited to special water treatment applications: for oxidation of iron, manganese, and phenolic and chlorphenolic compounds; for control of certain algae; and for control of certain taste and odor problems.

TABLE 6-2   Stability of NaOCl Solutions.

Available Chlorine Trade Percent	Chlorine, g/l	Half-Life Days, 77°F
3	30	1700
6	60	700
9	90	250
12	120	180
15	150	100
18	180	60

Although the bactericidal properties of chlorine dioxide are about equal to those of chlorine, it is rarely applied solely for the purpose of disinfection. High cost and the lack of a satisfactory differential residual test at the low concentrations normally maintained in water treatment contribute to this condition.

However, chlorine dioxide possesses properties which, in the light of recent developments, may expand its field of application. Chlorine dioxide does not combine with ammonia in water solution, a property which may be important in wastewater disinfection. Perhaps more important, chlorine dioxide does not react with organic materials to produce chloroform, a potential carcinogen. The properties of chlorine dioxide must be described in two separate categories: (1) gas, and (2) aqueous solutions. Chlorine dioxide is almost never used commercially as a gas because of its explosiveness. For wastewater processes, it is only used in aqueous solutions. One of the most important physical properties is its extreme solubility in water—it is five times more soluble than chlorine. Paradoxically, it is extremely volatile and can be easily removed from aqueous solutions by a minimum of aeration.

Chlorine dioxide is entirely harmless in aqueous solution. Although it is readily soluble in water, it does not react chemically with water as does chlorine. For example, it is easily expelled from solution in water by blowing a small amount of air through the solution. For this reason it is not stable while in solution in an open vessel. The strength of the solution deteriorates rapidly in these circumstances. Exposure of the gas to light results in photochemical decomposition, but aqueous solutions of ClO_2 will retain their strength for several months if properly stored in the dark. The rate of photodecomposition from an ultraviolet light source is low as compared to decomposition from its volatility, described above.

The most popular method of generating chlorine dioxide is the chlorine-chlorite process. A sodium chlorite solution containing 1.34 pounds of sodium chlorite is reacted for one minute in a glass tower filled with porcelain Raschig rings with a chlorine solution containing 0.5 pounds of chlorine to produce 1.0 pound of chlorine dioxide. To obtain maximum ClO_2 yield in the short reaction time the pH of the mixture should be in the range of 2–4. In most waters that require an HOCl concentration of not less than 500 mg/l.

Smaller installations that use hypochlorite feeding systems can also utilize the chlorine–chlorite process for ClO_2 generation. This is achieved by acidifying the two solutions with sulfuric acid. Sulfuric acid must never come in contact with solid sodium chlorite, as an explosive reaction results.

The chemistry of chlorine dioxide in water treatment is not completely understood. In many instances it has been thought of as a more power-ful oxidant than chlorine, but this is not so in the narrow pH ranges encountered in wastewater treatment.

Although chlorine dioxide does not belong to the family of "available chlorine" compounds (those chlorine compounds that hydrolyze to form hypochlorous acid), nevertheless, the oxidizing power of chlorine dioxide is referred to as having an "available chlorine" content of 263 percent. In effect this indicates that chlorine dioxide theoretically has about 2.5 times the oxidizing power of chlorine. However, the oxidizing capacity of chlorine dioxide is not all used in treatment practice because the majority of its reactions with substances in water only cause the chlorine dioxide reduction to chlorite; actually chlorine is a better oxidant than chlorine dioxide except under special conditions. One of these special conditions is the reaction of chlorine dioxide with phenols. In this instance chlorine dioxide utilizes its full oxidizing potential of 263 percent "available chlorine." Although chlorine dioxide is extremely soluble in water, it does not react chemically with water and does not dissociate or disproportionate. Another important reaction occurs in the presence of free available chlorine. The residual seems to disappear slowly, owing to the oxidation of the ClO_2 by the free chlorine to chlorate; this is a common situation in waterworks practice.

One chemical property of chlorine dioxide that is unique in water treatment is that it will not react with ammonia. However, since it has potentially a high oxidizing power it does react chemically in some way with organic matter. Waters having a chlorine demand from substances other than ammonia nitrogen will also display a chlorine dioxide demand, with a tendency to consume proportionally greater amounts of chlorine dioxide. It is generally accepted that the predominant reaction of chlorine dioxide in water treatment is the reduction to chlorite. Chlorine dioxide may be completely converted to chlorite within 35 minutes. The chemical relationship of chlorine to the chlorite ion may be of considerable importance in water treatment. The chlorite ion may be toxic to humans. Another important property of chlorine dioxide is that its efficiency is not impaired (as is that of HOCl) by a high pH environment. Since chlorine dioxide is made on site from chlorine and sodium chlorite, there exists the possibility to utilize existing chlorination equipment in any conversion to a chlorine dioxide system.

Comparative Chlorine Costs.   The cost of chlorine varies with the quantity used and the point of manufacture and the point of use. Average January, 1976 chlorine cost figures are given below:

Description	Approximate cost
Chlorine in tank cars	$0.05/lb Cl_2
Chlorine in 1 ton cylinders	0.11/lb Cl_2
Chlorine in 150 lb bottles	0.24/lb Cl_2
Sodium hypochlorite in 55 gal drums (14 percent available Cl_2)	0.58/lb Cl_2
Calcium hypochlorite (65 percent available Cl_2) in 50 lb drums	1.55/lb Cl_2
Chlorine dioxide (1.34 lb sodium chlorite @ $1.20/lb and 0.5 lb chlorine @ $0.11/lb)	1.67/lb ClO_2

On Site Hypochlorite Generation

Recently, increased consideration is being given to on-site generation of hypochlorite because:

Chlorine gas, although economical, is risky to store and use. This is true especially in populated areas where the release of a 50 ton tank, for example, could require the evacuation of a 5 square mile area. In order to eliminate the risks inherent in chlorine storage and use, some large cities, such as New York, Chicago, and Providence, have shifted to the use of concentrated solutions of sodium hypochlorite.

Municipal water and wastewater facilities have had difficulty in obtaining chlorine because of national shortages.

Purchase of sodium hypochlorite solutions of 15 percent (maximum concentration used) involves transport and storage of large quantities of water. A solution to the storage and safety problem is to use a system which can produce at the treatment plant sodium hypochlorite as needed.

There have recently been introduced several commercial systems which can produce hypochlorite electrolytically from sodium chloride brine or seawater. These systems are being evaluated in major installations.

The following systems are available.[15-20]

Sanilec System (Diamond Shamrock).
Pepcon System (Pacific Engineering).
Chloropac System (Englehard Industries).
Cloromat System (Ionics).

The Sanilec system is an electrolytic process for the on site generation of sodium hypochlorite solution. This system basically involves the direct electrolytic conversion of sodium chloride brine or seawater to sodium hypochlorite solution.

In actual operation, diluted brine (3 percent chloride) is metered into the Sanilec cell, where it cascades through a series of compartmented electrode packs which convert the sodium chloride to sodium hypochlorite:

$$NaCl + H_2O \xrightarrow{e^-} NaOCl + H_2\uparrow$$

The degree of salt conversion to sodium hypochlorite increases during passage through the cell, reaching a final concentration of 0.8 ± 0.1 percent chlorine equivalent. The cell electrode is ruthenium coated titanium. From the cell, the sodium hypochlorite solution is then pumped directly to the point of use or to a hypochlorite storage tank until needed. Being quite stable, this electrolytically produced hypochlorite can be stored with little or no decrease in effective concentration.

The raw materials for this electrolytic process are salt, electric power, and water. The following are reported requirements to produce one pound of chlorine equivalent with a Sanilec system:

3.5 lb of salt
2.5 kwh of electrical power (AC)
15 gal of water

When calcium and magnesium are present in the diluted brine, both will form precipitates which can foul the system in brine conversion. Consequently, it is desirable that both the salt and the water be as hardness free as possible. Thus, the use of evaporated salt, solar salt, or southern rock salt is recommended rather than northern rock salt. In locations where softened, deionized, or tap water of sufficient quality is not available, water softeners with automatic regeneration cycles are included in the system.

The Sanilec system is illustrated in Figure 6-7. Concentrated brine is automatically produced in a salt dissolver which is designed for pneumatic salt delivery by truck. Water, introduced through headers approximately 3 ft above the bottom of the tank, dissolves the salt while passing through the salt bed to collectors on the bottom. As the salt dissolves, dry salt shifts down to replace it. The brine is then transferred to a brine storage tank either by gravity or by a brine transfer pump. The water feed to the automatic brine makeup system and the liquid level in the brine storage tank are controlled by automatic liquid level controls located in each unit. Concentrated brine is pumped by a metering pump to the cell inlet, where it is diluted to the proper concentration (3 percent) with water and discharged into the cell where it is electrolyzed. Low level electrodes in the brine storage tank will alarm and stop the pump in the event of insufficient brine supply.

Hypochlorite is generated continuously or intermittently, depending

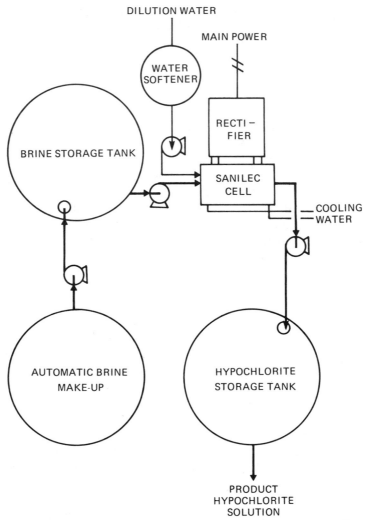

Fig. 6-7 Illustrative Sanilec hypochlorite generation system.[15]

upon the chlorination practices at the point of use. The main power circuit and transfer pumps are automatically activated or deactivated by electrical signals from level controls in the hypochlorite storage tank. No special maintenance attention is required other than periodic replacement of the electrodes, routine pump inspection, and supervision during salt deliveries. Safety features will alarm and shut down the cell to prevent possible malfunction. These are a low flow alarm on the inlet water dilution line, a thermal switch in the cell, and a high–low voltage control for the cell. Corrosion of nearby equipment from stray electric currents is prevented by grounding the inlet of the cell and allowing the cell effluent to discharge through a breaker box. Centrifugal pumps transfer the hypochlorite solution to a fiberglass holding tank with capacity for one day's supply. Upon signal from a chlorination cycle timer, an automatic valve on the tank outlet will open and discharge the hypochlorite by gravity into the intake of the condenser cooling water circulation pumps.

The Pepcon electrolytic cell for on site production of hypochlorite is shown in Figure 6-8. A metal tube, which serves as the cathode and container, is mounted on and supported by the cathode bus. The anode is a graphite rod electroplated with a dense, hard, impervious lead dioxide coating. The anode is inserted into the metal tubing and sealed in place with nonconducting supports and spacers. The graphite rod projects above the anode–cathode assembly and is electrically connected to the anode bus. The brine system is pumped into the cell at the bottom and out of the cell at the top. DC power is supplied from a suitable source.

The patented lead dioxide anode is chemically inert under power; it is relatively hard (about 5 on the MOH scale) and thus resists abrasion; and its electrical resistivity is low (40–50×10^{-6} ohm/cm). It is more conductive than mercury or graphite and lower in cost. Current efficiency in production of hypochlorite is 1.7 times that for graphite in 100 g/l sodium chloride

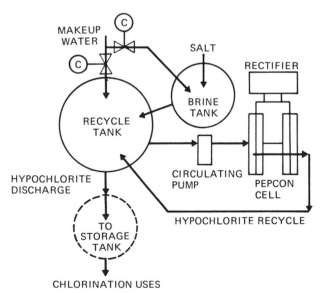

Fig. 6-8 **Pepcon hypochlorite generation system.**[17]

solution, using identical cell configuration and operating conditions (current density: 1 amp/in^2, 30°C).

The Pepcon cells are grouped into modules (generally 5000 amp). Fresh water and concentrated brine solutions are fed at predetermined rates to a recycle tank. A pump continuously recirculates the brine solution through the electrolytic cells where chloride ions are oxidized to hypochlorite. The hypochlorite solution overflows the recycle tank either directly to the stream being treated or to an intermediate storage tank. Seawater can also be used as the brine source.

The Chloropac System uses platinum plated titanium anodes. The bulk of experience with this system has been with seawater as the source of salt. About 1 lb equivalent chlorine is produced from each 120,000 gal salt water processed. Anode life of 5–7 years is claimed. This system is typically manufactured in cell modules of $\frac{1}{2}$ lb/hour or larger capacity. All materials are PVC or titanium and are corrosion resistant. The cells are arranged in series with as many as 20 cells in an electrical series. The system is usually operable from a central control panel and is designed for automatic operation. Automatic shutdown features are incorporated to secure the power in case of insufficient flow through any cell.

The Cloromat system uses a membrane cell developed by Ionics. The cell is illustrated in Figure 6-9. In the cell, brine solution is fed into the anode compartment where the chloride ion is discharged at the anode, forming chlorine gas:

$$2Cl^- \longrightarrow Cl_2\uparrow + 2e^-$$

At the cathode, water is electrolyzed and produces hydrogen gas and hydroxyl ions:

$$2H_2O + 2e^- \longrightarrow 2OH^- + H_2\uparrow$$

Current is carried through the electrolyte between the electrodes principally by the sodium ion (Na$^+$). A separator is placed between the electrodes. The purpose of the separator is twofold. First, it provides a barrier that prevents the hydrogen and chlorine gases from coming into contact with each other. Second, it provides a hydraulic barrier to prevent hydroxyl ions from migrating into the anode compartment.

In the Cloromat membrane cell the anode and cathode compartments are separated by a cation exchange membrane. This membrane must have good chemical resistance because it is exposed to caustic on one side and freshly generated chlorine on the other side. There is no direct hydraulic flow from the anode compartment to the cathode compartment. The only water that crosses the membrane is endosmotic water, which is associated with the ions being transferred. The anolyte liquor and the chlorine exit

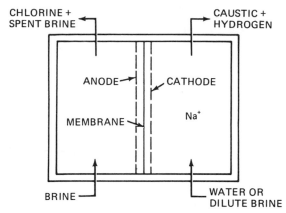

Fig. 6-9 **Membrane cell used in the Ionics, Inc. Cloromat system.**[16]

from a common port and are sent to a gas–liquid separator, where the product chlorine gas is separated from the depleted brine.

A complete Cloromat system is illustrated in Figure 6-10. Saturated brine is fed to the anode compartment. Chlorine gas is generated in the anode compartment and sent to a gas–liquid separator where the chlorine gas is separated from the depleted brine, which is discarded. Hydrogen

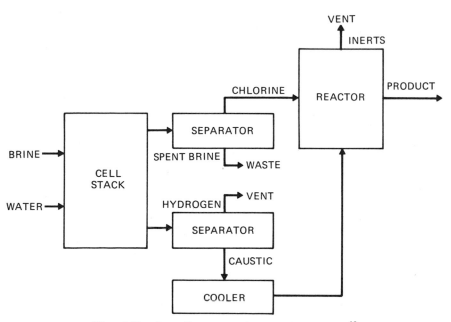

Fig. 6-10 **Overall system using membrane cells.**[19]

gas and hydroxyl ions are generated at the cathode. The effluent from this compartment goes to a gas–liquid separator in which the hydrogen is removed from the caustic solution. The hydrogen is diluted with air to a harmless level and vented to the atmosphere. Chlorine and caustic are fed to the reactor, with the caustic in slight excess. The reactor is water cooled to avoid decomposition of hypochlorite and formation of oxygen. Insoluble gases, mainly oxygen, are removed in a gas–liquid separator before the hypochlorite is sent to the product storage tank.

Brine feed for the system comes from the salt storage tank (lixator) which is used for both salt storage and production of saturated brine. Because of the crystal properties of sodium chloride, the bed of salt acts as a filter for the saturated brine. In applications where the hypochlorite is not used as it is produced, a storage tank may be provided for it.

The anode is an expanded titanium screen coated with a special metal-oxide coating. Anode life of 3 years is projected.[19] The cathode is constructed from a sheet of expanded mild steel. The membrane is Nafion, an ion exchange plastic membrane developed by DuPont. The membrane acts as a barrier in the cell, allowing sodium ions to pass into the cathode compartment but excluding chlorine gas and brine. It also prevents hydroxyl ions from migrating into the anode compartment, which would cause a loss in current efficiency by generation of oxygen gas. Membrane life of at least 2 years is projected.[19]

It is reported that the system produces hypochlorite with 8 percent available chlorine and requires the following to produce one pound of chlorine equivalent:

> 2.1 lb salt
> 1.7 kwh electrical power

These power and salt requirements are compared with those reported by other manufacturers below:

	Power, kwh/lb NaOCl	Salt, lb/lb NaOCl
Ionics Cloromat	1.7	2.1
Engelhard Chloropac	2.8	(seawater)
Pacific Engineering Pepcon	3.5	3.25
Diamond Shamrock Sanilec	2.5	3.5

Cost figures for one municipal wastewater installation for on-site chlorine generation are available as a result of construction bids and proposed maintenance contracts for the Orange County (California) Water District. Two

1000 lb/day Ionics units were bid for installation. The costs per pound of chlorine based on these data are summarized below:

	Cost/lb Cl_2
Amortization of capital costs	$0.032
Labor, O & M	0.021
Power, at 2¢/kwh	0.068
Salt (solar) at $20/ton	0.042
Total per pound of chlorine	$0.163

This is about 48 percent more than for chlorine delivered in 1 ton containers, but is only 37 percent of the cost of sodium hypochlorite delivered in truckload lots. Thus the safety in transport, storage, and feeding of chlorine which is provided by the use of sodium hypochlorite solution rather than gaseous chlorine can be provided at much lower cost by on site generation as compared to solution delivered in bulk by truck to the plant site.

The economics are attractive for on site generation if (1) seawater is available as a salt source or (2) the use of sodium hypochlorite rather than chlorine is dictated by safety considerations. To date, operating experience with the available systems has been good, and there are no known serious difficulties in operation or maintenance of existing installations. The reliability of on site generation systems appears to be as good as and probably better than that of chlorine supplies which may be subject to interruptions in production, allocation, or delivery.

The Importance of Mixing

A complete and uniform mixing of the chlorine with the wastewater to be disinfected is important. Many full scale chlorination facilities are inadequately designed with the chlorine applied directly to the chlorine contact basin, sometimes even without a diffuser. Flows in the contact basins are usually highly stratified, resulting in very little dispersion of the chlorine. Disinfection is one process for which any measurable short circuiting is completely ruinous to process efficiency. Attention to tank shape mixing and to proper baffling is critical.

Recent studies have evaluated the importance of rapid initial mixing in chlorination.[21] The researchers believed that prolonged segregation of the chlorine from the bacteria results in poorer performance. The bactericidal compounds are formed by the reaction of chlorine with nitrogenous and carbonaceous matter to form chlorine complexes. The speed of these reactions varies; some are quite fast while others are relatively slow. They concluded that the residuals initially formed are apparently much more bactericidal than the compounds formed later. They concluded that rapid initial mixing allows these more lethal residuals to come in contact with the bacteria and act as killing agents.[22].

Collins[21] also concluded that backmixing that occurs in a propeller mixed tank may contribute to poorer disinfection. It was noted that reactions may take place between the more bactericidal residuals initially formed upon entering the reactor and the more complex residuals which form over a longer time. Thus, these simple and more bactericidal compounds may be "destroyed" before they can act. Collins concluded that improved performance is provided by applying the chlorine to plug flow or tubular reactors designed to provide rapid initial mixing. Where mechanically stirred mixers are necessary, the chlorine solution should be applied to the stream ahead of the mixer. The two process streams (chlorine solution and water) would not be applied directly and separately to the mechanical mixer.

A subsequent study[22] was made on both a pilot scale and plant scale (2.4 mgd). To parallel the tubular reactor suggested by Collins, a diffusion grid was installed in the full scale plant in a 27 in. effluent pipe. The data from the plant scale study (which were consistent with the pilot scale tests) revealed that while rapid initial mixing provided improved performance, i.e., lower coliform survival ratios, at low values of Rt (R is the total chlorine residual, t is the contact time), a somewhat lesser degree of initial mixing produced the same results when t was increased with R maintained constant. Thus, in situations where a long pipeline is used to provide long contact time for disinfection, a reasonably good initial mixing device will probably produce the best obtainable results.

The advantages of the high initial kill which rapid initial mixing provides are of significant importance. For example, existing plants which have constant flow stirred tank reactors (CSTR) for chlorination could be improved by installing a chlorine injection device in the pipe, if one exists, leading to the chlorine contact unit. Use of tubular reactors also provides a much better residence time distribution than a CSTR. This is particularly important in chlorination because the coliform kill is so strongly time dependent. Use of a CSTR allows "short circuiting" of fluid particles with short chlorine contact times and thus a very high concentration of coliform organisms.

Chlorination Effect on Organics

There is much interest in the trace (concentrations less than 1 mg/l) organic compounds in drinking water, some of which may be the result of upstream wastewater chlorination. Much of this interest results from a 1974 study of New Orleans drinking water and the national publicity which followed. About the same time, other studies called attention to the presence in finished drinking water of some trihalomethanes (mostly chloroform) which were not found in the respective raw waters at the locations of study. Conclusions were drawn in both reports that the trihalomethanes were formed during chlorination.

Because of the interest in the various organic contaminants and the concern as to their significance, in 1975 the U.S. Environmental Protection Agency conducted a survey of 80 selected cities to measure the concentrations of six halogenated compounds in raw and finished water. The six included four trihalomethanes (chloroform), bromodichloromethane, dibromochloromethane, bromoform) suspected of being formed on chlorination, plus carbon tetrachloride and 1,2-dichloroethane, known contaminants at New Orleans, but not necessarily formed on chlorination. The occurrence of trihalomethanes in finished drinking water was demonstrated to be widespread and a direct result of the chlorination practice. No hard evidence was found in this regard with respect to 1,2-dichloroethane or carbon tetrachloride. Based on the survey results, the median finished water concentration of each compound would be about 21 μg/l of chloroform, 6 μg/l of bromodichloromethane, 1.2 μg/l of dibromochloromethane, and an amount less than the detection limit of bromoform. Although most of the finished waters tested demonstrated this decreasing order of concentration, this was not always the case. The range of chloroform concentrations was <0.1–311 μg/l; bromodichloromethane, none found (NF)–116 μg/l; dibromochloromethane, NF–100 μg/l; and bromoform, NF–92 μg/l.

The health significance of trihalomethanes produced during chlorination has not been completely evaluated at this time. In the future, if removal of trihalomethanes is deemed important for public health reasons, there are several approaches to reducing the potential contribution of wastewater discharges to trihalomethane concentrations in drinking water supplies. These approaches include: (1) reduction of precursor compound concentrations in wastewater by pH adjustment, coagulation, settling, filtration, and carbon adsorption; (2) changing disinfectant (to chloroamines, ozone, chlorine dioxide, etc.); (3) changing point of chlorine application; and (4) removing trihalomethanes after formation. At this time, changing the disinfectant without intensive research of public health ramifications could be a catastrophic step.

OTHER DISINFECTANTS

Ozone

Ozone has been used for the disinfection of water supplies since the beginning of the century when it was first applied to the treatment of water for the

City of Paris, France. There are now nearly 1000 installations in operation, primarily in Europe. There are 20 installations in Canada, where the largest is operating on drinking water supplied to the City of Quebec, treating flow rates up to 60 mgd.

With the advent of modern ozone generators, resulting in reduced installation and operating costs, and with the increased emphasis on higher degrees of wastewater treatment, ozone is receiving renewed attention in the wastewater treatment field. There are several ongoing or recently completed studies on the feasibility of ozone treatment for wastewater disinfection.

Because ozone is unstable, it cannot be stored, and it must be produced at its point of use by passing dry air or oxygen between two high potential electrodes to convert oxygen into ozone. Improvements in the technology of ozone production have bettered the reliability and economy of its generation. The development of dielectric, ozone resistant materials has simplified design and operation of ozone generators; in addition, new electronic designs have increased the efficiency of ozone production.

The potential advantages of using ozone are its high germicidal effectiveness, even against resistant organisms such as viruses and cysts; its ability to oxidize many organic compounds; and the fact that on decomposition, the only residual material is dissolved oxygen. No dissolved solids, such as chlorides, are added. Disinfecting action is effective over a wide temperature and pH range. Bactericidal and sporicidal action is more rapid with ozone than with chlorine, and only short contact periods are required.

The use of ozone does have disadvantages. Because it must be produced electrically as it is needed and cannot be stored, it is difficult to adjust treatment to variations in load or to changes in water quality with regard to ozone demand. The ozone process is less flexible than chlorine in adjusting for flow rate and water quality variations. As a result, ozone historically has been found most useful as a disinfectant for waters with low or constant ozone demand, such as groundwater sources. Moreover, although ozone is a highly potent oxidant, it is quite selective, and by no means universal in its action. Some otherwise easily oxidized substances, such as ethanol, do not react readily with ozone. Waters of high organic and algal content usually require thorough pretreatment to satisfy the ozone demand. The ozone–air mixture produced by necessary on-site generation is only slightly soluble in water, and production is complicated when temperature and humidity are high. No lasting residual disinfecting action is provided. Analytic techniques are not sufficiently specific or sensitive for ready and efficient control of the process. Electric energy requirements and capital and operating costs are higher than for chlorine.

There are three basic approaches to the generation and use of ozone in wastewater treatment: (1) generation from air; (2) generation from oxygen,

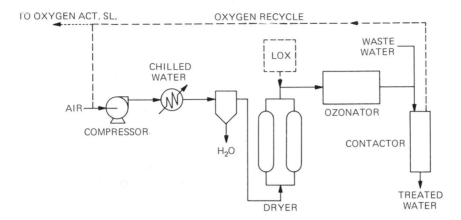

Fig. 6-11   Alternative ozone generation techniques.

with recycle of the oxygen which vents from the ozone contact basin to the ozone generation system; and (3) generation from oxygen in conjunction with an oxygen activated sludge system, with the oxygen venting from the ozone contactor being recycled as the source of oxygen for the activated sludge system. Figure 6-11 illustrates these alternatives schematically.

In the air feed system, air is the primary feed to the ozone generator. Air must be cleaned and dried to a $-60°F$ dewpoint to maximize generator output, since water vapor in the corona can adversely effect generation efficiency and may contribute to maintenance costs. The normal drying schemes involve compression and refrigerative drying to a constant dewpoint, followed by dessicant drying. Economic considerations in the drying system involve designing for the worst ambient condition as far as temperature and humidity of the air to the inlet of the dryer train. The system must also be designed to overcome any pressure drops through the dryers and ozonators in order to deliver the desired outlet pressure to the ozone contactor. Ozone is normally generated from air in the concentration range of 0.5–3.0 wt. percent. Power requirements for higher concentrations are prohibitive.

Another common system is one in which "pure" oxygen is fed to the ozonator. The source is indicated as LOX (liquid oxygen) in Figure 6-11. In this case, the oxygen rich contactor off-gases may be returned to the drying system for recycle. Makeup oxygen is added to the loop as required from an on-site LOX storage facility with vaporizer. With pure oxygen feed, ozone generators usually operate most efficiently at an output of 1.5–2.0 percent ozone. Production of up to 4 percent is technically feasible, but these higher concentrations entail increasingly severe economic pen-

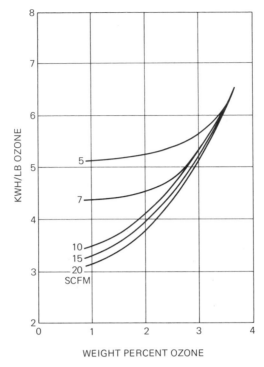

WEIGHT PERCENT OZONE

Fig. 6-12   Power requirements for ozone generation from a liquid oxygen source.[36]

alties. Figure 6-12 illustrates the effect of increasing the ozone concentration on power requirements as observed in a pilot facility at the Blue Plains plant.

The choice of using air or LOX as feed is an economic determination. Ozone generators produce 2–2.5 times as much ozone from oxygen per unit time and require only 50–40 percent of the power per pound that is used for generation from air. However, oxygen is not obtained without cost and is lost from system in several ways. The most obvious loss is through formation of ozone and subsequent loss during the oxidation processes which take place in the contactor. The second loss is as oxygen dissolved in the wastewater. At 25°C and 1 atm, the solubility of pure oxygen in water is 40 mg/l. In most situations it might be expected that dissolved oxygen (DO) in the effluent from an ozonation contactor will be 20–30 mg/l. This loss is a function of oxygen concentration, flow, water temperature and oxygen mass transfer efficiency in the contactor and is independent of ozone dosage.

The third major need for makeup oxygen comes from the requirement of maintaining a high (normally greater than 80 percent) concentration of oxygen in the recycle loop in order to maintain ozone generation efficiency. As the wastewater is treated with the high gas volumes used in ozonation, dissolved nitrogen is stripped from the wastewater and builds up in the recycle stream. Some method of purging the nitrogen diluted stream and adding oxygen to sustain a specified oxygen concentration is necessary. There are also other minor losses of oxygen such as the 0.5 percent per day normally allowed as boiloff from cryogenic storage vessels.

One more possibility exists which may be the most economical approach in many cases. This is the use of once through oxygen feed for ozone generation in conjunction with oxygen activated sludge treatment. In this system, oxygen from the on site supply for the activated sludge system is fed to the ozone generator. Since this oxygen is already clean and dry, no gas pretreatment equipment is necessary. The oxygen–ozone solution is then fed to the contactors, and the oxygen rich off gas used as a feed to the biological system (see Figure 6-11). The off gases from the oxygen activated sludge process contain little oxygen, so a highly efficient oxygen utilization is potentially achievable. However, as will become apparent later, in many instances the oxygen quantity needed for the required ozone dose for high degrees of disinfection may be significantly larger than that required by the activated sludge system. In this case, only a portion of the off gas from the ozone contactor is recycled to the activated sludge system and the rest of the oxygen is merely vented to the atmosphere. Under these circumstances, an inefficient use of the oxygen results, and the benefits of tying together the oxygen activated sludge system and ozone generation are reduced when compared to circumstances where the oxygen needs for ozonation and activated sludge treatment are in reasonable balance. A detailed evaluation of this balance is presented later.

Past research in the effectiveness of ozone as a germicide has been well summarized by Hann.[23] In general, ozone has been found to equal or exceed chlorine in its germicidal effects under a wide variety of circumstances. For example, ozone has been found to be many times more effective than chlorine in inactivating the virus of poliomyelitis. Under experimental conditions, an identical dilution of the same strain and pool of virus, when exposed to chlorine in residual amounts of 0.5–1.0 mg/l and to ozone in residual amounts of 0.05–0.45 mg/l was inactivated within 2 minutes by ozone, while $1\frac{1}{2}$–2 hours were required for inactivation by chlorine.

For many water reuse applications, it will be desirable to reduce coliform concentrations to less than 2.2/100 ml prior to recharge. Unfortunately,

TABLE 6-3 Reported Results—Wastewater Disinfection by Ozone.

Reference	Effluent Type	Ozone Dose, mg/l	Contact Time, min.	Coliform Most Probable Number/ 100 ml
24,25	unox. secondary	15	22	500 total 103 fecal
24,25	unox. secondary	5	?	114 fecal
26	microscreened secondary	6.6	8	480 total 100 fecal
26	secondary	15	5–22	774 total 153 fecal
26	unox. secondary	5	1	735 total 114 fecal
27	nitrified secondary	5	1.6	2200 total < 300 fecal
27	filtered secondary	5	1.6	257 total < 30 fecal
27	AWT	5	1.6	< 30 total < 30 fecal
27	secondary	7.5	?	20 total
26	secondary	5.4	?	0 fecal
26	secondary	4.8	?	1.3 fecal
26	secondary	4.2	?	5.7 fecal
26	secondary polishing pond	5	?	120-530 total < 4 fecal
28	secondary	50–60	10	33-1600 total
28	secondary	50–60	30	0-33 total
29	microscreened secondary	20	?	90 total
30	unox. secondary lime, filtered	20.6	11.6	< 3 total
30	three stage biological, lime, filtered	7	8.1	3 total
30	secondary, lime, coagulated	107	12.5	3 total
31	secondary	14	?	2 total

virtually none of the past research on wastewater disinfection by ozone has been oriented toward achieving a coliform concentration of less than 2.2/100 ml. Also, most of the work has been done with secondary effluents rather than AWT effluents. Table 6-3 summarizes recent data from several sources.

TABLE 6-4   Oxygen Available from Ozone Treatment for Disinfection.

Ozone dose = 15 mg/l
Flow = 5 mg/l
Ozone needs = 625 lb/day
Assume 3% Ozone
Oxygen + ozone feed = $\dfrac{625}{0.03}$ = 20,833 lb/day
Oxygen loss in raising DO of effluent by 35 mg/l = 1,751 lb/day
Oxygen in off gas from ozone contactor = 20,833 − 625 − 1,751 = 18,457 lb/day
Oxygen needs for activated sludge = 7,660 lb/day

Measurements of virus kill[32] in secondary effluent found that seeded coliphage f_2 was totally destroyed in five minutes in secondary effluent by 15 mg/l of applied ozone. Another study[33] of virus found that only 0.006 percent of poliovirus in secondary effluent survived one minute of contact with an initial ozone concentration of 5.05 mg/l. It was found that a threshold value for initial ozone concentration of about 1 mg/l ozone had to be surpassed before extremely high (greater than 99.99 percent) virus kills resulted.

Ozone has also been reported to be several times faster than chlorine in its germicidal effects on *Entamoeba histolytica*. For most cases, an ozone *residual* of 0.1 ppm for five minutes is adequate to disinfect water low in organics and free of suspended material.[23] Organic material does exert a significant ozone demand; therefore, the ozone dosage required to achieve this residual will be dependent upon the degree of pretreatment.

It appears that to achieve the very low coliform concentrations (less than 2.2/100 ml) that may be required in many AWT applications, ozone dosages of about 15 mg/l may be required.

To determine the potential merits of the combined use of oxygen activated sludge and ozone treatment, it is necessary to evaluate the relative quantities of oxygen required for each process to determine if they are in balance. Example calculations for a 15 mg/l ozone dose for a 5 mgd plant are shown in Table 6-4. The oxygen requirements for the activated sludge process are only 40 percent of the available oxygen in the off gas from the ozone contactors. Thus, a significant inefficiency in oxygen utilization results as some 11,200 lb/day of oxygen generated in the pure oxygen system would be wasted to the atmosphere. These 11,200 lb are, of course, available for recycle to the ozone generation system, but the advantage of combining the systems is largely lost because nitrogen, carbon dioxide, and water must now be removed from the recycled oxygen. If one were to increase the ozone to 4 percent, a balance with the activated sludge system needs would be struck at an ozone dose of 9 mg/l. These calculations indicate that, when standards call for 2.3 coliform/100 ml, the ozone oxygen requirements will exceed those of the activated sludge process substantially. When standards call for fecal coliform of 200/100 ml, the required oxygen for ozonation will be roughly in balance with the activated sludge needs, improving the benefits of the intertie.

Available Ozonation Systems.   W. R. Grace and Company of Columbia, Maryland produces equipment for on-site generation of ozone by the corona discharge technique. Grace uses an advanced ceramic dielectric with a high dielectric constant. This is applied to flat steel electrodes in an ultrathin layer by a special manufacturing process to ensure that no current leakage path exists. Grace's ozone generators utilize a patented solid state electronic circuit to provide high frequency power to the dielectrics at the low operating voltage required for long dielectric life. Grace units claim high efficiency due to a patented cell design which incorporates a narrow discharge gap, a thin dielectric with a high dielectric constant, cooling of both electrode surfaces, and uniform gas distribution in the corona. Grace equipment can be cooled using ambient air. Grace can supply modular generation units with capacities ranging from 2 to 2,500 lb/day per unit on oxygen feed. Grace gas preparation systems generally include a compressor, an aftercooler, a refrigerant dryer, a desiccant dryer, and various equipment items to remove contaminants from the ozonation process gas stream when oxygen recycle is utilized.

The Welsbach Corporation of Philadelphia, Pennsylvania is a pioneer in the manufacturer and supplier of ozonation equipment in sizes ranging from 3.7 lb/day to 9,400 ft^3/day (at atmospheric pressure).

The Cochrane Division of the Crane Company, King of Prussia, Pennsylvania, markets the C.E.O. ozonator of French manufacture, employing the M. P. Otto process. The generators are of the central air collector-flat electrode type. C.E.O. also makes an ozonizer with horizontal rectangular plates for service at atmospheric pressure or under higher pressures.

The PCI (Pollution Control Industries) Ozone Corporation of Stamford, Connecticut manufactures and supplies ozone equipment. They make two series of generators suitable for municipal and industrial service which have capacity ranges of (1) 2–20 lb/day from air and 4–40 lb/day from oxygen or from oxygen enriched air (40 percent O_2); and (2) 50–500 lb/day from dry air and 100–1000 lb/day from oxygen or oxygen enriched air (40 percent O_2). Design features of the PCI equipment include direct liquid cooling of the two electrodes and dielectric of the electric discharge field, leading to low discharge gas temperatures to avoid thermal decomposition of the ozone produced; high frequency (2000 cps) and low voltage (~10,000 V rms) operation using solid state proprietary frequency inverters.

Telecommunications Industries, Inc., Lindenhurst, New York, produces and markets equipment utilizing the Sonozone Process, which incorporates physical-chemical treatment with a combination of ultrasonics and ozone for disinfection. It is claimed that the action of the sonics and ozone exceeds the individual actions of the two processes. Sonics disrupts aggregation of yeast, bacteria, molds, and virus and thereby increases surface areas for oxidation by ozone. Units are available for treating flows of 0.01–30 mgd.

Costs. The cost estimates available in the literature for ozonation systems are nearly all based on dosages of 5 mg/l or less. For example, one estimate compares the cost of 5 mg/l ozone with 8 mg/l chlorine for a 64 mgd plant.[34] The 1971 cost of ozone (using pure oxygen) was 1.30/1000 gal as compared to 0.49 ¢/1000 gal for liquid chlorine and 0.90 ¢/1000 gal for sodium hypochlorite. The Blue Plains pilot plant report[35] estimates the annual cost to produce and provide ozonation with 750 lb/day of ozone at $88,000, including amortization. If one translates this to a 15 mg/l dose for 5 mgd, the equivalent cost is about 4 ¢/1000 gal. When one considers that the previous estimate for 5 mg/l was 1.3 ¢/1000 gal on a 64 mgd scale, then 4 ¢/1000 gal for 15 mg/l on a 5 mgd scale appears reasonable. Another check is offered by the estimate[36] that the cost for a 5 mg/l dose of ozone for a 10 mgd plant in which there is a perfect balance between an oxygen activated sludge system and the ozone system will be 1 ¢/1000 gal.

Iodine

Iodine possesses the highest atomic weight of the four common halogens, is the least soluble in water, with a solubility in water of 339 mg/l, and is the least hydrolyzed by water; it has the lowest oxidation potential, and reacts least readily with organic compounds. Taken collectively, these characteristics mean that low iodine residuals should be more stable and, therefore, persist longer in the presence of organic (or other oxidizable) material than corresponding residuals of any of the other halogens.

Iodine reacts with water in the following manner:

$$I_2 + H_2 \rightleftharpoons HIO + H^+ + I^-$$
$$HIO \rightleftharpoons H^+ + IO^-$$

At pH values over 4.0, the hypoiodous acid (HIO) undergoes dissociation, as shown in the second equation. At pH values over 8.0, HIO is unstable and will not form the hypoiodite ion, but will decompose according to the following equation:

$$3HIO + 2OH^- \rightleftharpoons HIO_3 + 2H_2O + 2I^-$$

This results in the formation of iodide (I^-) and iodate (in the form of HIO_3).

As the pH increases from 6.0 to 8.0, the percent of iodine as I_2 decreases while the percent of iodine as HIO increased. At pH 5.0 almost all the iodine is in the form of I_2. However, at pH 8.0, only 12 percent is present as I_2 while 88 percent is present as HIO and only 0.005 percent is present IO^-.

Some concern has arisen over the possible harmful effects of iodinated water on thyroid function, on possible sensitivities to iodine itself, and objectionable taste and odors. However, field analyses of human users of water treated with iodine have failed to disclose any adverse effects on health. Results from the study by Black and co-workers[37,38] indicated that few subjects, if any, were able to detect by either color, taste or odor a concentration of 1.0 mg/l elemental iodine in water. Black also indicated that the iodide ion cannot be detected in concentrations exceeding 5.0 mg/l. When the concentration of iodine was between 1.5 and 2.0 mg/l, many people were able to detect a taste but found it unobjectionable.

Since iodine does not combine with many organic compounds, the production of tastes and odors is minimized. Black states that iodine does not produce any detectable taste or odor in water containing several parts per billion phenol, whereas chlorine produces highly objectionable taste and odor under similar conditions.

Research into the bactericidal effects of iodine have shown only a slight difference between pH 5.0 and pH 7.0. At pH values of 9 or above, a marked reduction in efficiency occurs. Apparently, hypoiodous acid formed at higher pH values is less bactericidal than free iodine. At pH 7.0, it has been found[39] that 99.99 percent kill of the most resistant of 6 bacterial strains tested could be achieved in 5 minutes with a free iodine residual of 1.0 mg/l.

In general, it has been found that the bactericidal action of iodine resembled that of chlorine with respect to temperature and pH, but higher concentrations of iodine were required to produce comparable kills under similar conditions. However, toward higher organisms, and especially toward cysts and spores, iodine and bromine may have some advantages.[40] Rapid (5–10 minutes) destruction of cysts with iodine concentrations of 5–10 mg/l have been reported for several different cysts in different neutral waters.

Figure 6-13 demonstrates the relative efficiencies of hypochlorous acid and elemental iodine in the destruction of two Coxsackie viruses. With equal weight concentrations of disinfectant, the iodine took 200 times longer to produce the same amount of destruction.

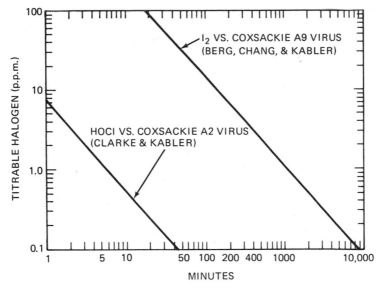

Fig. 6-13 Concentration–time relationship for 99 percent destruction of two Coxsackie viruses by HOCl and I$_2$ at about 5° C.[42]

HIO (hypoiodous acid), which results from the hydrolysis of I$_2$ as the pH of solution increases above 6, destroys viruses at a rate considerably faster than that achieved by I$_2$. HIO$_3$ and I$^-$ are essentially inert as viricides and I$_2$ has little or no viricidal activity.

In summary, iodine's great advantage over chlorine is that it does not react with ammonia or similar nitrogenous compounds. As a result, the bactericidal and cysticidal potency of I$_2$ and the viricidal effectiveness of HIO are not adversely affected by the presence of such nitrogenous pollutants, whereas the action of a given applied concentration of chlorine is badly affected either by exertion of demand or formation of halamines. Moreover, by appropriate control of dosage, pH, and other manageable conditions, aqueous iodine will occur about 50 percent as I$_2$ and 50 percent as HIO, thus providing a broad spectrum of germicidal capability. However, iodine seems unlikely to become a municipal disinfectant in any broad sense because of its high cost and restricted availability when compared to chlorine.

Bromine

As well summarized by Morris,[47] bromine exhibits chemistry in water qualitatively similar to that of chlorine—hydrolyzing to HOBr, ionizing to OBr$^-$, and reacting with ammonia to form bromamines. The major differences are that the same degrees of ionization for HOBr occur at pH values about one pH unit greater than for HOCl and that bromamines appear to be nearly as active germicidally as HOBr. At the same time, the bactericidal and viricidal effectiveness of HOBr has been found to be comparable to that of HOCl. As a result, solutions of free aqueous bromine exhibit a sharp decline in germicidal effectiveness with increasing pH only at pH greater than 8.5, whereas the effectiveness of free aqueous chlorine falls off sharply at pH greater than 7.5. In addition, the bactericidal and viricidal potency of bromine is reported to change in only a minor way in the presence of nitrogen in the form of ammonia, although aqueous bromine does undergo a rapid breakpoint process with ammonia, so that rapid bromine demand equal to about 20 times the ammonia-N may occur. However, the bromamine formed is not stable and does not present the fish toxicity problems of stable chloramines.

In spite of its advantages, it seems unlikely that bromine will become a general substitute for chlorine in municipal disinfection. Its much greater cost and scarcity are strong drawbacks. Moreover, there have been no thorough investigations on a plant scale that would bring out engineering problems.

Effluent Storage

Although not often thought of as a treatment process, storage of wastewater effluents can provide significant degrees of disinfection. The die off rate of viruses in the aquatic environment is a complex phenomenon which is dependent upon many factors.[48] These rates vary with the type of virus as well as water temperature, nature of the water, and chemical characteristics of the water. Under otherwise identical conditions, virus survival is inversely related to temperature. Also, the die off is directly related to the time the virus is in the water. Factors other than temperature and time have not been well defined. Studies have been made on virus survival in sewage stabilization ponds. Poliovirus type 3 was found to be reduced by 99.99 percent in 96 hours in one case and by 99.9 percent in 72 hours in another. Still another study found a 99.5 percent reduction in poliovirus type 1 in 72 hours. Virus reductions of 99.9 percent were measured in detention times of 80–160 hours with poliovirus type 1 in another set of experiments. A review of the loss of infectivity of enteric viruses when suspended in various aquatic environments showed the time required for a 99.9 percent reduction at water temperatures of 15–16°C varied from 7 to 56 days, but at temperatures of 20–25°C, the time range was 2–28 days, with most investigators reporting a 99.9 percent reduction in several types of enteric virus in 8 days or less.

SYSTEM REQUIREMENTS FOR DISINFECTION

The complete disinfection of wastewater is dependent upon optimizing the unit processes of chemical coagulation, sedimentation, and filtration (which in themselves provide substantial reductions in bacteria and viruses) to produce minimum water turbidities in order to assure the maximum contact between any remaining pathogens and the disinfectant added.

In AWT plant design utilizing effluent coagulation and filtration, provisions can be made so that a coagulant, alum, may be added continuously to the filter influent water as a filter aid following high lime treatment, so that maximum filterability is assured. At times, a polymer may also be applied at this point. Several investigators have pointed out the close relationship between virus removal and turbidity reduction. The mixed media filters at the South Lake Tahoe AWT Plant have under test produced water with turbidity as low as 0.02 JTU. Ordinarily they are operated to produce a turbidity of about 0.2 JTU. The excellent clarity obtainable with proper coagulation and filtration of the wastewater enhances chlorination efficiency so that viruses cannot escape chlorine contact by being encapsulated in particulate matter. Activated carbon treatment used for refractory organic removal removes chlorine (or ozone) demanding substances, further increasing the efficiency of disinfection.

The high pH of 11.0 when lime treatment of secondary effluent is used also benefits virus removal in addition to its bactericidal effects. Recent tests by EPA indicated that pH had little effect on virus at pH values of 10 or below. However, lime coagulation to raise the pH to 10.8–11.1 rapidly increased the rate of virus destruction. In addition to all of the above treatment considerations, AWT plant designs may also include breakpoint chlorination. The free chlorine residuals provided by this process are more effective disinfectants than the combined forms typically found in chlorinated wastewaters. Thus, even a higher degree of assurance of pathogen removal may be provided.

The South Lake Tahoe AWT plant records provide a full scale, long term reflection of the overall capability of a system utilizing lime coagulation of secondary effluent followed by recarbonation, granular filtration, carbon adsorption, and chlorination. The efficiency of the plant for bacterial removal is shown in Table 6-5. Of the 1402 samples examined between 1972–1975, 96.5 percent had coliform concentrations of less than 2.2/100 ml. The entire year of 1975 produced no detectable coliforms. Unfortunately, there are only limited data on virus in the Tahoe effluent. As shown in Table 6-6, the available data show no detectable virus in the final effluent.

TABLE 6-5   Summary of Bacteriological Tests at South Lake Tahoe, 1972–1975.

Time period	Total No.	Less Than 2.2	2.2	5.0	8.8	24	38	More Than 38
NUMBER OF SAMPLES WITH COLIFORM MOST PROBABLE NUMBER/100 ML OF								
1972								
January	31	25	3	1	1	0	0	1
February	29	23	5	0	1	0	0	0
March	30	22	6	1	0	0	1	0
April	30	28	1	0	1	0	0	0
May	31	31	0	0	0	0	0	0
June	29	29	0	0	0	0	0	0
July	31	29	0	0	2	0	0	0
August	31	31	0	0	0	0	0	0
September	30	30	0	0	0	0	0	0
October	31	30	0	0	1	0	0	0
November	30	25	0	0	2	1	2	0
December	31	31	0	0	0	0	0	0
1973								
January	31	26	0	1	0	2	2	0
February	28	28	0	0	0	0	0	0
March	31	31	0	0	0	0	0	0
April	30	30	0	0	0	0	0	0
May	31	31	0	0	0	0	0	0
June	30	30	0	0	0	0	0	0
July	31	24	3	2	2	0	0	0
August	31	31	0	0	0	0	0	0
September	30	30	0	0	0	0	0	0
October	31	31	0	0	0	0	0	0
November	30	30	0	0	0	0	0	0
December	27	26	0	0	1	0	0	0
1974								
January	26	26	0	0	0	0	0	0
February	28	28	0	0	0	0	0	0
March	31	31	0	0	0	0	0	0
April	30	30	0	0	0	0	0	0
May	31	31	0	0	0	0	0	0
June	29	29	0	0	0	0	0	0
July	31	31	0	0	0	0	0	0
August	31	31	0	0	0	0	0	0
September	30	28	0	0	2	0	0	0

TABLE 6-5 (*Continued*)

Time period	Total No.	Less Than 2.2	2.2	5.0	8.8	24	38	More Than 38
		NUMBER OF SAMPLES WITH COLIFORM MOST PROBABLE NUMBER/100 ML OF						
October	31	28	0	0	1	0	1	1
November	30	30	0	0	0	0	0	0
December	31	31	0	0	0	0	0	0
1975								
January	31	31	0	0	0	0	0	0
February	28	28	0	0	0	0	0	0
March	31	31	0	0	0	0	0	0
April	30	30	0	0	0	0	0	0
May	31	31	0	0	0	0	0	0
June	27	27	0	0	0	0	0	0
July	31	31	0	0	0	0	0	0
August	26	26	0	0	0	0	0	0
September	28	28	0	0	0	0	0	0
October	30	30	0	0	0	0	0	0
November	26	26	0	0	0	0	0	0
December	30	30	0	0	0	0	0	0
Totals	1402	1354	18	5	14	3	6	2

TABLE 6-6 Virus Tests on The South Lake Tahoe Effluent.

Date	Primary Effluent	Secondary Effluent	Column Effluent (U)	Final Effluent (C)
	VIRUS RECOVERED, PFU			
May 29	3	0	1	0
June 5	—	—	0	0
June 12	—	—	0	0
Aug. 20	3	18	NRD	0
Aug. 27	—	—	NRD	0
Sept. 11	—	—	0	0
Sept. 18	179	14	9	0
Sept. 25	NRD	430	0	0
Oct. 2	207	320	0	0

PFU = Plaque forming units; U = unchlorinated carbon column effluent; C = chlorinated carbon column effluent; NRD = no reliable data.
Tests performed by the EPA Laboratory in Cincinnati.

The Windhoek, South West Africa plant provides another example of full scale disinfection capabilities. In this project, the subject of several published papers,[49-53] reclaimed water is taken directly from maturation (or holding) ponds to the public water system and comprises as much as one-third of the total municipal water supply. The secondary effluent is reclaimed for reuse by chemical (lime) treatment, flotation, ammonia stripping, recarbonation, filtration, foam fractionation, and granular carbon adsorption. The water supply sources consist of an impoundment reservoir, 36 wells, and the reclaimed wastewater. The dam water is treated in a conventional plant while the well water is pumped directly to the distribution reservoirs. The reclaimed water is mixed with the purified dam water in the clearwell. The admixed streams are chlorinated to a free residual of 0.2 mg/l and then pumped to the service reservoir where they are mixed with the borehole water before final distribution. Although enteroviruses and reovirus entered the reclamation plant at levels as high as the $TCID_{50}$ (50 percent tissue culture infective dose) per liter, no virus could be recovered from the reclaimed wastewater.[53]

REFERENCES

1. Cooper, R. C., "Wastewater Contaminants and Their Effect on Public Health, A State-of-the-Art Review of Health Aspects of Wastewater Reclamation for Groundwater Discharge," State of California, November 1975.

2. Culp, R. L., "Disease Due to Nonpathogenic Bacteria," *Journal American Water Works Assoc.*, 1969, p. 157.

3. "Engineering Evaluation of Virus Hazard in Water," *Journal Sanitary Engineering Division, ASCE*, 1970, p. 111.

4. Gelderich, E. E., and Clarke, N. A., "The Coliform Test: A Criterion for Viral Safety of Water," Proceedings of the 13th Water Quality Conference, Urbana, Illinois (1971).

5. Committee Report, "Viruses in Water," *Journal American Water Works Assoc.*, 1969, p. 491.

6. Weibul, S. R.; Dixon, F. R.; Weidner, R. B.; and McCabe, L. J., "Waterborne-Disease Outbreaks, 1946–60," *Journal American Water Works Assoc.*, 1964, p. 947.

7. Kerr, R. W., and Butterfield, C. T., "Notes on the Relationship Between Coliforms and Enteric Pathogens," *Public Health Reporter*, 1943, p. 589.

8. Geldreich, E. E., "Applying Bacteriological Parameters to Recreational Water Quality," *Journal American Water Works Assoc.*, 1970, p. 113.

9. Mechelas, J. M., et al., "An Investigation into Recreational Water Quality," in *Water Quality Data Book*, Vol. 4. EPA 18040 DAZ 04/72, U.S. Gov't. Printing Office, Washington, D.C., 1972.

9a. Fair, G. M., and Geyer, J. C., *Elements of Water Supply and Waste-Water Disposal*, Wiley & Sons, New York, 1958.

10. Berg, G., *Removal of Viruses from Water and Wastewater*. Dept. of Civil Engineering, University of Illinois, 1971.

11. Wolf, H. W., et al., "Virus Inactivation During Tertiary Treatment," *Journal American Water Works Assoc.*, September, 1974, p. 526.

12. Robeck, G. G.; Clarke, N. A.; and Dostal, K. A., "Effectiveness of Water Treatment Processes in Virus Removal," *Journal American Water Works Assoc.*, 54, 1962, p. 1275.

13. Hudson, H. E., Jr., "High Quality Water Production and Viral Disease," *Journal American Water Works Assoc.*, 54, 1962, p. 1265.

14. Walton, G., Effectiveness Water Treatment Processes as Measured by Coliform Reduction, U.S. Public Health Service Publication 898 (1962).

15. "Sanilec Systems." Diamond Shamrock brochure.

16. "Cloromat." Ionics, Inc. brochure.

17. "Pep-Clor Systems." Pacific Engineering and Production Co. of Nevada brochure.

18. "Cheaper Sewage Treatment," *Chemical Week*, January 12, 1972, p. 52.

19. Michalek, S. A., and Leitz, F. B., "On-Site Generation of Hypochlorite," *Journal Water Pollution Control Federation*, September, 1972, p. 1697.

20. Ionics, Incorporated, "Hypochlorite Generator for Treatment of Combined Sewer Overflows." Report for EPA, Contract No. 14-12-490, March, 1972.

21. Collins, H. F., "Chlorination—How to Use It." Proceedings, Fifth Annual Sanitary Engineering Symposium, California Department of Public Health, May, 1970.

22. Stenquist, R. J., and Kaufman, W. J., "Initial Mixing Coagulation Processes," EPA Report EPA-R2-72-053, November, 1972.

23. Hann, V. A., "Disinfection of Drinking Water with Ozone," *Journal American Water Works Assoc.*, October, 1956, p. 1316.

24. McBride, T. J., and Taylor, D. M., "Can Ozone Ever Replace Chlorine," *Water and Wastes Engineering*, May, 1973, p. 29.

25. Venosa, D. A., "Ozone as a Water and Wastewater Disinfectant: A Literature Review," in *Ozone in Water and Wastewater Treatment*, Ann Arbor Science Publishers, 1972.

26. Clark, R. G.; Lowther, F. E.; and Rosen, H. M., "Disinfection of Municipal Secondary-Tertiary Effluent with Ozone: Five Recent Pilot Plant Studies," Paper presented at the First International Ozone Symposium and Exposition, Washington, D.C., December, 1973.

27. Petrasek, A. C., Jr., et al., "The Impact of Ozone on Water Quality for Various Wastewaters." Paper presented at the First International Ozone Symposium and Exposition, Washington, D.C., December, 1973.

28. Kinman, R. N., "Ozone in Water Disinfection," *Ozone in Water and Wastewater Treatment*, Ann Arbor Science Publishers, 1972.

29. Boucher, P. L., et al., "Use of Ozone in the Reclamation of Water from Sewage Effluent," *Institute of Public Health Engineers Journal*, April, 1968, p. 1.

30. Wynn, C. S., et al., "Pilot Plant for Tertiary Treatment of Wastewater with Ozone," U.S. Environmental Protection Agency Report EPA-R2-73-146, January, 1973.

31. Nebel, C., "Ozone Treatment of Secondary Effluents." Unpublished paper printed by the Welsbach Corp.

32. Pavoni, J. L., et al., "Virus Removal from Wastewater Using Ozone," *Water and Sewage Works*, December, 1972, p. 59.

33. Majumdar, S. B.; Ceckler, W. H.; and Sproul, O. J., "Communication-Inactivation of Poliovirus in Water by Ozonation," *Journal Water Pollution Control Federation*, 1974, p. 2048.

34. Yao, K. M., "Is Chlorine the Only Answer?" *Water and Wastes Engineering*, January, 1972, p. 30.

35. Wynn, C. S., et al., "Pilot Plant for Tertiary Treatment of Wastewater with Ozone." U.S. Environmental Protection Agency Report EPA-R2-73-146, January, 1973.

36. "Costs of Ozone Disinfection." W. R. Grace and Co., 1973.

37. Black, A. P.; Kinman, R. N.; Thomas, W. C., Jr.; Freund, G.; and Bird, E. D., "Use of Iodine for Disinfection," *Journal American Water Works Assoc.*, November, 1965, p. 140.

38. Cookson, J. T., Jr., "Mechanism of Virus Adsorption on Activated Carbon," *Journal American Water Works Assoc.*, January, 1969, p. 52.

39. Scarpino, P. V., Berg, G., Chang, S. L., Dahling, D., and Lucas, M., "A Comparative Study of the Inactivation of Viruses in Water by Chlorine." *Water Research*, p. 959 (1972).

40. Kabler, P. W., Chang, S. L., Clarke, N. A., and Clark, H. F., "Pathogenic Bacteria and Viruses in Water Supplies," 5th Sanitary Engineering Conference, U. of Illinois, Jan., 1963.

41. Kruse, C. W., Olivieri, V. P., and Kawata, K., "The Enhancement of Viral Inactivation by Halogens," *Water and Sewage Works*, p. 187 (June, 1971).

42. Clarke, N. A., and Kabler, P. W., "The Inactivation of Purified Coxsackie Virus
in Water by Chlorine," *American Journal of Hygiene*, p. 119 (January, 1954).

43. Stringer, R., and Kruse, C. W., "Amoebic Cysticidal Properties of Halogens,"
Proceedings of the National Specialty Conference on Disinfection, American
Society of Civil Engineers, New York, N.Y. (1970).

44. Black, A. P.; Thomas, W. C., Jr.; Kinman, R. N.; Bonner, W. P.; Keirn, M. A.;
Smith, J. J., Jr.; and Jabero, A. A., "Iodine for the Disinfection of Water,"
Journal American Water Works Assoc., January, 1968, p. 69.

45. Karalekas, P. C., Jr.; Kuzminski, L. N.; and Feng, T. H., "Recent Developments in the Use of Iodine for Water Disinfection," *Journal New England
Water Works Association*, June, 1970, p. 152.

46. McKee, J. E.; Brokaw, C. J.; and McLaughlin, R. T., "Chemical and Colicidal
Effects of Halogens in Sewage," *Journal Water Pollution Control Federation*,
August, 1960, p. 795.

47. Morris, J., "Chlorination and Disinfection—State of the Art," *Journal American Water Works Assoc.*, December, 1971, p. 769.

48. Clarke, N. A.; Berg, G.; Kabler, P. W.; and Chang, S. L., "Human Enteric
Viruses in Water: Source, Survival and Removability," in *Advances in Water
Pollution Research, Proc. Int. Conf., London, 1962*, Vol. 2, Pergamon Press,
London, 1964.

49. Van Vuuren, L. R. J., et al., "The Full-Scale Reclamation of Public Sewage
Effluent for the Augmentation of the Domestic Supplies for the City of Wind-
hoek." Paper presented at the Fifth International Water Pollution Research
Conference, July, 1970.

50. Van Vuuren, L. R. J., et al., "The Flotation of Algae in Water Reclamation,"
International Journal Air and Water Pollution, 1965, p. 823.

51. Cillie, G. G., et al., "The Reclamation of Sewage Effluents for Domestic Use."
Paper presented at the Third International Conference on Water Pollution
Research, 1966.

52. Stander, G. J., et al., "The Reclamation of Wastewater for Domestic and Industrial Use." Paper presented at the Biennial Health Congress of the Institute
of Public Health, October, 1966.

53. Nupen, E. M., "Virus Studies on the Windhoek Wastewater Reclamation
Plant (South-West Africa)," *Water Research*, 1970, p. 661.

7

Nitrogen removal

The removal or control of nitrogenous matter in wastewater has assumed greater significance in recent years as a means of protecting and preserving the environment. Some of the environmental problems associated with the various forms of nitrogen are as follows:

Nitrogen compounds are nutrients and may cause undesirable algal growths.

NH_3 can be toxic to fish and other aquatic life.

NH_3 and organic nitrogen in effluents can cause a dissolved oxygen demand in receiving waters.

NH_3 is corrosive to certain metals.

NH_3 can have detrimental effects on disinfection of water supplies.

NO_3 can be a health hazard.

Processes for the removal of nitrogen are not as well established as those for phosphate. The varied forms of nitrogen and the different chemistry associated with each presents an interesting challenge to the scientist and engineer to develop successful inexpensive removal processes. Precipitation reactions which have been so successful for phosphate removal are not readily adaptable to ammonia or nitrate removal because of the relatively high solubility of salts containing ammonia or nitrate. There are four processes which are being applied on a plant scale for nitrogen removal.

1. Ammonia Stripping
2. Selective Ion Exchange
3. Biological Nitrification—Denitrification
4. Breakpoint Chlorination

This chapter will review each of these processes.

AMMONIA STRIPPING

Of the available methods for removal of nitrogen from wastewater, ammonia stripping is the simplest and the easiest to control. However, ammonia stripping has two serious limitations which the designer must recognize: (1) the practical inability to operate the process at ambient air temperatures below 32° F, and (2) the deposition of calcium carbonate scale from the water onto the stripping tower fill, which results in loss of efficiency from reduced air circulation and droplet formation, and may eventually completely plug the tower. In warm climates, the temperature limitation does not apply. In some locations in cold climates, such as on a flowing stream or river, nitrogen removal may not be required during the winter season. The other problem—that of scaling—appears to be susceptible to control or elimination, and possible methods for doing this will be discussed later.

Under the right climatic conditions and with the proper precautions regarding scale prevention or removal, ammonia stripping is a practical, reliable method for nitrogen reduction. In cold climates, if nitrogen removal is required during freezing weather, then ammonia stripping may be used in warm weather, and a supplemental method such as ion exchange provided for use in other seasons.

The Ammonia Stripping Process

Ordinarily more than 90 percent of the nitrogen in raw domestic wastewater is in the form of ammonia or compounds from which ammonia is readily formed. If certain environmental conditions are provided in the secondary treatment process, the ammonia will be converted to nitrate.

However, this nitrification process can be eliminated by maintaining a relatively high organic loading on the secondary process. It is cheaper and easier to maintain the nitrogen in the form of ammonia than it is to convert it to nitrate. Thus, the removal of nitrogen in the form of ammonia offers some economic and operational advantages over the removal of nitrate-nitrogen.

In wastewater, either ammonium ions (NH_4^+), or dissolved ammonia gas (NH_3), or both may be present. At pH 7, only ammonium ions in true solutions are present. At pH 12, only dissolved ammonia gas is present, and this gas can be liberated from wastewater under proper conditions. The equilibrium is represented by the equation

$$NH_4^+ \rightleftharpoons NH_3 + H^+$$

As the pH is increased above 7.0, the reaction proceeds to the right.

Figure 7-1 illustrates this relationship at various temperatures. Two major factors affect the rate of transfer of ammonia gas from water to the atmosphere: (1) surface tension at the air–water interface, and (2) difference in concentration of ammonia in the water and in the air. Surface tension is at a minimum in water droplets when the surface film is being formed, and ammonia release is greatest at this instant. Little additional gas transfer takes place once a water droplet is completely formed. Therefore, repeated droplet formation and coalescing of the water assists ammonia stripping.

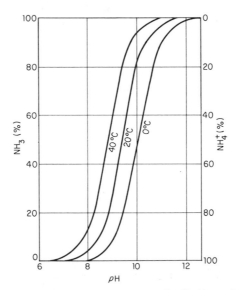

Fig. 7-1   Effects of pH and temperature on the distribution of ammonia and ammonium ion in water.

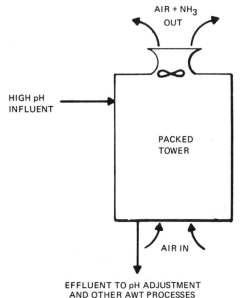

Fig. 7-2   Ammonia stripping process.

To minimize ammonia concentration in the ambient air, rapid circulation of air is beneficial. Air agitation of the droplets may also speed up ammonia release. The ammonia stripping process, then, consists of (1) raising the pH of the water to values in the range of 10.8–11.5, (2) formation and re-formation of water droplets in a stripping tower, and (3) providing air–water contact and droplet agitation by circulation of large quantities of air through the tower. Figure 7-2 schematically illustrates the process.[1]

Before addressing detailed design considerations, the general environmental impacts of the stripping process must be evaluated. It is obvious from Figure 7-2 that ammonia is being discharged into the atmosphere. Does the process solve a water pollution problem while creating an air pollution problem? What is the fate of the ammonia in the atmosphere? These questions must be satisfactorily addressed prior to the selection of air stripping for ammonia nitrogen removal.

Environmental Considerations

There are three major potential environmental impacts which must be evaluated if use of the ammonia stripping process is proposed: air pollution, washout of ammonia from the atmosphere, and noise. If these three concerns cannot be favorably resolved for a given situation, then the potential process advantages of simplicity and low cost may become only academic.

At an air flow of 500 ft^3/gal and at an ammonia concentration of 23 mg/l in the tower influent, the concentration of ammonia in the stripping tower discharge is about 6 mg/m^3. As the odor threshold of ammonia is 35 mg/m^3, the process does not present a pollution problem in this respect. Concentrations of 280–490 mg/m^3 have been reported to cause eye, nose, and throat irritation.[3] Concentrations of 700 mg/m^3 can have adverse effects on plants. Concentrations of 1700–4500 mg/m^3 must be reached before human or animal toxicities begin to occur. Ammonia discharged to the atmosphere is a stable material that is not oxidized to nitrogen oxides in the atmosphere.[2] Ammonia can react with sulfur dioxide and water to form an ammonium sulfate aerosol. However, for this consideration to be a limitation, the stripping tower would have to be located adjacent to a point source of sulfur dioxide.

The production and release of ammonia as part of the natural nitrogen cycle is about 50×10^9 tons per year. Roughly 99.9 percent of the atmosphere's ammonia concentration is produced by natural biological processes, primarily the bacterial breakdown of amino acids.[2,3] Although it is a relatively insignificant source, the burning of coal and oil produces measurable quantities of ammonia.[2] The background levels of ammonia in the atmosphere have been observed to vary from 0.001 mg/m^3 to 0.02 mg/m^3, with a value of 0.006 mg/m^3 being typical.[2]

Available diffusion technology can be used to estimate the atmospheric concentration of ammonia at any point downwind of the stripping tower.[4] Calculations were made for the Orange County, California stripping tower for low mixing conditions (wind speed 1 m/sec.). The resulting surface concentrations at the center of the downwind discharge zone including natural background levels were as follows.

Distance from Tower, ft	Surface Air Concentration of Ammonia, mg/m^3
300	5.2
1,000	1.6
1,600	0.6
3,200	0.2
16,000	0.0006

Background levels of ammonia are reached within 3 miles. No U.S. ammonia emission standards have been established by regulatory agencies because there are no known public health implications at concentrations normally encountered.

The American Conference of Governmental Industrial Hygienists recommended in 1967 an occupational threshold limit of 35 mg/m^3.[3] The permissible limit for ammonia in a submarine during a 60 day dive is 18 mg/m^3.[2]

The Navy's Bureau of Medicine and Surgery has recommended an ammonia threshold limit for 1 hour of 280 mg/m³. All of these values are above the 6 mg/m³ which will typically occur at the tower discharge. As noted above, no ambient air quality standards for ammonia exist for the United States. However, such ambient air standards exist for Czechoslovakia, the U.S.S.R., and Ontario, Canada, as shown below:[3]

Location	Basic Standard		Permissible	
	mg/m³	Averaging Time	mg/m³	Averaging Time
Czechoslovakia	0.1	24 hours	0.3	30 minutes
U.S.S.R.	0.2	24 hours	0.2	20 minutes
Ontario, Canada	3.5	30 minutes		

There is a large turnover of ammonia in the atmosphere, with the total ammonia content being displaced once per week on the average. Ammonia is returned to the earth through gaseous deposition (60 percent), aerosol deposition (22 percent), and precipitation (18 percent).[3]

Although not the most significant mechanism for removal of ammonia from the atmosphere, precipitation does provide one pathway for the return of atmospheric ammonia to bodies of water and to soil. In rainfall, the natural background ranges from 0.01 to 1 mg/l, with the most frequently reported values being 0.1–0.2 mg/l. The amount of ammonia in rainfall is directly related to the concentration of ammonia in the atmosphere. Thus, an increase in the ammonia in rainfall would occur only in that area where the stripping tower discharge increases the natural background ammonia concentration in the atmosphere. Calculations for the ammonia washout in the rainfall rate of 3 mm/hour (0.12 in./hour) have been made for the Orange County, California project with the following results:

Distance from Tower, ft	Peak Rainfall Ammonia Concentration, mg/l
300	60
1,000	18
1,600	11
3,200	5
16,000	0.5

The concentrations of ammonia in the rainfall would approach natural background levels within 16,000 ft of the tower. The ultimate fate of the ammonia which is washed out by rainfall within this 16,000 ft downwind distance depends on the nature of the surface upon which it falls. Most soils will retain the ammonia. That portion which lands on paved areas or directly on a stream will appear in the runoff from that area. Unless the stripping tower is located upwind in close proximity to a lake or reservoir, the direct return of ammonia to the aquatic environment by atmospheric washout should not make a significant contribution to the total ammonia discharged to the aquatic environment. However, this is a factor which must carefully be evaluated for each potential application.

There are three potentially significant noise sources in an ammonia stripping tower: (1) motors and fan drive equipment; (2) fans; and (3) water splashing. The following control measures are available:

Motors: proper installation, maintenance, and insulation.

Fans: reduction in tip speed and installation of exhaust silencers.

Water: shielding of the tower packing and air inlet plenum.

Based on sound level measurements from the tower at Lake Tahoe, the expected noise level at the tower is calculated to be about 64 dBA. This noise level can be reduced to 46 dBA at 600 ft from the towers by control measures. The Orange County project includes several specific noise control measures. Before construction of the plant, ambient nightime noise levels in the residential neighborhood around the Orange County plant were 40–45 dBA.

Design Considerations

The major factors affecting design and process performance include tower configuration, pH, temperature, hydraulic loading, tower packing depth and spacing, air flow, and control of calcium carbonate scaling. Detailed discussions of mass and enthalpy relationships and theoretical mathematical models of the stripping process are available in references 5, 6, and 7. However, these models are not normally used for stripping tower design, and an empirical design procedure is used.

Type of Stripping Tower.   There are two basic types of stripping towers now being used in full scale applications: countercurrent towers and crossflow towers (see Figure 7-3). Countercurrent towers (the entire airflow enters at the bottom of the tower, while the water enters the top of the tower and falls to the bottom) have been found to be the most efficient. In the crossflow towers, the air is pulled into the tower through its sides throughout the height of the packing. This type of tower has been found to be more prone to scaling problems.

pH.   The pH of the water has a major effect on the efficiency of the process. The pH must be raised to the point where all of the ammonium ion is con-

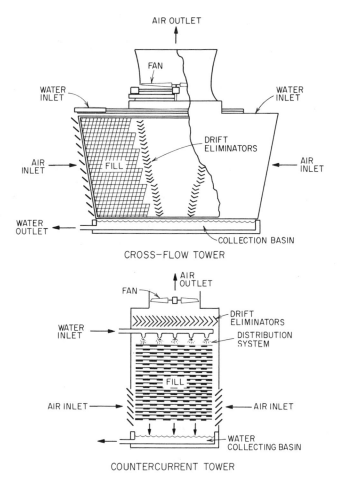

AIR OUTLET

FAN

WATER
INLET

WATER
INLET

DRIFT
ELIMINATORS

AIR
INLET

FILL

AIR
INLET

WATER
OUTLET

COLLECTION BASIN

CROSS−FLOW TOWER

AIR
OUTLET

FAN

DRIFT
ELIMINATORS

WATER
INLET

DISTRIBUTION
SYSTEM

FILL

AIR INLET

AIR INLET

WATER
COLLECTING BASIN

COUNTERCURRENT TOWER

Fig. 7-3 Types of stripping towers.

verted to ammonia gas. If phosphorus removal is required, the use of lime as the coagulant will generally enable the necessary pH elevation to be achieved concurrently with phosphorus removal. If pH elevation does not occur in some upstream process, the economics of the stripping process are adversely affected, since the costs of pH elevation must then be incurred solely for ammonia stripping.

Based on published data on the stripping process,[8-11] Slechta and Culp[12] evaluated the use of a small air conditioning tower, packed with redwood slats, as a pilot stripping tower. The packing of this tower was compact, having a depth of 2 ft and an area of 5 ft^2. The tower was equipped with a fan which drew 1520 cfm of air through the tower. This type of tower was used to evaluate the effect of aeration following pH adjustment, recognizing that this type of tower packing did not provide optimum air–water contact.

In each of several runs, the pH of 22 gal of chemically coagulated and filtered secondary effluent was adjusted with sodium hydroxide. This effluent was then recycled through the tower with frequent measurements of the ammonia-nitrogen content of the tower effluent being made. The pH of the wastewater (tower influent) was adjusted to values ranging from 8.0 to 10.8, and the flow to the tower was varied from 0.5 to 4 gpm (0.1–0.18 gpm/ft^2).

The efficiency of ammonia-nitrogen removal in the particular aeration tower used in these studies at various pH values is shown in Figure 7-4. Optimum ammonia removals at pH values above 9.0 were obtained at an aeration time of about 0.5 minute, or, with this particular tower, at an air–liquid loading of about 750 ft^3/gal. With the tower used in this study, the optimum aeration time corresponded to a tower depth of about 20 ft.

Approximately 94 percent of the nitrogen in the applied water was in the form of ammonia-nitrogen. Thus, the 98 percent ammonia-nitrogen removal obtained at a pH of 10.8 corresponds to a removal of about 92 percent of all nitrogen present. At a pH of 9.7, 93 percent of the ammonia-nitrogen, or 87 percent of all the nitrogen, was removed. Organic nitrogen was not affected by passage through the tower. Table 7-1 summarizes the ammonia-nitrogen content of the aeration tower influent and effluent at

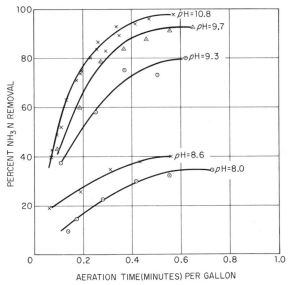

Fig. 7-4   Percentage ammonia removal vs. aeration time.[12]

**TABLE 7-1   Ammonia-Nitrogen Content of
Aeration Tower Influent and Effluent.**

	TOWER INFLUENT			TOWER EFFLUENT		
pH	Temp. (°C)	NH_3-N (mg/l)	pH	Temp. (°C)	NH_3-N (mg/l)	Percent NH_3-N Removal
8.0	22	28.8	8.0	16	18.2	37
8.6	21	26.6	8.1	17	15.8	41
9.3	23	26.2	8.6	18	5.3	80
9.7	21	30.0	9.1	16	2.1	93
10.8	22	25.5	10.1	17	0.6	98

various pH values. The values shown for tower effluent in Table 7-1 are those occurring at the end of each run shown in Figure 7-4.

Temperature.   A critical factor is the air temperature. The water temperature reaches equilibrium at a value near the air temperature in the top few inches of the stripping tower. As the water temperature decreases, the solubility of ammonia in water increases and it becomes more difficult to remove the ammonia by stripping. Ammonia partial pressure data are not available for concentrations as low as those in municipal wastewater. The data available for higher concentrations indicate, however, that a 10°C decrease in process water temperature would decrease the difference in partial pressures by about 40 percent. This decreases the driving force for ammonia stripping by this amount. The amount of air per gallon must be increased to maintain a given degree of removal as temperature decreases. However, it is not practical to supply enough air to offset major temperature decreases fully. For example, at 20°C, 90–95 percent removal of ammonia is typically achieved. At 10°C, the maximum practical removal efficiency drops to about 75 percent. Data collected in pilot tests by EPA at the Blue Plains plant in Washington, D.C. well illustrate the temperature effects and are shown in Figure 7-5.[13] In warm weather tests at pH 11.5 with inlet air and water temperatures averaging 25.5°C and 26°C, respectively, air stripping cooled the outlet water temperature by evaporation of the liquid within the tower to an average of 22.2°C. In a similar test with the inlet air temperature averaging 6°C and the inlet water temperature averaging 16°C, the air stripping cooled the outlet water to an average temperature of 5°C. Data from both the 22.2°C and 5°C conditions are shown in Figure 7-5. The decrease in efficiency from the warm to cold temperatures was approximately 30 percent over a wide range of air and water flows. Data on tem-

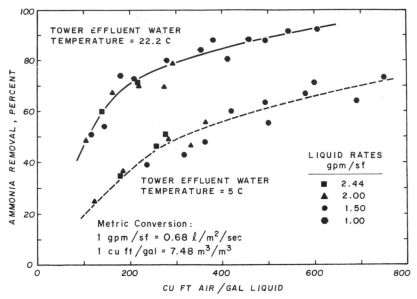

Fig. 7-5   Effect of temperature on ammonia removal efficiency observed at Blue Plains pilot plant.[13]

perature effects were also collected in studies at Lake Tahoe.[14] Figure 7-6 presents ammonia removal data collected with all operating conditions except temperature held constant.

The removal efficiency of the tower was substantially reduced as the temperature fell below 20° C, but a lower limit of removal efficiency was apparently reached as the temperature approached 0° C. A change in effluent temperature from 3° C to 9° C improved ammonia removal by only 5 percent, while a change in temperature from 9° C to 12° C resulted in an 8 percent increase in ammonia removal. Data collected from operation of the pilot tower during winter conditions indicate that an average lower limit of the process is in the range of 50–60 percent ammonia removal. From these tests, it appeared that a taller tower, or provisions for tower effluent water recirculation, would be beneficial under winter conditions. After 24 ft of packing contact, a limit for ammonia removal had not been reached at any temperature level.

When air temperatures reach freezing (or when the wet bulb temperature of the air within the tower reaches 0°C), the tower operation must generally be shut down because of icing problems. The very large volumes of air required for the stripping process make it impractical to heat the air in cold climates. Waste heat from potential on-site sources such as sludge incinerators is typically only a small percentage of that needed.

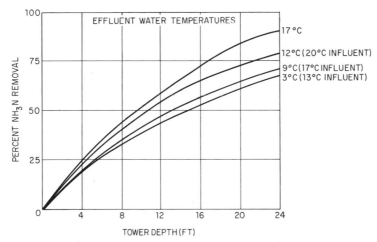

Fig. 7-6 Effect of water temperature on ammonia stripping (operating conditions— 2.0 gpm/ft,^{2} 480 ft^{3} air/gal).[14]

Hydraulic Loading. The hydraulic loading rate of the tower is an important factor. This is typically expressed in terms of gallons per minute applied to each square foot of the plan area of the tower packing. When the hydraulic loading rates become too high, the good droplet formation needed for efficient stripping is disrupted and the water begins to flow in sheets. If the rate is too low, the packing may not be properly wetted, resulting in poor performance and scale accumulation. Data collected in pilot tests at South Tahoe illustrate this relationship and are shown in Figure 7-7.[15] In optimum summer conditions, the pilot data indicate that a flow rate of 2 gpm/ft^{2} is compatible with efficient tower operation at 20–24 ft packing depths. There was little difference between the efficiency of the 20 and 24 ft towers at flow rates up to 7 gpm/ft,^{2} as shown by Figure 7-7. The rapid decrease in removals for loadings greater than 3 gpm/ft^{2} was due to the observed sheeting of water in the tower. Sheeting of the water decreases the number of droplets formed and consequently reduces the efficiency of ammonia stripping. Adequate flow distribution over the entire packing area is a critical factor. Full scale towers at Orange County, California and Pretoria, South Africa are based on tower loadings of 1–1.13 gpm/ft.^{2}

Tower Packing. The depth of tower packing required for maximum ammonia removal will depend on tower packing selected. Most stripping tower designs are based on the use of an open, cooling-tower-type packing (horizontal packing members spaced about 2 in. apart both horizontally and vertically) to minimize the power required to move adequate air quantities through the tower. If maximum removals are desired, tower packing depth

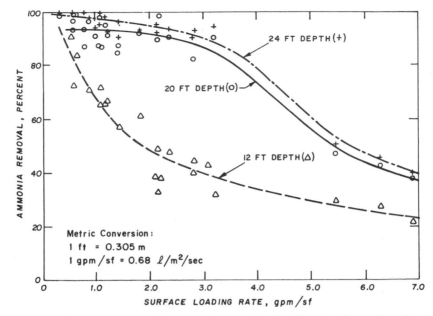

Fig. 7-7 Percentage ammonia removal vs. surface loading rate for various depths of packing.[15]

should be at least 24 ft with this type of packing, unless pilot plant data indicate that a lesser depth of a specific packing will accomplish the required removal. Packing with members spaced more than 2 in. apart may require greater depths, and pilot tests should be run to determine the required depth if greater spacings are proposed.

Both wood (Lake Tahoe) and plastic packings (Orange County) have been used in full scale towers. The smooth plastic surfaces appear to be one factor accounting for reduced calcium carbonate scaling at the Orange County facility. Plastic packing has an advantage in that it does not suffer from dilignification and consequent weakening that occurs with wood at elevated pH values.

Pilot studies at Orange County evaluated three different types of packing: $\frac{1}{2}$ in. diameter PVC pipe, triangular shaped splash bars, and vertical film packing like that used in cooling towers.[16] With vertical packing the water moves in a thin film down vertical sheets of packing rather than moving as droplets, as occurs in packing composed of horizontal splash bars. The film packing was found to provide only 50 percent or less ammonia removal and was eliminated from consideration early in the tests. Film packing fails to provide the repeated droplet formation and rupture needed for efficient stripping.

Since repeated splashing and droplet formation is a key parameter in ammonia stripping, a triangular shaped splash bar was tested. It was thought that it might provide two points of droplet formation, compared to only one for a round splash bar. It was observed that droplet formation throughout the tower still occurred at only one point. The water flowed down the sides of the triangle and around the corners, where it collected on the base and dripped from a single point. Air flow and pressure drop measurements were made on both the circular and triangular packings. The static pressure drop (24 ft packing depth) was 0.40–0.44 in. of water when the triangular packing was used, compared to 0.36–0.40 in. of water when the circular packing was used. No significant differences in ammonia removals were noted between the round and triangular shaped packings.

With a typical packing configuration using wood or plastic slats, the slats may be spaced 2 in. apart (center to center) on the horizontal and 1.5 in. apart vertically. This spacing is referred to as 1.5×2 in. Figure 7-8 shows that the spacing of the tower packing is important in determining the air requirements for ammonia stripping. The 1.5×2 in. packing has 2.66 more slats for droplet formation and coalescing than does a 4×4 in. packing. Although spacing the packing numbers closer than 1.5×2 in. would improve performance, the increased pressure drop would greatly increase power costs.

Tests at Orange County indicate that packings in which alternate layers of packing are placed at right angles, rather than a parallel position, main-

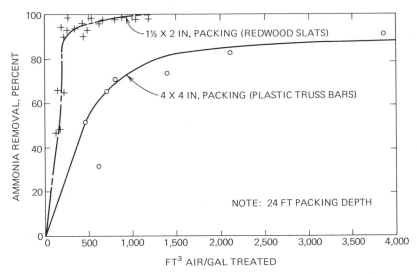

Fig. 7-8 **Effect of packing space on air requirements and efficiency of ammonia stripping.**[12]

tains better flow distribution and may be less susceptible to scale accumulation.

Air Flow.   Gas transfer relationships indicate that an increase in ammonia removal can be achieved by increasing the air flow for a given tower height. However, there is a practical limit on air flow rate because of the increase in air pressure drop with increasing flow rate. This results in higher capital investment for fans and increased power costs. Pressure drop increases exponentially with air flow rate. In general, air velocities of 550 cfm/ft^2 are considered to be the practical upper limit for countercurrent towers.

Figure 7-5 reflects the effects of the ratio of air to wastewater as observed at the Blue Plains pilot plant. These data are in general agreement with similar pilot data collected at South Lake Tahoe (Figure 7-8). For warm weather conditions, typical air requirements are about 300 ft^3/gal for 90 percent removal and 500 ft^3/gal for 95 percent removal. In cold weather conditions, the air requirements to achieve maximum tower efficiency increase substantially. Full scale data at Tahoe indicate that, for their packing design, air flows of about 800 ft^3/gal would be needed to achieve 90 percent removal at an air temperature of 4°C. However, reliance solely on the stripping process in cold weather conditions is usually not practical, and most designs are based on moderate to warm weather conditions. Typical air design quantities for 90 percent removal are as follows: Orange County, California—400 ft^3/gal; South Lake Tahoe—390 ft^3/gal; Pretoria, South Africa—338 ft^3/gal.

The required air quantities are usually provided by a fan located on top of the tower. Two speed fan motors may be used for better matching of air supplied to the actual requirements. Because of the low pressure drops with the types of packings typically used (less than 1 in. water), the horsepower requirements for the fans are not great for these large quantities of air. For the 15 mgd Orange County plant, the total installed fan brake horsepower is 1380 hp.

Scale Control.   A factor which may have an adverse effect on tower efficiency is scaling of the tower packing resulting from deposition of calcium carbonate from the unstable, high pH water flowing through the tower. Scaling potential can be minimized by maximizing the extent of completion of the calcium carbonate reaction in the lime treatment step. Using a high level of solids recycle in the clarification step will ensure more complete reaction. Another approach is to eliminate CO_2 from the air.

The original crossflow tower at the South Lake Tahoe plant has suffered a severe scaling problem. The severity of the scaling problem was not anticipated from the pilot studies, in which a countercurrent tower was used. As a result, the full scale, crossflow tower packing was not designed with access for scale removal in mind. Thus, portions of the tower packing are inaccessible for cleaning. Those portions which were accessible were readily cleaned by high pressure hosing.

The severity of the scaling problem has varied widely. Perhaps the most severe case is that reported at the Blue Plains pilot plant.[13] When operating at a pH of 11.5, a heavy scale of calcium carbonate formed on the crossflow tower packing (polypropylene grids). The scale was crystalline and hard, and could not be removed by a high pressure water hose. In contrast, several months of operation of the countercurrent pilot tower at Orange County at pH values above 11 resulted in only a thin coat of calcium carbonate scale on the 0.5 in. diameter PVC pipe packing. The thin coat of scale stabilized and did not continue to accumulate. The scale was friable and easily removed by water hosing. The Orange County pilot tower was later moved to South Tahoe, where several months of operation indicated no scale buildup. The differences in operation between the pilot tower at Tahoe and the full scale Tahoe tower were as follows: packing of plastic rather than wood, packing shape round rather than rectangular, countercurrent tower rather than crossflow, alternate layers of packing at right angles rather than parallel. The relative importance of these factors in eliminating the scaling noted in the full scale Tahoe tower is uncertain. The experience with the full scale (1 mgd) countercurrent tower at Pretoria, South Africa is similar to that with the full scale Tahoe tower, in that the scale formed can be readily flushed from the packing by a water jet. The feeding of a scale inhibiting polymer to the tower influent may also offer a means of scale control, and such provisions are made in the Orange County facility.

Ammonia Recovery or Removal From Off Gases.   One approach to overcoming the scaling and other limitations of the stripping process is currently being developed.[17] The process has the advantage of recovery of ammonia as a byproduct.

The improved process is shown diagrammatically on Figure 7-9. The process includes an ammonia stripping unit and an ammonia absorption unit. Both of these units are sealed from the outside air but are connected by appropriate ducting. The stripping gas, which initially is air, is maintained in a closed cycle. The stripping unit operates in the manner that is now being or has been used in a number of systems, with the exception that the gas stream is recycled rather than outside air being used in a single pass manner.

Most of the ammonia discharged to the gas stream from the stripping unit is removed in the absorption unit. Because of the favorable kinetics of the absorption reaction, the absorption unit may be reduced in size by about one third from that required for the stripping unit. The absorbing liquid is maintained at a low pH to convert absorbed and dissolved am-

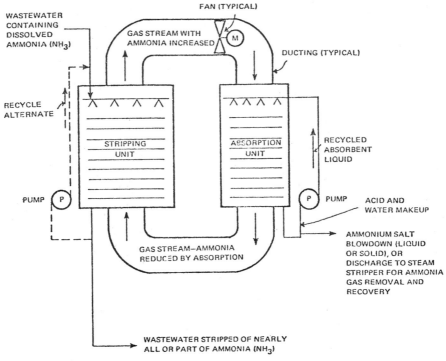

Fig. 7-9   **Ammonia removal and recovery process (ARRP).**[17]

monia gas to ammonium ion. This effectively traps the ammonia and also has the effect of maintaining the full driving force for absorbing the ammonia, since dissolved ammonia gas does not build up in the absorbent liquid. The absorption unit can be a slat tower or packed tower using sprays similar to those in the stripping unit, but will usually be smaller because of the kinetics of the absorption process.

The absorbent liquid initially consists of water with acid to obtain low pH (usually below 7). In the simplest case, as ammonia gas is dissolved in the absorbent and converted to ammonium ions, acid is added to maintain the desired pH. If sulfuric acid is added, as an example, an ammonium sulfate salt solution is formed. This salt solution continues to build up in concentration, and the ammonia is finally discharged from the absorption device as a liquid or solid (precipitate) blowdown of the absorbent. With shortages of ammonia based fertilizers, a saleable byproduct may result. Ammonium sulfate concentrations of 50 percent are obtainable.

Mist eliminators are necessary between the absorber and stripper to prevent carryover of the ammonia laden moisture from the absorber to the stripper effluent. Because of the headloss in the mist eliminator and absorber packing, total headloss for the air approaches 2 in. It is believed that the usual scaling problem associated with ammonia stripping towers will be eliminated by the improved process, since the carbon dioxide which normally reacts with the calcium and hydroxide ions in the water to form the calcium carbonate scale is eliminated from the stripping air during the first few passes. The freezing problem is eliminated by the exclusion of nearly all outside air. The treatment system will normally operate at the temperature near that of the wastewater.

This approach is being used for stripping of ammonia from selective ion exchange process regenerant. Because of the higher power requirements (as compared to single pass stripping), the use of this process may be limited to regeneration of the brines from the selective ion exchange process or other liquids with a relatively high (100 mg/l or more) ammonia concentration. Full scale designs of this application are underway for a 22.5 mgd plant serving the Upper Occoquan Sewage Authority, Virginia and for a 6 mgd plant for the Tahoe–Truckee Sanitation Agency, California as discussed in a later section of this chapter.

Stripping Ponds

The South Tahoe system has been modified to reduce the impact of temperature and scaling limitations encountered at the plant.[18] Basically, the modified process consists of three steps (see Figure 7-10): (1) holding in high pH, surface agitated ponds, (2) stripping in a modified, crossflow forced draft tower through water sprays installed in the tower, and (3) breakpoint chlorination. This system was inspired by observations in Israel of ammonia nitrogen losses from high pH holding ponds.[19]

Pilot tests at South Tahoe indicated that the release of ammonia from high pH ponds could be accelerated by agitation of the pond surface. In the modified Tahoe system, the high pH effluent from the lime clarification process flows to holding ponds. Holding pond detention times of 7–18 hours are used in the modified South Tahoe plant. The pond contents are agitated and recycled 4–13 times by pumping the pond contents through vertical spray nozzles into the air above the ponds. The holding pond detention time and number of recycles vary with plant flow with the time and cycles decreasing as plant flow increases. At least 37 percent ammonia removal in the ponds is anticipated, even in cold weather conditions. The pond contents are then sprayed into the forced draft tower. The packing has been removed from the tower and the entire area of the tower is equipped with water sprays. At least 42 percent removal of the ammonia in the pond effluent is anticipated, based on pilot tests, from this added spray in cold weather that includes recycling of the pond effluent through the tower to achieve 2–5 spraying cycles. The ammonia escaping this process is removed

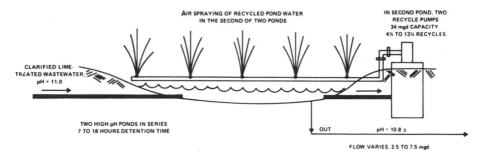

AIR SPRAYING OF RECYCLED POND WATER
IN THE SECOND OF TWO PONDS

IN SECOND POND, TWO
RECYCLE PUMPS
34 mgd CAPACITY
4½ TO 13½ RECYCLES

CLARIFIED LIME-
TREATED WASTEWATER,
pH = 11.0

TWO HIGH pH PONDS IN SERIES
7 TO 18 HOURS DETENTION TIME

OUT   pH = 10.8 ±

FLOW VARIES, 2.5 TO 7.5 mgd

New High pH Flow Equalization Ponds

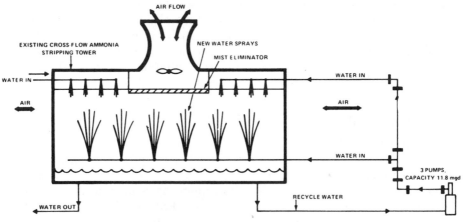

AIR FLOW

EXISTING CROSS FLOW AMMONIA
STRIPPING TOWER

NEW WATER SPRAYS

MIST ELIMINATOR

WATER IN   WATER IN

AIR   AIR

WATER IN

3 PUMPS,
CAPACITY 11.8 mgd

RECYCLE WATER

WATER OUT

Stripping Tower Modified with New Sprays

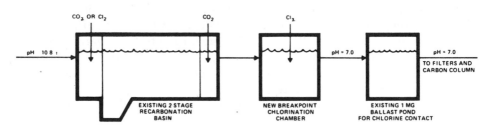

CO_2 OR Cl_2 CO_2 Cl_2

pH 10.8 ±   pH = 7.0   pH = 7.0

TO FILTERS AND
CARBON COLUMN

EXISTING 2 STAGE
RECARBONATION
BASIN

NEW BREAKPOINT
CHLORINATION
CHAMBER

EXISTING 1 MG
BALLAST POND
FOR CHLORINE CONTACT

Breakpoint Chlorination

Fig. 7-10   Ammonia stripping pond system.

by downstream breakpoint chlorination. It appears that stripping ponds offer an approach that takes advantage of the low cost and simplicity of the stripping process for removal of the bulk of the nitrogen, making breakpoint chlorination more attractive for complete removal of ammonia.

Plant Scale Ammonia Stripping Experience

South Lake Tahoe, California. At South Tahoe an experimental full scale ammonia stripping tower built to handle half of the total plant design flow of 7.5 mgd has been operated on an intermittent basis since 1969.[20] The stripping process was installed at South Lake Tahoe as an EPA research and demonstration installation and not as the result of any requirement to remove nitrogen from the plant effluent. In the absence of any need for full time operation, the purposes of this stripping tower have included: (1) demonstration of full scale tower efficiencies as compared to pilot plant test results; (2) determination of cold weather operating limitations and other operating problems, with investigation of solutions to these problems; and (3) collection of data for design purposes for future expansion of this process to full plant design capacity, as well as for use in planning similar facilities at other locations. The large tower capability to remove ammonia almost exactly duplicates the results of pilot plant operation, reaching 95 percent removal in warm weather.

The design data for the original packed tower are given in Table 7-2. It is a crossflow cooling type tower modified for ammonia stripping (see Figure 7-11). The overall dimensions of the tower are 32 ft × 64 ft × 47 ft high. Water at pH 11 was pumped to the top of the tower by either or both

**TABLE 7-2   Design Data for Full Scale Ammonia
Stripping Tower.**

Capacity: Nominal, 3.75 mgd
Type: Cross flow with central air plenum and vertical air discharge through
fan cylinder at top of tower.
Fill: Plan area, 900 sq ft
Height, 24 ft
Splash bars:
material, rough-sawn treated hemlock
size, ⅜ x 1½ in.
spacing, vertical 1.33 in.
horizontal 2 in.
Air Flow: Fan, two-speed, reversible, 24-ft diameter, horizontal

Water Rate		Air Rate	
gpm	gpm/ft²	cfm	cfm/gpm
1,350	1.0	750,000	550
1,800	2.0	700,000	390
2,700	3.0	625,000	230

Tower Structure: Redwood
Tower Enclosure: Corrugated cement asbestos
Air Pressure Drop: ½ in. of water at 1 gpm/ft²

Fig. 7-11 Ammonia stripping tower when under construction at South Tahoe.
(*Courtesy Cornell, Howland, Hayes & Merryfield*)

of two constant speed pumps. These pumps were backflushed two or three times daily to minimize buildup of calcium carbonate scale in the pump units. When the plant inflow was less than the rate at which the pumps were delivering water to the tower, some water was recycled from the tower effluent back to the pump suction well. This avoided the need for variable speed pump control, and at the same time provided some recirculation through the tower, which improved ammonia removal. At the top of the tower, the influent water entered a covered distribution box and overflowed to a distribution basin. The distribution basin is a flat deck with a series of holes fitted with plastic nozzles. Further distribution of the inflow was provided by diffusion decks immediately below the distribution basin. Three other diffusion decks were provided at 6 ft vertical intervals in the fill. The tower fill provided, theoretically, 215 successive droplet formations as the water passed down through the tower. The tower effluent fell into a concrete collection basin, which also formed the base for the tower structure. From the collection basin, the tower effluent passed through a Parshall measuring flume into the first stage recarbonation chamber, where excess pumpage returned through a flap gate into the tower pump sump to recirculate through the tower.

Air entered the tower through side louvers, passed horizontally through the tower fill and drift eliminators (or air flow equalizers), and entered a central plenum. At the top center of the plenum was a 24 ft diameter, six bladed, horizontal fan. Fan blades and fan cylinder are both made of glass reinforced polyester. The fan takes suction from the plenum and discharges to the atmosphere through the fan cylinder. The fan has a maximum capacity of about 750,000 cfm. It is equipped with a two speed, reversible, 75 HP motor.

The limitations of the packed tower at Tahoe are caused by the cold winter temperatures and the scaling of the tower packing. Because the scaling problem was not anticipated from the pilot studies, the access to the tower packing needed to remove the scale with water jets was not provided. As a result, portions of the original hemlock packing became hopelessly fouled with calcium carbonate scale, thereby decreasing effective tower packing area.

Because of the cold weather limitations of packed tower inherent in the Tahoe location and the reduced efficiency of the existing tower due to scaling of the tower packing, a research program was initiated at the South Lake Tahoe plant to develop alternative, low cost techniques which could be applied to the full scale plant while using as much of the existing facilities as possible. Although there are no current regulatory agency requirements for nitrogen removal for exported wastes, future requirements for effluent reuse or disposal in the Lake Tahoe basin will probably include nitrogen removal. At present (1976), the States of California and Nevada require that all wastewaters, regardless of the degree of treatment, be exported from Lake Tahoe basin. The South Tahoe effluent is exported to Alpine County, California where it is used to form the 1 billion gallon Indian Creek Reservoir. An excellent trout fishery has been established in this recreation reservoir, and it is necessary to control pH and ammonia concentrations to prevent fish toxicity. Thus, anticipated future regulatory agency requirements and current effluent reuse practices required that a system for full scale nitrogen removal be developed for the South Lake Tahoe plant. The stripping pond system described briefly in the preceding section is the result (see Figure 7-10).

Design data are given in Table 7-3. Because ammonia removals by stripping vary so much with temperature, and since low temperatures, high ammonia concentrations, and high flows never occur simultaneously at this location, design data and presented for two sets of conditions which are expected to occur: low flow (2.5 mgd), low water temperature (3°C), and low ammonia content; and high flow (7.5 mgd), high water temperature (22°C), and high ammonia content.

The two high pH ponds also serve to equalize flow to the modified stripping tower and breakpoint facilties. This reduces design capacity requirements for these two processes and improves their operating characteristics.

TABLE 7-3   Design Data and Estimated Nitrogen Removals For All-Weather Ammonia Stripping at South Tahoe, California

Description	Flow 2.5 mgd (0.11 m^3/sec) Water Temp. 3°C NH$_3$-N		Flow 7.5 mgd (0.33 m^3/sec) Water Temp. 22°C NH$_3$-N	
	Estimated reduction, percent	Remaining, mg/l	Estimated reduction, percent	Remaining, mg/l
Influent, pH = 11.0		15		35
Holding ponds				
detention = 7 hr			15	30
detention = 18 hr	10	13.5		
Air spraying in second pond				
turnovers = 4½			28	
turnovers = 13½	30	9.5		21.5
Stripping tower, air spraying with forced draft				
recycle turnovers = 1.6			23	16
recycle turnovers = 5	42	5.4		
Overall removal	64		40	
NH$_3$-N remaining to be removed by breakpoint chlorination, mg/l		5.4		16
Chlorine required for breakpoint chlorination, lb/day		1,130		10,000

The total surface area of these ponds is about 64,000 ft.^{2} The water depth varies from 3.0 to 7.25 ft. Detention time ranges from 7 hours at 7.5 mgd to 18 hours at 2.5 mgd. The second pond in the series is provided with surface agitation in order to increase the ammonia removal. The system of surface agitation consists of sprinkling with 34 mgd of recycled pond water through 2 pumps and a spray system consisting of about 75 vertical nozzles each delivering about 320 gpm. At the 2.5 mgd plant flow, the spray system recycles the pond water 13 times; at the 7.5 mgd flow, 4 recycles occur. The spray nozzles are 4 in. × 2½ in. female pipe reducers each fitted with a 2½ in. × 1½ in. bushing. Each spray orifice is about 1⅞ in. in diameter. Nozzles having interval vanes or other obstructions introduce air containing carbon dioxide to the water in the nozzle. This causes deposition and rapid buildup of calcium carbonate within the nozzle. Such nozzles are unsatisfactory because of the resulting plugging and flow restriction problems. Only nozzles with unobstructed, clear opening, and without internal vanes, should be used.

A major modification has been made to the existing stripping tower. The existing packing has been removed, and the entire area of the tower equipped with water sprays. The existing trays at the top of the tower distribute part of the flow, and four nozzle equipped headers in the bottom of the tower spray water upward into the tower. The pump capacity to the tower is 11.8 mgd. At cold weather plant flow rates of 2.5 mgd, this flow will provide a recycle rate of 5 through the sprays in the tower. The capacity and type of nozzle used in the tower is similar to the nozzles used in the ponds. Based on plant scale tests of this spray system, with the induced draft fan operating at high speed, it is anticipated that at least 42 percent of the ammonia in the pond effluent will be removed in the tower under cold weather operating conditions.

Chlorine may be added at two points in the process (Figure 7-10). The first point of application is in the primary recarbonation chamber at a pH of 11.0. Only enough chlorine is added to reduce the pH to about 9.6, thus eliminating the need for addition of carbon dioxide (CO_2) at this point. About 65 mg/l is required to reduce the pH from 11.0 to 9.6. The balance of the chlorine needed to reach the breakpoint for complete ammonia removal is added in a chamber immediately downstream from the secondary recarbonation chamber. At times, sufficient chlorine is added at this point to reduce the pH to 7.0 or less, so that no CO_2 will be required. A dose of approximately 160 mg/l of chlorine is required to reduce the pH from 11.0 to 7.0. When the breakpoint is reached with a lesser dose of chlorine, then some CO_2 must be added in the secondary stage of recarbonation to produce a pH of 7.0. About 10 mg/l of chlorine is required for each mg/l of NH_4^+-N to reach breakpoint. After breakpoint chlorination treatment, water from the ballast pond is pumped to the existing filters and carbon columns. The carbon columns remove any excess chlorine.

Orange County Water District, California.   The Orange County Water District (OCWD) at Fountain Valley, California has under operation a 15 mgd wastewater reclamation plant. The reclaimed water is pumped into a line of injection wells located on the land side of a line of seawater pumping wells to form a barrier against seawater intrusion into the freshwater acquifer.[21,22] The injection system also serves to replenish the supply of groundwater available for use. The OCWD water reclamation plant takes up to 15 mgd of trickling filter effluent from the secondary treatment plant of the Orange County Sanitation District. It is subjected to high lime treatment at pH 11.0 and clarification in a basin equipped with settling tubes. The clarified, high pH water is pumped to two countercurrent ammonia stripping towers. In the climate at this location, freezing temperatures are

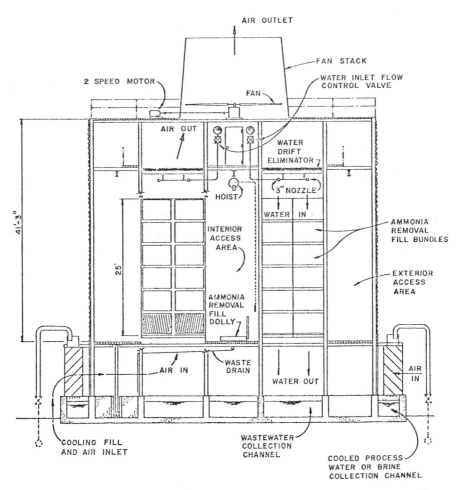

Fig. 7-12 Orange County, California ammonia stripping–cooling tower section.

not experienced. The design also fully allows for scaling problems encountered elsewhere and incorporates provisions for scale control, even though scaling was not a problem in pilot tests at this site and has not yet proven to be a problem in the full scale plant.

The plant is designed with two ammonia stripping–cooling towers, each equipped with six 18 ft diameter fans. An end section view of a tower is shown in Figure 7-12. The stripping sections are designed for a hydraulic loading of 1.0 gpm/ft^2 and an air flow of 400 ft^3/gal. Splash bar packing is used for ammonia stripping.

A prime design criterion was that the ammonia stripping package be accessible and removable for cleaning because scaling of the packing might reduce air flow and ammonia removal efficiency. Provisions have been made to feed a scale inhibiting polymer, if needed, to the tower influent.

The tower fill or packing is made from $\frac{1}{2}$ in. diameter Schedule 80 PVC pipe at 3 in. centers horizontally and with alternate layers placed at right angles and at 2 in. centers vertically. The fill was factory prefabricated in modules which are about 6 ft $\times$ 6 ft $\times$ 4 ft high. Each module is supported within its own steel or fiberglass frame, so that it is easily removable if necessary for cleaning. An overhead hoist and movable dolly are provided to assist in removal of packing for cleaning. However, the access corridors and the removable air baffle panels in the tower should make it possible to reach all of the fill in its normal operating position within the tower for hosing down to remove any excess calcium carbonate scale which may form. The ammonia remaining in the tower effluent is removed by breakpoint chlorination using about 10 mg/l of chlorine for each mg/l of ammonia nitrogen still present.

SELECTIVE ION EXCHANGE

Basis of the Process

The use of conventional ion exchange resin for removal of nitrogenous material from wastewaters has not proven attractive because of the preference of these exchangers for ions other than ammonium or nitrate. In addition, the regeneration of conventional ion exchange resins results in regenerant wastes which are difficult to handle. The nonselective nature of conventional resins is unfortunate because, as a unit process, ion exchange is easily controlled to achieve almost any desired product quality. The efficiency of the process is not significantly impaired at temperatures usually encountered, and ion exchange equipment can be automatically controlled, requiring only occasional monitoring, inspection, and maintenance.

A process has been developed using an ion exchanger which is selective for ammonium nitrogen.[23,24,25] The limitations of conventional resins given above may be largely overcome by using an exchanger which is selective for ammonium. The exchanger currently favored for this use is clinoptilolite, a zeolite, which occurs naturally in several extensive deposits in the western United States. It is selective for ammonium relative to calcium, magnesium, and sodium. The removal of the ammonium from the spent regenerant permits regenerant reuse. The ammonium may be removed from the regenerant and released to the atmosphere as ammonia (in certain situations) or nitrogen gas, or it may be recovered as an ammonium solution for use as a fertilizer. Figure 7-13 is a simplified schematic diagram of the process.

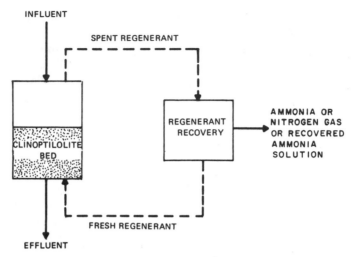

Fig. 7-13   Selective ion exchange process.

The wastewater is passed downward through a bed of clinoptilolite (typically 4–5 ft of 20 × 50 mesh particles) during the normal service cycle. When the effluent ammonium concentration increases to an objectionable level, the clinoptilolite is regenerated by passing a concentrated salt solution through the exchange bed. By removing the ammonia from the spent regenerant, the regenerant can be recycled for reuse, eliminating the difficult problem of brine disposal associated with conventional exchange resins. Some of the regenerant recovery techniques, as discussed in detail later, remove the ammonia as nitrogen gas, which is discharged to the atmosphere, while others remove and recover the ammonia in solution for potential use as a fertilizer.

The basic principles of ion exchange can be used to determine the capacity of clinoptilolite for ammonium; an excellent example of such a determination is available in reference 24. However, such calculations are lengthy and rather complex, and a simplified technique for quickly approximating the ammonium capacity of clinoptilolite for varying concentrations of competing cations is presented later in this chapter. Isotherms demonstrating the selectivity of clinoptilolite for ammonium over other cations have been reported in the literature.[25] An example is the comparison of Hector clinoptilolite and a strong acid polystyrene resin, IR 120 (Figure 7-14). If the equivalent fraction of ion in the solid phase is plotted against the equivalent fraction of ion in the solution three cases can be identified corresponding to $\alpha < 1$, $\alpha = 1$, and $\alpha > 1$; as explained below. Isotherms which are concave upward ($\alpha < 1$) are designated as being unfavorable to the uptake of ion A; those which fall along a rising diagonal ($\alpha = 1$) are termed linear and

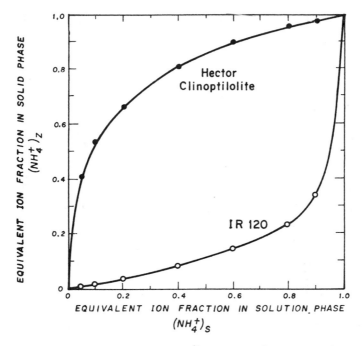

Fig. 7-14   Isotherms for the reaction $(Ca^{2+})_z + 2(NH_4^+)_N = 2(NH_4^+)_z + (Ca^{2+})_N$ with Hector clinoptilolite and IR 120.

exhibit no preference for ion A or B; and curves which are concave downward ($\alpha > 1$) are referred to as favorable isotherms, since the solid prefers ion A to ion B. Ion exchange operations almost always use systems in which the ion of concern has a separation factor (α) greater than unity during the service cycle. Figure 7-14 illustrates well that IR 120 prefers calcium to ammonium ions. Clinoptilolite, on the other hand, prefers ammonium to calcium ion and is of greater utility for ammonium ion removal from wastewaters containing calcium. The equilibrium isotherms for ammonia and other cations which are present as macrocomponents in wastewaters are shown in Figure 7-15. These isotherms illustrate that clinoptilolite is selective for ammonium relative to all of the listed ions except potassium. Plots of the NH_4^+ selectivity coefficients (K_A) vs. the solution concentration ratios of these cations are shown in Figures 7-16 and 7-17. The selectivity coefficient is defined as follows:

$$K_A^B = \frac{(A)_z^{nA}\,(B_N)^{nB}}{(B)_z^{nB}\,(A_N)^{nA}},$$

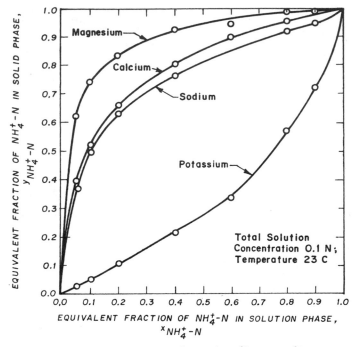

Fig. 7-15   Isotherms for exchange of NH_4^+ for K^+, Ca^{2+}, and Mg^{2+} on clinoptilolite.[23]

where

(A_N), (B_N) = normality of cations A and B in the equilibrium solution,

$(A)_z$, $(B)_z$ = equivalent fractions of cations A and B on the zeolite, and

nA, nB = the number of moles of A and B represented in the chemical equation for the exchange reaction of A and B.

Using these data in conjunction with published calculation techniques[24] it is possible to predict accurately the ammonium capacity of clinoptilolite in the presence of various concentrations of other cations.

Although the total ion exchange capacity of a material is by no means a complete description of its ion exchange properties, it is an indication of the applicability of the substance for process use. For example, New Jersey greensand, which was widely used in water softening before the development of organic exchangers, has a total exchange capacity of 0.17 meg/g (milliequivalents per gram). In comparison, the exchange capacities of strong acid cation exchangers are usually 4–5 meg/g. The total exchange capacity

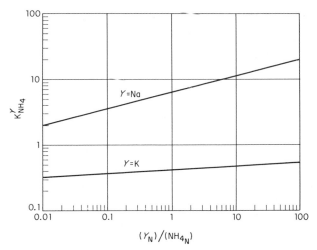

Fig. 7-16 Selectivity coefficients vs. concentration ratios of sodium or potassium and ammonium in the equilibrium solution with Hector clinoptilolite at 23°C for the reaction $(Y)_z + (NH_4^+)_N = (NH_4^+)_z + (Y)_N$.[25]

of clinoptilolite as measured by several different investigators ranges from 1.6 to 2.0 meg/g and is slightly lower than the average for zeolites. With typical cation concentrations encountered in municipal wastewaters, the capacity for NH_4^+ is typically 0.4 meq/g.

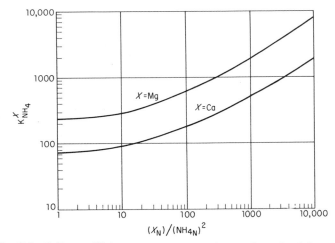

Fig. 7-17 Selectivity coefficients vs. concentration ratios of calcium or magnesium and ammonium in the equilibrium solution with Hector clinoptilolite at 23°C for the reaction $(X)_z + 2(NH_4^+)_N = 2(NH_4^+)_z + (X)_N$.[25]

Physical Properties of Clinoptilolite

The zeolites are classified as a family in the silicate group. They are hydrated aluminosilicates of univalent and bivalent bases which can be reversibly dehydrated to varying degrees without undergoing a change in crystal structure and are capable of undergoing cation exchange. The general composition of zeolites is given by the formula $(M,N_2)O\cdot A_2O_3\cdot nSiO_2\cdot mH_2O$, where M and N are, respectively, the alkali metal and alkaline earth ions present in the zeolite cavities.

Clinoptilolite is a common material found in bentonite deposits in the western United States. The largest known deposit of clinoptilolite in the United States is found in southern California within a deposit of bentonite called hectorite because of its proximity to Hector, California. The U.S. deposits are predominantly in the sodium form. Although clinoptilolite occurs widely, not all deposits produce a material of adequate structural strength to withstand the handling which occurs in a columnar operation. When crushed, sieved, and thoroughly washed with agitation to remove fines, clay, and other impurities, 20 × 50 mesh Hector clinoptilolite gives a wet attrition test of 3 percent.[26] The wet attrition test determines the amount of fines (less than 100 mesh) generated by 25 g of the granular zeolite during rapid mixing with 75 m of water on a paint shaker for 5 minutes. Commercial zeolites, such as erionite and chabazite, which are powdered, mixed with clay binder, extruded, and fined, will generally give a wet attrition test of about 6 percent, or twice that of the Hector clinoptilolite. Low wet attrition is important to minimize losses of clinoptilolite in an ion exchange column operation. Clinoptilolite (20 × 50 mesh) has been reported to have a wet particle specific gravity of 1.59 and a bulk density of 0.74–0.79 g/cm^3.

The instability of natural clays and zeolites toward acids and alkalis is known, as these materials are widely used in water softening. However, clinoptilolite is considerably more acid resistant than other zeolites. Very high strength (20 percent NaOH) caustic solutions produce significant chemical attack on the clinoptilolite. However, at the lower solution strengths encountered in systems which use a caustic regenerant, physical attrition is more significant than chemical attack.

Design Factors

The factors which have a major effect on process efficiency include: pH, hydraulic loading rate, clinoptilolite size, pretreatment, wastewater composition, and bed depth.

pH.   Within an influent pH range of 4–8, optimum ammonium exchange occurs. As the pH drops below this range, hydrogen ions begin to compete with ammonium for the available exchange capacity. As the pH values increase to about 8, a shift in the NH_3–NH_4^+ equilibrium toward NH_3 begins. Operation outside of the pH range of 4–8 results in a rapid decrease of exchange capacity and increased ammonium leakage.

Hydraulic Loading Rate.   Variations in column loading rates within the range of 7.5–20 bed volumes (BV)/hour (7.5 BV/hour is equivalent to 0.95 gpm/ft^3) have been shown to produce no significant effects on the ammonium removal efficiency of 20 × 50 mesh clinoptilolite.[25] Ammonium concentrations in the clinoptilolite effluent of 0.22–0.26 mg/l were produced throughout the above range in one set of tests.[23] When rates exceed 20 BV/hour the exchange kinetics suffer, as demonstrated by a significant leakage of ammonium early in the loading cycle.

Clinoptilolite Size.   Mine run clinoptilolite is typically 1–2 in. chunks which must be ground and screened to the size desired for column operation. As would be expected, the smaller the clinoptilolite size, the better the kinetics of the exchange reaction. This effect is illustrated by data that show that 20 × 50 mesh clinoptilolite kinetics begin to suffer at rates of 20–30 BV/hour, while 50 × 80 mesh kinetics do not suffer until rates of 40 BV/hour are reached. However, the improved rate of exchange is accompanied by the disadvantage of higher headloss. It appears that 20 × 50 mesh clinoptilolite (about the size of typical filter sand) offers an adequate compromise between acceptable headloss and exchange kinetics. At a loading of 15 BV/hour in a 3 ft deep bed (5.6 gpm/ft^2), the headloss is 2.1 ft with 20 × 50 mesh clinoptilolite. Lower headlosses could be obtained by lower rates. Use of deeper beds would result in greater headloss, i.e., 6 ft depth would have a headloss of 4.2 ft at 5.6 gpm/ft^2. These headloss values do not include losses in inlet and outlet piping or in the underdrain system.

Pretreatment.   To avoid excessive headloss, the clinoptilolite influent must be relatively free of suspended solids—preferably less than 35 mg/l. Tests with clarified and filtered raw wastewater indicate no problems with organic fouling. Biological growths which occurred were adequately removed in the regeneration cycle.[24]

Wastewater Composition.   As noted earlier, although clinoptilolite prefers ammonium ions to other cations, it is not absolutely selective, and other cations do compete for the available exchange capacity. Pilot tests conducted at several locations illustrate the effects of wastewater composition on the useful capacity of the clinoptilolite.[25] Tests at three locales that span a wide range of wastewater compositions are shown in Table 7-4.

TABLE 7-4 **Influent Composition for Selective Ion Exchange Pilot Tests at Different Locales.**

| Parameter, mg/l | Activated Sludge Plant Effluent | | Clarified Raw Wastewater Blue Plains (Washington, D.C.) |
	Tahoe, Carbon treated	Pomona[a]	
NH_4^+-N	15	16	12
Na^+	44	120	35
K^+	10	18	9
Mg^{2+}	1	20	0.2
Ca^{2+}	51	43	30
pH Range	7–8[b]	6.5–8.2[b]	7–9[b]
COD	11	10	50
TDS	325	700	250

[a]Approximately half of the runs at Pomona were made with carbon treated secondary effluent and the others with alum coagulated secondary effluent.
[b]pH units.

The equilibrium NH_4^+-N bed loading computed for each of the wastewater listed in Table 7-4 was 4.1 g/1, 3.9 g/1, and 4.3 g/1, respectively, for Tahoe, Pomona, and Blue Plains. Figure 7-18 presents equilibrium bed loading in an alternate way. The minimum bed volumes required to attain

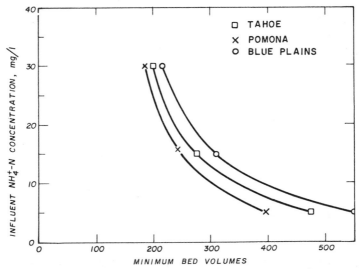

Fig. 7-18   Minimum bed volumes as a function of influent NH_4^+-N concentration to reach 50 percent breakthrough of ammonium.[1]

equilibrium NH_4^+-N loading are expressed as a function of the NH_4^+-N concentration in the influent wastewater. The minimum bed volume values given in Figure 7-18 normally represent the 50 percent breakthrough point (i.e., the point at which the effluent concentration is 50 percent of the feed concentration). The water lowest in competing cations (Blue Plains) had the greatest ammonium removal capacity. In the 10–20 mg/l influent NH_4^+-N range, the lower competing ion concentrations at Blue Plains resulted in a useful ammonium exchange capacity about 33 percent greater than that for Pomona with its higher TDS water. The lower degree of pretreatment at Blue Plains (i.e., no biological pretreatment) did not impair the effectiveness of the clinoptilolite for ammonium removal.

Length of Service Cycle.   To illustrate the permissible length of service cycle for a given wastewater, ammonia breakthrough curves for a single 6 ft deep bed of clinoptilolite are illustrated in Figure 7-19 for Tahoe tertiary effluent with flow rates varying from 6.5 to 9.7 BV/hour with 15–17 mg/l NH_4^+-N in the feed stream. These curves indicate a throughput value of 150 BV should be used for this wastewater for design for effluents requiring a high degree of ammonia removal. Although the effluent concentration had reached 2–3 mg/l at 150 BV, the average concentration produced to this point in the cycle was less than 1 mg/l. Breakpoint chlorination would be

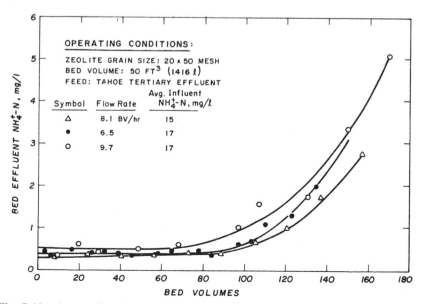

Fig. 7-19   **Ammonium breakthrough curves for a 6 ft clinoptilolite bed at various flow rates.**[1]

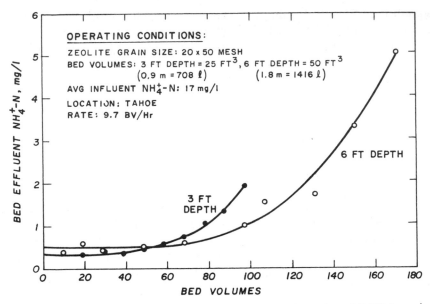

Fig. 7-20   Effect of bed depth on ammonium breakthrough at 9.7 BV/hour.[1]

more economical for removing the 1 mg/l residual, if required, than would provision of greater ion exchange column capacity. The average ammonium concentration for a breakthrough curve is obtained by integrating the area under the breakthrough curve and dividing by the total flow. For example, integrating the area under the curve for 8.1 BV/hour in Figure 7-19 indicates an average NH_4^+-N concentration of 0.67 mg/l for 150 BV.

Bed Depth.   The effect of bed depth on ammonia breakthrough at 9.7 BV/hour is illustrated in Figure 7-20. In general, the 3 ft bed of clinoptilolite was not as effective for ammonium removal as the 6 ft bed at the same bed volume rate. The shallow bed has a lower velocity because a 9.7 BV/hour flow in a 3 ft deep bed corresponds to 3.6 gpm/ft², while in a 6 ft deep bed it corresponds to 7.2 gpm/ft². The lower velocity might increase the likelihood of plugging of portions of the bed. Plugging would cause poor flow distribution and lower bed efficiency. Full-scale designs are using bed depths of 4 ft with a high degree of pretreatment (coagulation and filtration), which will minimize plugging of the clinoptilolite bed.

One Column vs. Series Column Operation.   Operation to the 150 BV throughput value (Figure 7-20) to maintain an average NH_4^+-N concentration at or below 1 mg/l uses only 55–58 percent of the zeolite's equilibrium capacity. The number of bed volumes throughput per bed can be increased while maintaining low NH_4^+-N effluent concentrations with semi-countercurrent operation, using two beds in series. Semi-countercurrent operation is achieved by first operating the columns in a 1–2 sequence and then placing column 2 into the lead position, after the first regeneration, with column 1 becoming the polishing column. A column is removed from the influent end when it becomes loaded, while a regenerated column is simultaneously added at the effluent end. This procedure in effect moves the beds countercurrent to liquid flow by continually shifting the more saturated beds closer to the higher influent concentrations. Beds can be loaded nearer to capacity with this procedure than with single column or parallel feed multicolumn operation. The most highly loaded column is always at the influent end, backed up by one (if there are only two in series) or more columns having decreasing loadings and NH_4^+-N concentrations at locations progressively nearer the end of the series. Removal of a column is not decided by applying a breakthrough criterion to the column's own effluent but by breakthrough at the end of the series. Tests have indicated that the ammonium loadings could be increased from 55–58 percent of the equilibrium capacity to 85 percent by using two columns in series.[24] Average throughputs for the Tahoe example discussed earlier increased from 150 to 250 BV/cycle. However, such a two column operation requires three columns (two on stream while the third is being regenerated) and more complicated valving and piping than a parallel column operation. Because of the added capital costs involved in a series system, all of the full scale systems currently under design or in operation utilize parallel single beds. By blending the effluents from several parallel columns, each of which is in a different stage of exhaustion, improved utilization of the available exchange capacity is also achieved.

Determination of Ion Exchanger Size.   In order to calculate the size of the ion exchange unit needed, the ammonium capacity of the clinoptilolite must be determined from the characteristics of the influent water. The ammonium capacity of clinoptilolite can be estimated from Figure 7-21 if the cationic strength of the wastewater is known. The data used to plot Figure 7-21 were determined in several experimental runs where the influent ammonium nitrogen concentration was 16.4–19.0 mg/l.[23] Although the curve is empirical and is a simplification of the complex effect of competing cation concentrations on ammonium capacity, it illustrates this effect and serves as a useful tool in sizing the exchange bed.

Assuming that the influent water has a cationic strength of 0.006 moles/l, the breakthrough ammonium capacity of the clinoptilolite will be approximately 0.25 meq/g for a 3 ft bed; the capacity to saturation will be approximately 0.44 meq/g. A greater effective ammonium capacity can be realized

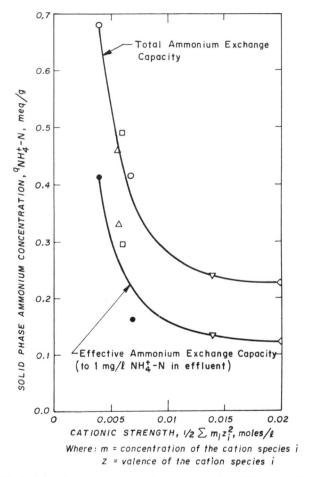

Fig. 7-21   Variation of ammonium exchange capacity with competing cation concentration for a 3 ft deep clinoptilolite bed.[23]

by increasing the depth of the zeolite bed. The use of a 6 ft bed would result in greater ammonia capacity per unit of exchanger, and while requiring a deeper structure would incur only nominal additional cost. Assuming that 3 ft of the zeolite bed will have an ammonium exchange capacity equal to 0.25 meq/g and that the remaining 3 ft will have a capacity equal to 90 percent of the saturation capacity, or 0.40 meq/g, the 6 ft bed will have an effective capacity of 0.32 meq/g (equivalent to 6.6 eq/ft^3 and 5.1 Kgr/ft^3).

The zeolite volume required to treat a 10 mgd waste flow at 15 BV/hour (1.9 gpm/ft^3) is 3650 ft^3. Assuming complete removal of ammonium, the throughput to ammonium breakthrough would then be 165 BV with a run length of 11 hours. Allowing 2 hours downtime per cycle for regeneration and rinsing, the zeolite volume would be increased proportionately to 4300 ft^3 to accommodate the total design flow. Using four units, each having the dimensions 12 ft × 15 ft × 6 ft deep, the total zeolite volume would be 4320 ft^3.[23]

Regeneration Alternatives

After about 150–200 BV of normal strength municipal waste have passed through the bed, the capacity of the clinoptilolite has been used to the point that ammonium begins to leak through the bed. At this point, the clinoptilolite must be regenerated so that its capacity to remove ammonium is restored. The zeolite is regenerated by passing concentrated salt solutions through the exchange bed when the ammonium concentration in the solid phase has reached the maximum desirable level. The volume of ammonium laden, spent regenerant is about 2.5–5 percent of the throughput treated prior to regeneration. By removing the ammonium from the spent regenerant, the regenerant may be recycled for reuse. The alternative approaches available for regenerant recovery are:

air stripping

steam stripping

electrolytic treatment

The ammonium retained on the clinoptilolite exchange sites may be eluted by either sodium or calcium ions contained in a regenerant solution. While the normal service cycle is downflow, regeneration is carried out by passing the regenerant up through the clinoptilolite bed.

High pH Regeneration. The approach originally studied for wastewater applications was to use a lime slurry (5 gm/l) as the regenerant so that the ammonium stripped from the bed during regeneration would be converted to gaseous ammonia, which could then be removed from the regenerant by air stripping.[25] It was found that elution with lime could be speeded up by the addition of sufficient NaCl to render the regenerant $0.1N$ with respect to NaCl.[25]

In addition to converting the ammonium ion to ammonia so it can readily be removed from the regenerant, the volume of regenerant required for complete regeneration has been found to decrease with increasing regenerant pH.[23] However, high pH regeneration was found to be accompanied by an operational problem of major significance:[24,25] precipitation of magnesium hydroxide and calcium carbonate occurs within the exchanger during the regeneration cycle. This leads to plugging of the exchanger inlets and outlets, as well as coating of the clinoptilolite particles. Violent backwashing of the clinoptilolite was found to be necessary to remove these precipitants from the clinoptilolite particles; this procedure resulted in increased mechanical attrition of the clinoptilolite. Chemical attrition also increases at elevated pH values.[23]

Substantial data have been collected on high pH regeneration.[23,24,25] However, the practical problem of scale control is a major limitation which can be overcome by using neutral regenerants. The use of closed loop regenerant recovery processes negates the disadvantage of higher regenerant volumes required at lower regenerant pH values.

Neutral pH Regeneration.   Two of the largest municipal wastewater installations under design which will use clinoptilolite are the Upper Occoquan (Virginia) regional plant (15 mgd) and the Tahoe–Truckee (California) Sanitation Agency plant (6 mgd), both of which will utilize a regenerant with a pH near neutral. The active portion of the regenerant will be a 2 percent sodium chloride solution. Calcium and potassium will be eluted as well as ammonia and will build up in the regenerant until they reach equilibrium. Approximately 25–30 BV are required before the ammonium concentration reached equilibrium.[27] Although greater regenerant volumes are required than with a high pH regenerant (10–30 BV), this is not a major disadvantage if the regenerant is recovered and reused in a closed loop system.

Variations in regenerant flow rates of 4–20 BV/hour do not affect regenerant performance. Higher rates result in less ammonia removed per volume of regenerant. Typical design values are 10 BV/hour, which insures efficient use of the regenerant while keeping headloss values at low levels. Provisions should be made for backwash at rates of 8 gpm/ft² and surface wash of the contactor prior to initiation of the regenerant flow.

Air Stripping of High pH Regenerant.   In the original pilot work on this approach to regenerant recovery, a stripping tower packed with 1 in. polypropylene saddles was used.[24,25] Because the regenerant volume is only a small portion of the total wastewater flow, it becomes feasible to heat the air used in the stripping process. The regenerant was normally recycled upflow through the zeolite bed at a flow rate of 4.8–7.1 gpm/ft² until the NH_3-N approached a maximum concentration. The regenerant was then recycled through both the zeolite bed and the air stripper until the NH_3-N was reduced to about 10 mg/l. The liquid flow rate to the stripper was normally 2 gpm/ft² with an air–liquid ratio of 150 ft³/gal.

Ammonia removal in the air stripper generally averaged about 40 percent per cycle at 25°C. Calcium carbonate scaling occurred on the polypropylene saddles, but the scale could generally be removed by water sprays. The headloss through the 1 in. pilot plant saddles caused the power requirements for the air stripping to be excessive. It was suggested for a full scale design that the ammonia stripping tower be sized to treat the contents of an elutriant tank in 8 hours, using two passes through the tower at 85 percent removal per pass at an air-to-water ratio of 300 cfm/gpm, and a loading of 3.5 gpm/ft^2.[24] The tower would be a modified cooling tower with low differential pressure across the tower, as discussed in the ammonia stripping portion of this chapter.

Air Stripping of Neutral pH Regenerant. The use of high pH regenerant is accompanied by scaling problems within the ion exchange beds. A regenerant with 2 percent sodium chloride as the active agent and pH nearer neutral has been used to overcome the scaling problem. This regenerant may also be recovered for reuse by air stripping. Figure 7-22 is a schematic diagram of such a system.[28] In this system, the stripping tower off gases are not discharged to the atmosphere but are instead passed through an absorption tower where the ammonia in the off gases is absorbed in sulfuric acid (see Figure 7-9 and related text).

In the system shown in Figure 7-22, batch countercurrent regenerant flow similar to that described above for the high pH regenerant is practiced to reduce the amount of regenerant which must be stripped per cycle. The first 11 BV of spent regenerant are discharged to the spent regenerant tank for stripping. The second and third 11 BV batches are stored and used as the firts 22 BV of regeneration flow in the next regenerant cycle. The last 6–11 BV batch of regenerant is mixed with the 11 BV of stripped regenerant for use as the final regenerant flow in the next cycle. Thus, although 40–44 BV of regenerant are passed through the exchanger per cycle, only 11 BV are actually renovated by air stripping per cycle.

Regenerant stored in the tanks varies in ammonium-nitrogen concentration from about 250 mg/l in the spent tank to 50 mg/l in the recovered regenerant tank. The intermediate tanks have intermediate concentrations. The regenerant pH varies from about 9.5 in the recovered regenerant tank to about 7.0 in the spent tank. As discussed earlier, higher pH values produce more efficient regeneration, but near neutral pH levels avoid problems with magnesium hydroxide precipitation in the bed during regeneration and attrition of the clinoptilolite caused by high pH. Media attrition has been insignificant in pilot studies under these pH conditions.

A typical ammonium elution curve is shown on Figure 7-23 with the background concentrations in each regenerant storage tank also shown. At the end of the cycle, the last portion of spent regenerant is discharged to the recovered regenerant tank. This has the effect of neutralizing the alkaline pH from the ARRP process. ARRP effluent is normally at a pH of 10.7–11.0.

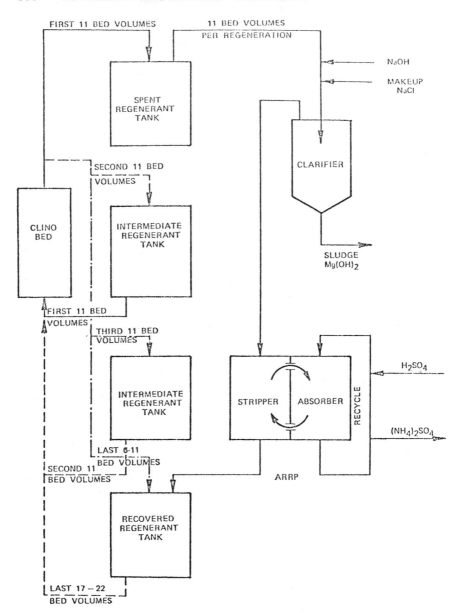

Fig. 7-22 Flow diagram of neutral pH regeneration system using air stripping.[1]

This is reduced to about 9.5 by recycling the last portion of spent regenerant. In this manner, pH is controlled without use of acid.

When spent regenerant is accumulated to a predetermined amount, the recovery portion of the process is activated. This system operates at a flow

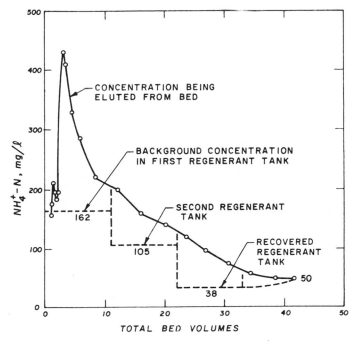

Fig. 7-23   Typical elution curve.[1]

rate of approximately $\frac{1}{13}$ of average plant flow since the regenerant concentration is about 13 times as concentrated as plant waste. Initially, sodium hydroxide is added to the spent regencrant to achieve a pH of about 11. Sodium chloride is also added because of some salt loss from the regenerant solution in sludge removal and bed rinsing. Following pH adjustment, the regenerant is clarified and any magnesium hydroxide formed is removed; the clarified regenerant at pH 11 is then pumped to the ARRP process for ammonia removal and recovery. The ARRP effluent flows to the recovered regenerant tank, where it is mixed with the last 6–11 BV of spent regenerant for pH adjustment prior to reuse. This system is being used in the design of plants for the Tahoe–Truckee Sanitation Agency (California) and for the Upper Occoquan (Virginia) regional plant.

Steam Stripping.   Steam stripping of regenerant is being practiced at the 0.6 mgd (0.026 mg^3/sec) Rosemount, Minnesota physical-chemical plant.[29,30] This process is economically feasible only with high pH regenerant.

The higher regenerant volumes resulting from the neutral regenerant approach are not economically treated by this approach. This process is feasible only if the regenerant volume requiring stripping is held to 4 BV per cycle, which is achievable with a high pH regenerant batch recycle system.

In this case, the necessary portion of the spent regenerant is stripped in a distillation tower in which steam is injected countercurrent to the regenerant. An air cooled, plate and tube condensor condenses the vapor for collection in a covered tank as a 1 percent aqueous ammonia solution which could be used as a fertilizer. A stripping tower depth of 24 ft and a loading of $7 \, gpm/ft^2$ are being used at Rosemount. Ceramic saddles are used rather than wooden slat packing because wood is not a suitable packing in a high pH–steam environment.

Heat exchangers are used to transfer heat from the stripped regenerant to the incoming, cold regenerant. Heat transfer to the incoming regenerant from the condenser used to condense the stripped regenerant may also be attractive. Provisions for scale control in the heat exchangers should be provided. The steam requirements have been estimated to be $15 \, lb/1000 \, gal$.

Electrolytic Treatment of Neutral pH Regenerant. In this approach, ammonium in the regenerant solution is converted to nitrogen gas by reaction with chlorine, which is generated electrolytically from the chlorides already present in the neutral pH regenerant solution. The regenerant solution is rich in NaCl and $CaCl_2$, which provide the chlorine produced at the anode of the electrolysis cell. A diagram of the regeneration system is presented in Figure 7-24. The regeneration of the clinoptilolite beds is accomplished with a 2 percent sodium chloride solution. The spent regenerant is collected in a large holding tank and then subjected to soda ash treatment for calcium

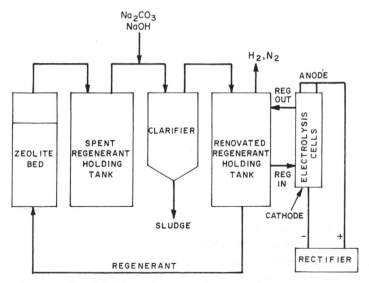

Fig. 7-24   Simplified flow diagram of electrolytic regenerant system.

removal. After the soda ash addition, the regenerant is clarified and trans-ferred to another holding tank, where the regenerant is recirculated through electrolysis cells for ammonia destruction.

During the regeneration of the ion exchange bed, a large amount of calcium is eluted from the zeolite along with the ammonia. This calcium tends to scale the cathode of the electrolysis cell, greatly reducing its life. Calcium may be removed from the spent regenerant solution by a soda ash softening process prior to passing the spent regenerant through the electrolytic cells. High flow velocities through the electrolysis cells are required in addition to a low concentration of $MgCl_2$ to minimize scaling of the cathode by calcium hydroxide and calcium carbonate.

In pilot tests of the electrolytic treatment of the regenerant at Blue Plains, about 50 watt-hours of power were required to destroy 1g of ammonia nitrogen.[24] When related to the treatment of water containing 25 mg/l NH_4^+-N, the energy consumed would be 4.7 kwh/1000 gal. Tests at South Tahoe also indicated that a value of 50 watt-hours per gram is reasonable for design.[27]

The electrolytic process also results in about 56 ft′ (1586 l) of hydrogen gas being evolved per pound of ammonia nitrogen destroyed. Provisions must be made to vent, burn, or otherwise adequately control the hydrogen gas evolved in the electrolytic process.

The major disadvantage of the electrolytic approach is the substantial amount of electrical energy required. The electrical requirements of the air stripping (ARRP) system are only about 10 percent of that required by the electrolytic process.

Plant Scale Selective Ion Exchange Experience

Rosemount, Minnesota. This new 0.6 mgd plant operated by the Metro-politan Sewer Board of the Twin Cities area (Minneapolis–St. Paul) provides independent physical-chemical treatment of a municipal wastewater.[29,30] This plant is the first full scale physical-chemical plant to be placed in operation in the U.S.A. Final effluent standards are as follows:

Parameter	Value
BOD_5, mg/l	10
Suspended solids, mg/l	10
COD, mg/l	10
Ammonia-nitrogen, mg/l	1
Phosphorus, total (P), mg/l	1
pH	8.5

TABLE 7-5   Rosemount (Minnesota) Ion Exchange Design Criteria.

Ammonium exchange columns (two trains of 3)	
Loading rate[a]	4.2 gpm/ft^2, 5.6 BV/hour
Clinoptilolite capacity	
per unit volume	0.3 lb/ft^3
per column	90 lb
Ammonia nitrogen loading rate	50 lb/day
Ammonia removal	95 percent
Clinoptilolite depth per column	6 ft
Clinoptilolite size	20 × 50 mesh
Normal operation	2 columns in series, 250 BV/cycle
Backwash rate	8 gpm/ft^2
Regeneration system	
Brine solution to columns[b]	
Hydraulic application rate	2.0 gpm/ft^2
Volume	4.5 BV
Strength	6% NaCl
Temperature	71°C
pH[c]	11
Brine solution regeneration	
Regeneration cycle length	5 hours
Hydraulic loading rate to steam stripping tower	7 gpm/ft^2
Tower depth	24 ft
Caustic soda added	3 lb/lb NH$_4^+$-N
Bed rinse[a]	
Rinse rate	300 gpm
Time	70 minutes
Ammonia recovery	
Aqueous ammonia strength	1%
Aqueous ammonia volume	1000 gpd
Ammonia stripper	
Steam @ 10 psig	3300 lb/hour
Throughput	53 gpm
Size: diameter	3 ft
height	18 ft

[a]Downflow.
[b]Upflow.
[c]Elevated with caustic soda.

The ion exchange treatment is preceded by chemical coagulation, sedimentation, filtration, and activated carbon adsorption. Selective ion exchange is accomplished by clinoptilolite in 2 of 3 downflow columns in series, each containing a 6 ft depth of clinoptilolite. When the ammonia-nitrogen reaches 1 mg/l in the effluent from the polishing column, the lead column is removed from service and regenerated (upflow) and the third column placed on line. The steam stripping process is used for clinoptilolite regeneration and ammonia recovery. Brine is stored at 77°C and is cooled to 27°C while passing through a heat exchanger on the way to the column. The waste brine leaving the column is passed through the other side of the heat exchanger to elevate its temperature to 71°C before entering the stripping process. Brine temperature in the storage tanks is controlled by steam supplied to internally mounted coils. Waste brine is collected in a mixed storage tank before being stripped. Soda ash is added to the storage tank to elevate the pH to 12; this procedure results in the precipitation of calcium carbonate and magnesium hydroxide. Mixing is discontinued 20 minutes after the pH has reached 12 to allow these precipitates to settle. The sludge is then pumped from the bottom of the waste brine storage tank. The waste brine is then pumped to the steam stripper at a rate of 53 gpm and a steam flow of 3000 lb/hour. The stripper operates at an equilibrium temperature of 104°C and the condenser at 38°C. The time required for a stripping–reclaiming operation is about 5 hours. Design criteria are summarized in Table 7-5.

Upper Occoquan Sewage Authority, Va.   This regional plant replaced 11 small secondary plants which discharge into tributaries of a water supply reservoir which serves as the raw water source for water treatment plants serving about 500,000 people in the Virginia suburbs of Washington, D.C.[31] The effluent eventually reaches a water supply reservoir in which nitrogen is believed to be one of the principal eutrophication factors. The effluent standards are shown below:

Parameter	Value
BOD_5, mg/l	1.0
COD, mg/l	10.0
Suspended solids	unmeasurable
Phosphorous, mg/l	0.1
Methylene Blue Active Substances (MBAS), mg/l	0.1
Turbidity, JTU	0.4
Coliforms, total/100 ml	2
Nitrogen, total	1.0 mg/l

The ion exchange process follows secondary treatment, chemical coagulation, filtration, and activated carbon adsorption. The selective ion exchange process was selected primarily because of its inherent reliability and efficiency and the minimal effect on total dissolved solids (TDS). When a 2–3 percent solution of salt is used for regeneration, elution of this salt remaining in the bed after regeneration at the start of the service cycle may result in an increase in TDS of about 50 mg/l. The increment would be greater with stronger regenerants. The TDS effect is much less than for the breakpoint process, however.

Carbon fines and other solids trapped in the clinoptilolite bed are removed by backwashing before the regeneration cycle. These backwash wastes are returned to the treatment process, typically to the chemical coagulation process, where the solids are removed in the precipitation process and trapped in the chemical sludge.

The selective ion exchange regenerant recovery was initially planned to be accomplished by breakpoint chlorination of the ammonium using electrolytic cells. A delay in the project of about one year occurred following final design before project funding developed. During this period, the ammonia removal and recovery process (ARRP) was developed by the design engineers. Based on pilot plant results, it was concluded that the annual operating cost for the plant could be reduced by $375,000 at a flow of 15 mgd. In addition, the electrical energy requirement would be only 10 percent of the electrolytic cell breakpoint chlorination process needs. Also, a byproduct would be obtained in the form of ammonium sulfate, a common chemical fertilizer. Based on this information, the Authority authorized a redesign of the regeneration facilities to incorporate the ARRP process.

Eight ion exchange beds operate in parallel and are separated into two independent trains, each with four beds and a common manifold. Each bed is a horizontal steel pressure vessel, 10 ft (3.05 m) in diameter by 50 ft (15.24 m) long containing a 4 ft (1.2 m) deep bed of clinoptilolite. Each

TABLE 7-6 Design Criteria for Selective Ion Exchange Process for Ammonium Removal at the Upper Occoquan Plant (Virginia)[1]

Flow Rate	15 mgd
Beds in service	4
Beds in regeneration	2
Beds—backup capacity	2
Flow per bed	3.75 mgd
Bed loading rate	10.82 BV/hour
	5.25 gpm/ft^2
Backwash rate	8 gpm/ft^2
Bed volumes to exhaustion	145
Average ammonia removal efficiency	95 percent
Average influent ammonia nitrogen concentration	20 mg/l
Average effluent ammonia nitrogen concentration	1 mg/l
Normal concentration of ammonia nitrogen at initiation of regeneration	2.5 mg/l
Clinoptilolite size	20 × 50 mesh
Clinoptilolite depth	4 ft

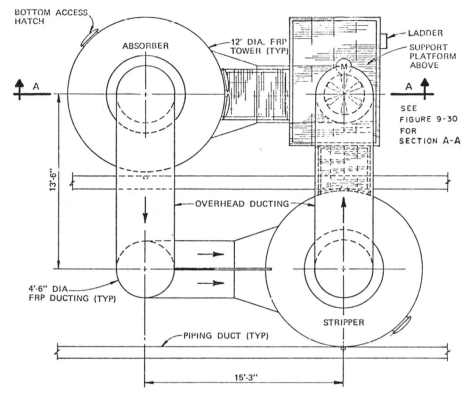

Fig. 7-25   Plan view of ARRP module, Upper Occoquan plant.[1]

parallel train is completely independent, including piping, instrumentation and control, and electrical supply. In addition, such backup facilities as key instrumentation and control are available in each train.

Table 7-6 is a summary of the design criteria for the ion exchange process at a flow of 15 mgd. The system is entirely automated, using automatic valves in a manner similar to most larger water treatment plant filtration facilities. Regeneration is initiated either on a run time basis, volume throughput basis, or manually. Backwashing is done before each regeneration. Backwash water is returned to the wastewater process (typically to the chemical coagulation process) or to the plant headworks.

The beds are regenerated with a 2 percent sodium chloride solution. The regeneration process is that shown in Figure 7-22. The regenerant recovery system consists of four 375,000 gal tanks, a regenerant pumping system and associated automatic valves, two 35 ft diameter clarifiers for magnesium hydroxide removal, and 18 ARRP modules. Figures 7-25 and 7-26 illus-

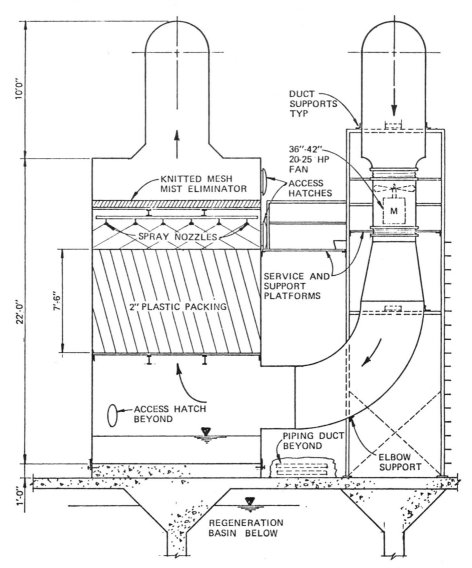

Fig. 7-26   Section of ARRP module, Upper Occoquan plant.[1]

trate the ARRP module design. The basic tower units are 12 ft diameter
fiberglass tanks. All materials will be fiberglass or PVC. Each tower has
a 25 hp fan. The air rate is approximately 34,000 cfm/tower at an air to
liquid ratio of 566 ft^3/gal. Tower air velocities are 300 ft/minute. Knitted
mesh mist eliminators prevent moisture carryover from tower to tower.
The total system headloss is about 2.5–3.0 in., of which about 1.5 in. is in

TABLE 7-7   Regeneration and Regenerant Recovery System Design Criteria at the Upper Occoquan Plant (Virginia)[1]

Regeneration system[a]	
number of regenerant tanks	4
size of each tank	375,000 gal
number of beds regenerated at once	2
number of regeneration cycles per day	3.58
regeneration bed volumes	39–44
Regenerant recovery system[a]	
recovery system flow rate	1080 gpm
operation time per day	16 hours
clarifiers	
number of units	2
diameter	35 ft
overflow rate	800 gpd/ft^2
Ammonia removal and recovery process[a]	
number of ARRP modules	18
liquid loading rate	760 gpd/ft^2
air to liquid loading rate	566 ft^3/gal
media height	7.5 ft
removal efficiency at 10°C	90 percent
at 20°C	95 percent

[a]At 15 mgd flow rate.

the media. The media is 2 in. diameter polypropylene plastic packing (Tellerette). A summary of the regeneration system design criteria is shown in Table 7-7. The average ammonia-nitrogen concentration from ion exchange beds will be about 1 mg/l. The organic nitrogen is expected to be 0.5–0.8 mg/l and the nitrate-nitrogen is expected to be from 0.1–0.2 mg/l. Thus, the total nitrogen leaving the ion exchange process will be about 1.6–2.0 mg/l. Since the discharge standard is 1.0 mg/l, additional nitrogen removal is necessary. This will be accomplished by breakpoint chlorination of the ion exchange effluent. A dosage of approximately 8–10 mg/l will result in nearly complete removal of ammonia nitrogen. The final effluent is then expected to have a total nitrogen concentration of less than 1 mg/l.

The estimated costs are shown in Table 7-8. Since the initial constructed capacity of 22.5 mgd may be operated at no more than 15 mgd because of state requirements for backup capacity, costs are shown on the basis of operation of two-thirds of the constructed capacity and at the full

TABLE 7-8   Estimated Costs of Selective Ion Exchange at the Upper Occoquan Plant (Virginia)[28]

Item	ESTIMATED COSTS, $/million gal	
	at 15 mgd	at 22.5 mgd
Operating and maintenance[a]		
Chemicals		
NaOH	$ 26.80	$ 26.80
NaCl	7.10	7.10
H_2SO_4	9.80	9.80
	$ 43.70	$ 43.70
Income from sale of $(NH_4)_2 SO_4$ @ $43/ton	$(12.60)	$(12.60)
Net chemical cost	$ 31.10	31.10
Power, 18 hp/million gal @ $0.0192/kwh	6.90	6.90
Labor	17.70	17.70
Total O & M	$ 55.70	$ 55.70
Capital[a]		
$4,470,000, 20 years @ 7%	$ 77.22	$ 51.59
Total annual cost[a]	$132.92	$107.29

[a]August, 1974 costs.

constructed capacity, which would be the more generally applicable circumstance.

BIOLOGICAL NITROGEN REMOVAL

Basis of Process

The nitrogen compounds found in raw sewage may be biologically oxidized to nitrates, provided that the proper aerobic environment is maintained in the biological treatment process. Should the nitrified effluent then be subjected to a period of anaerobiosis, the saprophytic bacteria which make up the bulk of the organisms found in a typical biological waste treatment process can utilize nitrates as electron acceptors. Under these conditions, nitrates and nitrites are reduced to nitrogen gas. This process effects a reduction in the nitrogen content of the wastewater as it escapes from solution. The simplicity of the structures required for such a process and the fact that no liquid or solid waste byproducts are created in the nitrogen removal process have led to many studies of such a system.

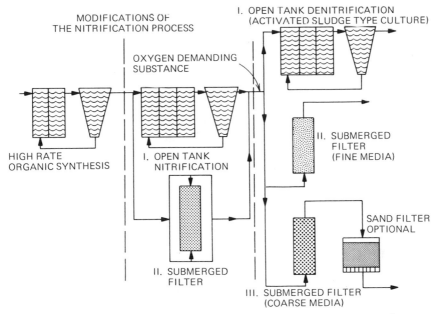

MODIFICATIONS OF
THE NITRIFICATION PROCESS

I. OPEN TANK DENITRIFICATION
(ACTIVATED SLUDGE TYPE CULTURE)

OXYGEN DEMANDING
SUBSTANCE

HIGH RATE
ORGANIC SYNTHESIS

I. OPEN TANK
NITRIFICATION

II. SUBMERGED
FILTER
(FINE MEDIA)

SAND FILTER
OPTIONAL

II. SUBMERGED
FILTER

III. SUBMERGED FILTER
(COARSE MEDIA)

Fig. 7-27   Alternative approaches to biological nitrogen removal.[1]

All of the system designs proposed for this process are based on the sequential steps of (1) oxidation of carbonaceous materials, (2) oxidation of nitrogenous materials, and (3) reduction of nitrates. Figure 7-27 summarizes the various approaches that have been proposed. In some cases, the carbonaceous and nitrogenous oxidation steps are combined in one stage activated sludge system rather than the two stages shown in Figure 7-27.

Nitrification Process Considerations

Biological nitrification is carried out by two groups of autotrophic bacteria, *Nitrosomonas* and *Nitrobacter*, which derive their energy from the oxidation of inorganic nitrogen compounds. These two groups of bacteria must operate sequentially to achieve nitrification because *Nitrosomonas* can oxidize ammonia to nitrite but cannot complete the oxidation to nitrate. *Nitrobacter* is limited to the oxidation of nitrite to nitrate.

The stoichiometric reaction for oxidation of ammonium to nitrite by *Nitrosomonas* is:

$$NH_4^+ + 1.5\ O_2 \longrightarrow 2H^+ + H_2O + NO_2^-$$

The reaction for oxidation of nitrite to nitrate by *Nitrobacter* is:

$$NO_2^- + 0.5O_2 \longrightarrow NO_3^-$$

Nitrosomonas obtains more energy per mole of nitrogen oxidized than *Nitrobacter*. Thus, significant nitrite concentrations are not observed in systems where both groups of bacteria are performing.

Nitrification processes fall into several major classifications: *Single stage*—the carbonaceous and nitrogenous oxidation steps are combined in a single reactor, rather than first achieving the carbonaceous oxidation in a separate reactor; *suspended growth*—nitrification is achieved in a mixed reactor (as in a conventional aeration tank); *attached growth*—nitrification is accomplished by organisms attached to a growth media (as in a trickling filter). Both suspended growth and attached growth systems are used for denitrification. Several factors basic to the nitrification process affect the design of any process configuration. Some of the critical factors are discussed below.

*p*H.   *p*H has been found to have a major effect on the rate of nitrification. Figure 7-28 presents one curve developed for activated sludge at 20° C.[32] Although other investigators have reported a wide range of values for optimum *p*H, it is widely recognized that as the *p*H moves to the acid range, the rate of nitrification declines.[1] This has been found to be true for both unacclimated and acclimated cultures, although acclimation tends to moderate *p*H effects. Nitrifiers can adapt to *p*H levels as low as 5.5.[33] Sudden decreases in *p*H (7.2–5.8) have proven to inhibit nitrification until *p*H increases, but not to have a residual toxic effect.[1] The nitrification process

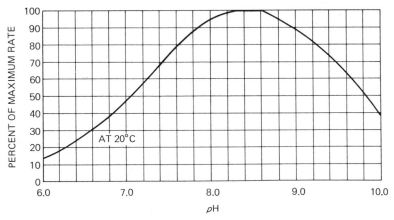

Fig. 7-28   Percentage of maximum rate of nitrification at constant temperature vs. *p*H.[32]

itself can depress the pH to undesirable levels. The overall nitrification reaction can be represented as follows:

$$NH_4^+ + 2O_2 + 2\ HCO_3^- \longrightarrow NO_3^- + 2H_2CO_3 + H_2O$$

Alkalinity is destroyed by the oxidation of ammonia and carbon dioxide (H_2CO_3 in the aqueous phase) is produced. Approximately 7.14 mg of alkalinity as $CaCO_3$ is destroyed per mg of ammonia nitrogen oxidized.

Since nitrification reduces the HCO_3^- level and increases the H_2CO_3 level, it tends to depress pH. This tendency is offset somewhat by stripping of carbon dioxide from the liquid by the process of aeration; the pH is thereby shifted upward. If the carbon dioxide is not stripped from the liquid, as in enclosed high purity oxygen systems, the tendency for pH depression is enhanced. Even in open systems where the carbon dioxide is continually stripped from the liquid, severe pH depression can occur when the alkalinity in the wastewater approaches depletion by the acid produced in the nitrification process. For example, if in a wastewater 20 mg/l of ammonia nitrogen is nitrified, 143 mg/l of alkalinity as $CaCO_3$ will be destroyed. In many wastewaters there is sufficient alkalinity initially

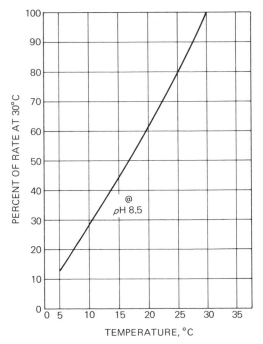

Fig. 7-29 **Rate of nitrification at all temperatures compared with the rate at 30°C.**[32]

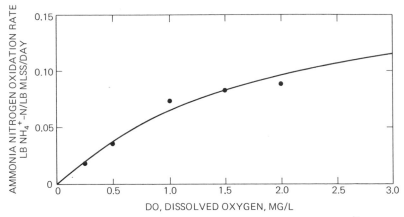

Fig. 7-30   **Effect of dissolved oxygen on nitrification rate.**[34]

present to leave a sufficient residual for buffering the wastewater during the nitrification process. Because of the effect of pH on nitrification rate, it is especially important that there be sufficient alkalinity in the wastewater to balance the acid produced by nitrification. Caustic or lime addition may be required to supplement moderately alkaline wastewaters.

Temperature.   The rate of nitrification is also strongly affected by temperature. Figure 7-29 presents this effect for activated sludge as measured

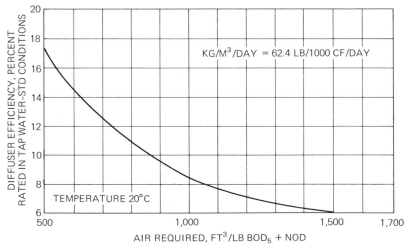

Fig. 7-31   **Relationship of aeration air requirements for oxidation of carbonaceous BOD and nitrogen.**[1]

by one observer.[32] Attached growth systems are not as sensitive to temperature variations as suspended growth systems.[1] Attached growth systems can compensate for colder temperatures by accumulation of thicker slime growths, while suspended growth reactors are limited in this regard by the solids thickening capabilities of the final clarifier.

Dissolved Oxygen. The concentration of dissolved oxygen (DO) has a significant effect on the rates of nitrifier growth and nitrification in biological waste treatment systems.[1] Figure 7-30 summarizes the results noted in one series of tests[34] for a single stage system. Nitrification can be achieved at DO levels of 0.5 mg/l, but at significantly lower rates than at higher DO levels. Figure 7-31 illustrates air requirements as a function of diffuser efficiency.

Ammonia Concentration. Nitrification rates are relatively unaffected by NH_4^+-N concentration greater than 2.5 mg/l.[1] As NH_4^+-N concentrations drop below 2.5 mg/l, the rate of nitrification decreases sharply.

Toxins. Certain heavy metals, complex anions, and organic compounds are toxic to nitrifiers. Substances toxic to unacclimated nitrifying organisms include:[1,35]

Organics	Inorganics
Thiourea	Zn
Allyl-thiourea	OCN^-
8-Hydroxyquinoline	ClO_4^-
Salicyladoxine	Cu
Histidine	Hg
Amino acids	Cr
Mercaptobenzthiazole	Ni
Perchlorethylene	Ag
Abietec acid	

It has been suggested[1] that 10–20 mg/l of heavy metals can be tolerated because of low ionic concentrations at pH values of 7.5–8.0. However, precipitated metals that concentrate in the sludge at these pH values can cause serious problems if the pH falls and the precipitate dissolves. Such conditions may occur in the sludge collection zone of the secondary clarifier, or if pH control systems fail. High concentrations of ammonia or nitrite can also be temporarily toxic to nitrifiers. An effective source control program is needed if there are potential dischargers of toxins.

Nitrification Suspended Growth System Design

Aeration System Configuration. The most efficient configuration of aeration basin for nitrification is plug flow because the rate of oxidation of ammonia is essentially linear. Completely mixed basins can achieve nitrification at comparable rates but are likely to produce an effluent slightly higher in ammonia because of increased short circuiting of the aeration basin influent. Variations of the conventional activated sludge process, such as contact stabilization, step aeration, and high rate, are not effective for nitrification.

Aeration Basin Sizing, Single Stage. Detailed design equations have been published for single stage systems[1] and should be consulted for design. For most applications (temperature of 18°C), provision of a sludge age of 7 days or more or an aeration basin loading ratio of less than 0.25 pounds of BOD$_5$ per day per pound of mixed liquor volatile suspended solids (MLVSS) in the aeration tank will provide nitrification. Single stage nitrification is achieved in the Flint, Michigan full scale plant[36] with these conditions: aeration @ 7 hours at Ave. Q, minimum DO = 2.0 mg/l; food to microorganism (F/M) $\leqslant$ 0.25; sludge age $<$ 6.5 days; final sedimentation 525 gpd/ft^2 average. Nitrification varied between 72 and 98 percent for average monthly values over 12 month period (low values in winter). A minimum solids retention time of 10 days at 5°C was reported[37] as required for nitrification.

Aeration Basin Sizing, Two Stage. The first stage activated sludge system is generally designed to produce an effluent BOD of 50 mg/l or less. The second stage aeration basin is often designed based on experimentally determined nitrification rates. Figure 7-32 summarizes nitrification rates observed at several locales in pilot studies. The nitrification rates shown in Figure 7-32 show that as temperature rises, nitrification rates increase. The BOD$_5$/TKN* ratio also can be seen to strongly influence the nitrification rates. The lower the BOD$_5$/TKN ratio the higher the nitrification rates. The lower ratio results in a higher percentage of organisms being nitrifiers. The effect of pH depression on nitrification rates is apparent from Figure 7-32.

In the absence of pilot data specific to a particular situation, Figure 7-32 can be used to approximate design nitrification rates based upon BOD$_5$/TKN ratio in the influent, the minimum temperature for nitrification, and the mixed liquor pH. Once the design value of the nitrification rate is

*Total Kjeldahl nitrogen.

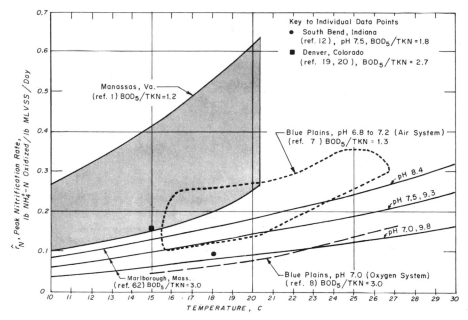

Fig. 7-32   Observed nitrification rates at various locations.[1]

established, the design can proceed in a manner similar to the F/M design approach adopted for activated sludge design. The total nitrogen load per day and the nitrification rate are used to establish the mass of solids that must be maintained in the nitrification reactor. The volume of the reactor is determined from the allowable mixed liquor solids level and the inventory of solids required. The allowable mixed liquor level is influenced primarily by the efficiency of solids–liquor separation. Figure 7-33 is an illustrative curve prepared to show the permissable volumetric loading of nitrification tanks at Marlborough, Massachusetts.[32] The following is a sample calculation for computing the tank size made from the Marlborough data:[32]

Design flow = 10 mgd
Average concentration to nitrification tanks = 15 mg/l
Minimum temperature = 10°C
Operating pH = 7.8
MLVSS concentration = 1,500 mg/l
NH₃ load
    Average = 10 × 8.34 × 15 = 1250 lb/day
    Maximum = 1250 × 1.5 = 1870 lb/day

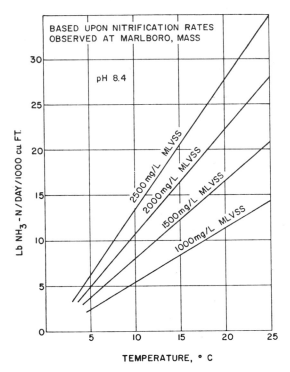

Fig. 7-33 Permissible nitrification tank loadings at Marlborough, Massachu-setts.[32]

Tank volume at 10°C, MLVSS = 1500 mg/l
From Figure 7-33, volumetric loading = 8.2 lb per 1000 ft^3
Tank volume = 1870/8.2 × 10^3 = 228,000 ft^3
Tank volume adjusted to pH 7.8 (see Fig. 7-28) = 228,000/0.88 = 226,000
Detention period = (260,000 × 24 × 7.48)/(10 × 10^6) = 4.65 hours

Aeration Capacity. Ammonia must be oxidized during the period it is in the nitrification reactor and sufficient oxygen must be available at all times if nitrification is to be achieved. Diurnal variations in ammonia concentrations can be significant and must be recognized in sizing the air supply. In some cases, diurnal variations in ammonia may be so great as to justify in-plant flow equalization facilities rather than incurring the added capital and operating costs of aeration tankage and facilities. The maximum hourly ammonia load should be used as the basis for sizing the air supply. Diffused air systems are well suited to modulation to meet variation in demand. Mechanical aerators sized for peak demands will overaerate the mixed liquor for most of the day.

The theoretical oxygen requirement for nitrification, neglecting synthesis, is 4.57 mg O_2/mg NH_4^+-N. An oxygen requirement sufficiently accurate to be used in calculations for aeration requirements is 4.6 mg O_2/mg NH_4^+-N. This oxygen demand for nitrification is significant; for instance, if 30 mg/l of ammonia nitrogen is oxidized by the nitrification system, about 138 mg/l of oxygen will be required.[1] Of course, in virtually all practical nitrification systems, oxygen demanding materials other than ammonia are present in the wastewater, raising the total oxygen requirements of nitrification systems even higher, as illustrated below. Calculation of the total oxygen requirements for the aeration system can be illustrated by the following example, which assumes that the peak BOD level occurs concurrently with the peak nitrogen load:

Aeration Basin Influent Quality:
    design flow = 10 mgd
    average BOD_5 load = 4170 lb/day
    average TKN load = 2100 lb/day
    peak/average TKN load ratio = 1.8
    peak/average BOD_5 load ratio = 1.5 (coincident with peak TKN load)

Average load condition:
    BOD_5 = 4170 lb/day
    NOD = 4.6 × 2,100 = 9660 lb/day
    BOD_5 × NOD = 13,830 lb/day

Peak hourly load condition:
    BOD_5 = 4170 × 1.5 = 6260 lb/day
    NOD = 9660 × 1.8 = 17,390 lb/day
    BOD_5 + NOD = 23,550 lb/day

Ratio peak hour to average load: 1.7

Of course, the volume of air required in a diffused aeration system will depend on the transfer efficiency of the specific system. Figure 7-31 shows the relationship between air requirements and transfer efficiency. In the above example, if we assume an efficiency of 12% and an air requirement of 725 ft^3/lb BOD_5 + NOD, the aeration requirements may be calculated as follows:

Average aeration requirements:
    725 × 13,830/1440 = 6,963 cfm

Peak aeration requirement:
    725 × 23,550/1440 = 11,860 cfm

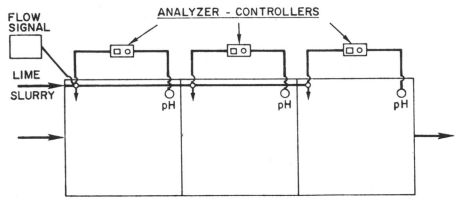

Fig. 7-34 *p*H control for nitrification system.[32]

*p*H control.   As noted earlier, nitrification will destroy about 7.2 lb of alkalinity per pound of NH₃-N oxidized. If the wastewater is deficient in alkalinity, alkaline feed and *p*H control must be provided. Sufficient alkalinity should be provided to leave a residual of 30–50 mg/l after complete nitrification. Either sodium hydroxide or lime may be used for this purpose. Lime is the predominantly used form because of its lower cost. For plug flow reactors, it may be prudent to place *p*H probes and chemical feed points at several points in the reactor (Figure 7-34).

Sedimentation.   The final clarifier overflow rate should not exceed a peak hourly rate of 1000 gpd/ft² and should have a minimum depth of 12 ft. The basin should be equipped with a surface skimming device. Because it is a time consuming task to rebuild an adequate population of the slow growing nitrifiers, adequate final sedimentation basin capacity should be available with one final clarifier out of service. Sludge return capacity of a maximum of 100 percent of the average daily flow should be provided. Rising sludge caused by denitrification has occasionally been a problem in nitrification systems, particularly with warm wastewaters. Control measures to prevent floating sludge which can be incorporated into plant design include:[1] provision of rapid sludge removal in sedimentation tank design to prevent sufficient contact time for bubble formation to occur; flexibility of influent feed points (e.g., allowing switching from step aeration to plug flow in warm weather periods) to provide the operator with options in process operation; provision for chlorination of the return activated sludge to control sludge bulking and to reduce the sludge volume index, to allow more rapid sludge withdrawal from the sedimentation tanks.

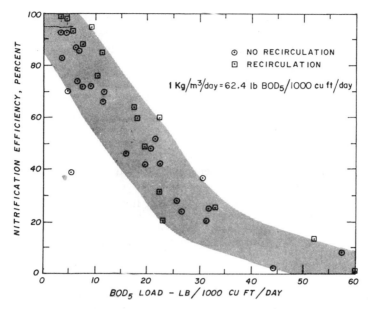

Fig. 7-35   Effect of organic load on nitrification efficiency of rock trickling filters.[1]

Nitrification by Attached Growth Systems

Rock Trickling Filters. Nitrification is usually achieved only at low hydraulic loading rates and warm temperatures. The nitrification process efficiency is difficult to maintain in a conventional rock trickling filter. Significant nitrification is not likely to occur at loading rates greater than 12 lb BOD/1000 ft^3/day.[38] Figure 7-35 summarizes data from several full scale trickling filter operations. If a nitrification system design is to be based on a trickling filter system, synthetic media filters offer a more effective design basis. The surface area of synthetic media is substantially greater than that of rock media (27 ft^2/ft^3 typical for synthetic media, vs. 12–18 ft^2/ft^3 for 3 in. rock).

Synthetic Media Trickling Filters. Combined carbon oxidation–nitrification can be achieved in a single stage system using synthetic media towers.[38,39] BOD loading is the limiting factor in this application. The following guidelines apply:

Influent Ammonia-Nitrogen, 25 mg/l
Media Area = 27 ft^2/ft^3 (94 percent void volume)

Forced draft provided

Maximum BOD loading $= 25$ lb/1000 ft^3/day

Recycle to maintain minimum flow of 0.5–1 gpm/ft^2 to keep media from drying out

For nitrification of secondary effluents in a two stage system, the contact time in the filter becomes limiting and the following guidelines apply:

Influent BOD$_5$, 50 mg/l

Influent Ammonia-Nitrogen, 25 mg/l

Media Area $= 27$ ft^2/ft^3 (94 percent void void volume)

Recycle to maintain constant flow

For media depth $= 21.5$ ft, the following flow rates may be used:

Nitrification Performance	Flow Rate @ Wastewater Temp. Shown	
	65° F	44° F
90 percent	0.5 gpm/ft^2	0.5 gpm/ft^2
85 percent	0.75 gpm/ft^2	0.65 gpm/ft^2
80 percent	1.0 gpm/ft^2	0.75 gpm/ft^2
75 percent	1.5 gpm/ft^2	0.85 gpm/ft^2

Rotating Biological Disks.   Rotating biological disks may also be used in single stage or two stage nitrification systems.[44] This process consists of a series of closely spaced disks (10–12 ft in diameter) which are mounted on a horizontal shaft and rotated while about one-half of their surface area is immersed in wastewater (Figure 7-36). The process has been used in Europe for several years. The disks are typically constructed of light-weight plastic. When the process is placed in operation, the microbes in the wastewater begin to adhere to the rotating surfaces and grow there until the entire surface area of the disks is covered with a 1/16–1/8 in. layer of biological slime. As the disks rotate, they carry a film of wastewater

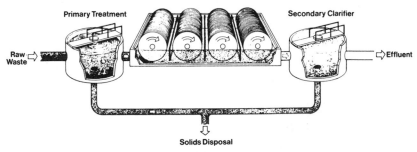

Fig. 7-36   Rotating biological disk process.

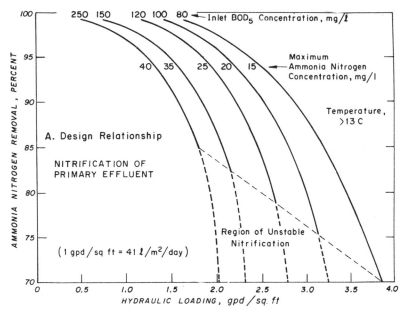

Fig. 7-37   Design criteria for single stage nitrification with rotating biological disks.[1]

into the air where it trickles down the surface of the disks, absorbing oxygen. As the disks complete their rotation, this film mixes with the reservoir of wastewater, adding to the oxygen in the reservoir and mixing the treated and partially treated wastewater. As the attached microbes pass through the reservoir, they absorb other organics for breakdown. The excess growth of microbes is sheared from the disks as they move through the reservoir. These dislodged organisms are kept in suspension by the moving disks. Thus, the disks serve several purposes: they provide media for the growth of attached microbial growth, they contact the growth with the wastewater, and they aerate the wastewater and suspended microbial growth in the wastewater reservoir. The speed of rotation is adjustable. The attached growths are similar in concept to a trickling filter, with the exception that the microbes are passed through the wastewater rather than the wastewater being passed over the microbes.

As the rotating disks operate in series, organic matter is removed in the first disk stages and subsequent disk stages are used for nitrification. Figure 7-37 summarizes the design criteria in terms of hydraulic loading (gallons per day per unit surface of available surface area). When the ammonia concentration exceeds the maximum ammonia concentration on the appropriate BOD curve, the curve for the ammonia concentration should be used. The region of unstable nitrification is a region of hydraulic loading

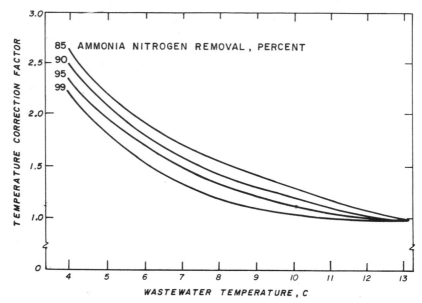

Fig. 7-38   **Temperature effects on single stage nitrification with rotating biological disks.**[1]

where a slight change in either the hydraulic loading or the influent BOD strength could result in a displacement of the nitrifying population. It is advisable to avoid this region even during daily peak flow conditions. The nitrifying capability of the disks has been found to be relatively constant in the temperature range of 15–26°C.[40] Temperature correction factors derived from pilot data are shown in Figure 7-38. These correction factors should be used to reduce the design by hydraulic loading determined in Figure 7-37 for any wastewater temperature lower than 13°C.

The rotating biological disk (RBD) process, may also be applied to nitrifying secondary effluents. In this application it may be possible to eliminate the downstream clarifier when the secondary effluent being treated has a BOD_5 and suspended solids less than about 20 mg/l. Under this circumstance the very low net growth occurring in the nitrification stage causes the process effluent suspended solids to approximately equal the influent solids level. If lower levels of suspended solids are required, the process could be followed directly by tertiary filtration without the need for intermediate clarification.[1] Figure 7-39 presents criteria for design of nitrification systems utilizing 4 stages of disks, the most commonly employed configuration, for nitrification of secondary effluents. It can also be employed for other numbers of stages using the relative capacities shown in the

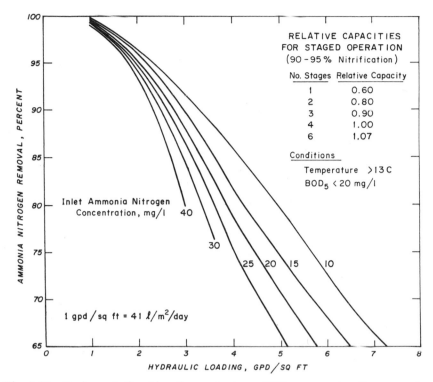

Fig. 7-39 Design relationships for a 4 stage RBD process treating secondary effluent.[1]

figure. The relative capacity factor should be applied to the hydraulic loading to obtain design values for situations where other than 4 stages are employed. Very few data are available for temperatures below 13°C. For applications to secondary effluents below 13°C, the recommendation has been made that the temperature correction factors developed for combined carbon oxidation–nitrification be applied for separate stage nitrification.[1]

Submerged Filters.   Submerged filters have recently been developed and applied for nitrification. The process consists of a bed of media (Figure 7-40) upon which biological growth occurs overlaying an inlet chamber, much as in an upflow carbon column or filter. Wastewater is distributed evenly across the bottom of the filter. The wastewater flow is upward, and a nitrifying biological mass is developed on the large surface area of the media.

Several types of media have been used successfully, including 1–1.5

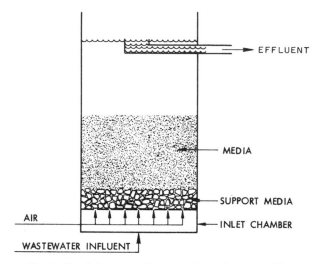

Fig. 7-40 Schematic diagram of a submerged filter.

in. stones, 0.5 cm gravel, 1.8 mm (effective size) anthracite and a plastic media. Approximate empty bed detention times to oxidize 20 mg/l NH₃-N with 1 in. stones were found to be as follows:[41]

Temperature	Detention Time
25° C	30 minutes
15° C	60 minutes
10° C	90 minutes
5° C	120 minutes

The media type affects the amount of surface available for nitrifier growth. Generally, the smaller the media, the higher the oxidation rate due to the higher surface area per unit volume. Occasional backflushing is needed to avoid clogging of the filter. The frequency will depend upon the media type and size and influent quality. Several means have been employed for supplying the necessary oxygen including: injection of air into the feed line entering the filter, distribution of air across the filter floor, injection of oxygen directly into the filter, and injection of oxygen into a reaction chamber prior to entry into the filter. Effluent recycle at ratio of 2–3:1 has been used to provide further oxygen. Filter effluent BOD and SS levels are affected by the type of aeration. Preoxygenation produces effluents of similar quality to those from tertiary multimedia filtration. Injection of air or oxygen into the filter causes continuous shearing of the biological

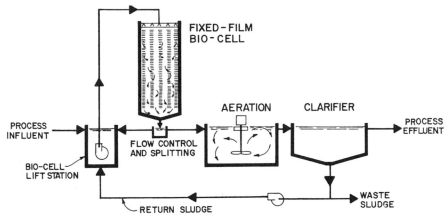

FIXED – FILM
BIO - CELL

AERATION   CLARIFIER

PROCESS
INFLUENT

PROCESS
EFFLUENT

FLOW CONTROL
AND SPLITTING

BIO-CELL
LIFT STATION

WASTE
SLUDGE

RETURN SLUDGE

Fig. 7-41   ABF process flow diagram.

film from the media, resulting in lower reductions of BOD and suspended solids. When very low levels of effluent solids are required, effluent filtration may be required in this latter case.

Activated Biofilter. The activated biofilter (ABF) process operates as a two stage biological system. Figure 7-41 shows the following major components of the ABF system: (a) a lift station, (b) a fixed film reactor or "biocell", (c) flow control and splitting, (d) a short detention aeration basin, (e) a clarifier, and (f) return sludge facilities. Influent waste is combined with biological solids returned from the final clarifier and biocell recycle to form a mixed liquor that is pumped at a constant rate to the biocell surface. This mixture is distributed over a horizontal media, creating an attached microorganism growth. Biocell underflow proceeds to a flow control and splitting arrangement where the flow is automatically proportioned. Part of it is recycled to the biocell lift station, and the remainder passes to the aeration basin. The second stage reactor is a completely mixed activated sludge unit designed to oxidize biodegradable organics further and to provide a flocculent mixed liquor prior to final clarification. The final clarifier is a conventional rapid sludge removal design. Settled biological solids are returned to the system and combined with process influent at the biocell lift station.

It has been reported[45] that this system can provide complete nitrification at average biocell loadings of 200–250 lb BOD/1000 ft^3/day (with peaks of 450 lb BOD/1000 ft^3/day) with an aeration basin detention time of 3–4 hours. Suggested design criteria are shown in Table 7-9.

TABLE 7-9 General Design Parameters for Treating Domestic Wastewater with ABF Nitrification Process.

Parameter	Units	Typical Value	Range
Effluent criteria			
5-Day BOD	mg/l	15	5–30
Suspended solids	mg/l	20	15–30
NH_3-N	mg/l	1.0	0.5–2.5
Biocell parameters			
Organic load	lb BOD_5/day/1000 ft³	200	100–350
Media depth	ft	14	5–22
BOD_5 Removal	percent	65	55–85
Hydraulic parameters			
Biocell recycle	[a]	1.5Q	0.5–2.0 Q
Sludge recycle	[a]	0.5Q	0.3–1.0 Q
Biocell flow	[a]	3.0Q	2.3–4.0 Q
Biocell hydraulic load	gpm/ft²	3.5	1.5–5.5
Aeration parameters[b]			
Detention time	hours	3.5	2.5–5.0
Organic load	lb BOD_5/day/1000 ft³	25	20–40
Ammonia load	lb NH_3-N/day/1000 ft³	10	5.0–15.0
F/M	lb BOD_5/day/lb MLVSS	0.13	0.1–0.2
MLVSS concentration	mg/l	3000	1500–4000
MLSS concentration	mg/l	4000	2000–5000
Carbonaceous oxygen[c]	lb O_2/lb BOD_5	1.4	1.2–1.5
Clarifier parameters			
Overflow rate	gpd/ft²	600	300–1200
Solids loading	lb/hour/ft²	1.0	0.5–2.0
Return sludge concentration	percent	1.5	1.0–3.0
Sludge Production	lb Volatile solids/lb BOD_5 removed	0.45	0.30–0.55

[a] Based on design average flow and secondary influent BOD_5 = 150 mg/l.
[b] Based on aeration BOD_5 loading after biocell removal.
[c] Total oxygen utilization = carbonaceous oxygen + 4.6 lb O_2/lb NH_3-N oxidized.

Denitrification Process Considerations

The biological process of denitrification involves the conversion of nitrate nitrogen to a gaseous nitrogen species, primarily nitrogen gas. As opposed to nitrification, a relatively broad range of bacteria can accomplish denitrification, including *Psuedomonas, Micrococcus, Achromobacter* and *Bacillus*.[1] Many bacteria can shift between using oxygen and nitrate (or nitrite) rapidly and without difficulty. Denitrification is achieved by contacting nitrified wastewater with biomass in the absence of oxygen. The apparent simplicity of the overall process is lessened by the fact that the reduction of nitrate proceeds too slowly to be practical without the addition of biologically degradable organic material to the anaerobic step. Several early investigators added raw sewage to the denitrification basin to speed the reaction, but this has the limitation of adding unoxidized nitrogen compounds and additional BOD to the final effluent. Most recent investigators have used methanol to accelerate the biological denitrification.

A method of estimating the quantity of methanol needed for denitrification based on the following considerations has been presented:[46]

First Step of Denitrification:

$$NO_3^- + \tfrac{1}{3}CH_3OH \longrightarrow NO_2 + \tfrac{1}{3}CO_2 + \tfrac{2}{3}H_2O$$

Second Step of Denitrification:

$$NO_2^- + \tfrac{1}{2}CH_3OH \longrightarrow \tfrac{1}{2}N_2 + \tfrac{1}{2}CO_2 + \tfrac{1}{2}H_2O + OH^-$$

Overall:

$$NO_3^- + \tfrac{5}{6}CH_3OH \longrightarrow \tfrac{1}{2}N_2 + \tfrac{5}{6}CO_2 + \tfrac{7}{6}H_2O + OH^-$$

Based on the above, one mole of nitrate requires at least five-sixth moles of methanol for complete denitrification, or 1.9 mg/l of methanol is required for each mg/l of nitrate-nitrogen. If the effluent contains dissolved oxygen, then it must be removed before denitrification can occur. This biological reaction can be accomplished by additional methanol as follows:

$$O_2 + \tfrac{2}{3}CH_3OH \longrightarrow \tfrac{2}{3}CO_2 + \tfrac{4}{3}H_2O$$

Each mg/l of dissolved oxygen requires 0.67 mg/l of methanol for its removal. Methanol must also be supplied to satisfy the requirements for bacterial growth. This quantity is an additional amount of about 30 percent over the stoichiometric amounts given in the above equations. From these considerations, the following formula may be used for estimating the total amount of methanol required:

$$C_m = 2.47N_0 + 1.53N_i + 0.87D_0$$

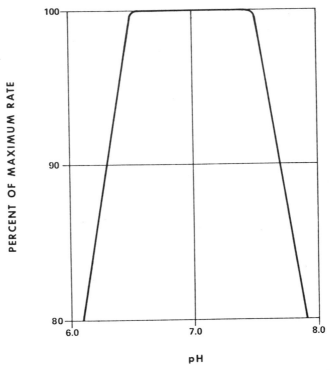

Fig. 7-42   Percent of maximum rate of denitrification vs. pH.[32]

where

C_m = required methanol concentration in mg/l
N_0 = initial nitrate-nitrogen concentration in mg/l
N_i = initial nitrite-nitrogen concentration in mg/l
D_0 = initial dissolved oxygen concentration in mg/l

The value of C_m calculated above is somewhat conservative in that it does not make any allowance to the residual BOD entering the denitrification step. A methanol to nitrate-nitrogen ratio of 3.0 has been suggested as a design guideline.[1] The hydroxide produced in the above reactions reacts with carbonic acid to produce bicarbonate alkalinity. A value for alkalinity production suggested for engineering calculations is 3.0 mg alkalinity as $CaCO_3$ produced per mg nitrogen reduced.[1] Since both the alkalinity concentration is increased and the carbonic acid concentration is reduced, denitrification at least partially reverses the effects of nitrification and raises pH. Denitrification only partially offsets the alkalinity loss caused by nitrification, since the alkalinity gain per mg of nitrogen is only one-half

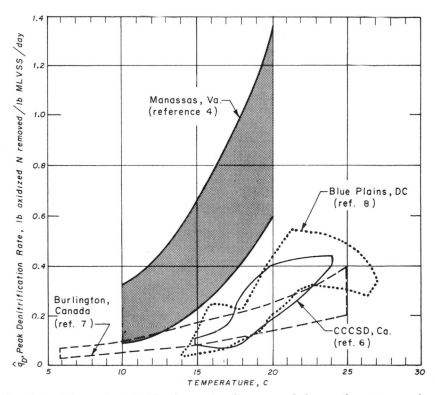

Fig. 7-43 Observed denitrification rates for suspended growth systems using methanol.[1]

the loss caused by nitrification. Methanol is the predominantly used electron donor in the U.S., but recent cost increases have stimulated consideration of alternative materials. Some of the alternatives cause greater sludge production than others. For instance, about twice as much sludge is produced per mg of nitrogen reduced when saccharose is used than when methanol is used. On the other hand, acetone, acetate, and ethanol produced similar quantities of sludge to that produced when methanol is employed.[47] Nitrogen deficient industrial wastes, such as brewery wastes, have been used[48] and are worthy of consideration when available. As with nitrification, pH and temperature have marked effects on the rate of denitrification.

*p*H. Figure 7-42 presents one observation[32] of the effects of pH on denitrification rate. Although other findings may differ somewhat, there is a consensus that denitrification rates are significantly depressed below pH 6.0 and above pH 8.0, with the highest rates occurring between 7.0 and 7.5.

Temperature.   Observations on the effects of temperature have been made at several locations and are summarized in Figure 7-43.

Denitrification Suspended Growth System Design

Mixing.   The suspended growth reactors used for denitrification are mixed mechanically to keep the biomass in suspension but only with the minimal energy input needed to keep the solids in suspension so as to minimize oxygen transfer. The reactors need not be covered. The plug flow configuration should be followed as closely as possible to minimize nitrate leakage.[32] Power requirements for mixing are typically 0.25–0.50 hp/1000 ft^3.[32]

Basin Sizing.   As can be seen from Figure 7-43, denitrification rates have varied substantially from one pilot test to another and it is desirable to conduct pilot tests on the specific wastewater involved to determine design values. Once the design rate has been selected, the permissable tank loadings for various MLSS concentrations can be plotted. The data from Manassas, Virginia offers an illustrative example.[32] Figure 7-44 shows the curves plotted at optimum pH for various MLSS values based upon the observed rates of oxidized nitrogen removed per pound of MLVSS per day. As is apparent from Figure 7-43, the rates observed at Manassas are higher than reported elsewhere and these data are used solely to illustrate a technique for calculating basin size. If pilot tests cannot be run, it would be prudent to select more conservative rates from Figure 7-43 than those observed at Manassas, where analytical techniques are suspected of causing the apparently higher rates.[1] The following calculations illustrate basin sizing for the Manassas data:

> Design flow = 10 mgd
> Average NO_3^--N + NO_2^--N concentration = 15 mg/l
> Minimum temperature = 10°C
> Expected operating pH = 7.3
> MLVSS = 2000 mg/l
> NO_3^--N + NO_2^--N loading
> average = 10 × 8.35 × 15 = 1250 lb/day
> peak = 1250 × 1.5 = 1870 lb/day
> Tank loading at 10°C, optimum pH (from Fig. 7-44) = 26.8 lb/1000 ft^3
> Tank volume at MLVSS = 2000, optimum pH = (1870/26.8) × 10^3 = 70,000 ft^3
> Detention period = (70,000 × 7.48 × 24)/10 mgd = 1.25 hour

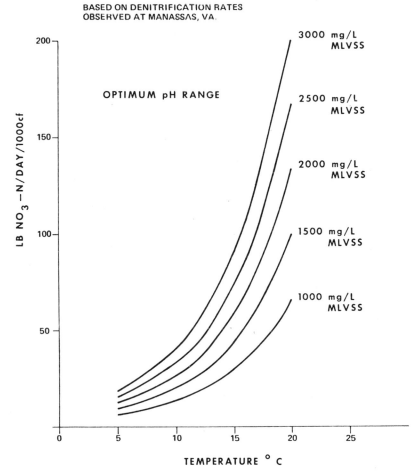

BASED ON DENITRIFICATION RATES
OBSERVED AT MANASSAS, VA.

Fig. 7-44 Permissible denitrification tank loadings.[32]

Such a system would have over twice the tankage needed at 20°C. For this reason, good design will allow for idle operation of part of the capacity during the warm months of the year.

Methanol Feed.   As noted earlier, about 3 mg/l of methanol is typically fed per mg/l of nitrate-nitrogen reduced. Detailed information on design of methanol storage and feeding systems is available.[1] The amount of methanol (or other oxygen demand source) fed must be closely controlled because an excessive feed would result in a residual BOD in the plant effluent.

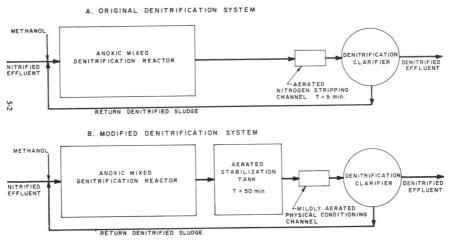

Fig. 7-45   **Process for removal of excess methanol.** [1]

A means of automatically pacing the feed to the incoming nitrate concentration should be provided. Flow pacing is generally not adequate because of variations in nitrate concentration. Unless specific measures are taken to provide for methanol removal, excess methanol addition will cause methanol to appear in the denitrification process effluent. In one instance, a methanol overdose caused an effluent BOD_5 of 106 mg/l.[49] Placing total reliance on the methanol feed control system to prevent methanol overdoses may be unrealistic in small plants. The provision of a methanol removal system as backup to provide failsafe operation may often be justifiable. Figure 7-45 shows one system used for removal of excess methanol. After denitrification, the mixed liquor passes to an aerated stabilization tank. In this tank facultative organisms oxidize any remaining methanol. A period of about 48 minutes has been found to be sufficient for methanol oxidation.[49] In attached growth denitrification systems, the provision of an aerated basin after the denitrification column would not ensure oxidation of excess methanol because there is an insufficient mass of facultative organisms in the column effluent to accomplish the biological oxidation of the carbon. Excess methanol removal systems applicable to attached growth systems have not yet been developed.[1]

Sedimentation.   Settling basins should have overflow rates less than 1000 gpd/ft^2 at peak hour and should be equipped with a surface skimming device with provisions made for returning the scum to the denitrification tank when desired.

Capability of returning sludge to the denitrification tank of up to at least

50 percent and preferably up to 100 percent of average flow should be provided.[32] Provisions should be made to transport sludge from this system to the nitrification system in the event that nitrifying sludge is unavoidably discharged into the denitrification system.

Waste sludge quantities will depend on the oxygen demand source fed to the system. For methanol, waste sludge quantities will be about 0.2 lb/lb methanol fed.[32] The provision of the aerobic reactor for excess methanol removal (Figure 7-45) has been found to minimize problems of rising sludge from denitrification.

Denitrification Attached Growth System Design

Submerged Filters.   Submerged filters using a variety of media have been used for denitrification. The most common aproach is to use relatively fine media (2–4 mm) in a downflow, packed bed. The concept combines two functions in one: (1) it serves the purpose of denitrifying the wastewater; (2) the column serves the purpose of effluent filtration. One commercial system[50] consists of 6 ft of uniformly graded sand 2–4 mm in size. Filtration rates recommended when removing 20 mg/l NO_3^--N from municipal wastewaters are 2.5 and 1.0 gpm/ft^2 for minimum wastewater temperatures of 21°C and 10°C, respectively.[1]

The filter is backwashed with 1 or 2 minutes of air agitation followed by 10–15 minutes of air and water scouring and finally 5 minutes of water rinse. Air and water backwash rates recommended are 6 cfm/ft^2 and 8 gpm/ft^2, respectively.[1] In addition, it has been found that nitrogen gas accumulates in the filter during a filter run. This imposes a loss of head on the filter and requires periodic removal of the trapped gas bubbles in the media. A "bumping" procedure may be used, consisting of a short backwash cycle. "Bumping" at backwash rates of 8–16 gpm/ft^2 for 1 or 2 minutes is required every 4–12 hours for the commercial system discussed above. During one study[51] air backwashing was found to cause a temporary partial inhibition of denitrification that was not present when only a water backwash was used. With an influent nitrate-nitrogen level of 15 mg/l, the effluent nitrate-nitrogen level was 10 mg/l $\frac{1}{2}$ hour after the air–water backwashing and reached 0 mg/l $7\frac{1}{2}$ hours after backwashing. In multiple filter installations, the effluents from the recently backwashed filter would be blended with other normally operating filters, so the impact of this nitrate leakage would be moderated.[1]

Mixed media filters (coal, sand, garnet) have also been used as downflow, packed beds for denitrification. Utilizing a 36 in. mixed media filter (3 in. of 0.27 mm garnet, 9 in. of 0.5 mm sand, 8 in. of 1.05 mm coal, and 16 in. of 1.75 mm coal), essentially complete denitrification of a highly nitrified wastewater has been achieved at filtration rates of 1.5 gpm/ft^2 at temperatures of 10°C, and at 3 gpm/ft^2 at temperatures of 20°C. At applied nitrate-nitrogen concentrations of 10 mg/l, filter run times between 16 and 24 hours to 8 ft of headloss were realized at a filtration rate of 3 gpm/ft^2. At higher applied nitrogen levels, filter runs were reduced in direct relation to nitrogen concentration.[1]

Media selected for downflow packed beds should be of such size that the media can be readily expanded by backwashing because severe plugging problems have occurred with coarser (0.5–1 in. diameter) rock media which cannot be backwashed.[52] Another approach to submerged filters is to use a coarse, plastic media (such as Raschig rings) in upflow columns. Such a system is in use at El Lago, Texas.[1] The relatively large rings provide a high void volume. As a consequence, biomass continuously sloughs from the media, minimizing the requirements for backwashing. However, the media does not build up the layers of biomass that would develop if the void fraction were smaller, as with sand. As a result, denitrification rates for this type of media are lower than for sand or rock. Backwashing, though infrequently used, is still required. At El Lago, Texas, the water backwash rate is 10 gpm/ft^2 coupled with an air backwash rate of 10 cfm/ft^2. Backwashing is routinely done every 4 weeks. Backwashing in this type of column is required to prevent the accumulated solids in the column from continuously sloughing into the effluent and causing high effluent suspended solids.

Still another approach to submerged filters for denitrification utilizes a fluidized sand bed (Figure 7-46).[53,54] Wastewater passes upward vertically through a bed of small media such as activated carbon or sand at a sufficient velocity to cause motion or fluidization of the media. The small media provides a large surface for growth of denitrifiers. In a pilot test, a clean bed of

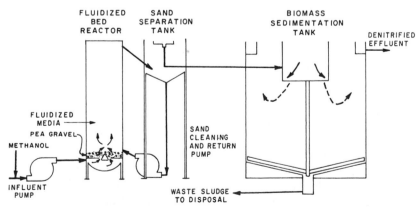

Fig. 7-46 Fluidized bed denitrification system.[54]

1.5 in. of pea gravel and 3 ft of silica sand with an effective size of 0.6 mm and a uniformity coefficient of 1.5 was placed in the column. During operation, the media became completely covered with denitrifier growth and the individual particles grew in size. The fluidized bed depth was 12 ft and the bed settled to a 6 ft depth when the flow was stopped. In a packed bed, this growth of particle size would result in high headloss, channeling, and a loss in efficiency.[1] In an expanded bed, however, there are sufficient voids between the sand particles to provide good liquid contact at modest headloss. This greater biological film development allows higher surface reaction rates (expressed per unit of media surface) than for any other type of column configuration. Since the surface contained in a unit volume is high, higher volumetric loadings are also possible as compared to any other column configuration. Empty bed detention time during the pilot tests was only 6.5 minutes. With proper methanol addition, 99 percent nitrate removal was achieved. The denitrifying filter is followed by a sand separation tank (Figure 7-46) sized at an overflow rate of 13,600 gpd/ft^2. The media settling in the sand separation tank is pumped back to the column, the pumping action shearing the denitrifiers from the media. This sheared biomass would pass through the column and sand separating tank and then settle in the biomass sedimentation tank. If very low levels of suspended solids are required, tertiary filtration would be required. This fluidized bed technology has also been applied for BOD removal and nitrification on pilot scale with promising results.[58]

Plant Scale Biological Nitrogen Removal Experience

Figure 7-27 illustrated the potential combinations of unit processes to provide biological nitrogen removal. Optimum performance of the overall biological system will result in an effluent total nitrogen concentration of about 2 mg/l when treating typical municipal wastewaters.[32] The following discussion summarizes plant scale experience.

Washington, D.C.   The 309 mgd Blue Plains plant in Washington, D.C. represents the largest planned biological removal system in the U.S. The denitrification portion of the process has been planned but the construction of this portion has been postponed while water quality requirements are further evaluated. The system (Figure 7-47) design is based on the three sludge system which utilizes three mixed reactors in series for carbonaceous oxidation, nitrogeneous oxidation, and denitrification.[55] The goal is to produce an effluent nitrogen concentration of 2.4 mg/l. The plant will utilize five stage nitrification tanks to minimize short circuiting. Lime is added to the reactors to maintain the desired pH value for nitrifica-

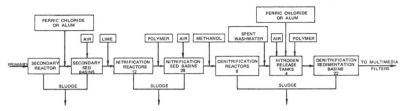

Fig. 7-47 Schematic flow diagram of three sludge system for Blue Plains wastewater treatment plant.[55]

tion. The detention period in the nitrification reactors will average 3.9 hours. The nitrification sedimentation tanks are designed for a peak hydraulic loading of 1210 gpd/ft^2. The denitrification system will employ eight covered, plug flow, 44 ft deep, suspended growth reactors, based on removal of 0.0425 lb NO$_3^-$-N/day/lb MLVSS, four nitrogen release tanks, and 22 sedimentation basins. Methanol will be added to the reactor influent from a feeding system which is automatically controlled from continuous nitrate and plant flow measurements at a ratio of 4.5 lb methanol/lb NO$_3^-$-N. Design permits shutdown of four of the reactors during the summertime. The nitrogen release tanks will remove supersaturated nitrogen gas to avoid rising sludge problems in the sedimentation basins and will provide an additional aeration period for removal of excess methanol. They will also provide for mixing of alum or ferric chloride added for removal of remaining phosphate. The effluent from the denitrification system will be filtered through multimedia filters and chlorinated prior to discharge. Sludge processing facilities will include gravity thickening of primary sludge, flotation thickening of secondary and advanced treatment sludges, vacuum filtration, and sludge incineration.

Central Contra Costa Sanitary District, California. This 30 mgd plant integrates chemical and biological treatment to their mutual advantage. The process flow diagram is shown in Figure 7-48. The goal of the project is to produce water for sale to industry for cooling and process water with 1 mg/l phosphorus and 5 mg/l total nitrogen. Primary treatment follows lime addition and preaeration and is followed with a separate stage nitrification step. The use of lime in the primary removes the bulk of the organic carbon before nitrification, resulting in a very stable oxidation of ammonia to nitrate. Addition of lime also enhances the removal of organic nitrogen, phosphorus, heavy metals, and viruses. Biological denitrification follows nitrification. Multimedia filtration will also be provided prior to distribution of reclaimed water to industry.

About 400 mg/l of lime, as Ca (OH)$_2$, is added along with enough ferric

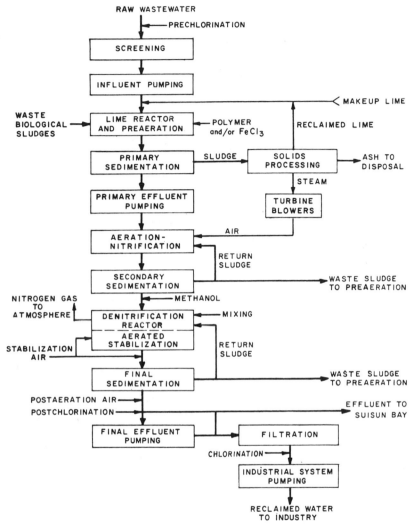

Fig. 7-48 Liquid process flow diagram, CCCSD water reclamation plant (California). [1]

chloride (about 14 mg/l) to maintain the primary basin pH at 11.0. The primary clarifiers are designed at 780 gpd/ft^2 average overflow rate. There is no recarbonation stage between the primary clarification and nitrification stages. External carbon dioxide (CO_2) is added directly to the first pass of the aeration–nitrification tanks as needed. External requirements are minimal, as the chief source of CO_2 is the CO_2 generated in the biological process. Carbon dioxide is derived from the oxidation of both organic carbon and ammonia, an example of lime clarification–nitrification compatibility. The nitrification pH is maintained in the 7.0–8.5 range optimal for nitrification. The nitrification basins provide 6.8 hour detention at average dry weather flow. The nitrification clarifiers are designed at 720 gpd/ft^2. average overflow rate and have a center feed, peripheral discharge arrangement with sludge removal by vacuum-type sludge collectors. Sludge return rate is controlled by the blanket level in each sedimentation tank. Solids retention time is controlled by use of waste activated sludge flowmeters and return sludge concentration. Uncovered reactors are employed for denitrification. Two parallel denitrification tanks consist of nine completely

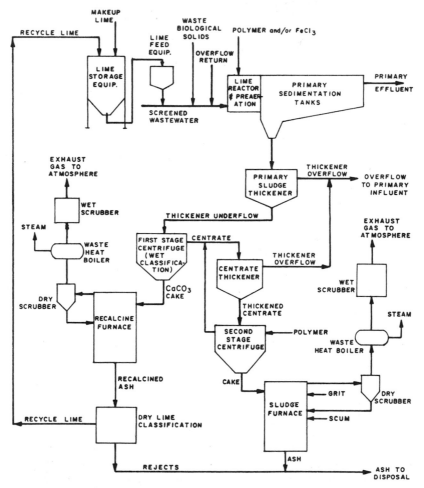

Fig. 7-49   Solids flow diagram at the CCCSD water reclamation plant (California).[1]

mixed reactors in series with an average detention time of 102 minutes. This arrangement allows approximation of a plug flow hydraulic regime. Each reactor is equipped with a low shear, turbine-type mixer to keep the mixed liquor solids in suspension. The last four reactors in the series are equipped with spargers for aeration. With this arrangement, the first five cells in the series can be used for denitrification and the last four cells can be used for denitrification or aerated stabilization for methanol removal depending on whether or not aeration is used. The volume devoted to each function can be varied to meet seasonal loads and temperatures. The denitrification reactors are designed to operate at 0.1 lb NO_3^--N removed/lb MLVSS/day at MLSS levels ranging from 3000 to 4000 mg/l. The denitrified mixed liquor channel between the denitrification tanks and the final sedimentation tanks is aerated for further stabilization. Final sedimentation tanks are similar to the nitrification sedimentation tanks.

All waste sludge is eventually cycled into the primary sedimentation tanks and appears in the primary underflow. Sources of sludge include the suspended solids associated with the raw wastewater, the solids that are wasted to the primary sedimentation tanks from the subsequent biological treatment stages, and the inorganic sludges that are precipitated by chemical action. The solids flow diagram is shown in Figure 7-49. The lime is recovered and reused by using wet classification of the sludges in series with dry classification of the recalciner output.

The process produces an effluent which typically has less than 2 mg/l total nitrogen in the denitrified effluent, a BOD of less than 5 mg/l, and suspended solids less than 8 mg/l. The process performance was quite stable in the pilot tests.[49] The full scale plant was scheduled to go into service in 1976.

Tampa, Florida. A 60 mgd plant utilizing pure oxygen aeration in a two stage nitrification system followed by columnar denitrification is under design for Tampa, Florida.[56,57] The warm sewage temperatures (average 80°F, above 70°F 95 percent of the time) are beneficial to the biological processes. The columnar denitrification process will utilize 2.9 mm effective size sand in 6 ft deep, open gravity flow beds loaded at 2 gpm/ft^2. Based on pilot tests, the anticipated effluent quality is 1.5 mg/l total nitrogen (1 mg/l of organic nitrogen, 0.4 mg/l NH_3-N, and 0.1 mg/l NO_3^--N). Sand larger than 2.9 mm was found to give poorer nitrogen removal while finer sands led to clogging problems.

BREAKPOINT CHLORINATION

At the present time (July 1976) there are no full scale plants in operation that utilize breakpoint chlorination as the principal method of nitrogen removal. However, there have been several pilot and laboratory scale tests and these results are available. The breakpoint process is capable of achieving near complete removal of ammonia. The process is relatively insensitive to changes in temperature and can rapidly adjust to changes in flow. For these reasons the breakpoint process is being used, or proposed for use, as a final polishing step in conjunction with other nitrogen removal methods.

Process Description

Ammonia present in aqueous solution can be removed by oxidation with chlorine, but the exact nature of the reaction and the end products is somewhat uncertain.

Chlorine gas reacts with water, as described in Chapter 6, to form hypochlorous acid and hypochlorite:

$$Cl_2 + H_2O \longrightarrow HOCl + H^+ + Cl^- \qquad (7\text{-}1)$$

$$HOCl \rightleftharpoons H^+ + OCl^- \qquad (7\text{-}2)$$

$$\frac{[H^+][OCl^-]}{[HOCl]} = K_i \qquad (7\text{-}3)$$

The ionization constant K_i varies with temperature; a $20°C$, $K_i = 3.3 \times 10^{-8}$ moles/l. The distribution of hypochlorous acid (HOCl) and hypochlorite ion (OCl⁻) varies with pH as shown in Figure 7-50. At pH 7.5, about 50 percent of the chlorine concentration will be HOCl and 50 percent OCl⁻. As the pH increases, the OCl⁻ concentration increases while the HOCl concentration decreases.

Ammonia reacts with hypochlorous acid in dilute aqueous solution to form choloramines:

$$NH_4^+ + HOCl \longrightarrow NH_2Cl \text{ (monochloramine)} + H_2O + H^+ \qquad (7\text{-}4)$$
$$NH_2Cl + HOCl \longrightarrow NHCL_2 \text{ (dichloramine)} + H_2O \qquad (7\text{-}5)$$
$$NHCl_2 + HOCl \longrightarrow NCl_3 \text{ (trichloramine or nitrogen trichloride)}$$
$$+ H_2O \qquad (7\text{-}6)$$

Chlorine (Cl_2), hypochlorous acid (HOCl), and hypochlorite ion (OCl⁻) are referred to as *free chlorine residual* and the chloramines are called *combined chlorine residual*.

It has been observed in several studies that chlorine added to water containing ammonia reacts as shown in Figure 7-51. The combined chlorine residual increases as chlorine is added up to a mole ratio of 1:1 (weight ratio of 5:1, $Cl_2:NH_4$-N), indicated as point A in Figure 7-51. Additions of chlorine above 5 mg/l per mg/l NH_4-N result in a decrease in the chlorine residual to a minimum, the "breakpoint", at a mole ratio of about 1.5:1 (weight ratio

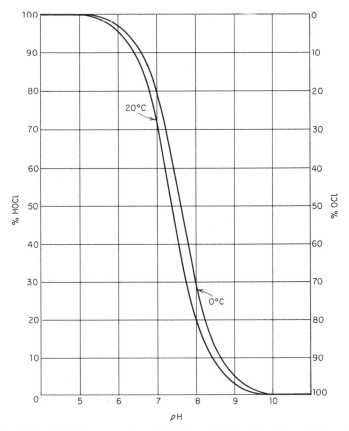

Fig. 7-50   Relative amounts of HOCl and OCl at various pH levels.

of 7.6:1, Cl_2:NH_4-N), indicated as point B on Figure 7-51. At the break-point, essentially all oxidizing chlorine is reduced and all ammonia is oxidized. Further additions of chlorine result in free chlorine residuals.

Chlorine dosage required to reach the breakpoint in wastewater treatment is almost always in excess of the stoichiometric ratio of 7.6:1 because of other chlorine demanding reactions. The actual reactions that take place in reaching the breakpoint are still not clearly understood. Sawyer and McCarty noted in 1967: "Theoretically it should require 3 moles of chlorine for the complete conversion of 1 mole of ammonia to nitrogen trichloride (trichloramine). The fact that only 2 moles of chlorine is required to reach the breakpoint indicates that some unusual reactions occur. Nitrous oxide, nitrogen, and nitrogen trichloride have been identified among the gaseous products of the breakpoint reaction."[59]

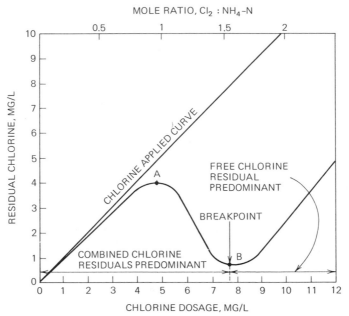

Fig. 7-51   Typical breakpoint chlorination curve.

The U.S. EPA conducted studies in breakpoint chlorination of synthetic buffered solutions and unclarified and lime clarified raw and secondary wastewaters.[60,61] These studies identified the major products of the breakpoint reaction as a function of chlorine dosage, pH, and temperature. These tests identified nitrogen gas as the principal breakpoint reaction product, with relatively small amounts of nitrate and nitrogen trichloride produced; oxygen (O_2), nitrous oxide (N_2O), nitric oxide (NO) and nitrogen dioxide (NO_2) were not present. It was suggested that monochloramine, equation (7-4), was oxidized to nitrogen gas according to the following reaction:

$$2NH_2Cl + HOCl \longrightarrow N_2 + H_2O + 3H^+ + 3Cl^- \qquad (7-7)$$

Combining equations (7-4) and 7-7), the overall reaction to reach the breakpoint is expressed as follows:

$$NH_4^+ + 1.5HOCl \longrightarrow 0.5N_2 + 1.5H_2O + 2.5H^+ + 1.5Cl^- \quad (7-8)$$

It requires one mole of chlorine (Cl_2) to produce one mole of HOCl as shown in equation (7-1), therefore, the weight ratio of $Cl_2:NH_4$-N required to reach the breakpoint is:

$$\frac{1.5 \times 71\ (Cl_2)}{1 \times 14\ (NH_4\text{-N})} = 7.6, \text{ or } 7.6:1$$

The other two products, besides nitrogen gas, found in the EPA studies were nitrogen trichloride (NCl_3) and nitrate (NO_3^-). The formation of NCl_3 can be accounted for as shown in equation (7-6), and the formation of nitrate might be explained by a mechanism suggested by Palin:[62]

$$NHCl_2 + H_2O \longrightarrow NH(OH)Cl + H^+ + Cl^- \qquad (7\text{-}9)$$
$$NH(OH)Cl + 2HOCl \longrightarrow NO_3^- + 2Cl^- + 3H^+ \qquad (7\text{-}10)$$

According to these reactions the chlorine dose required to form nitrogen trichloride is a weight ratio of 15.2:1 and nitrate is a ratio of 20.3:1. Other research in New York confirmed that the gaseous end product of the breakpoint reaction is virtually 100 percent N_2.[63]

Research by Wei and Morris[64] concluded that equation (7-4) does occur as shown but that equation (7-7) does not occur in the way shown. Instead, laboratory tests by Wei and Morris indicated that one of the initial reactants, HOCl, is partially regenerated at a later stage and these researchers proposed that the overall reaction leading to the breakpoint is as follows:

$$NHCl_2 + 0.5H_2O \longrightarrow 0.5N_2 + 0.5HOCl + 1.5H^+ + 1.5Cl^- \quad (7\text{-}11)$$

The breakpoint reaction would then be represented by equations (7-4), (7-5), and (7-11), resulting in equation (7-8) and, or course, the identical stoichiometric ratio of 7.6:1 for $Cl_2:NH_4\text{-}N$. Therefore, regardless of the actual sequence of reactions to reach the breakpoint, there is currently general agreement that the stoichiometric ratio of $Cl_2:NH_4\text{-}N$ is 7.6:1.

Other conclusions from the EPA research in breakpoint chlorination of wastewaters include:[60,61]

1. The chlorine demand required to reach breakpoint decreased, and approached the stoichiometric amount of oxidation of NH_4 to N_2 (7.6:1), as the degree of wastewater pretreatment increased.

$Cl_2:NH_4\text{-}N$ weight ratio required to reach breakpoint	Wastewater source
10:1	Raw
9:1	Secondary effluent
8:1	Lime clarified and filtered secondary effluent

2. The alkalinity of the wastewaters maintained the pH between 6.5 and 7.5. The concentrations of nitrate and nitrogen trichloride in this pH range were always less than 1 mg/l. Near complete removal of $NH_4\text{-}N$ was obtained with only a slight reduction of organic nitrogen within 2 hours' contact time.

3. Highly treated wastewater exhibited breakpoint performances similar to synthetic buffered waters.

EPA studies on synthetic samples containing 20 mg/l NH_4-N indicated the following:

1. The breakpoint reaction was complete within one min in the pH range of 6 to 8.
2. The minimum chlorine dosage required to reach breakpoint occurred in the pH range of 6–7.
3. The amount of nitrate produced at the breakpoint increased from about 1.5 percent of the NH_4-N at pH 5 to about 10 percent at pH 8.0.
4. The amount of nitrogen trichloride produced at the breakpoint decreased from approximately 1.5 percent of the NH_4-N at pH 5 to 0.25 percent at pH 8.0.
5. Temperature variation of 5–40°C at pH 6.0 and 2 hours contact time did not change the required chlorine dose or the amounts of the products.

Several studies have investigated the possibility of adding only enough chlorine to form monochloramines and then removing the monochloramines on activated carbon.[63,65,66] Some advantages would be realized if monochloramine could be removed by activated carbon. The theoretical Cl_2: NH_4-N ratio of 100 percent ammonia removal would drop from 7.6:1 for breakpoint to about 5:1 for the formation of monochloramine. The results of these studies are somewhat inconclusive but the complete removal of ammonia by this process could not be achieved.

There are at least two other factors to be considered in the breakpoint process:

1. The use of chlorine gas produces acid. Equations (7-1) and (7-8) show that oxidation of 1 mole of NH_4-N produces 4 moles of H^+. Stoichiometrically it requires 14.3 mg/l of bicarbonate alkalinity (as calcium carbonate) to neutralize the acid produced during the oxidation of 1 mg/l NH_4-N.
2. Chloride ion is added to the product water during the breakpoint process and the concentration will increase by at least 7.6 mg/l for each mg/l of ammonia oxidized. The TDS concentration may increase or decrease depending upon the initial alkalinity concentration.

Sodium hypochlorite can be used in the breakpoint process instead of chlorine gas. The reaction of sodium hypochlorite in water will tend to raise the pH instead of lowering it:

$$\text{NaOCl} + H_2O \longrightarrow \text{HOCl} + Na^+ + OH^- \qquad (7\text{-}12)$$

The hypochlorous acid formed disassociates and establishes equilibrium with hypochlorite ion as shown in equation (7-2). The breakpoint reaction with sodium hypochlorite is represented as follows:

$$1.5NaOCl + NH_4^+ \longrightarrow 0.5N_2 + 1.5NaCl + 1.5H_2O + H^+ \quad (7\text{-}13)$$

The weight ratio of $NaOCl:NH_4\text{-}N$ required to reach the breakpoint is $8.0:1$. Each mg/l of $NH_4\text{-}N$ oxidized increases the sodium concentration by at least 2.5 mg/l and the chloride by 3.8 mg/l. Stoichiometrically, it requires 3.6 mg/l of bicarbonate alkalinity (as calcium carbonate) to neutralize the acid formed during the breakpoint chlorination of 1 mg/l $NH_4\text{-}N$ with NaOCl.

It is possible to generate sodium hypochlorite at the point of use. Several commercial systems are described in Chapter 6. Also, at least one of the on site generation systems is capable of maintaining the chlorine and caustic produced in separate streams. An on site generation system of this type would not add sodium to the treated water.

Design Considerations

As previously noted, there are currently no full scale plants in operation that utilize breakpoint chlorination as the principal method of nitrogen removal. The EPA Nitrogen Control Manual[1] describes the following two full scale plants that will use breakpoint chlorination for nitrogen removal.

1. *Sacramento, California.* Effluent from this new 125 mgd plant will discharge into the Sacramento River. About half of the design flow will be treated by breakpoint chlorination on an intermittent basis to meet a total nitrogen limitation in the discharge during certain times of the year.

2. *Montgomery County, Maryland.* Effluent from this new 60 mgd AWT plant will discharge into the Potomac River above the raw water intakes for the Washington, D.C. water treatment plants. The effluent total nitrogen limit is 2 mg/l. An on-site chlorine generation system, of the type shown in Chapter 6, Figures 6-9 and 6-10, will be used to supply sodium hypochlorite. The design of this AWT nitrogen removal system is described in detail in the EPA manual.[1] The major design criteria are as follows:

- Nitrogen removal is accomplished by adding a sodium hypochlorite solution to the wastewater at a dosage slightly in excess of the stoichiometric requirement.
- The wastewater first passes through two in-line mechanical mixers in series (minimum $G = 1000$ fps/ft). The sodium hypochlorite is added to the first mixer along with sodium hydroxide if needed for pH adjustment. The second mixer is provided for redundancy in the event the first mixer

malfunctions. The second mixer is normally operated to provide additional mixing.

- Wastewater then flows to breakpoint reactors, closed concrete tanks (minimum retention, 20 minutes), where it is air mixed to complete the chemical reaction. Flow is distributed over the first half of the length of each basin through a multiport header. Distribution of flow in this manner minimizes the decrease in pH caused by the reaction of sodium hypochlorite with ammonia-nitrogen, and minimizes the formation of nitrogen trichloride. Air is also diffused into the wastewater over the bottom of the second half of the basin to provide a means of stripping any nitrogen trichloride from solution. Additional air is introduced in the space between the liquid surface and the basin cover to further dilute any nitrogen trichloride that might be present in order to preclude the development of the explosive concentration that occurs at 0.5 percent by volume. The diffusion of air into the breakpoint reactor contents also strips gaseous nitrogen and carbon dioxide from solution. Removal of CO_2 will result in a desirable increase in pH.

- Exhaust gases from the reactors are ducted to a gas treatment system. The treatment system consists of thermal decomposition of the nitrogen trichloride (by means of a burner and heat exchanger system to heat the gas stream to 300° F) and caustic scrubbing of any other noxious gases, including chlorine.

- The breakpoint process is controlled by pacing the sodium hypochlorite feed rate to the influent flow and influent ammonia nitrogen concentration. The pH is also monitored and controls the addition, when necessary, of sodium hydroxide to maintain an optimum pH for the breakpoint reaction.

- Ammonia nitrogen in the breakpoint effluent is monitored to determine efficiency of the process. Free and combined chlorine residuals and pH are also continuously monitored in breakpoint reactor effluent.

Considering the pH sensitivity of formation of NO_3^- and NCl_3, the EPA Nitrogen Manual[1] gives the following two specific recommendations for the design of full scale breakpoint chlorination facilities:

1. Add pH adjustment chemical to the chlorine solution prior to application of the chlorine solution to the breakpoint chlorination process influent. It is believed that this will prevent pockets of high or low pH that could form excessive concentrations of NO_3^- and NCl_3.
2. Operate the breakpoint process at pH 7 to reduce the formation of NCl_3.

Successful operation of the breakpoint chlorination process depends on careful control of the pH and chlorine dosage. An operation and control sys-

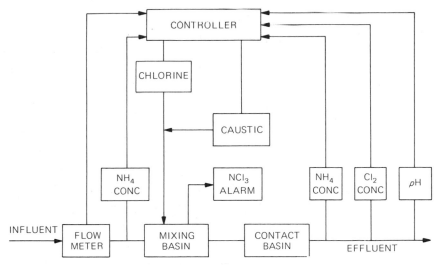

Fig. 7-52   Breakpoint chlorination system.

tem should include: (1) measurement of ammonia concentration in process influent, (2) addition of the appropriate amount of chlorine, (3) measurement of ammonia and chlorine residual in the process effluent, and (4) maintenance of pH near 7. A schematic of a system with these necessary elements is shown in Figure 7-52.

Although the breakpoint reaction is believed to be complete in less than one minute, complete and uniform mixing of the chlorine is important. Also, a well designed plug-flow-type contact basin following mixing is desirable. The importance of adequate mixing is discussed in Chapter 6.

ECONOMICS OF NITROGEN REMOVAL PROCESSES

Chapter 11 presents very general cost estimating guides for preliminary estimates. This section presents more detailed estimating guides for capacities of 0.01–10 mgd for cases where nitrogen concentrations have been defined and more specific estimates are desired. The following estimates are based upon work conducted by Culp/Wesner/Culp under EPA Contract 68-03-2186 and are based upon fourth quarter 1975 cost levels.

Breakpoint Chlorination

The basic design criteria used were as follows:

Provide 30 seconds of rapid mixing, $G = 900$

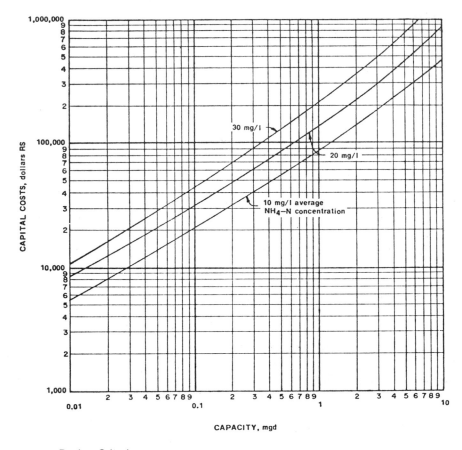

CAPITAL COSTS, dollars RS

CAPACITY, mgd

Design Criteria:
Mixing–reactor basin, 30 sec detention, G = 900
Peak NH₃-N concentration = 2× average NH₃-N concentration
Chlorine feed capacity = 10× peak NH₃-N concentration × average flow

Fig. 7-53 Capital costs for breakpoint chlorination system, including related yardwork, engineering, legal, fiscal, and financing costs during construction.

Peak NH_3 concentration = 2 × average NH_4-N concentration

Chlorine Feed capacity = 10 × peak NH_4-N concentration @ average flow

Costs of chlorine contact facilities were not included because such facilities would normally be provided for disinfection purposes even without the need to remove nitrogen.

Costs were estimated for average NH_4-N concentrations of 10, 20 and 30 mg/l. Chlorine usage in the various size facilities is as follows:

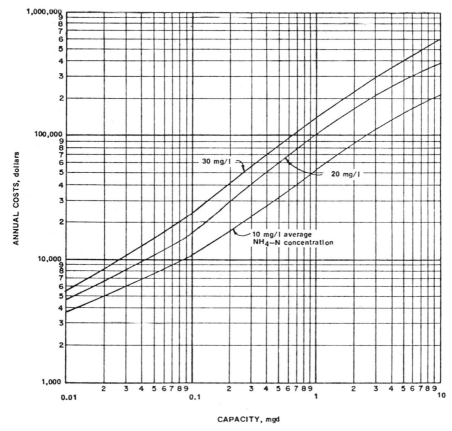

CAPACITY, mgd

Chlorine Costs = 11¢/lb, ton cylinders 0.01 - 1 mgd; 5¢/lb, tank cars, 10 mgd
Labor = $9/hr
Power =$0.02/kwh

Fig. 7-54 Operation and maintenance costs for breakpoint chlorination system, exclusive of chemicals for *p*H control.

Capacity	10 mg/l NH₄-N	20 mg/l NH₄-N	30 mg/l NH₄-N
0.01 mgd	1.5 ton/year	3.1	4.5
0.1	15	31	45
1	150	310	450
10	1500	3100	4500

Costs of chlorine were based on use of 1 ton cylinders for quantities up to 450 tons/year (11¢/lb) and on tank cars over 450 tons/year. Between 1 and 10 mgd, the demurrage cost on the rail cars will result in an effective chlorine cost in excess of 5¢/lb and a gradually decreasing cost down to 5¢/lb at 10

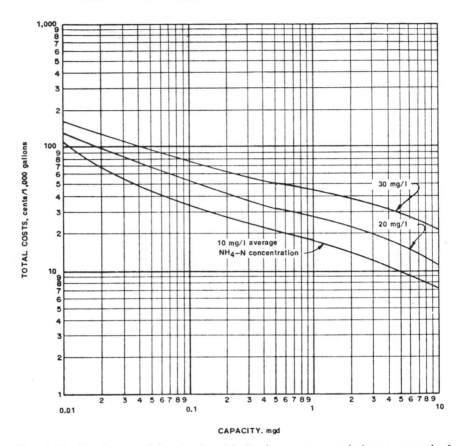

Fig. 7-55 Total costs of breakpoint chlorination system; capital costs amortized @ 7 percent over 20 years.

mgd. To determine operation and maintenance costs, a labor rate of $9/hour and a power cost of $0.02/kwh were used. The same unit costs for labor and power were used for all alternatives. Capital costs were amortized at 7 percent over 20 years to determine total annual costs (for all alternatives) which are summarized as follows:

TOTAL ANNUAL COSTS (¢/1000 gal)

	10 mg/l NH₄-N			20 mg/l NH₄-N			30 mg/l NH₄-N		
	Cap	O & M	Total	Cap	O & M	Total	Cap	O & M	Total
0.01 mgd	14	105	119	21	134	155	28	152	180
0.1 mgd	6	30	36	9	46	55	12	67	79
1 mgd	2.4	15.5	18	4.2	28	32	5.7	38	44
10 mgd	1.1	6.3	7.4	2.2	11.7	13.9	3.8	16.4	20.2

Cost curves are shown in Figures 7-53–7-55.

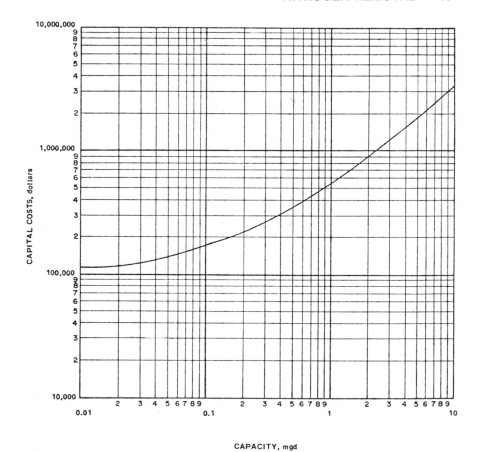

CAPACITY, mgd

Fig. 7-56 Capital costs of selective ion exchange system with regenerant recovery by closed loop stripping tower, including related yardwork, engineering, legal, fiscal, and financing costs during construction.

Selective Ion Exchange

The costs for this process are based on use of clinoptilolite exchange media in gravity structures with recovery of the regenerant in closed loop stripping towers. A minimum of four exchangers was provided for each capacity. A 4 ft deep clinoptilolite bed loaded at 5.25 gpm/ft^2 was used. Costs include the exchange structure, backwash facility, influent pumping, clarification-softening facility for the spent regenerant, and closed loop stripping tower for regenerant recovery. Capital costs are essentially unaffected by ammonia concentration, but regeneration frequency and operating costs increase as ammonia concentration increases. Costs for the closed loop tower modules are based on the estimated cost of such units for the Upper Occoquan, Virginia plant. Influent pumping costs are based on 15 ft total dynamic

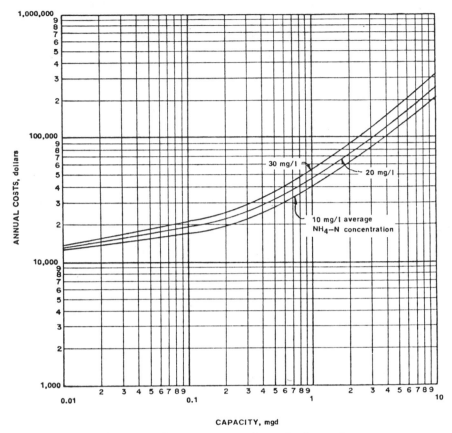

Fig. 7-57   Operation and maintenance costs of selective ion exchange system; labor fixed @ $9/hour, power @ $0.02/kwh.

head with regenerant recovery pumping at 35 ft total dynamic head (TDH). The composition of total annual costs is as follows:

TOTAL ANNUAL COSTS (¢/1000 gal)

		O & M			Total		
		mg/l NH₄-N			mg/l NH₄-N		
	Amortized Capital	10	20	30	10	20	30
0.05 mgd	330	382	403	430	712	733	760
0.1	45	48.5	52.5	57.2	93	97	102
1	15.2	10.9	12.9	14.9	26	28	30
10	8.6	5.5	7.3	8.9	14.3	15.9	17.5

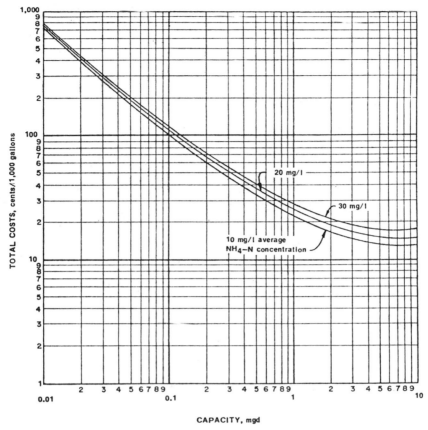

Fig. 7-58   Total costs of selective ion exchange system; capital costs amortized @ 7 percent over 20 years.

Cost curves developed are presented in Figures 7-56–7-58.

Ammonia Stripping

Capital costs are based on a tower loading rate of 1 gpm/ft^2 with a plastic tower packing with 24 ft packing depth. Capital and operation and maintenance costs include influent pumping (50 ft TDH). The costs *do not* include the costs of elevating the pH of the wastewater to an adequate level for stripping nor the cost of susbsequent downward pH adjustment following stripping. The costs of the stripping process to provide a given percentage removal of ammonium-nitrogen are independent of influent concentration (at a given temperature). The total annual costs are as follows:

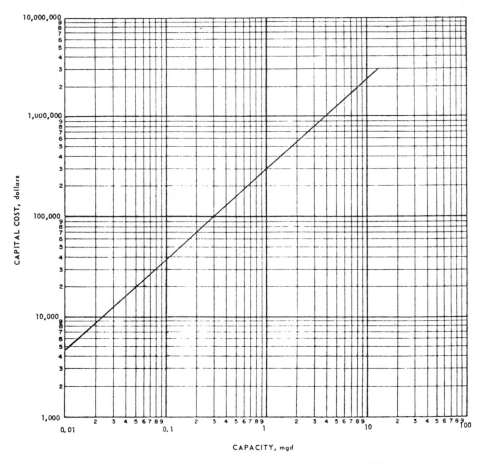

Fig. 7-59 Capital costs of ammonia stripping system, including related yard-work, engineering, legal, fiscal, and financing costs during construction, and ex-cluding cost of *p*H adjustment. Based on 1 gpm/sf of tower packing, 24 ft packing depth. Includes influent pumping (TDH = 50 ft).

TOTAL ANNUAL COSTS (¢/1000 gal)

	Amortized Capital	O & M	Total
0.01 mgd	20.3	63	83
0.1	17.8	16	34
1	9.9	3.8	13.7
10	6.9	3.6	10.5

Cost curves are presented in Figures 7-59–7-61.

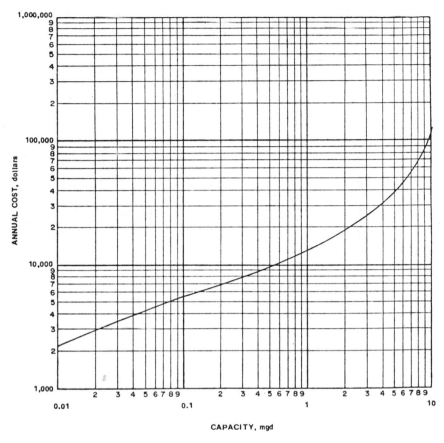

Fig. 7-60 Operation and maintenance costs of ammonia stripping system, excluding cost of *p*H adjustment; labor fixed @ $9/hour, power @ $0.02/kwh.

Nitrification, Mixed Reactor

The design of the single stage system is based on a mean cell residence time of 10 days to achieve nitrification in a single stage activated sludge system ($F/M = 0.20$). The following table presents the results of the cost calculations. The effects of different average ammonium nitrogen concentrations (10, 20, and 30 mg/l) were estimated based upon providing 4.6 lb O_2/lb NH_4-N. Peak hourly ammonium-nitrogen concentrations of twice the average were assumed with peak hourly BOD concentrations of 1.5 times the average. An oxygen transfer efficiency of 2 lb/hp-hour was used. Costs include aerobic sludge stabilization but not sludge dewatering or disposal.

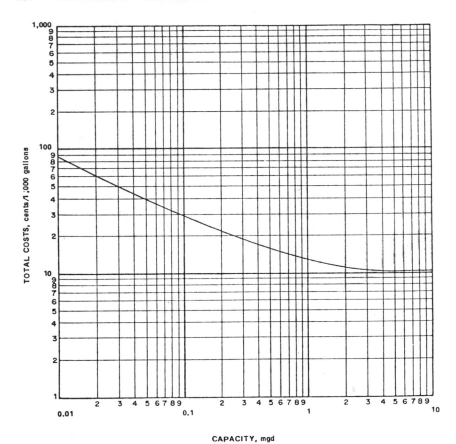

Fig. 7-61   Total costs of ammonia stripping system; capital costs amortized @ 7 percent over 20 years.

	for 20 mg/l NH₄-N		
	1 mgd	5 mgd	10 mgd
Amortized capital	139	369	588
Labor	52	99	147
Power	14	52	105
Maintenance materials	13	26	38
Chlorine	3	17	33
Total	222	562	911
Costs/1000 gal (operating @ capacity)	$0.61	$0.31	$0.25
for 30 mg/l NH₄-N	0.63	0.32	0.26
for 10 mg/l NH₄-N	0.59	0.30	0.24

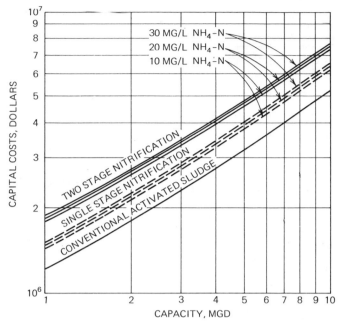

Fig. 7-62 Capital costs of biological treatment system with aerobic sludge stabilization, including yardwork, engineering, legal, fiscal, and financing costs during construction, but exclusive of sludge dewatering and disposal.

Two-stage nitrification costs were based on a 3 hour detention second stage aeration basin following a first stage activated sludge process, with $F/M = 0.355$ and a mean cell residence time of 5 days in the first stage. The same basic assumptions used for the single stage nitrification system on ammonium concentrations oxygen transfer, etc. were used. The costs, including the first stage, are as follows:

	Two-Stage Nitrification, for 20 mg/l NH₄-N		
	1 mgd	5 mgd	10 mgd
Amortized capital	167	441	695
Labor	64	120	178
Power	14	53	107
Maintenance materials	14	30	44
Chlorine	3	17	33
Total	263	660	1,057

(*Continued*)

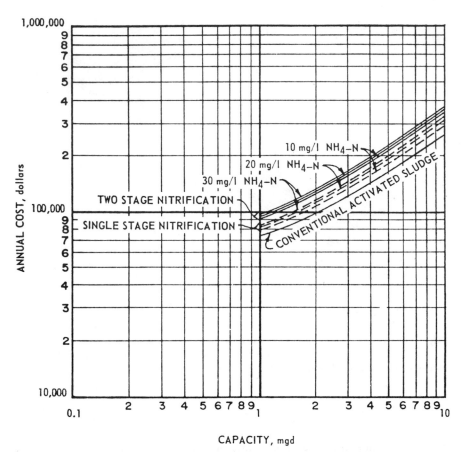

Fig. 7-63 Operation and maintenance costs of biological treatment system with aerobic sludge stabilization, exclusive of sludge dewatering and disposal; labor fixed @ $9/hour, power @ $0.02/kwh.

	Two-Stage Nitrification, for 20 mg/l NH₄-N		
	1 mgd	5 mgd	10 mgd
Costs/1000 gal (operating @ capacity)	$0.72	$0.36	$0.29
for 10 mg/l NH₄-N	0.70	0.35	0.28
for 30 mg/l NH₄-N	0.74	0.37	0.30

Curves 7-62–7-64 summarize the costs for conventional activated sludge

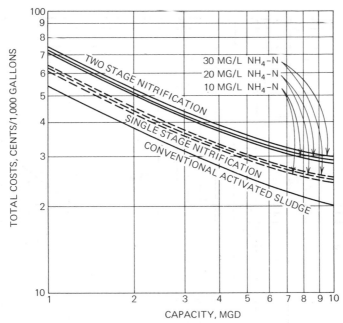

Fig. 7-64 Total costs of biological treatment system with aerobic sludge stabilization, exclusive of sludge dewatering and disposal; capital costs amortized @ 7 percent over 20 years.

single stage, and two stage nitrification. The incremental cost to achieve nitrification may be summarized as follows:

INCREMENTAL COSTS[a] FOR
NITRIFICATION WITH MIXED REACTORS

	¢/1000 gal		
	1 mgd	5 mgd	10 mgd
Single-Stage Nitrification			
10 mg/1 NH₄-N	5	4	4
20 mg/1 NH₄-N	7	5	5
30 mg/1 NH₄-N	9	6	6
Two-Stage Nitrification			
10 mg/1 NH₄-N	16	9	8
20 mg/1 NH₄-N	18	10	9
30 mg/1 NH₄-N	20	11	10

[a]Costs above those required for conventional activated sludge and aerobic sludge digestion.

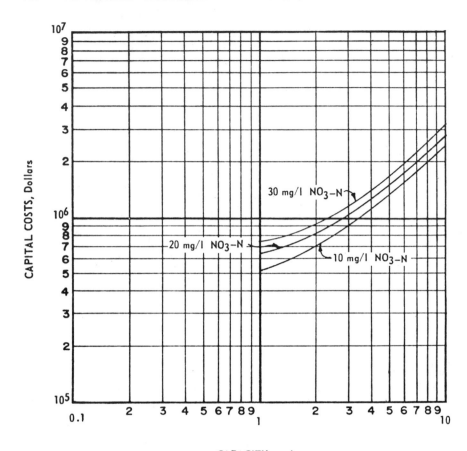

Nitrate Removed Rate	0.1 lb NO_3-N/lb MLVSS/day
MLVSS	1500 mg/l
Aerated Stabilization Reactor	50 minutes detention
Clarifier	700 gpd/sft
Methanol Feed Rate	3 × NO_3-N

Fig. 7-65   Capital costs of denitrification, mixed reactor system.

Denitrification

The design basis for denitrification in a mixed, uncovered reactor is summarized on Figure 7-65. The anoxic denitrification reactor is followed by an aerobic stabilization reactor for removal of any excess methanol (see Figure 7-45). Solids are then removed in a clarifier and recycled to the denitrification

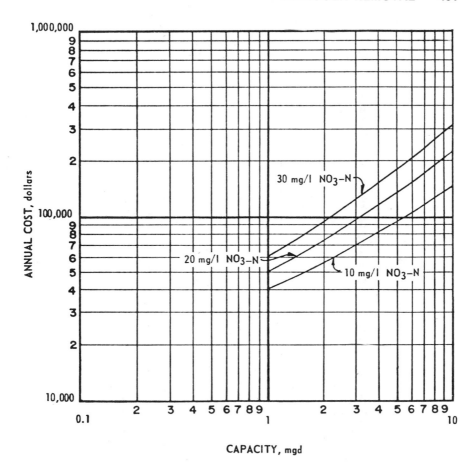

Fig. 7-66 Operation and maintenance costs of denitrification, mixed reactor system; labor fixed @ $9/hour, power @ $0.02/kwh, methanol @ $150/ton.

reactor. The costs for denitrification may be summarized as follows (methanol cost of $150/ton used):

	¢/1000 gal		
	1 mgd	5 mgd	10 mgd
Mixed Reactor Denitrification			
10 mg/l NO_3^--N	26	12	10
20 mg/l NO_3^--N	32	16	14
30 mg/l NO_3^--N	36	19	17

Curves 7-65–7-67 summarize the costs of denitrification.

TABLE 7-10 Comparative Costs (Capital, Operation and Maintenance) of Alternative Nitrogen Removal Processes. (¢/1000 gal)

	1 mgd			5 mgd			10 mgd		
	10 mg/l[b]	20 mg/l	30 mg/l	10 mg/l	20 mg/l	30 mg/l	10 mg/l	20 mg/l	30 mg/l
Single-stage nitrification[a]	5	7	9	4	5	6	4	5	6
Single-stage nitrification[a] and denitrification	31	39	45	16	21	25	14	19	23
Two-stage nitrification[a]	16	18	20	9	10	11	8	9	10
Two-stage nitrification[a] and denitrification	42	50	56	21	26	30	18	23	27
Breakpoint chlorination	18	32	44	10	17	26	7	14	20
Selective ion exchange	26	28	30	14	16	18	14	16	17.5
Ammonia stripping	14	14	14	10	10	10	8	8	8

[a]Incremental costs above conventional activated sludge required to nitrify.
[b]Ammonium nitrogen concentrations.

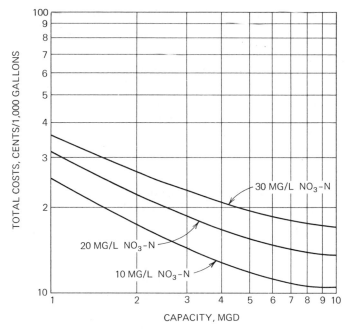

Fig. 7-67 Total costs of denitrification, mixed reactor system; capital costs amortized @ 7 percent over 20 years.

Comparative Costs

Table 7-10 summarizes the costs presented in the preceding paragraphs for all alternatives discussed. The limitations of such generalized cost estimates must be kept in mind, and process selection for a specific plant must be based upon the conditions encountered at that plant. However, Table 7-10 does serve to indicate the general level of costs for the alternative processes and their sensitivity to variations in nitrogen concentrations.

REFERENCES

Ammonia Stripping

1. "Process Design Manual for Nitrogen Control." U.S. Environmental Protection Agency, Office of Technology Transfer, October, 1975.

2. "Nitrogenous Compounds in the Environment." EPA Report SAB-73-001, December, 1973.

3. Miner, S., "Preliminary Air Pollution Survey of Ammonia." U.S. Public Health Service, Contract No. PH22-68-25, October, 1969.

4. Pasquill, F., *Atmospheric Diffusion*. D. Van Nostrand Co., Ltd., London, 1962.

5. Roesler, J. F.; Smith, R.; and Eilers, R. G., "Mathematical Simulation of Ammonia Stripping Towers for Wastewater Treatment." U.S. Department of Interior—FWPCA, Cincinnati, Ohio, January, 1970.

6. Roesler, J. F.; Smith, R.; and Eilers, R. G., "Simulation of Ammonia Removal From Wastewater," *Journal Sanitation Engineering Division*, ASCE, 97, No. SA 3, 1971, pp. 269–286.

7. Perry, J. H., *Chemical Engineering Handbook*, McGraw-Hill Book Co., New York, N.Y.

8. Kuhn, P. A., "Removal of Ammonium Nitrogen from Sewage Effluent." Unpublished MS Thesis, University of Wisconsin, Madison, Wisconsin, 1956.

9. Prather, B. V., "Wastewater Aeration May Be Key To More Efficient Removal of Impurities," *Oil and Gas Journal*, November 30, 1959, p. 78.

10. Prather, B. V., "Chemical Oxidation of Petroleum Refinery Wastes." Paper presented at the Thirteenth Industrial Wastes Conference, Oklahoma State University, Stillwater, Oklahoma, November, 1962.

11. Prather, B. V., and Gaudy, A. F., "Combined Chemical, Physical and Biological Processes in Refinery Wastewater Purification." Paper presented at the 29th Midyear Meeting of the American Petroleum Institute's Division of Refining, St. Louis, Missouri, May, 1964.

12. Slechta, A. F., and Culp, G. L., "Water Reclamation Studies At The South Public Utility District," *Journal Water Pollution Control Federation*, 1967, p. 787.

13. O'Farrell, T. P.; Bishop, D. F.; and Cassel, A. F., "Nitrogen Removal by Ammonia Stripping." EPA Report 670/2-73-040, September, 1973.

14. Smith, C. E., and Chapman, R. L., "Recovery of Coagulant, Nitrogen Removal, and Carbon Regeneration in Wastewater Reclamation." Final Report by South Tahoe Public Utilities District to FWPCA, Demonstratoin Grant WPD-85, June, 1967.

15. South Tahoe Public Utility District, "Advanced Wastewater Treatment as Practiced at South Tahoe." EPA Report 17010ELQ 08/71, August, 1971.

16. Wesner, G. M., and Argo, D. G., "Report on Pilot Waste Water Reclamation Study," Orange County Water District Report, July, 1973.

17. Kepple, L. G., "Ammonia Removal and Recovery Becomes Feasible," *Water and Sewage Works*, April, 1974, p. 42.

18. Gonzales, J. F. and Culp, R. L., "New Developments in Ammonia Stripping," *Public Works*, 104, No. 5, 1973, p. 78; and No. 6, 1973, p. 82.

19. Folkman, Y., and Wachs, A. M., "Nitrogen Removal Through Ammonia Release From Ponds." Proceedings, 6th Annual International Water Pollution Research Conference, Tel Aviv, Israel, June 18–23, 1972.

20. Culp, R. L., and Moyer, H. E., "Wastewater Reclamation and Export at South Tahoe," Civil Engineering, June, 1969, p. 38.

21. Wesner, G. M., and Culp, R. L., "Wastewater Reclamation and Seawater Desalination," Journal Water Pollution Control Federation, 44, No. 10, 1972, pp. 1932–1939.

22. Wesner, G. M., and Baier, D. C., "Injection of Reclaimed Wastewater Into Confined Aquifers," Journal American Water Works Assoc., 62, No. 3, 1970, pp. 203–210.

Selective Ion Exchange

23. Koon, J. H., and Kaufman, W. J., "Optimization of Ammonia Removal by Ion Exchange Using Clinoptilolite." Environmental Protection Agency Water Pollution Control Research Series No. 17080 DAR 09/71.

24. Battelle Northwest and the South Tahoe Public Utility District, "Wastewater Ammonia Removal by Ion Exchange." Environmental Protection Agency Water Pollution Control Research Series No. 17010 ECZ 02/71, February, 1971.

25. Battelle Northwest, "Ammonia Removal from Agricultural Runoff and Secondary Effluents by Selective Ion Exchange." Robert A. Taft Water Research Center Report No. TWRC-5, March, 1969.

26. Mercer, B. W., "Clinoptilolite in Water Pollution Control," The Ore Bin, published by Oregon Dept. of Geology and Mineral Industries, November, 1969, p. 209.

27. Prettyman, R., et al., "Ammonia Removal by Ion Exchange and Electrolytic Regeneration." Unpublished Report, CH2M/Hill Engineers, December, 1973.

28. Suhr, L. G., and Kepple, L., "Design of a Selective Ion Exchange System for Ammonia Removal." Paper presented at the ASCE Environmental Engineering Division Conference, Pennsylvania State University, July, 1974.

29. "Physical-Chemical Plant Treats Sewage Near the Twin Cities," Water and Sewage Works, September, 1973, p. 86.

30. Larkman, D., "Physical-Chemical Treatment," Chemical Engineering, Deskbook Issue, June 18, 1973, p. 87.

31. Culp, G. L.; Culp, R. L. and Hamann, C. L., "Water Resource Preservation by Planned Recycling of Treated Wastewater," Journal American Water Works Assoc., 65, 1973, p. 641.

Biological Nitrogen Removal

32. Sawyer, C. N.; Wild, H. E., Jr.; and McMahon, T. C., "Nitrification and De-nitrification Facilities, Wastewater Treatment." Report prepared for the EPA Technology Transfer Program, August, 1973.

33. Haug, R. T., and McCarty, P. L., "Nitrification with the Submerged Filter." Report prepared by the Department of Civil Engineering, Stanford University for the Environmental Protection Agency, Research Grant No. 17010 EPM, August, 1971.

34. Nagel, C. A., and Haworth, J. G., "Operational Factors Affecting Nitrification in the Activated Sludge Process." Paper presented at the 42nd Annual Conference of the Water Pollution Control Federation, Dallas, Texas, October, 1969.

35. Painter, H. A., "A Review of Literature on Inorganic Nitrogen Metabolism in Microorganisms," *Water Research*, 4, No. 6, 1970, pp. 393–450.

36. Beckman, W. J., et al., "Design and Operation of a Combined Carbon Oxidation–Nitrification Activated Sludge Plant" (at Flint, Michigan plant).

37. Sutton, P. M.; Murphy, K. L.; Jank, B. E.; and Monaghan, B. A., "Efficacy of Biological Nitrification," *Journal Water Pollution Control Federation*, 47, 1975, p. 2665.

38. Brown and Caldwell Engineers, "Case Histories of Nitrification and Denitrification Facilities." Paper presented at EPA Technology Transfer Seminar, Orlando, Florida, May 7–9, 1974.

39. Richardson, S., "Pilot Plants Define Parameters for Nitrification on Plastic Trickling Filters." Paper presented at the 46th Annual Conference of the Water Pollution Control Federation, Cleveland, Ohio, October, 1973.

40. "Rotating Biological Disk Wastewater Treatment Process—Pilot Plant Evaluation." Report by the Department of Environmental Sciences, Rutgers University, prepared for the Environmental Protection Agency, Project No. 17010 EBM, August, 1972.

41. Haug, R. T., and McCarty, P. L., "Nitrification With the Submerged Filter." Department of Civil Engineering, Stanford University, Technical Report No. 149, August, 1971.

42. Haug, R. T., and McCarty, P. L., "Nitrification With the Submerged Filter, *Journal Water Pollution Control Federation*, 44, November, 1972, p. 2086.

43. McHarness, D. D.; Haug, R. T.; and McCarty, P. L., "Field Studies of Nitrification with Submerged Filters," *Journal Water Pollution Control Federation*, 1975, p. 281.

44. Antonie, R. L., "Nitrification of Activated Sludge Effluent: Bio-Surf Process," *Water and Sewage Works*, November, 1974, p. 44.

45. Owen, W. F.. and Slechta, A. F., "Organic Removal or Nitrification with a Combined Fixed/Suspended Growth Biological Treatment System." Paper presented at the 48th Annual Conference, WPCF, Miami Beach, Fla., 1975.

46. St. Amant, P. P., and McCarty, P. L., "Treatment of High Nitrate Waters," *Journal American Water Works Assoc.*, 1969, p. 659.

47. McCarty, P. L.; Beck, L.; and St. Amant, P., "Biological Denitrification of Wastewaters by Addition of Organic Materials." In *Proc. of the 24th Industrial Waste Conference, May 6, 7, and 8, 1969. Purdue University, Lafayette, Indiana, 1969.*

47. McCarty, P. L.; Beck, L.; and St. Amant, P., "Biological Denitrification of Wastewaters by Addition of Organic Materials." In *Proc. of the 24th Industrial Waste Conference, May 6, 7, and 8, 1969.* Purdue University, Lafayette, Indiana, 1969.

48. Wilson, T. E., and Newton, D., "Brewery Wastes as a Carbon Source for Denitrification at Tampa, Florida." Paper presented at the 28th Annual Purdue Industrial Waste Conference, May, 1973.

49. Horstkotte, G. A., et al., "Full-Scale Testing of a Water Reclamation System," *Journal Water Pollution Control Federation*, 1974, p. 181.

50. Savage, E. S., and J. J. Chen, "Operating Experiences with Columnar Denitrification." Dravo Corporation, Pittsburgh, Pennsylvania, 1973.

51. Kapoor, S. A., and Wilson, T. E., "Biological Denitrification on Deep Bed Filters at Tampa, Florida." Unpublished paper, Greeley and Hansen, Engineers, Chicago, Illinois.

52. Ehreth, D. J., and Barth, E., "Control of Nitrogen in Wastewater Effluents." EPA, Technology Transfer, March, 1974.

53. Ecotrol, Inc., "Biological Denitrification Using Fluidized Bed Technology." August, 1974.

54. Jeris, J.; Beer, C.; and J. A. Mueller, "High Rate Biological Denitrification Using a Granular Fluidized Bed," *Journal Water Pollution Control Federation*, 46, No. 9, 1974, pp. 2118–2128.

55. Bishop, D. F., et al., "Advanced Waste Treatment System at the Environmental Protection Agency District of Columbia Pilot Plant." In *Water—1971*, American Institute of Chemical Engineers Symposium Series, Vol. 68, No. 124, 1972, p. 11.

56. Wilson, T. E.; Kapoor, S. K.; and Newton, D., "Pilot Studies for Advanced Waste Treatment at Tampa, Florida." Paper presented at the 46th Annual Conference of the Water Pollution Control Federation, Cleveland, Ohio, October, 1973.

57. Newton, D., and Wilson, T. E., "Oxygen Nitrification Process at Tampa." Paper presented at Applications of Commercial Oxygen of Water and Waste-

water Systems, Water Resources Symposium Number Six, University of Texas at Austin, November 13–15, 1972.

Breakpoint Chlorination

58. Jeris, J. S.; Owens, R.; Hickey, R.; and Flood, F., "Biological Fluidized Bed Technology—BOD and Nitrogen Removal in Less Than One Hour." Paper presented at the 48th Annual WPCF Conference, October, 1975.

59. Sawyer, C. N., and McCarty, P. L., *Chemistry for Sanitary Engineers*, 2nd Edition. McGraw–Hill Book Co., New York, 1967.

60. Pressley, T. A., et al., "Ammonia Removal by Breakpoint Chlorination," *Environmental Science and Technology*, Vol. 6, July, 1972, pp. 622–628.

61. Pressley, T. A., et al., "Ammonia-Nitrogen Removal by Breakpoint Chlorination." EPA Report 670/2-73-058, September, 1973.

62. Palin, A. T., "A Study of the Chloro Derivatives of Ammonia and Related Compounds, with Special Reference to Their Formation in the Chlorination of Natural and Polluted Waters," *Water and Water Engineering*, 54, 1950, pp. 151–159, 189–200, 248–256.

63. Stasiuk, W. N., et al., "Nitrogen Removal by Catalyst Aided Breakpoint Chlorination," *Journal Water Pollution Control Federation*, 26, August, 1974, pp. 1974–1983.

64. Wei, I. W., and Morris, J. C., "Dynamics of Breakpoint Chlorination." Paper presented at 165th National Meeting, American Chemical Society, Dallas, Texas, April 8–13, 1973.

65. Bauer, R. C., and Snoeyink, V. L., "Reactions of Chloramines with Active Carbon," *Journal Water Pollution Control Federation*, 45, November, 1973, pp. 2290–2301.

66. Barnes, R. A., et al., "Ammonia Removal in a Physical-Chemical Wastewater Treatment Process," EPA-R2-72-123, Noevember, 1972.

8

Chemical sludge handling

INTRODUCTION

In advanced wastewater treatment, chemical sludges are produced primarily in processes for removal of suspended solids or phosphorus. Nitrogen removal processes do not generate significant quantities of sludge. Common chemicals used include lime, alum, iron salts, and polymers. The nature of sludges resulting from the chemical coagulation of sewage depends on the nature of the coagulant used and the point of chemical application, whether it be to raw wastewater, primary effluent, or secondary effluent.

Lime sludges can be recalcined and reused. In treatment of waters for which sodium aluminate is a good coagulant, the alkaline method for alum recovery may be used. In the event that sodium aluminate is not a good coagulant for the wastewater under consideration, then the acid method for alum recovery may be used in those instances where the treatment goal is suspended solids removal. On the other hand, the acid recovery method for alum is not applicable if one of the treatment goals is phosphorus removal. Iron salts are not subject to recovery and reuse at present.

In every case where chemical treatment is employed, there are waste chemical sludges or mixtures of biological and chemical sludges requiring disposal. Liquid sludges may be taken directly to disposal by land spreading or in landfill. Generally there is some processing prior to disposal. Alum sludges are often treated in anaerobic digesters. There are no known adverse effects on biological digestion and there is no release of phosphorus from the sludge to the supernatant. There may be some enhancement in the dewatering and drying of biological sludges containing alum. In tertiary treatment, alum and iron coagulants generally produce gelatinous floc which is difficult to dewater. Lime coagulation produces a sludge which readily thickens and dewaters in most cases. The quantity and nature of organics present in the chemical sludge may alter its dewatering characteristics substantially. For example, the presence of large quantities of activated sludge solids may make the dewatering of lime sludges much more difficult.

Methods and equipment ordinarily used in handling and processing sludges from primary and secondary treatment are generally applicable to chemical sludges or chemical-biological sludge mixtures. Gravity or flotation thickening may be used. Chemical sludges may be dewatered by use of drying beds, lagoons, centrifuges, vacuum filters, filter presses, or horizontal belt filters. The dewatered cake may then be disposed of by land spreading or burial in landfill, or subjected to heat drying or incineration. Incineration ash may be used to make concrete block or brick, or as a road subgrade stabilizer, or it can be disposed of as a soil conditioner or in landfill.

Sludge Production

In chemical precipitation, the pounds of sludge produced per million gallons of wastewater treated varies considerably. Data on production of sludges for chemical treatment at various points in a secondary plant are summarized in Tables 8-1–8-4.

Methods for Estimating Sludge Production

It is difficult to obtain an accurate gravimetric measurement of sludge quantities in laboratory tests owing to loss of solids during decanting, and so forth. It is possible, however, to estimate quantities of sludge from the chemistry involved and the data collected from the jar tests.

TABLE 8-1   Additional Sludges to be Handled with Chemical Treatment Systems: Primary Treatment for Removal of Phosphorus[25]

Sludge Production Parameter	Conventional Primary	Lime Addition to Primary Influent	Lime Addition to Primary Influent	Aluminum Addition to Primary Influent	Iron Addition to Primary Influent
Level of chemical addition (mg/l)	0	350–500	800–1600	13–22.7	25.80
Percent sludge solids					
Mean	5.25	11.1	4.4	1.2	2.25
Range	5.0–5.5	3.0–19.5	2.1–5.5	0.4–2.0	1.0–4.5
lb/mg					
Mean	788	5,630	9,567	1,323	2775
Range	600–950	2500–8000	4700–15,000	1200–1545	1400–4500
gal/mg					
Mean	4,465	8,924	28,254	23,000	21,922
Range	3,600–5,000	4663–18,000	16,787–38,000	10,000–36,000	9000–38,000

TABLE 8-2   Additional Sludge to be Handled with Chemical Treatment Systems: Phosphorus Removal by Mineral Addition to Aerator[25]

Sludge Production Parameter	Al^{3+} Addition to aerator		Fe^{3+} Addition to aerator	
	Conventional Secondary	With Al^{3+} Addition	Conventional Secondary	With Fe^{3+} Addition
Level of chemical addition (mg/l)	0	9.4–23	0	10–30
Percent sludge solids				
Mean	0.91	1.12	1.2	1.3
Range	0.58–1.4	0.75–2.0	1.0–1.4	1.0–2.2
lb/mg				
Mean	672	1,180	1,059	1,705
Range	384–820	744–1,462	218–1,200	1,100–2,035
gal/mg				
Mean	9,100	13,477	10,650	18,650
Range	7,250–12,300	7,360–20,000	10,300–11,000	6,000–24,000

TABLE 8-3   Additional Sludge to be Handled with Chemical Treatment Systems: Phosphorus Removal by Mineral Addition to Secondary Effluent.[25]

Sludge Production Parameters	Lime Addition	Alum Addition	Iron Addition
Level of chemical, addition (mg/l)	268–450	16	10–30
Percent sludge solids			
Mean	1.1	2.0	0.29
Range	0.6–1.72	—	—
lb/mg			
Mean	4,650	2,000	507
Range	3,100–6,800	—	175–781
gal/mg			
Mean	53,400	12,000	22,066
Range	50,000–63,000	—	6,000–36,000

TABLE 8-4   Sludge Characteristics—Chemical Treatment of Raw Sewage.

	Primary Sludge	Lime, low pH	Lime, high pH	Aluminum	Iron
% Sludge solids	5.25	11.1	4.4	1.2	2.25
Wt. solids,					
lb/mg of treated sewage	788	5,630	9,567	1,323	2,775
Volume sludge,					
gal/mg of treated sewage	4,465	8,924	28,254	23,000	21,922

Average of Data from Blue Plains, Lebanon, Taft Center, Salt Lake City.

The basic equations required for these calculations may be simplified as follows:

$$3PO_4^{3-} + 5Ca^{2+} + OH^- \longrightarrow Ca_5OH(PO_4)_3\downarrow$$
$$Mg^{2+} + 2OH^- \longrightarrow Mg(OH)_2\downarrow$$
$$Ca^{2+} + CO_3^{2-} \longrightarrow CaCO_3\downarrow$$

$$CaCO_3 \xrightarrow[\Delta]{1800°F} CaO + CO_2\uparrow \text{ (incineration)}$$

$$CaO + H_2O \longrightarrow Ca(OH)_2$$
$$Al^{3+} + PO_4^{3-} \longrightarrow AlPO_4\downarrow$$
$$Al^{3+} + 3OH^- \longrightarrow Al(OH)_3\downarrow$$

$$2Al(OH)_3 \xrightarrow[\Delta]{1400°F} Al_2O_3 + 3H_2O \text{ (incineration)}$$

$$Fe^{3+} + PO_4^{3-} \longrightarrow FePO_4\downarrow$$
$$Fe^{3+} + 3OH^- \longrightarrow Fe(OH)_3\downarrow$$

$$2Fe(OH)_3 \xrightarrow[\Delta]{1400°F} Fe_2O_3 + 3H_2O \text{ (incineration)}$$

$$\Sigma \text{ coagulant in} = \Sigma \text{ coagulant out}$$

Tables 8-5–8-7 describe the computations used to estimate the quantities of sludge produced from EPA Technology Transfer Manual, "Physical-

Table 8-5   Estimate of Lime Sludge Quantities.

Raw sewage suspended solids	250 mg/l
Raw sewage volatile suspended solids	150 mg/l
Raw sewage PO_4^{3-}	11.5 mg/l as P
Raw sewage total hardness	170.5 mg/l as $CaCO_3$
Raw sewage Ca^{2+}	60 mg/l
Raw sewage Mg^{2+}	5 mg/l
Effluent PO_4	0.3 mg/l as P
Effluent Ca^{2+}	80 mg/l
Effluent Mg^{2+}	0

Lime dosage	400 mg/l as $Ca(OH)_2$ or
	216 mg/l as Ca^{2+}
From equation 1	$Ca_5OH(PO_4)_3$ formed is 1 mole per 3 moles P
	$\dfrac{11.2}{30.97} = 0.365$ mole P removed
	Therefore $\dfrac{0.365}{3}$ or 0.122 mole $Ca_5OH(PO_4)_3$ are formed; fw is 502
	Therefore weight is $0.122 \times 502 = 61$ mg/l as $Ca_5OH(PO_4)_3$
From equation 2	$Mg(OH)_2$ formed is 1 mole per mole Mg^{2+}
	$\dfrac{5}{24.31} = 0.206$
	Therefore $0.206 \times 58.31 = 12$ mg/l as $Mg(OH)_2$
From equation 12	Ca^{2+} in = Ca^{2+} out; Ca^{2+} in = $60 + 216 = 276$
	Ca^{2+} content of $Ca_5OH(PO_4)_3$ formed = $5 \times 40 \times 0.122 = 24$ mg/l
	Ca^{2+} lost in effluent = 80 mg/l
	Therefore Ca^{2+} not accounted for = $276 - (80 + 24) = 172$ mg/l
From equation 3	$CaCO_3$ formed is 1 mole per mole Ca^{2+}
	Therefore $\dfrac{172}{40} = 4.3$ moles $CaCO_3$; fw = 100
	So weight of $CaCO_3$ = 430 mg/l

Sludge composition

Sludge species	Total weight	Ash
		Pounds per million gallons
Raw sewage solids	250 mg/l = 2,080 pounds per million gallons	832
$Ca_5OH(PO_4)_3$	61 mg/l = 510 pounds per million gallons	510
$Mg(OH)_2$	12 mg/l = 100 pounds per million gallons	100
$CaCO_3$	430 mg/l = 3,600 pounds per million gallons	2,020
Total	6,290 pounds per million gallons	3,462

Chemical Wastewater Treatment Plant Design" (August, 1973). The total quantities of raw and chemical sludges produced are as follows:

Lime at 400 mg/l $[Ca(OH)_2]$ = 6290 lb/million gal
Alum at 200 mg/l $[Al_2(SO_4)_3 \cdot 14H_2O]$ = 2648 lb/million gal
Ferric chloride at 80 mg/l $[FeCl_3]$ = 2662 lb/million gal

Table 8-6   Estimate of Alum Sludge Quantities.

Raw sewage suspended solids	250 mg/l
Raw sewage volatile suspended solids	150 mg/l
Raw sewage PO_4^{3-}	11.5 mg/l as P
Raw sewage total hardness	170.5 mg/l as $CaCO_3$
Raw sewage Ca^{2+}	60 mg/l
Raw sewage Mg^{2+}	5 mg/l
Effluent PO_4	0.3 mg/l as P
Effluent Ca^{2+}	60 mg/l
Effluent Mg^{2+}	5
Effluent Al^{3+}	0

Alum dosage	200 mg/l as $Al_2(SO_4)_3$ $14H_2O$ – fw = 594
From equation 6	$AlPO_4$ formed is 1 mole per mole of P
	$\dfrac{11.2}{30.97}$ = 0.365 mole P removed
	Therefore 0.365 mole of $AlPO_4$ are formed; fw is 122
	Therefore weight is 0.365 × 122 = 44 mg/l
From equation 12	Al^{3+} in = Al^{3+} out; Al^{3+} in = 18.1 mg/l
	Al^{3+} content of $AlPO_4$ = 0.365 × 27 = 9.9 mg/l
	Al^{3+} not accounted for = 18.1 – 9.9 = 8.2 mg/l
From equation 7	$Al(OH)_3$ formed is 1 mole per mole Al^{3+}
	Therefore $\dfrac{8.2}{27}$ = 0.31 mole $Al(OH)_3$; fw = 78
	So weight of $Al(OH)_3$ is 0.31 × 78 = 24 mg/l

Sludge composition

Sludge species	Total weight	Ash
		Pounds per million gallons
Raw sewage solids	250 mg/l = 2,080 pounds per million gallons	832
$AlPO_4$	44 mg/l = 368 pounds per million gallons	368
$Al(OH)_3$	24 mg/l = 200 pounds per million gallons	133
Total	2,648 pounds per million gallons	1,333

Sludge Thickening

The purpose of thickening is to reduce the sludge volume to be stabilized, dewatered, or hauled. It increases the solids content by partial removal of the liquid. A conventional gravity thickner is shown in Figure 8-1.

Thickening by gravity is simple and inexpensive, but it may not produce as highly concentrated sludges as other thickening processes. Gravity thickening is essentially a sedimentation process similar to that which occurs in all settling tanks, but in comparison with the initial waste clarification stage the thickening action is relatively slow.

Thickener operating costs are minimal because of their relatively low

Table 8-7   Estimate of Iron Sludge Quantities.

Raw sewage suspended solids	250 mg/l
Raw sewage volatile suspended solids	150 mg/l
Raw sewage PO_4^{3-}	11.5 mg/l as P
Raw sewage total hardness	170.5 mg/l as $CaCO_3$
Raw sewage Ca^{2+}	60 mg/l
Raw sewage Mg^{2+}	5 mg/l
Effluent PO_4	0.3 mg/l as P
Effluent Ca^{2+}	60 mg/l
Effluent Mg^{2+}	5
Effluent Fe^{3+}	0

$FeCl_3$ dosage	80 mg/l
From equation 9	$FePO_4$ formed is 1 mole per mole P
	$\dfrac{11.2}{30.97} = 0.365$ mole P removed
	Therefore 0.365 mole of $FePO_4$ are formed; fw = 151
	Therefore weight is 0.365 × 151 = 55 mg/l
From equation 12	Fe^{3+} in = Fe^{3+} out; Fe^{3+} in = 28 mg/l
	Fe^{3+} content of $FePO_4$ = 0.365 × 55.8 = 20.4 mg/l
	Fe^{3+} not accounted for = 28 − 20.4 = 7.6 mg/l
From equation 10	$Fe(OH)_3$ formed is 1 mole per mole Fe^{3+}
	Therefore $\dfrac{7.6}{55.8} = 0.136$ mole $Fe(OH)_3$; fw = 107
	So weight of $Fe(OH)_3$ = 0.136 × 107 = 15 mg/l

Sludge composition

Sludge species	Total weight	Ash
		Pounds per million gallons
Raw sewage solids	250 mg/l = 2,080 pounds per million gallons	832
$FePO_4$	55 mg/l = 460 pounds per million gallons	460
$Fe(OH)_3$	15 mg/l = 122 pounds per million gallons	105
Total	2,662 pounds per million gallons	1,397

horsepower (15 hp for 200 ft diameter), low maintenance requirements due to their slow peripheral speed (25 ft/minute average), and the low operating labor expenses that are experienced with automated units.

Gravity thickening usually exhibits hindered settling phenomenon because of the relatively concentrated nature of the sewage and industrial wastewater solids. There are four basic zones in a gravity thickening system: (1) a clarification zone at the top containing the relatively clear supernatant liquid, (2) a settling zone characterized by a constant rate of solids settling, (3) a compression zone characterized by a decreasing solids settling rate, and (4) a compaction zone where the settling rate is very low.

Continuous thickening units for suspensions with hindered settling

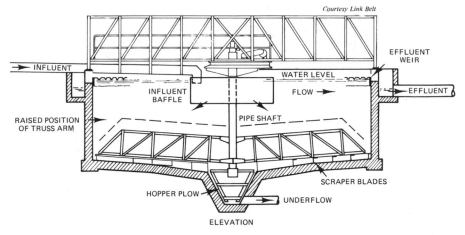

Courtesy Link Belt

Fig. 8-1   Gravity thickener.

characteristics can be designed on the basis of data collected during batch settling tests. A solid–liquid interface settling curve developed from laboratory tests can be used to determine the surface area required for clarification and the surface area required to thicken the sludge to a particular solids concentration. The degree to which waste sludges can be thickened depends on many factors; among the most important are the type of sludge being thickened and its volatile solids concentration. The initial solids concentration also has an effect on the degree of concentration achieved. In general, it has been found that optimum results are achieved when the feed solids concentration is between 0.5 and 1 percent. Within this range, sludge compaction and overhead clarity are optimized. Hydraulic and surface loading rates are also of importance.

In secondary treatment plants, current practice in the U.S. calls for the use of overflow rates of 400–800 gpd/ft^2. Excessively low flow rates can lead to odor problems. If the sludge flow to the thickener is far below the design rate, pumping of secondary effluent to the thickener may be practiced to minimize odors. The following solids loading rates are generally used:

Primary sludge: 22 lb/ft^2/day
Primary + trickling filter sludge: 15 lb/ft^2/day
Primary + waste activated sludge: 8–12 lb/ft^2/day
Waste activated sludge: 4 lb/ft^2/day

In the thickening of sludge from lime clarification of raw wastewater, incoming sludge containing 3 percent solids has been thickened to 6–9 percent solids at loading rates of 9–17 pounds per day per square foot

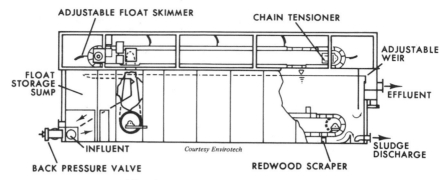

Fig. 8-2   Dissolved air flotation unit.

(ppd/ft^2). In the thickening of sludge from high lime tertiary treatment incoming sludge of 1–2 percent solids has been thickened to 8–20 percent solids at solids loading of 200 ppd/ft^2. Gravity thickeners have also been used in handling spent lime rejects in the centrate discharged by a centrifuge classifying thickened spent lime sludge from tertiary treatment. With an incoming solids content of 3 percent the thickener produced 5–7 percent solids at a loading of 15 ppd/ft^2 and at 24 hours detention, which is the time required to achieve maximum compaction of the sludge. Most continuous thickeners are designed with a side water depth of 10 ft.

Figure 8-2 shows a dissolved air flotation unit. In general, air flotation thickening can be employed whenever particles tend to float rather than sink. These procedures are also applied if the materials have a long subsidence period and resist compaction for thickening by gravity. They have been used successfully for thickening of primary, waste activated, and mixtures of primary and waste activated sludges. They have also been used to thicken mixtures of alum sludge or lime sludge with biological sludges. At the Wyoming, Michigan wastewater treatment plant, which uses ferric chloride and polymer in primary treatment, the primary–chemical sludge mixture is processed in a flotation thickener loaded at 20 ppd/ft^2. In primary and secondary treatment plants, dissolved air flotation units are loaded at 1–6.5 pph/ft^2 (pounds per hour per square foot) and at surface overflow rates of 0.3–1.7 gpm/ft^2. Rates of 2 pph/ft^2 and 0.8 gpm/ft^2 are commonly used in design.

Sludge Dewatering

As previously mentioned, the same dewatering devices used for primary and secondary sludges may be applied to the dewatering of chemical sludges and mixtures of chemical and biological sludges. The devices include:

drying beds, lagoons, vacuum filters, centrifuges, filter presses, and horizontal belt filters.

The selection of the dewatering method to be used is related to the sludge handling processes which precede and follow. Lagoons usually receive sludges directly without prethickening or other conditioning. Solids removed from lagoons are usually not suited for heat drying or incineration, and are taken directly to land spreading or landfill disposal. Sludge destined for drying beds may be thickened and, again, is not suitable for heat drying or incineration; it is usually disposed of on land or in landfill. In general, sludge which is to be dewatered by vacuum filtration, centrifuging (solid bowl), filter pressing, or horizontal belt filtration is pretreated by thickening or other conditioning, and may be disposed of by heat drying, incineration, land spreading, or in landfill.

The combination of sludge processes selected should best satisfy the objectives of high solids capture, minimum cost, optimum sludge moisture content, reliability, and ease of operation.

The technology and design of all available dewatering methods is constantly under development. Each type, therefore, should be given careful consideration. The applicability of a given method should be determined on a case by case basis with the specifics of any given situation being carefully evaluated, preferably in pilot tests.

Detailed descriptions of sludge dewatering processes and equipment are given in the EPA Technology Transfer Manual for Sludge Treatment and Disposal, October, 1974. Much of the design and operational data relates to primary and secondary sludges. The rest of this discussion is aimed primarily at chemical sludges, although the data available are quite limited.

The dewaterability of lime sludge by vacuum filtration can be measured by specific resistance determinations and filter leaf tests. Specific resistance (R) is a measure of the ease with which water may be drained from a given volume of sludge when subjected to a driving force such as gravity or vacuum. Hence a sludge with a high specific resistance would be difficult to dewater. If three or more specific resistance tests are performed at different pressures, a plot of specific resistance versus pressure on bilogarithmic paper will result in a straight line. The slope of the straight line is dimensionless and represents the coefficient of compressibility (σ). A sludge with a high coefficient of compressibility indicates that the filter cake is easily compressed and therefore reduced in volume. Typical values of specific resistance and coefficient of compressibility are shown in Table 8-8. From these values it can be seen that wastewater sludges are generally more difficult to dewater than water treatment sludges. Water treatment sludges are also more compressible.

TABLE 8-8 Typical Specific Resistance and Co-efficient of Compressibility Values for Various Sludges.

Sludge	Specific Resistance @ 38.1 cmHG (sec²/g)	Coefficient of Compressibility
Primary wastewater	$15-70 \times 10^9$	—
Digested wastewater	$14-70 \times 10^9$	0.51–0.74
Water treatment	$1-10 \times 10^9$	0.8–1.3

The results of tests using thickened lime sludge show that at 38.1 cmHg, the specific resistance is 24.0×10^9 sec²/g, which is within a range typical of wastewater sludges. The coefficient of compressibility is 1.056, which is characteristic of water treatment sludges. Consequently, the lime sludge can be expected to display dewatering characteristics similar to both types of sludge, that is, moderately difficult to dewater but easy to compress.

Although vacuum filter yield rates can be determined from specific resistance measurements, filter leaf tests provide a more realistic method of specifying operational characteristics of the vacuum filter. Not only are test results more accurate with respect to yield rate, cake moisture, and filtrate clarity but they offer a visual demonstration of cake texture, cracking, and dischargeability from the filter medium.

For example, one series of tests by Martel and co-workers[12b] on sludge obtained in the lime clarification of raw wastewater was performed on Amherst sewage. The lime sludge had good dewatering characteristics even without the use of conditioning chemicals. Tests indicated that thickener concentrations in the 6–9 percent range could be expected at thickener solids loading rates of 9–17 ppd/ft². At a typical thickener underflow concentration of 7 percent the filter yield was about 2.7 ppd/ft². The solids content of the cake ranged from 26 to 32 percent. It was estimated that filtrate returned to the treatment process would represent 0.2 percent of the total flow, which should not have a detrimental effect on the lime clarification process despite the high COD content of 1220 mg/l.

The low filter yield rates obtained in this study are consistent with those obtained by other investigators. For a primary lime sludge (3.6 percent solids) flocculated at pH 11.0 Parker and co-workers[15a] reported a filter yield rate of 1.5 pph/ft². From the data presented by Burns and Shell,[5b] a 7 percent sludge would produce a 2.4 pph/ft² filtration rate. They also conducted tests with anionic polyelectrolyte flocculation aid (Dow Chemical, AP-30) which increased vacuum filter yields by 30–70 percent.

This same Amherst sludge was subjected to centrifugation with production of cake containing about 30 percent solids and a clear centrate without the use of chemical aids. At a lime dosage of 380 mg/l as CaO and a pH of 11.5, the residual phosphorus was less than 1.0 mg/l. Sludge production amounted to 4000 lb/million gal, which is approximately four times the amount from plain primary sedimentation. It was estimated that two stage recarbonation would produce an additional 2250 lb/million gal of calcium carbonate sludge.

Bower and DiGiano[5a] at the University of Massachusetts compared alum coagulation of raw wastewater to lime coagulation for phosphorus removal. Measurements were made of specific resistance and coefficient of compressibility for sludges produced at two lime doses, 300 mg/l and 400 mg/l, and at an alum dose of 175 mg/l (pH 5.9). The values found for the specific resistance of the two lime sludges were very similar. Those of the alum sludge were approximately three times as great, as shown in Table 8-9. This means that, other factors being equal, it would take three times as long to dry the alum sludge on the sand beds and, with mechanical dewatering devices, 1.73 times as long when using a vacuum filter or centrifuge.

Additional data on dewatering of lime, alum, and iron sludges from raw wastewater by vacuum filtration are presented in Tables 8-10 and 8-11.

TABLE 8-9   Sludge Dewatering Characteristics.

Sludge	Specific Resistance (sec^2/g)			Coefficient of compressibility
	12.6 cmHg	25.4 cmHg	41.6 cmHg	
Lime, 300 mg/l	2.52×10^9	5.08×10^9	7.47×10^9	0.909
Lime, 400 mg/l	2.72×10^9	4.77×10^9	7.50×10^9	0.858
Alum, 175 mg/l pH 5.9	7.96×10^9	13.83×10^9	20.15×10^9	0.778

TABLE 8-10   Physical-Chemical Sludge Dewatering Data, Salt Lake City

Sludge Type	Vacuum Filter Area, ft^2/mgd	Cake Moisture, %
Iron	62	80
Aluminum	84	80
Lime (low pH)	29	60

**TABLE 8-11   Cleveland Westerly
Vacuum Filter
Performance**

Lime sludge	pH 10.5
Loading	3–19 pph/ft^2
Cake solids	19–36%
Solids capture	90–98%

Sludge Disposal

Sludge disposal is perhaps the most important factor governing the choice of chemical coagulants. Unfortunately, the least is known about this particular facet of sludge handling.

Alum and iron sludges can normally be added to existing anaerobic digesters. The higher digester loadings resulting from additional sludge protection usually will not be detrimental to operation unless an organic overloading condition exists. Release of soluble phosphorus from the sludge during digestion is minimal. Final disposal of the digested sludge can be on land or by dewatering and incineration.

Alum, iron, and lime sludges can be disposed of directly onto land. At warm temperatures alum and iron sludges may need lime treatment to prevent odors.

In large systems, sludge thickening or dewatering prior to lagooning or incineration should be considered. Here the type of sludge becomes important. Alum and iron sludges are much more difficult and expensive to thicken or dewater than are lime sludges. The data in Table 8-12 illustrate the low solids concentrations to be handled.

Sludge incineration, particularly for large cities, has the advantage of

**TABLE 8-12   Probable Sludge
Concentrations.**

Chemical Coagulant	Percent Solids
Gravity thickening:	
Alum and iron	2–5
Lime	10–25
Dewatering:	
Alum and iron	10–20
Lime	20–40

TABLE 8-13   Physical-Chemical Solids Reduction by Incineration.

| | Dry Weight | |
| | Before Incineration | After Incineration |
Coagulant	Pounds per Million Gallons	
Alum	2,648	1,333
Iron	2,662	1,397
Lime	6,920	3,462

converting organic solids to ash, and thereby reducing the weight and volume of solids. Alum, iron, and lime sludges can be incinerated. The relative amounts of water and solids described earlier control the incinerator size. Table 8-13 illustrates the weight reduction achieved by incineration.

When lime recovery systems are employed, recycling inert solids necessarily appear as a part of the reclaimed coagulant feed. If recalcination for coagulant reuse is employed, each cycle of coagulant recovery will increase the total dry solids to be processed by the amounts shown in Table 8-14. Following this line of reasoning, unless blowdown of inerts from the system occurs, regardless of plant size, coagulant recovery systems must in time approach an infinite capacity. Purely as a coarse approximation, the following equation can be used to illustrate this point:

$$\text{Feed} = CaCO_3 + \text{organics} + [\text{inerts} \times (C - 1)]$$

where $CaCO_3$, organics, and inerts are in pounds per million gallons, and C is the number of cycles, starting with the initial feed as No. 1. Note

TABLE 8-14   Theoretical Buildup of Inerts in a Recycling Coagulant-Recovery System.

Constituent	Increase of Inerts per Cycle, Pounds per Million Gallons
Ash (from raw sewage solids)	832
Hydroxyapatite	510
Magnesium hydroxide	100
Total inerts per cycle	1,442

TABLE 8-15   Incinerator Feed Rates Theoretically Required for a Nonblowdown Coagulant-Recovery System.

Cycles	Feed, Pounds per Million Gallons, Dry Solids
1	6,290
5	11,440
10	18,640
20	33,040

that the equation is usable only for more than one cycle. Table 8-15 illustrates for the example what would occur at the 5th, 10th, and 20th cycle of such a system.

Clearly such a buildup of inerts as indicated in Table 8-15 is unacceptable in the design of solids handling systems. This problem has spurred research into better techniques of separating or classifying chemical sludges one from another. Several techniques for reducing the buildup of inert solids within a coagulant recovery system are available, including the following:

Direct blowdown of unprocessed sludges
Blowdown of dewatered chemical sludges
Classification of solids content
Chemical treatment of unprocessed sludges
Indirect blowdown of recovered coagulant
Combinations of the above methods

Regardless of the methodology employed for blowdown of unwasted constituents, some fraction of inert materials will be present as a recycle in any solids handling system employing coagulant recovery and reuse. Therefore, the design engineer must be able to determine what the fraction is, as well as its characteristics, prior to design of a proper solids handling system. This determination is made most easily by calculation of mass balance under conditions when equilibrium is reached in the system. Under the conditions shown in Table 8-12 equilibrium would occur when blowdowns of inerts are:

2,080 lb/million gal organics
510 lb/million gal hydroxyapatite
100 lb/million gal magnesium hydroxide

Continuing the example, assume a coagulant recovery system employing the following unit processes:

Centrifugal dewatering and classification
Recalcination
Dry blowdown of 25 percent of calciner output

Calculate the theoretical centrifuge feed, cake output, calciner output, and blowdown of solids required and enumerate by type, assuming the following test results are available:

30 percent of hydroxyapatite is wasted in centrate
25 percent of magnesium hydroxide is wasted in centrate
25 percent of organics are wasted in centrate
10 percent of calcium carbonate is lost in centrate
10 percent of ash is wasted in centrate
25 percent of calciner output is blowdown

The solution is as follows:

Apatite to waste $= 0.3\,x + 0.25(0.7\,x)$

$$510 = 0.3\,x + 0.175\,x = 0.475\,x$$

$$x = \frac{510}{0.475} \cong 1075 \text{ lb apatite/million gal reports}$$
$$\text{in centrifuge feed}$$

Magnesium hydroxide to waste $= 0.25\,y + 0.25(0.75\,y)$

$$100 = 0.25\,y + 0.19\,y = 0.44\,y$$

$$y = \frac{100}{0.44} \cong 227 \text{ lb Mg(OH)}_2\text{/million gal reports in}$$
$$\text{centrifuge feed}$$

Organics to waste $= 2080$ lb/million gal wasted in two forms; i.e., ash and organics

Organic equivalent as ash $= 0.40(2080) = 830$ lb/million gal

Centrate wastage $= 0.25 \times 830 = 208$ lb/million gal

622 lb/million gal remain and are wasted as ash

Ash to waste $= 0.1\,z + 0.25(0.9\,z) + 0.25(622)$

$$622 = 0.1\,z + 0.225\,z + 156 = 0.32\,z + 156$$

$$z = \frac{466}{0.325} \cong 1440 \text{ lb ash/million gal reports in}$$
$$\text{centrifuge feed}$$

Calcium carbonate to waste $= 0.1 \, t + (0.25)(0.9 \, t) = 0.325 \, t$

$$t \cong 3600 \text{ lb/million gal (from Table 8-13)}$$

Calcium carbonate wasted $= 3600 \times 0.325$
$$= 1170 \text{ lb/million gal}$$

Using the above equations Tables 8-16, 8-17, and 8-18 may be constructed. The example assumes that there will be a net positive blowdown of the inert solids itemized in Table 8-18. The inerts *cannot* be recycled. Table 8-19 compares solids handling and lime requirements for a solids handling system with and without lime recovery. Tables 8-16 through 8-19 are from EPA Technology Transfer Manual, "Physical Chemical Wastewater Treatment Plant Design," August, 1973).

The arrangement of the calculations required to determine equilibrium values for chemical sludges in the manner illustrated provides the design engineer with a concise tabulation of the amounts of each type of sludge under any condition he may choose to investigate. This tabulation, in turn, allows an orderly economic evaluation to be made. The designer may choose to evaluate several alternative methods of solids handling,

TABLE 8-16   Theoretical Feed, Centrate, and Cake Content at Equilibrium in a Coagulant-Recovery System. (Values in pounds per million gallons, dry solids.)

Component	Sludge				
	$CaCO_3$	$Ca_5OH(PO_4)_3$	Organics	Ash	$Mg(OH)_2$
Centrifuge feed	3,600	1,075	2,080	1,440	227
Centrate	360	323	520	144	57
Cake	3,240	752	1,560	1,296	170

TABLE 8-17   Theoretical Calciner Output at Equilibrium in a Coagulant-Recovery System. (Values in pounds per million gallons, dry solids.)

Component	Product			
	CaO	$Ca_5OH(PO_4)_3$	Ash	$Mg(OH)_2$
Calciner output	1,820	752	1,920	170
Blowdown (25 percent)	455	187	480	43
Remainder to reuse	1,365	565	1,440	127

TABLE 8-18 Components of Inerts Actually Wasted with Theoretical Inerts Wastage Required at Equilibrium in a Coagulant-Recovery System.

Inert	Source of Wastage			Theoretical Required Table V-1 Total
	Centrate	Blowdown	Total	
$Ca_5OH(PO_4)_3$	323	187	510	510
$Mg(OH)_2$	57	43	100	100
Ash	208 + 144	480	832	832

TABLE 8-19 Solids Handling and Lime Requirements With or Without Lime Recovery at Equilibrium.

Component	With Lime Recovery	Without Lime Recovery
	Pounds per Million Gallons	
Sludge from primary clarifier	8,422	6,920
Sludge to be disposed of, assuming incineration	1,442	3,462
Makeup lime requirements (CaO)	1,135	2,500

ranging from no recovery to sophisticated recovery systems; he can, therefore, make a sound decision. In addition, the designer is assured that adequate capacity is provided for the system's needs. Weak points in the system can then be evaluated, and standby capacity or redundancy can be added as may be required or deemed advisable.

EXAMPLES OF CHEMICAL SLUDGES AND METHODS FOR HANDLING

Chemical Treatment for Phosphorus Removal by Mineral Addition to Raw Wastewater

Lebanon, Ohio. The primary sludge produced by alum addition received lime treatment before final disposal as landfill or on farmland. Before lime treatment, the sludge was similar in character to biological sludge and had a solids concentration of from 2 to 3 percent. It had very poor settling and thickening characteristics and was malodorous.

Lime treatment consisted on mixing the sludge with a 25 percent lime slurry until a pH of 11.5 was reached. The average lime requirement was

400 lb of lime as $Ca(OH)_2$ per ton of dry solids. A contact time of 30 minutes was maintained before the sludge was allowed to flow to sand beds for drying. Seventy percent of the total volume of sludge applied to the drying beds was removed as drainage in the first 3 days, and the sludge cake could be lifted off the beds in 3 weeks. Elimination of odors and complete pathogen kill was accomplished with lime treatment.

Lake Odessa and Grayling, Michigan; Cleveland, Lebanon, and Mentor, Ohio.   At these locations information related to iron sludges and sludge handling was obtained. It was found that the sludge resulting from the addition of iron for phosphorus removal can be handled by anaerobic digestion, vacuum filtration, incineration, a combination of these three, or lime treatment. Each of the first three methods has been used during plant scale studies of iron addition. The last method was used on a research study during a plant scale addition of iron.

At Lake Odessa, anaerobic sludge digestion is practiced and digester supernatant was returned to the primary unit during a test period. The total phosphorus concentration in the supernatant decreased during the study from an initial concentration of 100–200 mg/l to 23 mg/l during the latter stages of the study period, and it was concluded that phosphorus release was not occurring during digestion. A crystalline phosphate precipitate found in the raw sludge from Grayling and the digested sludge from Lake Odessa was identified by X-ray diffraction techniques as vivianite, $Fe_3(PO_4)_2 \cdot 8H_2O$. Analyses by X-ray diffraction also showed that the two sludges contained 20–24 percent iron.

The chemical sludge resulting from iron addition at Mentor, Ohio was digested anaerobically and dewatered on vacuum filters. The volume of sludge produced during the iron and lime treatment was never more than double that of periods without chemical treatment. Iron–lime primary sludge averaged 8.5–10.0 percent solids in November and December, 1969. The anaerobic digesters operated with normal gas production and pH. The larger volumes of sludge shortened digester residence time, but stabilization did occur. Typical concentrations of iron and phosphorus in the various sludges are given in Table 8-20.

At Mentor, Ohio, the iron–lime digested sludge had good dewatering characteristics. The digested sludge with 10 percent solids was dewatered on a vacuum filter to an average of 20 percent solids.

At Cleveland, Ohio, anaerobic digestion, vacuum filtration, and incineration were utilized to dispose of the 28 tons/day of solids produced by $FeCl_3$ addition to the primary units. Polymer conditioning of both raw and digested sludge resulted in filter yield rates of up to 10 lb/ft^2/hour. No problems were encountered in incinerating the cake.

TABLE 8-20   Typical Iron and Phosphorus Analysis During
Sludge Handling at Mentor, Ohio.

	Primary Sludge Analysis		
	Raw, mg/l	Digested, mg/l	Supernatant, mg/l
Total iron	3750	1020	3950
Soluble iron	105	26	trace
Total phosphorus (as P)	1140	393	1240
Soluble phosphorus	81.5	32.7	6.6

TABLE 8-21   Sludge from Ferric Chloride
Treatment of Raw Sewage at
Lebanon, Ohio.[25]

	Before Lime	After Lime
Sludge solids, %	1.9	2.4
volatile portion, %	70	57
dissolved solids, %	0.17	0.17
Vacuum filter yield[a]		
90 mg/l $FeCl_3$ dose	1.06	1.57
45 mg/l $FeCl_3$ dose	1.57	2.40
Filter cake moisture content[b]		
90 mg/l $FeCl_3$ dose	4.10	4.28
45 mg/l $FeCl_3$ dose	3.75	3.75

[a]lb sludge solids/ft^2/hr.
[b]lb water/lb sludge solids.

Lime treatment of ferric chloride sludge at Lebanon, Ohio was designed on the basis of 8700 gpd of sludge. Lime addition to the sludge to a pH of 11.5 required 244 lb of lime, as $Ca(OH)_2$, per ton of dry sludge solids. After 30 minutes of contact at this pH, the sludge was pumped to drying beds. Samples of limed and unlimed sludge were analyzed, and corresponding sludge characteristics are given in Table 8-21. The analyses show that lime treatment increased sludge solids and filter yields but had little or no effect on filter cake moisture content.

Escanaba, Michigan.   A complete mix, activated sludge plant at Escanaba, Michigan has been constructed to replace a combination trickling filter–conventional activated sludge plant. Provisions were made in the design to meet the state's requirement of 80 percent phosphorus removal. Effluent from the plant will flow to Little Bay DeNoc of Lake Michigan. The storage facilities for FeCl₃ are designed to provide a 30 day supply at a feed rate of 27 mg/l Fe^{3+} to a plant flow of 2.2 mgd. Sludge from the primary basins will be pumped to the existing anaerobic digesters, having a total capacity of 77,700 ft³. Digested sludge will be pumped to drying beds with a total area of 37,090 ft². Waste activated sludge will be digested aerobically.

Wyoming, Michigan.   An expansion of the Wyoming, Michigan wastewater treatment plant is being designed for 80 percent phosphorus removal and increased organic removal. The plant will include new chemical mixing and flocculation basins, two primary settling beasins, a chemical feed and storage building, secondary aeration basins (trickling filters are to be used for roughing), final settling basins, and sludge handling facilities. Phosphorus will be removed by addition of ferric chloride to the new chemical mixing basins, followed by polymer addition, flocculation, and sedimentation in the expanded primary facilities. Provision will be made to feed ferric chloride and polymer to the final clarifiers for increased phosphorus or solids removal. All chemical tanks and feed equipment will also be designed to allow the use of liquid alum, ferrous chloride, ferrous sulfate, and sodium hydroxide.

Plant components associated with phosphorus removal are to be designed on the following basis. The chemical mix and flocculation basins will have a combined detention time of 20 minutes at the average design flow of 19 mgd. At design flow the two existing and two new primary basins will have overflow rates of 604 gpd/ft², weir rates of 15,100 gpd/ft, and detention times of 2.07 hours.

Capacity of the sludge handling facilities will be more than doubled. The present units consist of two holding tanks with mixers, two 250 ft² vacuum filters, and one 6 hearth incinerator rated at 1500 lb dry solids/hour. The expanded sludge processing facilities have been designed on the following basis:

Suspended solids		
Plant load, lb/day	63,450	
Overall removal (95% assumed), lb/day		60,290
Biochemical oxygen demand		
Plant load, lb/day	48,260	
Removed by aeration basins, lb/day	7,850	
Waste activated sludge, lb/day[a]		3,925
Total		64,215
Solids added by lime and ferric chloride conditioning (15%)		9,635
Total solids loading, lb/day		73,890
Total solids loading, lb/hr		3,080

[a] *Waste activated sludge production based on 0.5 lb solids generated per lb BOD removed.*

Additional equipment which will be provided to handle this load includes two air flotation thickeners, two sludge holding tanks, two vacuum filters, and one incinerator. The two air flotation thickeners will each have an effective area of 1170 ft². The sludge will be conditioned with lime or ferric chloride prior to flotation thickening at a rate of 20 lb/ft²/day. The two holding tanks will allow 4 days storage capacity for the additional sludge. Each of the two new 500 ft² vacuum filters will have a capacity of 4.2 lb/ft²/hour, and one will be on a standby basis. The 6 hearth incinerator will handle 2100 lb dry solids/hour. Lagoons will be provided for disposal of the ash.

Handling Requirements for Sludges Produced by Mineral Addition to Raw Wastewater

Application of phosphorus removal chemicals during primary treatment will result in increased amounts of primary sludge because of the greater capture of suspended solids and the precipitated metal phosphate and hydroxide solids. Consequently, the amount of secondary sludge will generally decrease, and the ratio of primary to secondary sludge will increase.

Sludge conditioning characteristics will also be changed. Instead of requiring cationic polymers or ferric chloride and lime for conditioning, sludges resulting from mineral addition to the primary settler will probably best be conditioned with an anionic polymer. Filter yields should remain about the same as those for sludges from conventional treatment without mineral addition. Addition of polymer to the raw wastewater may reduce the polymer demand for conditioning the sludge.

Anaerobic digestion will be unaffected by the addition of chemically precipitated sludge. The higher digester loadings resulting from additional sludge production will not normally be detrimental to operation unless an organic overloading condition exists. Release of soluble phosphorus from the sludge during digestion is considered to be minimal. In situations requiring total phosphorus removal greater than 90 percent, it may be desirable to treat sidestreams such as digester supernatant and filtrate separately with metal salts before recycle to the head of the plant.

In comparing a conventional secondary treatment plant with the same plant having chemical treatment in the primary tank for phosphorus removal, the addition of chemical treatment results in better solids capture, improved BOD conversion, and greatly increased phosphorus removal. The calculated ratio of primary to secondary sludge is more than double that of conventional treatment, and in actual operation would result in improved sludge conditioning. The addition of chemical treatment also reduces the load to the secondary, which should result in some improvements in plant operation. Chemical treatment increases the total sludge mass about 24 percent, of which about 16 percent is due to formation of the insoluble iron–phosphate–hydroxide sludge.

Handling Requirements for Sludge Produced by Mineral Addition to the Activated Sludge Process

Pennsylvania State University. In studies conducted at Pennsylvania State University by Long and co-workers,[12a] indications were that sludge production in the chemical-biological process was significantly greater than in biological treatment only. This is to be expected because of the sludge produced by mineral addition. For a 1 year alum test period, the amount of waste sludge from the chemical-biological process was 2.1 times by weight that from the biological process alone. On the basis of volatile solids, the ratio was 1.56:1. However, high alum doses (mole ratios of Al:P of 2:1 or above) were used in the study and the sludge age of the undosed control was greater than that for the dosed system.

Manassas, Virginia. Changes were found at Manassas in sludge density and additional solids production with aluminate additions. During these studies the average weekly primary effluent phosphorus concentration ranged between 9.3 and 13.1 mg/l. From this figure it can be concluded that up to stoichiometric requirements, additional solids production will be about 4 lb/lb Al^{3+}. Above stoichiometric requirements, production of additional solids decreases. This observation can be explained by the predominant formation of $AlPO_4$ (molecular weight, 122) at less than stoichiometric Al additions and formation of $Al(OH)_3$ (molecular weight, 78) above stoichiometric Al additions. There is an improvement in sludge density (which is inversely related to the sludge volume index) with increasing aluminum dosages. Whether or not this density improvement is enough to compensate for the additional solids that must be handled, and thereby eliminate the need for additional sludge disposal facilities, is dependent upon the original sludge density and the amount of phosphorus removed and aluminum added. Mineral additions do produce a dense, stable sludge.

Sludge settling characteristics change as a function of the operating solids level. Sludge settling characteristics were determined in 1 l graduates. Manassas experience indicates that the settled sludge does have a tendency to classify (i.e., separate into biological and chemical fractions), although not to such an extent as to harm system performance. Therefore, sludge collection systems which have a tendency to classify sludges into heavy and light fractions, i.e., those that work with hydrostatic or vacuum principle, probably should be avoided.

Experience at Manassas has found no resolubilization of phosphorus with the waste organic–chemical sludge in anaerobic digestion. At Manassas, comparative bench studies of settled high rate activated sludge dewaterability with and without aluminum additions showed that dry cake filter yields will substantially increase with aluminum additions, but there is a tendency to produce a wetter cake. Similar high yield rate observations after aluminum additions have been observed by Smith and co-workers.[23a] This increase in dewaterability can probably be explained by the fact that with efficient chemical utilization, aluminum phosphate, not aluminum hydroxide, is the inorganic precipitate being dewatered.

Sludges resulting from aluminum addition are amenable to anaerobic digestion. In conjunction with laboratory scale aluminum addition to activated sludge units at the University of Missouri at Rolla, laboratory scale anaerobic digesters were studied to determine the fate of the aluminum and phosphorus. The digesters were fed 70 percent activated sludge and 30 percent primary sludge. Detention time in the digesters was 15 days. The phosphorus which had been precipitated with aluminum was concentrated in the digester sludge and not released to the supernatant. Tests confirmed that the aluminum was also retained in the digester sludge. Comparing the digesters which were fed aluminum sludge with a control digester, using the parameters of volatile acids and gas production, the digester performance was equivalent.

Aluminum additions will result in the least amount of additional solids for ultimate disposal, probably 30–50 percent of that found with iron and 5–20 percent of that found with two stage lime treatment.

Aluminum–organic sludges may be incinerated or spread on land. When the sludge is applied to the soil, leachates should not be a problem because of the natural phosphorus exchange capacity of soils and the low solubility of the aluminum phosphate precipitate. Preliminary tests by Dotson showed no plant inhibitory effect with applications of organic–aluminum sludges up to 1 in. per week, although some initial seed germination inhibition may be encountered at higher applications. Generally, it is felt that the soil building and fertilizing benefits derived from organic sludge application more than compensate for any deleterious effects from the inorganic aluminum sludge.

Phosphorus Removal By Lime Addition in Tertiary Treatment

The sludge from a lime treatment system may be handled in two general ways. It may be thickened, dewatered, and disposed of or it may be thickened, dewatered, and recalcined to recover lime for reuse. For small plants, recalcination will not be economically competitive with disposal and should not be considered unless there are restrictions on disposal which make that alternative difficult. For plants over 10 mgd, recalcination of lime sludge may be practical, but dewatering and burial at nearby sites may be competitive. Recalcination may also be practical at plants somewhat smaller than 10 mgd, depending on the cost of purchased lime, distance to disposal sites, and other local conditions.

Data indicate that the volume of sludge from lime treatment will vary from 1.5 percent to several percent of feed volume. Sludge concentration will probably be in the range of 1–5 percent. The actual weight of sludge will vary with the chemical composition and amount of suspended solids in the feed water. Values have been observed in the range of 4–7 lb/1000 gal. Sludge production from the first stage of a two stage system or from a single stage system can be estimated approximately by weighing the dried sludge from jar tests run at the planned operating pH. Additional sludge produced in the second stage can be assumed to be the $CaCO_3$ formed from calcium concentration reduction during recarbonation. Calcium content before recarbonation can be obtained from the above mentioned jar tests and after recarbonation can be assumed to be 40 mg/l as Ca.

Design information for the Tahoe sludge handling system is given in the material which follows. If the lime is not to be recovered from the sludge, the underflow from the thickener can be placed directly on drying beds for final dewatering, or it can be dewatered by centrifuge or vacuum filter.

Lime Recalcining and Reuse

General Considerations.   Lime recalcination is an old art in many industrial applications. It is also used to some extent in water works practice, principally in lime–soda water softening plants. The recalcining and reuse of lime in wastewater treatment plants is relatively new, but has found a number of successful applications in the past few years.

Phosphorus removal by lime coagulation requires massive doses, in the range of 100–500 mg/l as CaO, of lime. This high rate of lime addition produces large quantities of sludge. Roughly, the volume of lime sludge produced is about $1\frac{1}{3}$ times that of organic sludges resulting from secondary treatment, which gives some perspective to the magnitude of the problem. If the lime sludge emanating from a wastewater treatment plant can be disposed of satisfactorily on land or buried, then this may be the cheapest arrangement. In the absence of this expedient for lime sludge disposal, there is a real problem. One obvious solution is to recalcine and reuse the lime within the plant. The recalcining process is quite simple. It consists of heating the calcium sludge to a temperature of about 1850° F thus driving off water and carbon dioxide leaving only the calcium oxide or quicklime. In all cases there will be some sludge to be disposed of, but the total volume will be greatly reduced. Some lime sludge must be wasted periodically to avoid a buildup of inerts. The amount to be wasted is an economic balance between solids handling capacity and desired characteristics of the reclaimed lime. In soft water areas, some makeup lime will be required to replace the calcium which leaves the plant dissolved in the effluent water. In treating wastewaters high in calcium hardness an excess of recalcined lime may be produced. In small wastewater treatment plants (say 5–10 mgd), the cost of reclaimed lime may be only slightly less than the cost of new lime, but recalcining still may be warranted because it also saves the cost of disposing of the lime sludge. In large plants (over 10 mgd), recalcining probably can be justified by the lower cost of the reclaimed lime as compared to that for new lime.

At the South Tahoe Water Reclamation Plant, lime is used for clarification, phosphorus removal, and pH elevation for ammonia stripping. The spent lime sludge has been recalcined and reused in the process continuously for a period of more than eight years. During that time more than 8000 tons of recalcined lime have been produced. The system at Tahoe is typical of facilities for use in wastewater treatment plants, and will be used to illustrate process and design principles.

Tahoe Lime Recovery System.   The lime system at Tahoe is designed for a plant wastewater flow of 7.5 mgd, with average lime dosages of 400 mg/l. Very briefly, the system consists of lime mud thickening to a solids content of 8–20 percent, further dewatering in a concurrent flow solid bowl centrifuge to 30–60 percent solids, recalcining in a multiple hearth furnace at a temperature of 1850° F, and return to reuse in the process. About 25–35 percent makeup lime is required. Figure 8-3 is schematic representation of the solids handling system for lime.

Lime Sludge Volume.   If it is assumed that the underflow from the lime clarifier is 1 percent solids, then at 7.5 mgd flow and a lime dosage of 400 mg/l about 2,500,000 lb or 290,000 gal of liquid sludge per day is expected. This is about 200 gpm. If the solids content of the lime sludge is only 0.5 percent under the worst conditions, then the maximum flow might be 400 gpm. A single, variable speed, horizontal centrifugal pump with a capacity of 450 gpm is installed at Tahoe for this service. The suction line under the floor of the chemical clarifier is glass lined. The discharge line is plain cast iron, and is cleaned periodically using a polyurethane pig, forced through the line by water pressure applied behind it.

Lime Mud Thickener.   The lime mud thickener is designed for a dry solids loading of not more than 200 lb/day/ft^2, and a surface overflow rate of 1,000 gal/ft^2/day. The unit installed is 30 ft in diameter by 8 ft deep

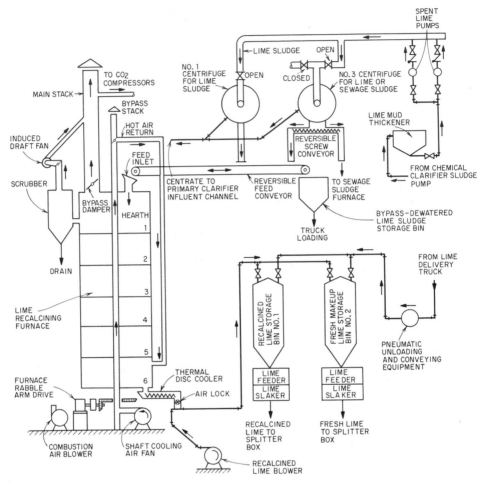

Fig. 8-3   Solids handling—lime sludges. (*Courtesy Clair A. Hill & Assocs.***)**

with a Dorr-Oliver thickener mechanism of the bottom scraper type with variable speed drive. The thickener will allow withdrawal of lime sludge up to a solids content of about 20 percent, which has about the same consistency as toothpaste. Lime sludge thicker than this is very difficult to pump. The suction line to the pump for thickened sludge is glass lined, and is extended to the surface of the ground on the side opposite the pump suction line to permit rodding of the line beneath the bottom of the thickener basin. There are two Wemco 3 in., Model C, recessed impeller, horizontal centrifugal pumps with variable speed drives for delivering sludge from the thickener to the centrifuges. Each of these pumps has a maximum capacity of 40 gpm. Ordinarily, sludge is withdrawn from the lime mud thickener at a consistency of about 8 percent solids, but under test the pumps have handled 20 percent lime solids satisfactorily.

Dewatering Lime Sludge. The thickened lime mud is further dewatered in a Bird Machine Company 24 × 60 in. concurrent flow centrifuge. The maximum hydraulic capacity is 100 gpm. The maximum expected total solids feed to the lime centrifugal is 1650 lb/hour. The minimum solids concentration in the feed is 8 percent and the maximum 20 percent. The maximum expected flow rate of feed to the machine is 36 gpm. In dewatering lime sludge, the centrifuge can produce a cake with 50–55 percent solids and can capture more than 90 percent of the solids. However, at Tahoe the lime centrifuge is used to classify or separate the phosphorus solids from the calcium solids, so that the phosphorus leaves the machine in the centrate, while the calcium is discharged in the cake. This is possible because of the greater specific gravity of the calcium solids. This type of classifying operation works very well, and is a good way to rid the lime recirculation system of phosphorus. However, it has two disadvantages: it produces a wetter cake (about 40 percent solids) and a cloudier centrate than if the centrifuge were operated for best dewatering. Ideally there should be two centrifuges in series. The first would be operated to classify with the calcium cake going to the lime furnace. The phosphorus laden centrate would go to the second centrifuge, which would be set to produce a centrate of high clarity. The phosphorus cake would go to the sewage sludge incinerator. This type of series operation was first used at Tahoe on a trial basis. It worked so well that it has now become the principal method of operation.

The dewatered lime sludge is fed to the lime recalcining furnace by a belt conveyor. By reversing the conveyor, dewatered lime can be delivered to a truck loading bin outside the building for wet disposal.

Lime Recalcining Furnace. The lime recalciner is a 14 ft 3 in. diameter, 6 hearth B-S-P furnace. It is operated at high temperatures, and there are several auxiliaries. The hearth temperatures are as follows: hearth No. 1, 800° F; No. 2, 1250° F; No. 3, 1850° F, No. 4, 1850° F; No. 5, 1850° F; and No. 6, 750° F. It is important not to let the temperature drop too low on the top hearth in order to avoid clinker information, which is due to slow drying and balling of the lime cake. The lime furnace will deliver a maximum of 20,400 lb/day of recalcined lime plus impurities. The stack gas is wet scrubbed in a Sly Company jet impingement unit. There is no smoke or steam plume and no odor in the off gases.

At the bottom outlet from hearth No. 6 of the lime furnace there is a lime grinder installed to break up large pieces of recalcined lime. It is protected by a coarse bar screen at the inlet. Following the grinder, the reclaimed lime passes through a thermal disk cooler (water cooled). The lime is then conveyed to a recalcined lime storage bin of 35 ton capacity by a pneumatic system. This equipment consists of a stainless steel air lock discharging device, air compressor, and other accessories. It has a rated capacity of 0.75 tons/hour. There is also a pneumatic conveying system with dust collector for new makeup lime, and a second 35 ton storage bin for makeup lime.

Lime Feeding and Slaking.   The makeup lime storage bin and the recalcined lime storage bin each discharge through gravimetric lime feeders and slakers. The feeders each have a capacity of 1500 lb of CaO per hour. The lime slakers are Wallace & Tiernan paste-type slakers.

Control and Instrumentation.   The furnace controls are similar to those previously described for the organic sludge incinerator. In addition, the lime feeders may be selectively controlled either manually or automatically, using recording pH meters at the point of lime application.

When the CaO content of the recalcined lime becomes less than 70 percent, some dewatered lime sludge is wasted to incineration and the proportion of new makeup lime to recalcined lime fed to the wastewater is increased.

Costs.   The capital costs for the complete lime recalcining system for a plant capacity of 7.5 mgd and a lime dosage of 400 mg/l is $516,000, based on the 1969 FWPCA STP Construction Cost Index of 127.1. This includes lime mud handling, thickening, dewatering, and recalcining. At plant capacity, the operation costs are about $1.90/ton for lime sludge dewatering, and about $18.10/ton for recalcining, or a total operation cost of $20.00/ton of recalcined lime.

Recalcining Primary Lime Sludges.   The large quantities of inert materials found in the sludges resulting from lime coagulation of raw sewage make lime recovery and reuse more difficult. One plant is now under construction at Contra Costa, California, where a system for such lime recovery and reuse has been developed which uses a combination of wet and dry classification techniques. At Contra Costa, a mixture of lime (to pH 11) and ferric chloride (about 14 mg/l) are added to the raw sewage as coagulants. Following primary clarification, the wastewater receives biological treatment, including biological nitrogen removal.

The lime sludges are thickened and then passed through series centrifugation for classification in a manner similar to that described for the Tahoe project. As shown in Figure 8-4, pilot data indicate that 90 percent

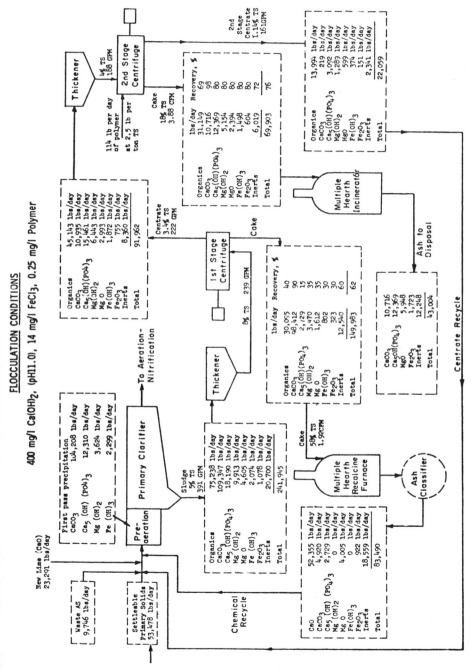

Fig. 8-4 Lime recovery from primary sludges at the Contra Costa, California waste treatment project.

calcium capture is possible in the first stage centrifuge, with rejection of 40–85 percent of the inert materials. Following recalcination in a multiple hearth furnace, the ash discharge from the furnace will be passed through a dry classification device for further purging of inert materials. The dry classification device makes a separation with air based on particle size, while the preceding centrifugal wet classification is based on particle weight. Based upon pilot tests, the use of these two classification techniques in series offers a means of providing adequate purging of inerts to permit lime reuse when coagulating raw sewage. The dry classification process is a development of Envirotech.

Recalcining with Fluidized Bed Furnaces. Although the Lake Tahoe and Contra Costa waste treatment plants are both utilizing multiple hearth furnaces, fluidized bed furnaces may be used and have for many years been used successfully for recalcining in water treatment plants.

Lime mud filter cake is fed into a paddle mixer along with dry recycled fines and quench water. The mixture then goes to a cage mill disintegrator, where precooled calciner stack gas at $1000°F$ dries and disintegrates the moist solids. The resultant fine carbonate is conveyed by the exhaust gas to a cyclone separator. Large fraction discharge from the cyclone is split, a portion being recycled to the mixer and a portion to the calciner feed bin.

Calcination takes place in a two compartment, fluidized bed furnace. The upper fluid bed of the reactor is used for low temperature calcination $(1500–1600°F)$ of calcium carbonate and the pelletization of calcium oxide. The lower fluid bed cools the calcined product. In both fluid beds the solid particles are supported on a rising column of air so that the solids behave in much the same fashion as a liquid. Fresh solids added to the bed are quickly and uniformly distributed. The beds are held in a constant state of agitation and suspension so that heat transfer is instantaneous and uniform.

Lime produced by this sytem is in the form of pelletized particles, 6–20 mesh in size. These uniform spheres are soft burned, dust free, and highly reactive. The chief advantage offered over the multiple hearth approach is the pelletized product rather than the lime dust obtained with the multiple hearth.

Alum Sludges

General Considerations. Alum is a good coagulant for many wastewaters. It is useful not only for clarification, but also for phosphorus removal. The alum dose for equal phosphorus removal will usually be about one half that required with the use of lime; the precise dose must be determined for each individual wastewater. Alum costs about twice as much as lime per pound, so that chemical costs for phosphorus removal are about equal with the use of either lime or alum. Alum sludges may be disposed of satisfactorily in anaerobic digestion tanks along with sewage sludges. The phosphorus remains with the sludge and is not returned in the supernatant liquor. Alum sludges and alum–sewage sludge mixtures are exceedingly difficult to dewater. There are two basic methods for recovery of aluminum from settled alum sludges: the alkaline method and the acid method. Under circumstances which favor the use of alum over lime, these will be of interest.

The Alkaline Method of Alum Recovery. Aluminum hydroxide is an amphoteric substance which can be dissolved in either an acidic or a basic solution. Lea and co-workers[12] suggested an alkaline method for alum recovery. The aluminum hydroxide is dissolved by raising the pH of the alum sludge to 11.9 with sodium hydroxide. The reaction converts the aluminum hydroxide to sodium aluminate and also returns the phosphates into solution. Calcium chloride is then added to react with the phosphates to form insoluble calcium phosphate. The reactions are as follows:

$$Al(OH)_3 \cdot PO_4 + 4NaOH \xrightarrow{pH\ 11.9} Na_3PO_4 + 2H_2O + 3OH + NaAlO_2$$

$$NaAlO_2 + 2Na_3PO_4 + 3CaCl_2 \longrightarrow NaAlO_2 + Ca_3(PO_4)_2 \downarrow + 6NaCl$$

Lea and co-workers found in pilot studies that the reclaimed sodium aluminate and fresh alum were equally effective in removing phosphates from sewage. Contrary to these findings, several investigators have found sodium aluminate to be ineffective for wastewater coagulation. Sawyer found that sodium aluminate was unsuitable for phosphate removal on Boston sewage unless supplemental caustic or alum or sulfuric acid were added to respectively raise or lower the pH into a favorable range. Sawyer also found that in the pH range of 10–11, considerably more sodium aluminate was required when alum alone was used. In addition, Sawyer found that sodium aluminate was not effective in removing the complex polyphosphates at high pH levels. Slechta and Culp[22] found that sodium aluminate was not an effective coagulant for coagulation of wastewater at South Lake Tahoe. Rose[17] also reports that sodium aluminate was not as effective as alum in wastewater coagulation.

The first step in the evaluation of the feasibility of alkaline alum recovery is to determine the effectiveness of sodium aluminate as a coagulant by jar tests with the particular wastewater involved. If the aluminate appears to be a practical coagulant, then alkaline alum recovery should be considered. The following laboratory techniques will enable evaluation of the feasibility of alkaline alum recovery:

1. Coagulate 5 gal of wastewater with appropriate alum dosage. Flocculate and settle. Decant clear liquid and divide sludge into 200 ml samples.
2. Treat sludge samples with varying quantities of sodium hydroxide to obtain pH values of 10–12.5. After the sludge is dissolved, determine the aluminum and phosphate concentrations in the resulting solution.
3. Add the stoichiometric amount of calcium chloride required to precipitate the phosphate.
4. Allow the calcium phosphate precipitate to settle. Decant the clear liquid, filter the remaining sludge, and combine the filtrate with the decanted liquid. Measure the aluminum and phosphate concentration in the liquid and the volume of recovered liquid. Calculate the percentage of alum recovery and cost of chemical required to recover alum.

The pH of the alum sludge must be raised to 12–12.5 with sodium hydroxide to maximize alum recovery. Alum recovery of 90–95 percent has been reported in the literature. Although the aluminum hydroxide is converted to sodium aluminate at pH values of 10–10.5, a side reaction occurs when the calcium chloride is added which results in the precipitation of a substantial portion of the aluminum ion unless the pH is above 12.

The studies by Slechta and Culp[22] showed for the wastewater studied, that the chemical cost for alum recovery was about \$62/ton recovered while the cost of fresh alum was \$57/ton. In addition to the fact that the recovered alum was a poor coagulant, the fact that the cost of the recovered alum exceeded that of fresh eliminated alkaline alum recovery with sodium hydroxide and calcium chloride as a realistic alternate for the South Lake Tahoe plant.

An alternative alkaline recovery scheme uses lime as a source of the hydroxyl and calcium ions required to dissolve the aluminum hydroxide and precipitate the phosphate. However, alum recovery is usually substantially lower when using only lime. Maximum recoveries of about 35 percent have been reported.

The Acid Method for Alum Recovery. Aluminum hydroxide can also be dissolved by decreasing the pH by the addition of sulfuric acid to the alum sludge. Roberts and Roddy[16] have reported on the successful recovery and reuse of alum at the Tampa, Florida, water treatment plant. The alum sludge from the clarifier is thickened and conveyed to a reactor where sulfuric acid is added. The resulting alum solution is used as a coagulant for the raw water.

Slechta and Culp[22] added sulfuric acid to the alum sludge resulting from alum coagulation of secondary effluent to obtain final pH values of 1–3. The resulting alum recovery is shown in Figure 8-5, and is in agreement

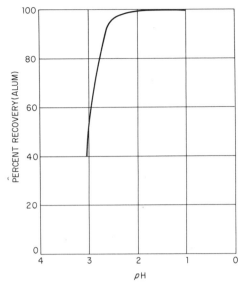

Fig. 8-5 Acid recovery of alum. (*From Slechta and Culp*[22])

with theory which indicates that aluminum hydroxide completely dissolves between pH 2.5 and pH 3. The acid recovered alum proved to be an effective coagulant and it was estimated that its cost would be about one-third the cost of fresh alum. However, the acid recovery also dissolves the phosphates which, if left in the alum solution, would be recycled to the effluent. If phosphate removal is the goal of alum addition, then such recycle of the phosphate would be intolerable. If coagulation for clarification only is the goal of alum addition, then acid recovery offers an attractive alternate for consideration.

Slechta and Culp[22] also investigated methods of removing the phosphates from the acid recovered alum. Use of anion exchange resigns and activated alumina were investigated but found to be impractical because of prohibitively high costs.

At the October 1975 Conference of the Water Pollution Control Federation, Cornwell and Zoltek[5c] report that a procedure was developed for the economical recovery of aluminum when the aluminum was used as a coagulant for phosphorus removal from secondary effluent in domestic wastewater treatment. The sludge was first thickened in order to obtain a solids concentration of about four times that of the raw sludge. The sludge was then reacted with sulfuric acid to dissolve the aluminum and phosphate. In order to reach a pH of 2.0, two moles of H_2SO_4 per mole Al^{3+} were needed. The supernatant was separated from the residual organic sludge by either sedimentation or filtration. In order to separate the acidified aluminum from the phosphate, a solvent extraction process was developed. A kerosene solution of alkyl phosphates was contacted with the aluminum phosphate solution by rapid mixing. The alkyl phosphates reacted with the aluminum, causing the aluminum to become kerosene soluble. At optimal extraction conditions, the reaction was found to follow an equilibrium equation such that the alkyl phosphate reacted as a dimer. The value of the equilibrium constant was 46.4 ± 0.95 (pK 1.67 ± 0.1). The kerosene and water were separated in a settler following rapid mix. The aluminum rich kerosene was contacted with H_2SO_4, causing the aluminum to transfer from the kerosene into the acid. The recovered aluminum sulfate was reused as a coagulant for phosphorus precipitation. The

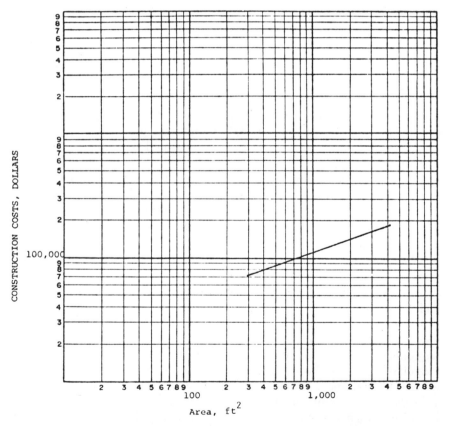

Fig. 8-6 Gravity thickening construction costs, 1975.

aluminum free kerosene was recycled to the extraction stages. The overall aluminum recovery using sedimentation for sludge separation was between 89 and 93 percent. In a full scale operation, very little capital investment would be required and the process is equally applicable to both large and small treatment plants. For a treatment plant using 200 mg/l alum for phosphate removal, the cost of sludge handling with the recovery process was reduced to 25 percent of the cost of sludge handling without the recovery process.

Numerous water plants in Japan have used acid alum recovery with vacuum filtration of the waste solids to produce a cake of about 30 percent solids, although instances of the sludge being sticky and difficult to dewater have been reported. Solid–liquid separation in the acidified sludge can be improved by the use of the proper polymer. Use of 20–40 mg/l of

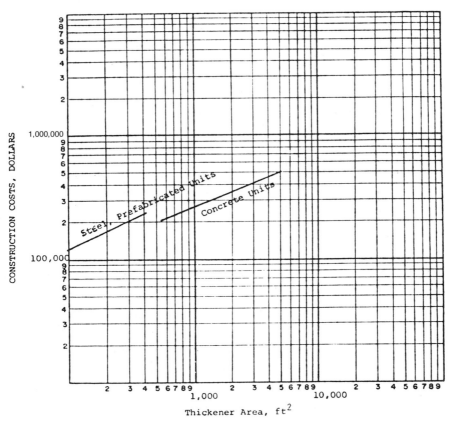

Fig. 8-7 Flotation thickening construction costs, 1975.

cationic polymers have been reported to improve this separation sub-stantially. Alum recovery of about 60 percent is reported in many of these cases in Japan. Filter processing has been reported to produce a 40–45 percent solids cake following acidification, with 80–93 percent alum re-covery achieved. The Japanese experience indicates that optimum plant scale recovery is obtained using a pH of 2 obtained by feeding about 1.2 times the stoichiometric requirements of sulfuric acid. The waste solids are reported to be about 15 percent solids prior to final dewatering. If iron and manganese are present in the alum sludge, they will also be re-covered along with the alum, necessitating the blowdown of some of the recovered alum to keep the iron and manganese content of the finished water in a satisfactory range.

A recent laboratory evaluation of the acid alum recovery process showed that the acidification of the alum sludge decreased the specific resistance

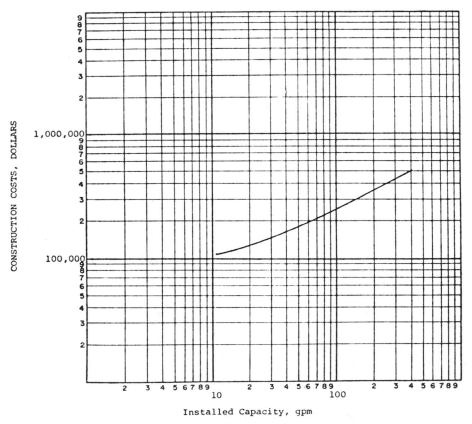

Fig. 8-8   **Vacuum filtration capital costs—construction costs, 1975.**

of the sludge (increasing the sludge's filterability) from $3–5 \times 10^8$ sec^2/g to $1.5–2.6 \times 10^7$ sec^2/g when comparing lime–polymer treated alum sludges and cationic–polymer treated acidified sludges. Cake solids of 28–36 percent were obtained following acidification, while solids of 16–25 percent were obtained with the conditioned alum sludge.

Costs

This section presents information which is intended for use in preparing rough preliminary estimates of installed costs involved in various unit sludge handling processes. These processes include gravity thickening, flotation thickening, vacuum filtration, centrifuging, filter pressing, multiple hearth recalcining, and sludge drying beds. Costs are based on fourth-quarter 1975 levels and do not include engineering, fiscal, or legal costs.

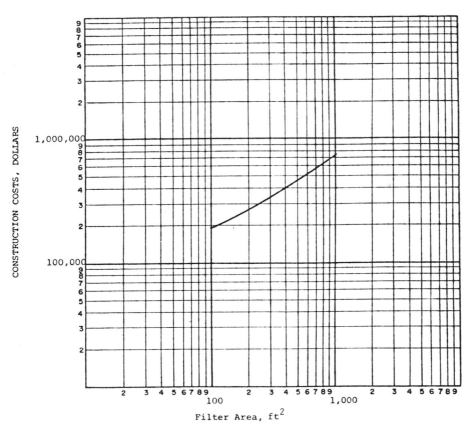

Fig. 8-9   Centrifuging construction costs, 1975.

Costs include an allowance for contractor's overhead and profit, electrical costs, and 15 percent contengency. These costs were developed by Culp/Wesner/Culp, Clean Water Consultants, under EPA Project 68-03-2186, "Conceptual Design, Process Performance, and Cost Analysis of Conventional and AWT Projects," 1975–76.

Gravity Thickening.   Structure related costs are based on the use of circular reinforced concrete basins. Equipment (thickener mechanism and related drive and motor) costs were obtained from manufacturers. Figure 8-6 summarizes the results. When total thickener area requirements exceed 3850 ft^2, it is assumed multiple units will be used to provide flexibility of operation.

Flotation Thickener.   Steel prefabricated units are available with areas up to about 450 ft^2. Costs of these units were obtained from manufacturers

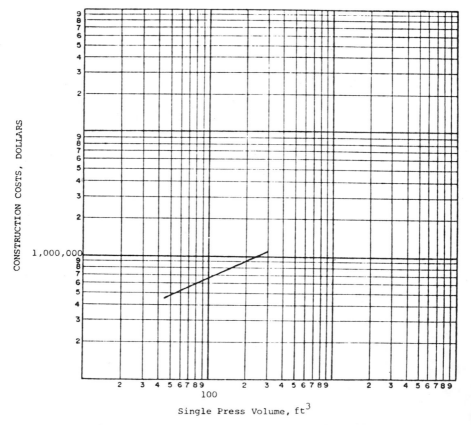

Fig. 8-10   Pressure filtration construction costs, 1975.

and an allowance (25 percent) for contractor's overhead and profit applied to calculate manufactured equipment costs. Larger areas are provided in concrete basins. Equipment costs were provided by manufacturers and the costs of the reinforced concrete basins were added. Manufactured equipment costs for the larger basins include the flotation cell equipment, air compressor, and controls. Figure 8-7 presents the results.

Vacuum Filtration. Manufactured equipment costs were obtained from manufacturers and include the basic filter, sludge flocculators chemical mix tanks and feeders, associated vacuum and filtrate pumps, internal piping, housing, etc. Figure 8-8 summarizes the results. The materials of construction are based on resistance to acid conditions.

Centrifuging. Manufactured equipment costs were obtained from manufacturers. Costs of overall centrifuge systems at two plants were also ana-

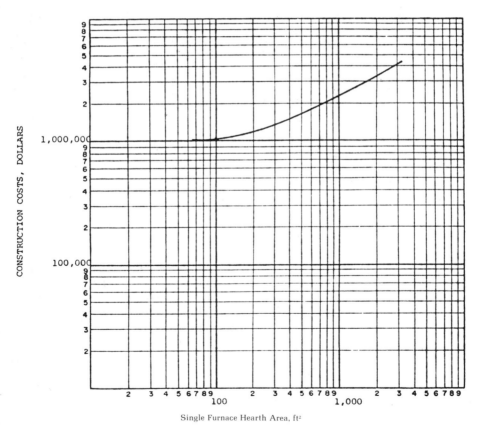

Single Furnace Hearth Area, ft²

Fig. 8-11   Multiple hearth recalcining construction costs, 1975.

lyzed. Costs are again presented in terms of *installed* capacity rather than firm capacity, which provides the designer flexibility in selecting the number and size of machines. Costs (Figure 8-9) include the centrifuge, associated pumps, cake conveyors, and housing.

Pressure Filtration. Equipment costs, including installation labor, were obtained from manufacturers. The manufacturers used 30 percent of the equipment costs as the estimated cost of installation labor. Housing is included. The largest single press currently available has a capacity of 278 ft³. Figure 8-10 summarizes the results.

Multiple Hearth Recalcining. Equipment costs include the furnace, housing, controls, scrubbing system, dry cyclone, and auxiliaries on an installed basis. Figure 8-11 summarizes the results.

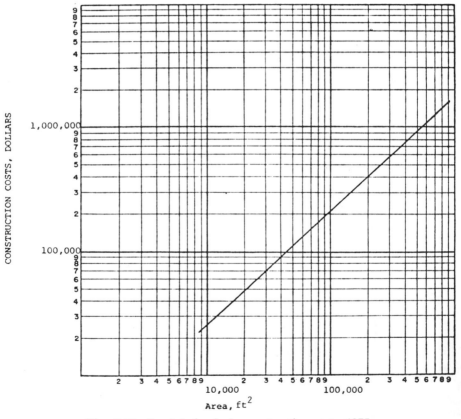

Fig. 8-12   Sand drying beds, construction costs, 1975.

Sand Drying Beds. The curve (Figure 8-12) includes excavation, piping for sludge distribution, sand and gravel drainage beds, and underdrain collection piping.

REFERENCES

1. Adrian, D.D., and Smith, J.E., Jr., "Dewatering Physical-Chemical Sludges." Paper presented at Conference on Application of New Concepts of Physical-Chemical Wastewater Treatment, Vanderbilt University, September, 1972.

2. Albrecht, A.E., "Disposal of Alum Sludges," *Journal American Water Works Assoc.*, January, 1972, p. 46.

3. Aultman, W.W., "Reclamation and Reuse of Lime in Water Softening," *Journal American Water Works Assoc.*, 1969, p. 640.

4. Bird Machine Co., South Walpole, Mass., Operating Manual, 1966.

5. Black, A.P., and Eidsness, F.A., "Carbonation of Water Softening Plant Sludge," *Journal American Water Works Assoc.*, 1957, p. 1343.

5a. Bowen, S.P., and DiGiano, F.A., "Evaluation of Process Design Parameters for Phosphorus Removed from Domestic Wastewaters By Chemical Clarification," Dept. of Civil Engr., Univ. of Mass., Report No. Env. E. 52-75-6 (1975).

5b. Burns, D.E., and Shell, G.C., "Physical Chemical Treatment of a Municipal Wastewater Using Powdered Carbon," EPA-R2-73-264 (1973).

5c. Cornwell, D.A., and Zoltek, J., "Recycling of Alum Used for Phosphate Removal in Domestic Wastewater Treatment," presented at Water Pollution Control Conference, Miami Beach, Florida (Oct., 1975).

6. Crow, W.B., "Techniques and Economics of Calcining Softening Sludges—Calcination Techniques," *Journal American Water Works Assoc.*, 1960, p. 322.

7. Eisenhauer, D.L., et al., "Design of an Integrated Approach to Nutrient Removal," *Journal Sanitation Engineering Division, ASCE*, February, 1976, p. 37.

8. Horstkotte, G.A., et al., "Full-Scale Testing of a Water Reclamation System," Journal *Water Pollution Control Federation*, 1974, p. 181.

9. Issac, B.C.G., and Vahidi, I., "The Recovery of Alum Sludge," *Proceedings Society Water Treatment and Examination*, 10, 1961, p. 1.

10. Jordan, V.J., and Scherer, C.H., "Gravity Thickening Techniques at a Water Reclamation Plant," *Journal Water Pollution Control Federation*, 1970, p. 180.

11. LaMer, V.K., and Smellie, R.H., Jr., "Flocculation, Subsidence, and Filtration of Phosphate Slimes. 1. General," *Journal Colloid Science*, 1956, p. 704.

12. Lea, W.L.; Rohlich, G.A., and Katz, W.J., "Removal of Phosphates from Treated Sewage," *Sewage and Industrial Wastes*, 1954, p. 261.

12a. Long, D. A., Nesbitt, J. B., and Kountz, R. R., "Soluble Phosphate Removal in the Activated Sludge Process—A Two Year Plant Scale Study", Presented at the 26th Annual Purdue Industrial Waste Conference, Purdue University, LaFayette, Indiana (May, 1971).

12b. Martel, C.J., et al, "Pilot Plant Studies of Wastewater Chemical Clarification using Lime," Dept. of Civil Engineering, Univ. of Mass., Report No. ENV.E. 46-75-2 (1975).

13. Mulbarger, M.C., et al., "Lime Clarification, Recovery, Reuse, and Sludge Dewatering Characteristics," *Journal Water Pollution Control Federation*, 1969, p. 2070.

14. Nelsen, F.G., "Recalcination of Water Softening Plant Sludge," *Journal American Water Works Assoc.*, 1944, p. 1178.

15. Nichols Engineering Co., Bulletin No. 238 A, Sludge Furnaces.

15a. Parker, D.S., et al., "Lime Use in Wastewater Treatment: Design and Cost Data," USEPA 600/2-75-038 (October, 1975).

16. Roberts, J.M., and Roddy, C.P., "Recovery and Reuse of Alum Sludge at Tampa," *Journal American Water Works Assoc.*, 1960, p. 857.

17. Rose, J.L., "Removal of Phosphorus by Alum." Paper presented at the FWPCA Seminar on Phosphate Removal, Chicago, Illinois, June, 1968.

18. Sawyer, C.N., "Some New Aspects of Phosphates in Relation to Lake Fertilization," *Sewage and Industrial Wastes*, 1952, p. 768.

19. Schmid, L.A., and McKinney, R.E., "Phosphate Removal by a Lime–Biological Treatment Scheme," *Journal Water Pollution Control Federation*, 1959, p. 1259.

20. Sebastian, F., and Sherwood, R., "Clean Water and Ultimate Disposal," *Water and Sewage Works*, August, 1969.

21. Sharman, L., "Polyelectrolyte Conditioning of Sludge," *Water and Wastes Engineering*, 4, 1967, p. 8.

22. Slechta, A.F., and Culp, G.L., "Water Reclamation Studies at the South Tahoe PUD," *Journal Water Pollution Control Federation*, 1967, p. 787.

23. Smith, C.E., "Use and Reuse of Lime in Removing Phosphorus From Wastewater," Annual Meeting of the National Lime Assoc., Phoenix, Arizona, April, 1969.

23a. Smith, J.E. Farrel, J.B., and Dean, R.B., Unpublished Data from the Advanced Waste Treatment Laboratory, Office of Research and Monitoring, Environmental Protection Agency, Cincinnati, Ohio (Jan., 1971).

24. Tomas, C.M., "The Use of Filter Presses for the Dewatering of Sewage and Waste Treatment Sludges," Paper presented at the Forty-second Annual Conference of the WPCF, Dallas, Texas, October, 1969.

25. "Process Design Manual for Sludge Treatment and Disposal," U.S. Environmental Protection Agency, Technology Transfer (October, 1974).

9

Demineralization

INTRODUCTION

Removal of common salts added to municipal wastewater through domestic use is generally not required for most current methods of disposal in the United States. At the time of this writing (July 1976) there are no large scale municipal wastewater treatment plants which utilize demineralization in the treatment process. Several pilot scale and demonstration plants have been or are being operated on municipal wastewater. Processes for wastewater demineralization that are discussed in this chapter include reverse osmosis (RO), electrodialysis (ED), and ion exchange (IX).

Although there are no large scale demineralization facilities which operate routinely on municipal wastewater, there has been a considerable amount of laboratory and pilot scale work and at least one large scale plant is under construction. A 50,000 gpd RO plant treats secondary effluent at the El Dorado Irrigation District, California.

The Orange County (California) Water District has a 5.0 mgd RO plant under construction. This RO plant will treat effluent from the District's AWT plant.

The incremental increase from one residential use of wastewater is summarized in Chapter 1. The data indicate that on the order of 300 mg/l of inorganic salts are added to the water by one domestic use. Incremental additions substantially above 300 mg/l occur in some locations because of industrial discharges or, as is common in many areas, because of the discharge of brine from home regenerated water softeners.

Because the experience in wastewater demineralization is somewhat limited, the progress in brackish water supply demineralization is also reviewed in this chapter. During the past few years there has been considerable research effort directed toward demineralization of municipal wastewaters. There has also been considerable recent work directed toward the use of demineralization processes for the treatment of industrial wastewater. Again, most of this work has utilized the RO process. There have been papers in the recent literature on demineralization of plating rinse waters, cheese whey, petrochemical effluents, wastes from photographic processing, cooling water, food processing, pulp and paper manufacturing effluents, and others.

Brackish Water Demineralization

Although demineralization of municipal and industrial wastewaters has thus far been limited to small scale or pilot plants, there are rather extensive installations of demineralization processes in industrial water treatment. Also, relatively recently, demineralization has been used for municipal water treatment and it appears that this use is increasing.

ED was the first process to be fully developed for desalting brackish municipal water supplies. There are at least six municipal plants in operation in the United States utilizing ED to desalt brackish water. The largest municipal desalting plant in the United States (3 mgd) began operation during 1975 at the Foss Reservoir in Oklahoma and utilizes the ED process. There is a 5 mgd ED plant in operation in Benghazi, Libya. The ED plant with the longest history of operation on a municipal water supply is located in Buckeye, Arizona.[1] The plant at Buckeye has a capacity of 650,000 gpd and was constructed in 1962.

RO has been intensively developed in the last few years, and the first municipal desalting application utilizing this process was a 150,000 gpd plant constructed in Greenfield, Iowa in 1971.[2] A 24,000 gpd plant was installed in 1971 to demineralize brackish well water supplied to a trailer park.[3] The largest RO plant in the United States serves the Ocean Reef Club Community on the northern tip of Key Largo, Florida. This plant was originally constructed in 1971 with a capacity of 350,000 gpd and was subsequently expanded to a total of 930,000 gpd.

Design data and operating experience were recently reported for 11 commercial membrane desalting plants, including four ED and seven RO facilities.[4] Data from the 11 plants reviewed are summarized in Table 9-1. As noted in the table the size range of the plants varies from 2500 gpd to 2.0 mgd and feedwater TDS ranges from 1300 to 7000 mg/l.

A large brackish water desalting plant is now planned for improving the salinity in the lower Colorado River.[5] A 100 mgd membrane demineralization plant is planned to treat irrigation return flows originating in the Wellton Mohawk Irrigation District near Yuma, Arizona. These return flows periodically have a TDS in excess of 4000 mg/l. Test work of both RO and ED plants is underway and a full scale facility could either be an ED or RO plant, or possibly a combination of the two.

The IX process has been used for many years in the United States for softening municipal water supplies, and there are several multi-mgd plants which have operated successfully; however, the process had not been applied to demineralizing municipal water supplies until recently, when a 500,000 gpd plant was constructed in Burgettstown, Pennsylvania.[6] This

TABLE 9-1   Brackish Water Membrane Demineralization Plants.

Plant	Type	Year Installed	Capacity, mgd	TDS, mg/l Feed	TDS, mg/l Product	Brine, percent of feed
Siesta Key, Florida	ED	1969	2.0	1300	485	15
Gillette, Wyoming	ED	1972	1.5	1840	775	20
Sanibel Island, Florida	ED	1973	1.2	2930	500	9
Sorrento Shores, Florida	ED	1973	0.070	2785	500	50
Grand Bahama Hotel	RO	1971	0.288	3000	300	25
Rotonda West, Florida	RO	1972	0.500	7000	350	50
Ocean Reef, Florida	RO	1971	0.930	6300	300	50
Schuster's, St. Croix	RO	1973	0.070	2300	NA[a]	30–50
Cinnamon Bay, St. John	RO	1970	0.010	5000	NA	30–50
Colonial Manor, Tortola	RO	1972	0.0025	5000	NA	30–50
Wittberg Gardens, St. Thomas	RO	1971	0.0025	1500	NA	30–50

[a]Not available

plant began operation in late 1972 and treats a potable water supply which has been contaminated by acid mine drainage. It is the only plant in the United States utilizing the IX process for demineralizing a muncipal water supply.

Municipal Wastewater Demineralization

Although there are no large demineralization facilities which operate routinely on municipal wastewater, a considerable amount of laboratory and pilot scale work has been conducted. Most of the recent test work has been with the RO process; however, there has also been some interest in ED and IX. There has been very little work with phase change processes, such as evaporation or freezing.

Selection of a particular demineralization process will depend upon several factors, including: amount of salt to be removed, desirability of removing organics and microorganisms, facilities for brine disposal, and availability of power and chemicals. A potential problem common to both membrane and IX processes is fouling of resin and membrane surfaces by colloidal and dissolved material present in municipal wastewater.

Fouling and scaling of the membrane surfaces has been the major problem in ED and IX treatment of wastewater. In one study, membrane fouling was attributed to the fraction of COD represented by MBAS. In another study, microscopic examination of the fouling material showed the presence of bacteria, protozoa and higher forms. Treatment by activated carbon adsorption is apparently a requirement for municipal wastewater to be demineralized by ED or IX.

RO units have achieved stable operation on wastewater without pre-treatment by activated carbon adsorption, but regular cleaning of the membranes is required. The principal problem in the treatment of municipal wastewater is membrane fouling, which can greatly reduce the capacity of the units. The primary foulants are believed to be colloidal material and dissolved organics. Fouling generally decreases with increasing degrees of wastewater pretreatment.

RO membranes remove a high percentage of almost all inorganic ions, turbidity, organic material, bacteria, and viruses present in municipal wastewater. The dissolved solids in municipal wastewater can be easily reduced to 50–100 mg/l in one stage. Power requirements for commercially available RO systems are about 7–9 kwh/1000 gal product water. This power requirement does not change appreciably over the range of dissolved solids generally present in municipal wastewater. Advantages of the reverse osmosis process include: (1) it removes a high percentage of organic material, turbidity, bacteria, and viruses, as well as inorganic material; (2) removal efficiencies and power consumption remain nearly stable over the range of dissolved solids present in most municipal wastewaters; and (3) waste brine contains only salts removed plus a small amount of acid used for pH control.

Typical removals of inorganic salts from municipal wastewater by ED range from 25 to 40 percent of dissolved solids per stage of treatment. Higher removals require treatment by multiple stages in series. Less than 20 percent of the organics remaining in activated carbon treatment secondary effluent are removed by electrodialysis. Power required for ED is about 0.2–0.4 kwh/1000 gal for each 100 mg/l dissolved solids removed, plus 2–3 kwh/1000 gal for pumping feed water and brine. Advantages of the ED process include: (1) well developed technology, including equipment and membranes; (2) efficient removal of most inorganic constituents; and (3) the fact that waste brine contains only salts removed plus a small amount of acid used for pH control.

The IX process has been used for many years in the treatment of industrial waters and in municipal softening plants. The equipment and operating procedures are well developed. However, the process has not been studied as extensively as other processes for brackish water or wastewater desalting.

It is possible easily to reduce the dissolved solids in municipal wastewater to about 50 mg/l by ion exchange. There is a direct relationship between operating costs and salts removed; therefore, as salt removal percentages increase, so do chemical operating costs. Most organic material is not removed by resins normally used for inorganic removal; however, special resins are now available which do adsorb organics. An equivalent amount of regenerant chemicals must be used for all inorganic ions removed from the wastewater. Wastes requiring disposal are regenerant chemicals which include strong acids and bases as well as the salts removed from the wastewater.

Advantages of the ion exchange process include: (1) well developed and proven technology; (2) availability of large scale equipment which has a history of reliable operation; (3) highly efficient removal of common inorganic salts and minimal problems with surface fouling in the processing of carbon treated municipal wastewater.

REVERSE OSMOSIS

Process Description

A natural phenomenon known as osmosis occurs when solutions of two different concentrations are separated by a semipermeable membrane such as cellophane. Water tends to pass through a semipermeable membrane from the more dilute side to the more concentrated side, thus producing equal TDS concentrations on both sides of the membrane. The ideal osmotic membrane permits passage of water molecules but prevents passage of ions such as sodium and chloride. For example, if a solution of sodium chloride in water is separated from pure water by means of a semipermeable membrane, water will pass through the membrane in both directions, but it will pass more rapidly in the direction of the salt solution. At equilibrium, the quantity of water passing in either direction is equal, and the pressure is defined as the osmotic pressure of the solution having that particular concentration of TDS. This principle is illustrated in Figure 9-1.

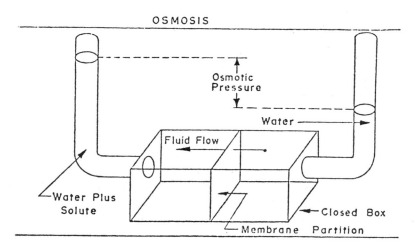

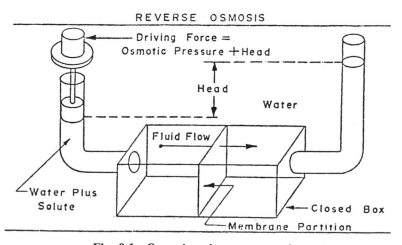

Fig. 9-1   Osmosis and reverse osmosis.

The magnitude of the osmotic pressure depends on the concentration of the salt solution which is related to the solution's vapor pressure and temperature. By exertion of pressure on the salt solution, the osmosis process can be reversed. When the pressure on the salt solution is greater than the osmotic pressure, fresh water diffuses through the membrane in the direction opposite to normal osmotic flow—hence the name for the process, reverse osmosis (see Figure 9-1).

The osmotic pressure of a solution increases with the solution concentration. A rule of thumb, which is based on sodium chloride, is that the osmotic pressure increases by approximately 0.01 psi for each mg/l. This approximation works well for most natural waters. However, high molecular weight organics produce a much lower osmotic pressure. For example, sucrose gives approximately 0.001 psi for each mg/l. Several methods are available for measuring the osmotic pressure. It can be calculated from the depression of the vapor pressure of the solution, by depression of the freezing point, and by an equivalent of the ideal gas law equation.

Many materials have been studied for possible use as membranes for water and wastewater purification and related separation and concentration procedures. The most widely used membrane developed to date is a modified cellulose acetate film. The techniques for preparing these membranes were discovered by Loeb and Sourirajan in 1959 at the University of California, Los Angeles. Their discovery constituted a milestone in technology of membrane separation by providing the means for casting membranes free of imperfections, consisting of an asymmetric structure which combines a very thin, dense skin on a porous support structure— a combination which provides very high solute rejection while permitting relatively high water flow.

The modified cellulose acetate membrane currently in general use is approximately 100 μ thick (0.004 in.). As noted, it is asymmetric, having on one surface a relatively dense layer, approximately 2000 Å (0.2 μ) thick, which serves as the rejecting surface. The remainder of the film is a relatively spongy, porous mass, containing approximately two-thirds water by weight; this portion must generally be kept wet at all times.

The semipermeable membrane acts as a filter to retain the ions, such as sodium and chloride, on the brackish water side, while permitting pure or nearly pure water to pass through the membrane. The properties of a membrane that permit water molecules to pass through but will not permit the flow of salt ions are not clearly understood. It is believed that it is not simply a molecular filtering action, even though individual water molecules are smaller than most of the ions of concern.

Normally RO membranes must be cast in an extremely clean environment under closely controlled conditions. After manufacture, the membranes must normally be kept moist and cannot be stored dry. However, Desalination Systems Corporation has developed a membrane which can be stored dry.

The behavior of semipermeable RO membranes can be described by two basic equations. The product water flow through a semipermeable membrane may be expressed as:

$$F_w = A\,(\Delta p - \Delta\pi)$$

where

F_w = water flux
A = water permeability constant
Δp = pressure differential applied across the membrane
$\Delta\pi$ = osmotic pressure differential across the membrane

The salt flux through the membrane may be expressed as:

$$F_s = B(C_1 - C_2)$$

where

F_s = salt flux
B = salt permeability constant
$C_1 - C_2$ = concentration gradient across the membrane

The water permeability and salt permeability constants are characteristic of the particular membrane which is used and the processing which it has received.

These equations show that the water flux is dependent upon the applied pressure, while the salt flux is not. As the pressure of the feedwater is increased, the flow of the water through the membrane should increase while the flow of salt remains essentially constant. It follows that both the quantity and the quality of the product water should increase with increased driving pressure.

The water flux increases as the available pressure differential increases; and the water flux decreases as the salinity of the feed increases (because the osmotic pressure contribution increases with increasing salinity). As more and more water passes through the membrane, the salinity of the feedwater becomes higher and higher; the osmotic pressure contribution of the concentrate becomes correspondingly higher, and this results in a lower water flux with increasing percentage water recovery. Also, since the salinity of the feed concentrate increases with increasing product water removal and the membrane rejects essentially a fixed percentage of the salt, the quality of the product water decreases with increasing percentage water recovery.

Operating plants carry out the RO principle in several different process designs and types of membrane configurations. There are essentially four types of membrane systems which have been used: (1) spiral wound, (2) hollow fine fiber, (3) tubular, and (4) plate and frame. The first three types are in commercial production and are currently in use in operating plants.

Spiral Wound RO Membranes.   A sketch of a spiral wound membrane and module assembly is shown in Figure 9-2. The spiral wound RO module was developed by Gulf Environmental Systems (now Fluid Systems Division, UOP) under contract to the U.S. Office of Saline Water. It was conceived

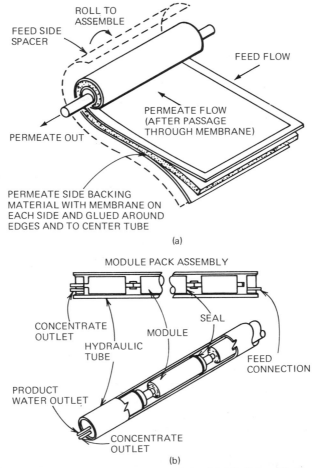

Fig. 9-2   Spiral wound reverse osmosis module. (a) Module. (b) Pressure vessel assembly. (*Courtesy Fluid Systems Div. UOP, San Diego, Calif.*)

as a method of obtaining a relatively high ratio of membrane area to pressure vessel volume. The membrane is supported on both sides of a backing material and sealed with glue on three of the four edges of the laminate. The laminate is also sealed to a central tube, which is drilled. The membrane surfaces are separated by a screen material which acts as a brine spacer. The entire package is then rolled into a spiral configuration and wrapped in a cylindrical form with tape as an outer wrap. Feed flow is parallel to the central tube while the permeate flow is through the membrane toward the central tube. Several brackish water desalting plants and wastewater desalting plants are in operation at the Orange County Water District, El Dorado Irrigation District, Pomona Municipal Water District, and the City of Escondido, all in California.

Hollow Fiber RO Membranes.   The hollow fiber type of membrane was developed by DuPont and Dow Chemical. The 150,000 gpd municipal desalting plant at Greenfield, Iowa was manufactured by DuPont. The fine fibers as manufactured by DuPont are about the size of a human hair, with an inside diameter of about 0.002 in. and an outside diameter of about 0.004 in. The aromatic polyamide fibers manufactured by DuPont are illustrated in Figure 9-3. In these very small diameters, fibers can withstand enormous pressures.

In an operating process the fibers are placed in a pressure vessel with one end sealed and the other end protruding outside of the vessel. The

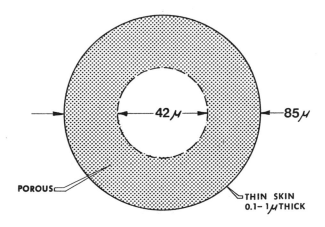

*ASYMMETRIC AROMATIC POLYAMIDE

Fig. 9-3 Hollow fiber (asymmetric aromatic polyamide) reverse osmosis membrane as manufactured by DuPont. Transverse section.

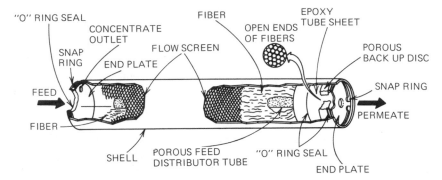

Fig. 9-4   Hollow fiber reverse osmosis module as manufactured by DuPont.

salt water is under pressure on the outside of the fibers and product water flows inside the fiber to the open end. A DuPont module is illustrated in Figure 9-4. For operating plants, the membrane modules are assembled in a configuration similar to the spiral wound unit.

Tubular RO Membranes.   Tubular membrane processes operate on much the same principle as the hollow fine fiber, except that the tubes are much larger, on the order of 0.05 in. ID. Westinghouse and several other firms manufacture tubular systems.

The helical tubular module shown in Figure 9-5 was developed by Philco-Ford in Newport Beach, California. This system has since been acquired by Occidental Petroleum. The basic module configuration is a nominal 1 in. diameter tubular membrane wound onto a support spool in a multiple layer helical coil. The membrane is supported by a network of polyester filaments. Composite metal and plastic fittings are attached to each end of the membrane tube and its polyester pressure support. The helical tubular membrane is completely enclosed by a plastic shroud. The shroud is intended to prevent damage to the membrane and provide a means for contamination free product collection.

Plate and Frame RO Units.   This design makes use of a rigid plate with the membranes mounted on opposite sides and sealed to the plate. Pressure is applied to the salt water on the outer sides of the plate, and the product water is forced through the membranes into the interior of the plate, which is at low pressure. The interior of the plate must, therefore, be either porous or have hollow channels through which the product water can flow to a collecting point, where it passes to the outside of the pressure vessel. Plate materials may consist of solid plastic plates with grooved channels, porous

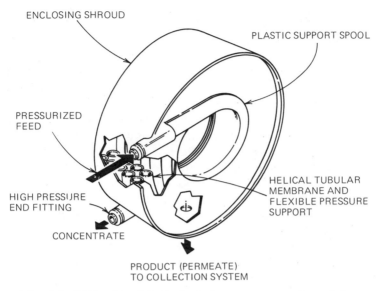

ENCLOSING SHROUD

PLASTIC SUPPORT SPOOL

PRESSURIZED
FEED

HIGH PRESSURE
END FITTING

CONCENTRATE

HELICAL TUBULAR
MEMBRANE AND
FLEXIBLE PRESSURE
SUPPORT

PRODUCT (PERMEATE)
TO COLLECTION SYSTEM

Fig. 9-5   Philco-Ford's helical tubular reverse osmosis module.

fiberglass materials, or reinforced porous paper. The plate and frame approach is not an efficient use of membrane surface area. Some of the early
experimental work in the treatment of wastewater by RO was with plate
and frame units, but the other RO module types (spiral, hollow fiber, and
tubular) are the only systems commercially available.

Early RO Studies

The application of RO to treating municipal wastewater was first studied
late in 1962 by Aerojet-General Corporation at Azusa, California. A bench-
scale study was conducted on filtered municipal secondary effluent, using
3 in. diameter flat plate test cells operated at pressures of 750 and 1500 psi.
Results of this study demonstrated that high quality water could be produced from secondary effluent by the RO process.[7]

A plate and frame pilot RO unit was operated by the Orange County Water
District in 1967 for treating municipal wastewater.[8,9] The results of this
test generally corroborated the earlier bench scale study and produced
high quality water. A number of operational problems were encountered,
and since this type of unit is not currently commercially available, the
results of this early pilot study are not discussed in detail.

Pomona, California

Spiral wound and tubular membrane modules were pilot tested at Pomona.[7,10] Two spiral wound units manufactured by Gulf, and tubular units manufactured by Havens and Universal Water, were operated over a period of several months. The Gulf modules had capacities of 5000 gpd and 12,000 gpd and were similar to those shown in Figure 9-2. The first tests at Pomona were with the spiral wound modules treating activated sludge effluent. Based on these first tests, it was concluded that spiral wound modules could process secondary effluent with high organic and inorganic removals, but that some form of pretreatment (or additional treatment of the secondary effluent) was necessary to prevent excessive clogging of the concentration side spacers from suspended solids. Data for the 12,000 gpd Gulf unit, after 9000 hours or approximately 1 year of operation, are summarized in Table 9-2.

Tests of tubular membranes were conducted at Pomona with a Universal Water Corporation unit. The feed and product water quality for this unit are shown in Table 9-3. The first run with this unit was terminated after 1000 hours of operation. During this time the total headloss across the system increased from an initial value of 90 psi to 110 psi. Following the shutdown after 1000 hours, the modules were flushed for 1 hour with a 0.75 percent solution of Biz (Proctor and Gamble). This technique increased the flux from 18.0 to 20.6 gpd/ft^2, compared to an original flux of 23.9 gpd/ft^2.

TABLE 9-2   Typical Water Quality, Pomona, California.

Constituent	Concentration, mg/l Influent	Effluent	% Removal
COD	8.7	1.0	88.5
NH_3-N	10.1	1.1	89.2
NO_3-N	4.9	2.4	51.0
PO_4-P	10.9	0.2	98.4
TDS	750	59	92.1

Notes
1. 12,000-gpd pilot plant manufactured by Gulf (now Fluid Systems Division, UOP).
2. Influent: municipal wastewater treated by activated sludge system and granular carbon adsorption.
3. Averages based on 40 grab samples (June–September 1969).
4. Average water recovery, 75%.

TABLE 9-3   Feed and Product Water Quality for Universal Water Corporation RO Unit, Pomona, California.

Chemical Analysis	No. of tests	Feed		Product		Brine		Percentage reduction
		avg.	range	avg.	range	avg.	range	
Total COD	10	9.4	7.2–12.7	0.3	0.0–1.0	14.6	11.1–19.0	96.8
Dissolved COD	9	8.3	6.3–12.0	—	—	12.7	9.7–18.6	—
NO_3^--N	9	3.6	0.1–12.0	0.7	0.0–2.2	5.5	0.1–19.0	80.6
NH_3-N	10	11.7	2.5–18.1	0.7	0.2–1.4	18.9	3.5–36.5	94.0
$PO_4^{3-}-P$	8	10.1	8.8–11.7	0.12	0.0–0.4	15.5	12.9–18.9	98.8
TDS	10	619	540–711	39	20–72	1018	633–1859	93.7
Ca^{2+}	4	39.6		<1		55.4		>97.5
Mg^{2-}	3	24.5		2.0		39.3		91.8
K^+	4	12.1		0.8		17.9		93.4
Na^+	4	111		7.6		156		92.8
SO_4^{2-}	3	227		0		383		100.0
Cl^-	3	85		8		153		90.6

Notes: 1. All analyses on grab samples; 2. feed samples taken after acidification with H_2SO_4, hence values of SO_4^{2-} shown in table include sulfate contributed by acid addition; 3. all values shown are in mg/l.

Gulf Environmental Systems Company
(Now Fluid Systems Division, UOP)

The tests at Pomona employing spiral wound modules with modified cellulose acetate membranes produced generally good removals of both inorganic salts and organics as measured by COD. Membrane fouling was identified as the most important operational problem and subsequent tests were directed at developing methods of fouling prevention and control, and maintaining performance. Both physical and chemical methods were investigated for cleaning the spiral wound RO modules while they were operated on activated sludge and carbon treated activated sludge effluent. The results of these studies have been reported in several publications by researchers with Gulf Environmental Systems Company.[11-14]

The pilot spiral wound RO units used in the earlier studies at Pomona were modified to incorporate provisions for chemical and physical cleaning. Five different RO feed streams were investigated: (1) primary effluent with and without sand filtration; (2) sand filtered activated sludge effluent; (3) chemically clarified primary effluent with sand filtration; (4) chemically clarified primary effluent with sand filtration and activated carbon treatment; and (5) activated carbon treated activated sludge effluent. There were four test objectives in this phase of the work:

1. Comparison of operation of RO modules on activated sludge effluent, with and without treatment by activated carbon adsorption.
2. Comparison of different chemical cleaning techniques, while operating the RO units on activated sludge effluent.
3. Operation of RO units on chemically clarified primary effluent with and without treatment by activated carbon adsorption.
4. Simulation of an actual RO plant operation by running one unit on activated sludge effluent at constant product flow.

Average product water fluxes in the range of 8–20 gpd/ft^2 were maintained by regular chemical cleaning. Three cleaning solutions in which the active constituent was either the enzyme detergent Biz, ethylenediamine tetraacetic acid (EDTA), or sodium perborate, performed well in the field tests. Test results indicated that enzymes per se were not the specific active materials in the detergent based compound. To prevent membrane damage due to exposure to high pH, all cleaning solutions were adjusted to a pH value of less than 8.0. Periodic depressurization also aided in maintaining the product water flux. Total inorganic solute rejections, determined by conductivity measurements during the activated sludge effluent tests, remained high throughout the test period and in some cases appeared to increase slightly above the initial value.

Tables 9-4, 9-5, and 9-6 show the average influent and product water concentrations and percent reductions for activated sludge effluent, carbon treated activated sludge effluent and chemically clarified primary effluent.

TABLE 9-4 Reduction of Constituents in Activated Sludge
Effluent, Pomona, California RO Pilot Study,
Gulf Environmental Systems Company.

	CONCENTRATION, mg/l				
	Feed[a]		Product		Percentage
Constituent	avg.	range	avg.	range	reduction
Calcium	92	62–160	0.33	<0.1–1.0	>99.6
Magnesium	20.6	14–31	0.1	Nil–0.18	>99.5
Sodium	177	99–310	10.6	7–18.2	94.0
Potassium	20.6	11–35.2	1.1	Nil–2.3	94.7
Ammonia-N	20.6	8.4–34	1.0	Nil–1.5	95.1
Chloride	159	94–315	8.6	6–15	94.6
Sulfate	422	220–852	1.9	Nil–6	>99.5
Phosphate	29.8	10.1–95	0.1	Nil–0.24	>99.7
Total COD	45.9	25.6–83.3	1.8	Nil–5	>96.1
Dissolved COD	38.1	16.9–72.3	1.5	Nil–5	>96.1
TDS	1123	751–1977	54	27–107	95.2

[a]Feed is sand filtered activated sludge effluent and contains recycled brine.

TABLE 9-5 Reduction of Constituents
in Activated Carbon
Treated Activated Sludge
Effluent, Pomona, Cali-
fornia RO Pilot Study, Gulf
Environmental Systems
Company.

	AVERAGE CONCENTRATION, mg/l		
Constituent	Feed	Product	Percentage reduction
Calcium	104	0.46	99.6
Magnesium	22.7	0.08	99.6
Sodium	198	14.7	92.6
Potassium	20	1.4	93
Ammonia-N	14.1	0.47	96.7
Chloride	221	15	93.2
Sulfate	379	1	99.7
Phosphate	59	0.06	99.9
Total COD	16.1	1.4	91.3
Dissolved COD	11.6	0.8	93.1
TDS	1206	57	95.3

TABLE 9-6 Reduction of Constituents in Chemically Clarified Primary Effluent, Pomona, California.

	CONCENTRATION, mg/l				
	Feed		Product		Percentage reduction
Constituent	avg.	range	avg.	range	
Calcium	91	54–114	0.4	<0.1–1.2	>99.6
Magnesium	19.8	13.3–26	0.13	<0.1–0.3	>99.3
Sodium	133	92–188	10	6.3–17	92.5
Potassium	21.3	17.2–29	1.8	0.7–2.6	91.5
Ammonia-N	46.8	37.5–57	4.6	2.8–7.5	90.2
Chloride	277	209–378	26.6	14–44	90.4
Sulfate	325	256–388	0.7	Nil–3	>99.8
Phosphate	1.7	0.27–3.1	0.01	0.1–0.08	>99.4
Total COD	70	24–109	24	15.7–33.4	65.6
Dissolved COD	64	22.2–103	22	15.7–32.3	65.6
TDS	963	757–1170	65	49–92	93.3

[a]Feed contained recycled brine.

These data show that the TDS removal from all feed streams was 93 percent or greater, and that the total COD in activated sludge effluent and carbon treated activated sludge effluent was reduced to less than 2 mg/l.

In all the test runs, some of the membrane water permeability could not be recovered by cleaning. This permanent decline could represent either a pressure increase or a water flux decrease, depending on whether the unit was being operated at constant flux or constant pressure. In general, permeability declines were greater for operation on primary effluent than for secondary effluent feeds. Activated carbon pretreatment improved performance in all cases.

The Gulf researchers have also conducted studies on the fouling of RO membranes for the Office of Saline Water, and the results of these studies were utilized in the work at Pomona.[15]

It was the general conclusion of the Gulf investigators, based on the studies at Pomona, that spiral wound RO units can be successfully operated on primary and activated sludge wastewater effluents with only moderate pretreatment and chemical cleaning. It was further concluded that activated carbon pretreatment is unnecessary for successful operation on sand filtered activated sludge effluent, or chemically clarified sand filtered primary effluent.

Aerojet-General Corporation

Aerojet-General has conducted studies on desalting seawater and brackish water under contract to the Federal Office of Saline Water. The company has also had several contracts with the Environmental Protection Agency for the study of salt removal from municipal wastewater.

A laboratory study utilized conventional cellulose acetate membranes in flat plate cells and tubular configurations.[16] Municipal wastewater for the test was collected from Treatment Plant No. 1 at the County Sanitation Districts in Orange County, California and the Pomona (California) Sewage Treatment Plant. Municipal wastewater treated in the test equipment ranged from raw sewage and digester supernatant to activated carbon treated secondary effluent. The flat plate membranes and the 0.25 in. diameter tubular membranes exhibited very similar wastewater constituent rejections. The average results for the tubular membranes are shown in Table 9-7.

In all the tests except one, higher quality feed water produced higher product water fluxes. The magnitude of the stabilized product water flux appears to be quite dependent on the feedwater quality. Although the sewages used in the test program were from two different sources and had quite different characteristics, no difference in RO performance was apparent. As an example, secondary sewage from Pomona, was, except for nitrate, total phosphate, and MBAS (all relatively low concentration constituents), markedly superior in quality to the winter season secondary sewage from Orange County; yet typical results obtained with these two feeds were not noticeably different.

Hemet, California

A pilot demonstration program on the use of RO in reducing the concentrations of TDS and refractory substances present in secondary sewage effluent was conducted at Hemet, California from March 6, 1970 to June 25, 1976.[17] It was proposed that the reclaimed wastewater would be used for groundwater recharge in the Hemet area. Other objectives in the Hemet study were:

1. Determination of efficiencies and costs of pretreatment operations.
2. Comparison of the performance of a selected group of RO units manufactured by major firms.

The nominal capacity of the pretreatment system was 150,000 gpd and consisted of: (1) chemical clarification with alum in a sludge blanket type clarifier, (2) filtration in pressure units with single media, 0.45–0.55 mm sand; (3) granular carbon adsorption with 8 × 12 mesh carbon; (4) diatomaceous earth filtration in a 30 in. diameter unit; and (5) chlorination with sodium hypochlorite.

There were six different RO units tested:

1. *Aerojet-General Corporation.* Tubular type unit with a design capacity

TABLE 9-7 Average Wastewater Constituent Rejections and Product Water Quality for 0.25 in. Diameter Tubes at 700 psig—RO Laboratory Study, Aerojet-General Corporation.

	EC^a	Total COD	NH_4^+-N	Organic N	NO_2^--N	NO_3^--N	Total PO_4^{3-}-P	MBAS
				Rejections, %				
Pomona Wastewater								
Carbon treated secondary sewage	88.9	90.2	88.3	92.3	93.7	67.4	98.6	94.0
Alum treated secondary sewage	89.7	89.8	84.5	93.0	40.5	34.5	99.2	—
Primary sewage	90.2	87.8	90.7	84.3	100	46.9	98.6	91.7
Orange County Wastewater								
Alum treated secondary sewage	85.6	86.1	79.9	77.0	88.1	24.4	92.5	93.6
Secondary sewage	92.4	91.5	89.7	91.1	76.6	54.2	98.8	94.8
Primary sewage	89.3	94.4	65.6	84.2	39.6	83.3	99.0	—
Raw sewage	89.1	60.0	77.1	89.7	92.6	51.9	98.2	94.1
				Product Water Quality, mg/l^b				
Pomona Wastewater								
Carbon treated secondary sewage	204	8.15	2.33	0.250	0.000	0.170	0.17	0.030
Alum treated secondary sewage	237	4.31	1.49	0.640	0.018	6.14	0.060	—
Primary sewage	232	21.3	1.71	1.65	0.000	0.990	0.094	0.19
Orange County Wastewater								
Alum treated secondary sewage	452	12.4	6.08	1.51	0.000	1.27	0.051	0.072
Secondary sewage	245	13.5	3.06	0.660	0.011	0.510	0.12	0.071
Primary sewage	245	6.23	6.07	1.55	0.032	0.092	0.051	—
Raw sewage	523	84.7	5.59	2.00	0.004	0.550	0.12	0.080

^{a}Electrical conductivity, μmhos/cm at 25°C, figures in the "Rejections" section are percentages.
^{b}Adjusted to product quality of total output of plant operating at 80 percent recovery.

of 8000 gpd; tubes were 0.56 in. ID by 14 ft 3 in. long. Startup March 9, 1970, shutdown December 21, 1970; 2031 hours of operation.

2. *American Standard* (*ABCOR*). Design capacity 10,000 gpd at 90 percent TDS removal; tubular unit with 0.5 in. ID membrane on inside with turbulence promoters. Startup June 3, 1970, shutdown June 25, 1971; 6485 hours of operation.

3. *DuPont*. Design capacity 10,000 gpd at 75–90 percent recovery. The B-5 modules were replaced by B-9 modules during the test program. B-5 modules are 15 in. diameter by 10.5 ft long; and B-9 modules are 5.5 in. diameter by 47 in. long.

4. *Gulf*. Capacity 10,000 gpd at 60 percent recovery. Startup March 9, 1970, shutdown June 25, 1971; 10,016 hours of operation.

5. *Raypak*. Capacity 3,000 gpd at 50 percent recovery; tubular unit with 0.9 in. ID tubes. Startup May 6, 1971, shutdown June 3, 1971; six weeks of operation.

6. *Universal Water*. Capacity 10,000 gpd. Tubular modules with 0.4 in. ID tubes; membrane on inside of tube. Startup March 9, 1970, shutdown June 25, 1971; 10,275 hours of operation (94.3 percent of available time).

In early 1971, a staged reduction in the treatment of the secondary effluent supplied to the RO units began. First carbon adsorption was eliminated, then chemical clarification and, finally, the RO units were supplied straight chlorinated secondary effluent.

First, the activated carbon treatment was removed and the RO units were operated for about two months on chemically clarified, filtered, and chlorinated secondary effluent. After this change all units experienced an accelerated flux decline. Water quality data with and without activated carbon treatment are shown in Table 9-8.

Next, the chemical clarification unit was taken out of operation and the RO units were operated for about six weeks on filtered and chlorinated secondary effluent. Additional flux decreases were observed during this period. Average water quality data are shown in Table 9-9.

The RO units were then operated on chlorinated secondary effluent for about five weeks. Water quality data collected during this final part of the test program are shown in Table 9-10.

Investigators at Hemet believe that there are two and perhaps three types of mechanisms involved in membrane fouling:

1. *Calcium deposits*: scaling by tricalcium phosphate could be removed with an EDTA solution adjusted to pH 7.

2. *Organic slimes*: these could be removed with a solution of an enzymatic detergent (Biz); this treatment was moderately effective.

3. *Scaling*: this refers to scaling due to the particulates in the wastewater, as opposed to $Ca_3(PO_4)_2$ scale.

TABLE 9-8   Typical Water Quality—With and Without Activated Carbon Pretreatment, Reverse Osmosis Pilot Plant, Hemet, California.

Constituent	Feed, mg/l	DuPont Product, mg/l	DuPont % rejection	Gulf Product, mg/l	Gulf % rejection	Universal Product, mg/l	Universal % rejection
Activated Carbon Pretreatment							
Total COD	7.6	0.8	90	1.9	75	0.5	93
Dissolved COD	3.9	0.4	90	1.1	72	0.4	90
TDS	726	100	86	95.1	87	45.0	94
Nitrate	5.2	0.6	88	—	—	2.6	50
Total phosphorus	6.2	0.6	90	0.1	98	0.1	98
Calcium	69.5	4.5	93	—	—	0.1	—
Magnesium	21.0	1.4	93	—	—	—	—
Turbidity, JTU	0.35	0.2	43	0.2	54	0.7	51
Hardness (as CaCO₃)	244	18.1	93	5.0	98	4.6	98
Without Activated Carbon Pretreatment							
Total COD	35.2	1.5	96	3.4	90	3.1	91
Dissolved COD	26.2	1.8	93	1.1	96	2.5	90
TDS	718	103	86	52	93	25.1	96
Nitrate	9.7	1.6	83	3.8	61	1.7	82
Total phosphorus	7.9	0.7	91	0.04	99	0.2	99
Calcium	63.9	5.2	92	0.4	99	0.5	99
Magnesium	16.4	1.6	90	0.3	98	0.3	98
Turbidity, JTU	3.0	0.4	87	0.2	93	0.2	93
Hardness (as CaCO₃)	236	16.5	93	0.2	99	2.8	99

Feedwater is activated sludge effluent treated by chemical clarification and filtration, with and without additional treatment by granular activated carbon.

TABLE 9-9   Typical Water Quality—Filtered Secondary Effluent, Reverse Osmosis Pilot Plant, Hemet, California.

Constituent	Feed, mg/l	DuPont Product, mg/l	DuPont % Rejection	Gulf Product, mg/l	Gulf % Rejection	Universal Water Product, mg/l	Universal Water % Rejection
Total COD	47.4	1.6	97	2.1	96	0.9	98
Dissolved COD	25.2	3.1	88	1.2	95	0	100
TDS	800	89	89	41	95	22	97
Nitrate	0.1	—	—	—	—	—	—
Total phosphorus	13.4	1.3	91	0.04	99	0.05	99
Calcium	61	3.4	94	0.3	99	0.4	99
Magnesium	16.4	1.5	91	0.3	98	0.4	97
Turbidity, JTU	2.1	0.4	80	0.5	80	0.6	67
Hardness (as CaCO₃)	228	10.2	96	1.8	99	2.4	99

TABLE 9-10   Typical Water Quality—Secondary Effluent, Reverse Osmosis Pilot Plant, Hemet, California.

Constituent	Feed, mg/l	Gulf Product, mg/l	Gulf % rejection	Universal Water Product, mg/l	Universal Water % rejection	DuPont Product, mg/l	DuPont % rejection
Total COD	60	5.5	91	4.8	92	2.1	97
Dissolved COD	32	5.2	84	3.5	89	0.2	99
TDS	731	198	73	80	91	29	96
Nitrate-N	4.1	3.4	17	1.4	66	0.6	85
Ammonia-N	13.6	—	—	—	—	0.1	99
Organic-N	3.9	—	—	—	—	0.1	97
Sulfate	159	90	43	4	97	1	99
Total phosphorus	11.7	3.0	74	0.3	97	3.5	70
Calcium	65.5	9.3	86	0.6	99	0.4	99
Magnesium	₁8.8	3.1	84	0.4	98	0.3	98
Total hardness	242	37	85	4	98	2	99
Turbidity, JTU	10.0	0.8	92	0.4	96	0.4	96

Lebanon, Ohio

The early pilot work at Lebanon was with a plate and frame RO unit.[7] Subsequently, several commercially available RO modules were tested at the Lebanon pilot plant, including tubular and hollow fiber configurations. The tubular assembly used was manufactured by Aerojet-General Corporation, and the study results have been reported.[18]

The pilot plant had a nominal capacity of 4000 gpd and consisted of three modules with 32 tubes in each module. The tubes were 0.56 in. ID and 14 ft long. Four different types of municipal wastewater were treated: (1) lime clarified raw wastewater; (2) secondary effluent; (3) primary effluent; and (4) primary effluent with turbulence promoters in the module. During most of the study the unit was operated 24 hours a day, 5 days a week. Based on this pilot study, Feige and Smith concluded the following:

1. A tubular RO unit can successfully treat municipal wastewater when membrane cleaning techniques are routinely practiced.
2. The frequency of tube breakages was a significant factor contributing to total downtime experienced during each run. Without exception, the tubes fractured near the end fittings. Failures decreased when daily pressurization–depressurization steps were discontinued.
3. Flux decline, probably due to membrane compaction, took place when new membrane tubes were first used.
4. Product water fluxes between 8 and 12 gpd/ft^2, and averaging 11

gpd/ft², could be maintained with regular chemical cleaning of the tubes when both lime clarified, dual media, filtered raw wastewater and unfiltered secondary effluent were treated.

5. Product water fluxes between 6 and 10 gpd/ft² and averaging 9.5 gpd/ft² could be maintained with frequent chemical cleaning of the tubes when primary effluent was treated.

6. Several chemical cleaners, including the commercial bioenzyme Biz, sodium perborate, and a nonenzyme threshold inhibitor were used for flux restoration. Biz was the most effective. The most successful bioenzyme cleaning occurred at the natural solution pH of 9.5.

7. Although product water fluxes were improved with chemical cleaning, the highest restored flux was 14 gpd/ft², about 78 percent of that obtained with tapwater.

8. Flux increases were evident after weekend shutdown periods. This suggests that relaxation of the membrane is a natural flux restoring characteristic.

9. Membrane rejection performance for both inorganic and organic constituents was consistently good for all waste streams tested. TDS removals averaged 93 percent; total organic rejection averaged 93 percent; total organic rejection averaged 94 percent for COD and 92 percent for TOC. Total COD in the feed water ranged from 66 to 90 mg/l; COD in the RO product water was: lime clarified raw wastewater, 8.4 mg/l; secondary effluent, 2.4 mg/l; primary effluent, 2.7 mg/l.

10. The insertion of turbulence promoters into each tube resulted in an effective 20 percent velocity increase and 42 percent flux increase when primary effluent was treated.

11. Chemical cleaning of the tubes helped reduce differential pressure increases experienced during each of the tests. The greatest improvement occurred during treatment of primary effluent, when cleaning with Biz reduced the operating differential pressure by about 30 psi.

Other RO Studies

A RO laboratory scale unit was used as part of a 15,000 gpd biological-chemical-physical pilot plant in a 30 day test program.[19] RO was chosen because of its ability to remove organics, bacteria, and virus as well as inorganic salts.

A laboratory study at the University of Washington evaluated the properties of a number of commercially available cellulose acetate membranes.[20] It was found that coliform bacteria and bacteriophage, of approximately the same shape and size as enteric viruses, did not permeate through a porous cellulose acetate membrane. Also, studies at the University of

Washington considered the permeation through a porous cellulose acetate membrane of a number of organic compounds found in municipal wastewater effluent.[21,22] The permeation of aqueous solutions or dispersions containing only one chemical species were also investigated. The permeation of aqueous solutions or dispersions of mixtures of chemical species were tested to determine the presence of synergistic or antagonistic actions. Reductions in the 80–99 percent range were obtained for those chemical species existing primarily in the colloidal aggregate, micelle, or macromolecular form. When the vapor pressure was appreciably greater than that of water the reduction was found to be in the 50–80 percent range. The RO membrane tested effected large percentage removals of most organic materials present in activated sludge and trickling filter effluents; however, compounds which have significantly higher vapor pressure than however, compounds such as phenol, which have significantly higher vapor pressure than water, appeared in significant quantities in the RO product water.

ELECTRODIALYSIS

Process Description

In the electrodialysis (ED) process, water flows between alternately placed cation permeable (C) and anion permeable (A) membranes, as illustrated in Figure 9-6. A direct electric current provides the motive force for the ion migrations through the membranes. Many alternating cation and anion

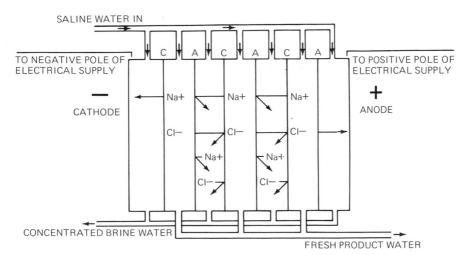

Fig. 9-6 Electrodialysis demineralization process.

membranes, each separated by a plastic spacer, are assembled into membrane stacks. The spacers (about 0.04 in. thick) contain the water streams within the stack and direct the flow of water through a tortuous path across the exposed face of the membranes. Membrane thicknesses generally range between 0.005 and 0.025 in.

Physically, the process takes the form of a plate and frame assembly similar to that of a filter press. The spacers determine the thickness of the solution compartments and also define the flow paths of the water over the membrane surface. Several hundred membranes and their separating spacers are usually assembled between a single set set of electrodes to form a membrane stack. End plates and tie rods complete the assembly.

Ion selective membranes are essentially IX resins in sheet form. Some membranes are made by mixing IX resins with a polymeric binder and casting or extruding a sheet from the mixture. Membranes are also manufactured by methods that exactly parallel the production of IX resins; e.g., styrene and divinylbenzene are copolymerized in sheet form, and the resulting structure is then chemically treated to give the sheet IX properties.

The ionic properties of an IX membrane are identical to those of the resin on which they are based. A cation exchange membrane, for example, may be thought of as an assembly of anions with which are associated mobile and exchangeable cations. When a membrane of this type is placed between two salt solutions and subjected to the passage of a direct electric current, most of the current will be carried through the membrane by cations, and hence the membrane is said to be cation selective. Selectivities are typically greater than 90 percent. When the passage of current is continued for a sufficient length of time, the solution on the side of the membrane that is furnishing the cations becomes partially desalted, and the solution adjacent to the other side of the membrane becomes more concentrated. It should be noted that these desalting and concentrating phenomena occur only in thin layers of solution immediately adjacent to the membrane, and not in the bulk of the solution. These phenomena are analogous to those that occur in the reverse osmosis process at the membrane solution interfaces.

Ion selective membranes in commercial electrodialysis equipment are commonly guaranteed for as long as 5 years, and experience has demonstrated an effective life of over 10 years. There is no reason to believe that an ion selective membrane should have a life any shorter than an ion exchange resin of similar composition.[23]

Passage of water between the membranes of a single stack, or stage, usually requires 10 to 20 seconds, during which time the entering minerals in the feedwater are removed. The actual percentage removal achieved varies with water temperature, type and amounts of ions present, flow rate of the water and stack design. Typical removals per stage range from 25 to 60 percent. Practical systems currently employ one to six stages.

Of the above parameters, water temperature is the most important single factor affecting removals per stage. Mineral removal increases at the rate of slightly more than 1 percent/°F. ED membranes are stable up to 100°F, and may be operated at or near this temperature when relatively low grade thermal energy is available at low cost, as in power and boiler plants.

According to Faraday's law, 100 mg/l of dissolved ionized solids can be removed from 1000 gal of water by 200 amp-hours of direct electric current. Since 1–2 volts per membrane or cell pairs are generally required in modern ED plants to achieve practical removal rates when feedwater mineral concentrations are less than 1000 mg/l, the amount of energy required to remove 100 mg/l of dissolved minerals from 1000 gal of water is 0.2–0.4 kwh. To this must be added about 2–3 kwh/ 1,000 gal for pumping the product and waste streams through the stack.[24]

The most commonly encountered problem in ED operation is scaling (or fouling) of the membranes by both organic and inorganic materials. Alkaline scales are troublesome in the concentrating compartments when the diffusion of ions to the surface of the anion membrane in the diluting cell is insufficient to carry the current. Water is then electrolyzed and hydroxide ions pass through the membrane and raise the pH in the cell. This increase is often sufficient to cause precipitation of materials such as magnesium hydroxide or calcium carbonate, and limits the allowable current density. The accumulation of particulate matter increases the electrical resistance of the membrane, thus interfering with the proper function of the stack, and in the long run may damage or destroy the membranes. This condition can be offset by feeding acid to the concentrate water stream to maintain a negative Langelier index.

Ionics, Inc. has developed a new type of ED unit which does not require the addition of acid or other chemicals for scale control. This new system reverses the DC current direction and the flow path of the diluting and concentrating streams every 15 minutes. The electrodes reverse by switching the polarity of the cathodes and the anodes. The stream flow paths also exchange their source every 15 minutes. Motor operated valves, controlled by the timers, switch the streams so that the flow path that was previously the diluting stream becomes the concentrating stream, and the flow path that was previously the concentrating stream becomes the diluting stream. Small scale units have operated successfully on brackish waters but have not been tested on wastewater.

Early ED Studies

From December 1962 through July 1963, secondary sewage effluent from the City of Clinton, Massachusetts was pumped "more or less continuously" through two laboratory scale ED stacks.[25] Membranes used were 9 in. by 10 in. The secondary effluent was treated by cellulose cartridge filters, and the supply to one stack was also treated by granular activated carbon adsorption. No other pretreatment was used. The typical quality of wastewater supplied to the two ED stacks is shown in Table 9-11.

Results of the study were described in terms of electrical resistances, limiting current densities, selective removals or separations, and membrane physical properties. Fouling of the membrane surfaces and calcium

TABLE 9-11 Typical Quality of Wastewater Influent to Electrodialysis Stacks, Ionics Laboratory Study.

Constituent	Concentration (mg/l, except as noted)	
	Stack No. 1[a]	Stack No. 2[b]
Sodium	39.0	40.0
Potassium	16.5	16.5
Ammonium	17.0	16.5
Hardness	14.5	12.6
Chloride	27.3	30.0
Bicarbonate	117.1	118.3
Nitrate	7.5	5.5
Sulfate	43.0	26.0
Phosphate	33.5	41.0
Total ions	315.4	306.4
ABS	10.0	<0.01
COD	73.10	19.5
pH	7.0	7.0
Resistivity, ohm-cm	3000	3000
Turbidity, SiO_2	42.0	26.0
Temperature, °C	18.0	18.0
Organic N	3.0	2.0

[a]Secondary effluent treated by cartridge filter.
[b]Secondary effluent treated by cartridge filter and granular carbon sorption.

carbonate deposition in the concentrating compartments increased the electrical resistance in the stack. Deposition of organic material in the diluting compartments increased the thickness of the boundary layer films adjacent to the membranes which resulted in an increase in electrical resistance. Membrane fouling, thought to be from the deposition of organics, caused erratic resistance data in stack no. 1 (feedwater to stack no. 1 was not treated by carbon adsorption). These erratic resistance data were attributed to buildup and subsequent discharge or sloughing of organic deposits. Throughout the study the electrical resistances of stack no. 2, which received carbon treated water, were generally below those in stack no. 1.

At the conclusion of the tests the two ED stacks were disassembled and the membranes subjected to standard test procedures to determine if any irreversible changes in membrane properties had occurred. The cation membranes in both stacks appeared not to have suffered any deterioration, and capacities of the cation membranes did not decrease throughout the test period. The anion membranes from stack no. 1 (which received non-carbon treated wastewater) showed the effects of fouling by an increase in resistivity and a decrease in capacity. It was postulated that the sorption of ABS or other organics was responsible for this fouling. Samples of the membranes from both stacks were tested before and after elution of ABS. The membrane resistivities were restored nearly to their original values by the elution of ABS and capacities were increased, but not to their original values.

Pomona, California

A 15,000 gpd ED unit constructed by Ionics was operated by the Sanitation Districts of Los Angeles County.[10] The unit contained 65 anion and cation membrane pairs, 18 in. by 20 in. in surface area. A flow of approximately 12 gpm in the feed stream was split, with 10 or 11 gpm passing through the dilute stream and becoming product water. The remaining 1–2 gpm of the feed stream was mixed with 10–11 gpm of brine recirculation flow to provide equal flow rates and equal pressures on both sides of the membranes. The pH in the concentrate stream was maintained at 3.5 by the injection of sulfuric acid. Typical removals achieved in the unit are shown in Table 9-12. These data show that the average TDS reduction was about 34 percent and that the process was more selective for nitrogen compounds and less selective for phosphate. The removal of soluble organic material as measured by COD was about 15 percent. The membranes were routinely cleaned every second day to maintain high performance. The cleaning was accomplished by flushing with an enzyme detergent, tapwater, and acid

TABLE 9-12 Water Quality—ED Pilot Study, Pomona, California.

Constituent	Concentration, mg/l		Percentage removal
	Influent	Effluent	
COD	9.4	8.0	14.9
NH_3-N	9.0	5.1	43.4
NO_3^--N	6.2	3.1	50.0
PO_4^{3-}-P	10.1	7.8	22.8
TDS	705	465	34.0

Notes:
1. 15,000-gpd pilot plant manufactured by Ionics, Inc.
2. Influent municipal wastewater treated by activated sludge and granular carbon sorption.
3. Averages based on 10 samples (June–August, 1969).
4. Analyses run on 24 hour composite samples.
5. Average water recovery 85 percent.

with the unit shut off. This method of cleaning permitted at least four months of operation without requiring manual cleaning of the membranes.

Orange County, California

An Ionics Mark III ED unit was used to treat chemically clarified secondary (trickling filter) effluent.[8] The unit used in the test contained 20 pairs of membranes.

The test was divided into two phases:

1. Operation with and without activated carbon pretreatment.
2. Operation with more contact time in the granular activated pretreatment unit. (At the end of the second phase the unit was operated for a short time without carbon pretreatment).

Typical quality of the water supplied to the ED unit for each phase is shown in Table 9-13.

During the first phase the unit was operated for 192 hours with activated carbon pretreatment, followed by 60 hours without carbon pretreatment. The COD and MBAS concentrations in the feed to the ED unit increased with time, because of exhaustion of the carbon. TDS removal averaged about 27 percent; however, without carbon pretreatment, the removal percentage decreased.

TABLE 9-13   Influent Water Quality —ED Pilot
Plant, Orange County, California.

Constituent	First phase	Second phase
TDS (mg/l)	1160–1400	950–1450
Chloride (mg/l)	290–360	275–390
Turbidity (JTU)	0.1–6.0	0.3–4.0[b]
pH (units)	6.5–7.3	6.1–7.1[c]
COD (mg/l)—with carbon	22–44	0–48
COD (mg/l)—without carbon	28–46	50–92
MBAS (mg/l)—with carbon	0.1–0.9	0.1–0.7
MBAS (mg/l)—without carbon	1.6–5.3	0.6–1.5

[a]Influent was municipal wastewater with 15–20% industrial wastes treated by trickling filters, chemical clarification, and dual media filtration; with and without additional treatment by granular carbon sorption as shown.
[b]A value of 12 JTU was measured during a plant upset at 460 hrs.
[c]A value of 4.4 units was measured at 500 hrs. This was due to a high alum dose.

In the second phase the unit was operated for 920 hours with carbon pretreatment and about 24 hours without carbon pretreatment. For the first 600 hours, the unit removed about 30 percent of the influent TDS; the electrical resistance remained nearly constant. Membrane resistance did increase at times, but wiping the membranes restored the unit to its original condition. After 600 hours some permanent membrane fouling began to appear, and physical wiping of the membranes would no longer restore the unit to its original condition. Between 600 and 920 hours of operation the removal of TDS dropped to about 22 percent.

After 920 hours the carbon pretreatment was removed, and within a day the performance had dropped to near zero, indicating rapid membrane fouling. A new set of membranes was then installed to determine if they would foul as quickly without carbon pretreatment as the old membranes. The new membranes were also fouled very rapidly by the non–carbon treated water.

When the unit was removing about 30 percent of TDS, there was also a small removal of COD and MBAS. Removal of chloride was slightly higher than the removal of TDS, and the removal of sulfate slightly lower. The concentrations of MBAS in the waste stream were no higher than in the influent; therefore, it was believed that the reduction of MBAS was through deposition on the membranes. There was also a high MBAS concentration found in the membrane flushing solutions.

As a result of this study it was concluded that membrane fouling resulted primarily from the fraction of COD represented by MBAS and that relatively high COD concentrations could be tolerated without fouling if the MBAS concentration was less than 0.3 mg/l. It was also concluded that after adequate pretreatment of the feed water, including carbon adsorption, operating characteristics and costs for an ED plant treating reclaimed water could be expected to parallel those same characteristics and costs for a similar plant treating brackish water of the same mineral content.

Santee, California

A 50,000 gpd ED pilot plant was operated from March 1970 through June 1971.[26] The unit was a two stage Ionics Mark III with 95 membrane pairs in each of the two stacks. The membrane surface was 18 in. by 40 in. and the membrane was 0.023 in. thick. The spacers were 0.04 in. thick. Feedwater to the ED unit was municipal wastewater which received secondary treatment (activated sludge), chemical clarification with lime in a solids contact clarifier, dual media filtration, and granular carbon adsorption.

The pilot study was divided into three operational periods as follows:

1. *March, 1970 to August 28, 1970 (240 days)*. No fouling or scaling prevention methods were used during this time.
2. *August 28, 1970 to January 12, 1971*. The ED plant was shut down 89 days in this 110 day period because of modifications under construction in the pretreatment system.
3. *January 12, 1971 to June 25, 1971 (176 days)*. Several membrane cleaning and rejuvenation methods were used during this period.

During the first period of operation the membrane stack was disassembled and hand cleaned when the amperage dropped below, or the influent pressure rose above, the manufacturer's (Ionics) recommended operating values. During this period the membrane stack was hand cleaned eight times, averaging 18 days between cleanings.

During the third period of operation all efforts were made to retard membrane fouling and scaling without stack disassembly. Sulfuric acid flushes, detergent flushes, air scouring, and shutdowns were used to maintain the amperage and the influent pressure at the recommended values. Using these methods the unit operated for 138 days before hand cleaning the membrane stack was necessary.

The influent and effluent concentrations of various constituents for the entire study period are shown in Table 9-14. These data indicate that the average TDS removal in this two stage ED process was about 50 percent. The removals of calcium and chloride were significantly more than 50 percent, while the removals of phosphorus, nitrate-nitrogen, and COD were significantly less than 50 percent.

TABLE 9-14 Influent and Effluent Water Quality, ED Pilot Plant, Santee, California.

Constituent	CONCENTRATION, mg/l (except as noted)				Average removal, percent
	Influent		Effluent		
	Range	Avg.	Range	Avg.	
Chloride	200–276	243	44–127	78	67.9
Sulfate	300–490	379	160–380	225	40.6
Bicarbonate	78–384	209	40–174	103	50.7
Nitrate-N	0.02–22.5	3.4	0.07–25.0	2.3	32.4
Phosphate-P	0.03–3.53	0.98	0.00–2.54	0.76	22.4
Total hardness as $CaCO_3$	140–430	378	92–244	146	61.4
Calcium	46–163	111	24–129	38	65.8
Magnesium	0.0–43	24	1.5–22.4	12	50.0
Potassium	14–25	20	4–14	11	45.0
Sodium	196–255	230	80–180	124	46.1
Ammonia	0.03–14.1	3.8	0.00–6.8	1.9	50.0
Turbidity, JTU	0.0–5.5	1.2	0.1–2.5	1.1	8.3
Color, units	5–20	7	5–15	5.6	20.0
Total COD	0.0–34.0	15.3	0.0–21.8	10.4	32.0
COD—filtrate	0.0–24.3	13.1	0.0–14.7	8.8	32.8
Iron	0.0–0.25	0.12	0.00–0.32	0.10	16.7
TDS	1000–1580	1245	400–990	630	49.4

[a]Municipal wastewater treated by activated sludge, chemical clarification, dual media filtration, and granular activated carbon sorption.

Complete hydraulic and membrane stack shutdowns of 1–2 hours were found to be somewhat effective in decreasing the membrane resistance. The effects of flushing with sulfuric acid appeared to be dependent upon the degree of fouling and the composition and location of the deposition. As might be expected, acid flushing had more effect on carbonate deposits than on organic fouling. There also appeared to be a level of fouling when acid flushes were not able to restore the membranes to their original condition. Flushing with enzymatic detergents was found to show some success in restoring the membrane resistance, and air injection or air scrubbing was found to be somewhat successful in lowering influent pressures.

Nine membranes required replacement during the 1.5 year study period. Two cation membranes were replaced during the third month of operation. These were the heavy membranes which came in direct contact with the cathode waste streams, and they were scaled with a carbonate compound.

Two anion membranes located near the cathode were replaced during the fourth month of operation. They were also scaled with a carbonate compound. Anion membranes were replaced during the nineteenth and thirteenth months because of a slight scaling of the carbonate compound.

The deposit on the membranes, as observed during hand cleaning, had the same general appearance throughout the study period. It was a yellowish colored, slimy material which adhered to the membrane but was easily wiped off with the palm of a hand or a wet rag. After hand cleaning on June 24, 1971 (after 175 days of operation) samples of the deposit on the diluting stream sides of both the anion and cation membranes were analyzed; the results are shown in Table 9-15. The analysis of the deposit from the anion membranes was not considered to be as representative as the deposit from the cation membrane because of the method of sample collection. The data in Table 9-15 indicate that there was considerable organic matter in the scale, and that the cation membrane deposit contained a higher proportion of inorganic matter than did the anion membrane deposit.

TABLE 9-15   Composition of Scale Deposits Found on ED Membranes, ED Pilot Plant, Santee, California.

Constituent	RELATIVE CONCENTRATIONS, mg/l	
	Diluting side of anion membrane	Diluting side of cation membrane
Sodium	148	358
Potassium	9.8	35
Calcium	103	88
Magnesium	31.7	16.6
Ammonia	3.0	12.9
Total Keldjahl-N	26.1	36.8
Iron	1.2	2.8
Aluminum	106	7.0
Copper	0.8	0.9
Chromium	0.1	0.1
Manganese	0.2	0.2
Zinc	0.3	1.6
Sulfate	380	440
Chloride	160	360
Bicarbonate	150	244
Phosphate-P	44.2	6.5
TDS	1050	1510
COD	393	395
TOC	102	82
Cadmium	0.1	0.1

Lebanon, Ohio

The Federal Water Pollution Control Administration (now Environmental Protection Agency) operated an ED pilot plant at Lebanon, Ohio.[27] Municipal secondary effluent was treated by diatomaceous earth filtration and granular activated carbon adsorption before it entered the ED unit. The ED unit was an Ionics Mark III with 18 in. by 40 in. membranes with an effective area of 2960 cm^2. At the beginning of the study the ED stack contained 150 pairs of membranes and produced 50 gpm of product water. The number of membrane pairs was reduced to 125 and the product water to 42 gpm for better diatomaceous earth filter operation. Part of the concentrate stream leaving the stack was recirculated to the stack with makeup water to equalize the pressures across the membranes. Sulfuric acid was fed to the concentrate and cathode streams and pH was maintained in the range of 3–5. The pilot plant was operated five days per week. Water temperature during the test varied from 12 to 24°C.

Membrane fouling was encountered throughout the tests. It resulted in increased membrane resistance and, therefore, decreased demineralization. During the first 106 hours of operation, little fouling occurred, but during test runs made after the initial period, increasingly serious fouling occurred.

Shutting down the unit for a day or more was found to be somewhat successful in restoring demineralization efficiency. Examination of the membranes after the unit was badly fouled showed that the dilute stream side of the anion membranes was coated with a slimy material.

It was believed that, since the fouling material was present as an easily removable layer on the membrane surface, rather than penetrating the membrane, colloidal material was the likely cause of the scaling. Soluble organics would have penetrated the membranes and been more difficult to remove. The carbon adsorption effluent produced a feedwater with a total organic carbon concentration averaging less than 2 mg/l for all runs, and the investigators believed that this type of pretreatment was adequate for preventing serious fouling by soluble organics.

Use of acid as a disinfectant at the beginning of and during runs appeared to decrease fouling, but not enough to make operation practical at turbidities above 0.1 JTU. Although there was no good correlation between turbidity and fouling rate, it was concluded that turbidities less than 0.1 JTU combined with periodic disinfection were the best method for controlling fouling.

Permanent damage to the membranes from fouling, as measured by an increase in membrane resistance, did not occur. The membrane resistance after more than 1400 hours of operation increased about 5 percent. The investigators considered this to be good membrane performance, since much of the 1400 hours of operation was under severe fouling conditions.

Ion selectivity data from eight sets of feed and product samples are

TABLE 9-16   Ion Removal Selectivity, ED Pilot Plant, Lebanon, Ohio.

Ion	Fraction of ion removed by fraction of all ions removed	
	Avg.	Range
Bicarbonate	0.77	0.63–0.95
Chloride	1.25	1.07–1.51
Nitrate	1.22	1.02–1.48
Phosphate	0.72	0.25–1.03
Sulfate	1.11	0.21–2.00
Ammonium	1.15	1.05–1.3
Calcium	1.25	1.01–1.48
Magnesium	0.98	0.71–1.35
Potassium	1.14	0.77–1.35
Sodium	0.79	0.52–0.99
Calcium and magnesium	1.13	1.05–1.28

summarized in Table 9-16. This data show that the membranes were not highly selective for any particular ion. Highest selectivities were for calcium and chloride, with the lowest for sodium and phosphate.

Other ED Studies

By the end of 1971 the method of manufacturing salt in Japan had completely changed from the salt field method to the IX membrane method. All salt manufacturing plants in Japan now produce brine by means of ED with IX membranes. The permeation of several trace metal ions from the diluting compartment through the IX membranes into the concentrating compartment during seawater concentration by ED was investigated.[28] The toxic heavy metals studied were cadmium, lead, mercury, and chromium. It was found that, during the concentration of seawater by ED, membranes which had a low permeability to divalent ions (cadmium, mercury, and chromic ions) had a low rate of permeation. The (percentage removal) lead permeability for this type of membrane was about the same as that for sodium ion; and chromate ions were passed more easily (higher percent removal) than chloride ions.

ION EXCHANGE

Process Description

The IX process has been used for many years to soften hard water and to demineralize water for various industrial uses. Several natural minerals, called zeolites, were found to have exchange characteristics suitable for the softening of water, which involves the exchange of calcium and magnesium for sodium. In 1904 the first artificial aluminosilicates were produced and these so-called artificial zeolites had capacities of two to three times those of naturally occurring materials. These zeolites were suitable for water softening reactions, but are not capable of resisting acids, and, therefore, cannot be used in the exchange of hydrogen ions. The first acidproof cation exchange materials were produced in the 1920s by the sulfonation of coal. The first organic IX resins were successfully synthesized in 1936.

IX resins are, by definition, insoluble solids containing fixed cations or anions capable of reversible exchange with mobile ions of the opposite sign in the solutions with which they are brought into contact. The resins which are commercially available in the United States are manufactured by four companies:

1. Rohm and Haas Company
2. Diamond Shamrock Chemical Company
3. Dow Chemical Company
4. Ionak (Division Permutit and Sybron Corporation)

Dow and Permutit can supply either the resins or a complete IX treatment system. Rohm and Haas and Diamond Shamrock supply only the resins for systems which are designed and manufactured by others.

In desalting of water by the IX process, the salts in the water react with specially formulated resins that are reactive but insoluble. One kind of resin takes up sodium ions and other cations and releases hydrogen ions. A second type of resin takes up chloride ions and other anions and releases hydroxyl ions. The hydrogen ions and hydroxyl ions combine to form small amounts of water. When the resins are exhausted, they must be regenerated for the process to be economically feasible. An acid such as sulfuric acid is used to supply hydrogen ions to the cation resin, and a base such as sodium hydroxide is used to supply hydroxide ions to the anion resins.

Commercial resins are broadly classed as (1) strong acid, (2) weak acid, (3) strong base, and (4) weak base. It is possible to use several combinations of the various cation and anion exchange resins to demineralize water. The reactions that occur when the water passes through a strong acid cation exchange resin in the hydrogen form and a typical strong base anion unit operating in the hydroxide form are illustrated in Figure 9-7. The reactions in weak acid and weak base resins are illustrated in Figure 9-8.

An IX contacting system can be operated downflow or upflow with single beds or mixed beds. A typical two bed IX system is shown in Figure 9-9. Another IX system which was developed recently is the moving bed unit, illustrated in Figure 9-10.

HYDROGEN-CYCLE CATION EXCHANGE

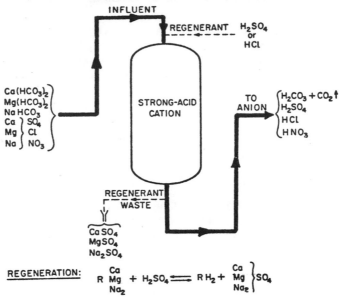

REGENERATION:

$$R \begin{matrix} Ca \\ Mg \\ Na_2 \end{matrix} + H_2SO_4 \rightleftharpoons RH_2 + \begin{matrix} Ca \\ Mg \\ Na_2 \end{matrix} \Big\} SO_4$$

HYDROXIDE-CYCLE ANION EXCHANGE (STRONG-BASE)

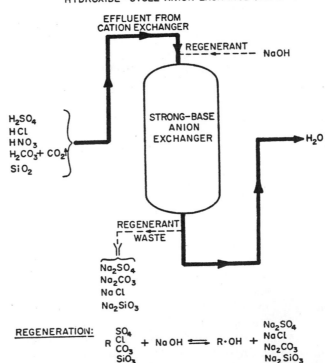

REGENERATION:

$$R \begin{matrix} SO_4 \\ Cl \\ CO_3 \\ SiO_3 \end{matrix} + NaOH \rightleftharpoons R \cdot OH + \begin{matrix} Na_2SO_4 \\ NaCl \\ Na_2CO_3 \\ Na_2SiO_3 \end{matrix}$$

Fig. 9-7 Strong acid and strong base ion exchange reactions.

490

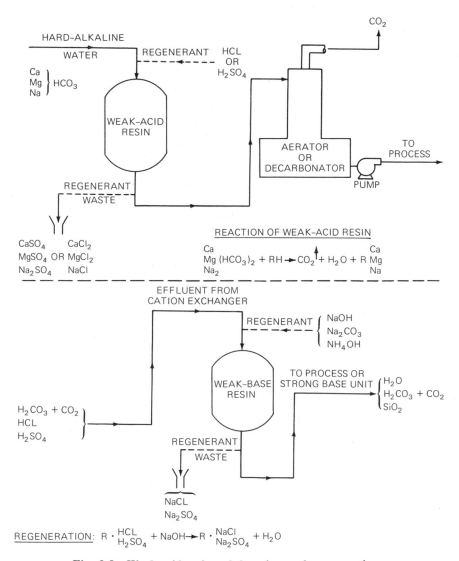

Fig. 9-8 Weak acid and weak base ion exchange reactions.

IX systems have been tested for the removal of specific ions from both water and wastewater. A moving bed system is being used for the removal of nitrate from groundwater in New York.[29] IX for the removal of ammonia from wastewater is discussed in Chapter 6.

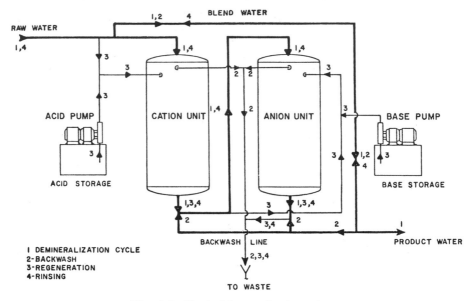

Fig. 9-9   Typical ion exchange system.

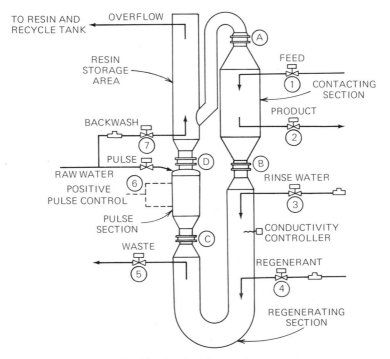

Fig. 9-10   Moving bed ion exchange system.

Pomona, California

An IX pilot plant was operated at Pomona, California for about three years.[30] Activated sludge effluent was treated in downflow carbon adsorption contactors and supplied to the IX pilot plant, which consisted of four columns of resin in series: primary cation, primary anion, secondary cation, and secondary anion. The dimensions and resin volumes of each IX column are shown below:

	Primary cation column	Primary anion column	Secondary cation column	Secondary anion column
Diameter, in.	20	20	13	10
Height, ft	6.0	4.7	4.8	4.7
Resin volume, ft³	8.0	1.6	3.2	1.6

The cation exchange column was filled with a strong acid resin in the hydrogen form and the anion units contained weak base resins in the hydroxide form. The cation columns were regenerated with a 4 percent sulfuric acid solution and the anion columns with a 4 percent aqueous ammonia solution.

Under what were considered to be typical operating conditions, the TDS of the feed and product water averaged about 610 mg/l and 72 mg/l, respectively. The concentrations of various constituents at successive stages of the demineralization process are shown in Table 9-17. The data

TABLE 9-17   Average Water Quality, Ion Exchange Pilot Plant, Pomona, California.

Constituent	Carbon column effluent, feed	Primary cation effluent	Primary anion effluent	Secondary cation effluent	Secondary anion effluent, product
Calcium	53	2.0	1.7	1.1	0.60
Magnesium	17	0.59	0.56	0.38	0.00
Sodium	126	61	59	16	15
Potassium	14	7.3	7.1	1.9	1.9
Ammonia, as N	20	9.6	9.2	4.0	3.8
Sulfate	72	72	3.6	3.6	1.3
Nitrate, as N	2.9	2.8	1.6	1.5	0.35
Chloride	135	132	84	83	14
Orthophosphate, as PO$_4^{3-}$	27	27	15	14	0.25
Total alkalinity, as CaCO$_3$	218		51		0.25
pH	7.4	2.7	5.7	2.8	5.8
Conductivity, μmho/cm	1040	1390	390	1040	100
TDS	610	298	198	104	72
COD	10.0				3.7

All concentrations in mg/l of ion, except conductivity and pH in Table 9-17 show that in addition to the 89 percent removal of TDS, the system removed about 63 percent of the COD, over 98 percent of the phosphate, 86 percent of the nitrate nitrogen, and 75 percent of the ammonia nitrogen.

After 500 cycles of operation there was no discernible decrease in the efficiency of the resins. Carbon adsorption was an effective pretreatment for organic removal, and fouling of the resins was not a problem. Operation of the system at leakages (equivalents of cations or anions in product water divided by equivalents of cations or anions in feed water) of about 5 percent produced regeneration efficiencies of 85 percent and 90 percent for the cation and anion exchangers, respectively. Regeneration efficiency is defined as the total equivalents of ions removed divided by the total equivalents of regenerate applied.

University of California, Berkeley

Sanks and Kaufman reported the results of studies at the University of California at Berkeley on the demineralization of brackish water for municipal water supply,[31] and the partial demineralization of secondary sewage effluent.[32] The general purpose of the latter investigation was to determine the feasibility of using IX for the removal of the mineral portion of the use increment in sewage.

The IX system used consisted of strong acid (Duolite C-25 Diamond Shamrock Company; or Amberlite 200 Rohm and Haas Company) and weak base resins (Amberlite IRA-93, Rohm and Haas) in fixed beds with a degasifier between the beds to remove carbon dioxide. The resins were chosen arbitrarily from manufacturer's literature, and no attempt was made to determine the optimum resin.

The feedwater was obtained from the City of Richmond standard rate activated sludge plant. Normally, the activated sludge effluent contained a total COD of about 80 mg/l, a turbidity of 40 JTU, and a threshold odor of 13. Approximately 600 gal of secondary effluent were collected for each series of runs. After analysis for the principal ions, several salts were added to increase the TDS to 1200 mg/l and chlorine and mercuric chloride were introduced to prevent decomposition.

Two series of runs were made in which the salted secondary effluent was applied directly to the ion exchange beds. IX removed 40–49 percent of the total COD, 85–94 percent of the turbidity, 58–83 percent of the color, 43–94 percent of the odor, and 69–74 percent of the dissolved minerals. However, the CCE increased from a feed value of 1.24 mg/l by 15–108 percent. The cation exchange efficiency was 82 percent for Duolite C-25 and 73 percent for Amberlite 200. The anion exchange efficiency of IRA-93 declined from 88 percent in the tenth run to 76 percent in the sixteenth, with a corresponding increase in leakage. Compared to the product obtained with flocculated and filtered secondary effluent, the performance was very poor and the product water was considered unacceptable.

Because the performance of the two cation exchangers, as measured by sustained operating exchange capacity and leakage, was similar, only Duolite C-25 resin was used. The feed was salted secondary effluent that was treated by chemical clarification, sand filtration, and carbon adsorption. The exchange efficiency was 83 percent for C-25 resin and 86 percent for IRA-93. The results are shown in Table 9-18.

TABLE 9-18 Ion Exchange Feed and Product Water Quality.[32] (All values mg/l except as noted.)

Constituent	Feed[a]	IX Effluent
Sodium	148	81
Potassium	13	9
Calcium	97	0.4
Magnesium	69	0.2
Ammonium	3	5
Sum cations	330	96
Chloride	194	117
Sulfate	270	10
Bicarbonate	254	12
Nitrate	112	29
Nitrate	0	0
Phosphate (PO_4^{3-})	3	2
Sum anions	833	170
Dissolved Solids		
Calculated	1163	266
Total, 105° C	1096	272
Fixed, 600° C	846	212
COD, suspended	1	2
COD, dissolved	8	9
Color, APHA units	0	1
Turbidity, JTU	0.4	1.3
Threshold odor	2	2

[a]Activated sludge effluent treated by chemical clarification, filtration and granular activated carbon adsorption.

Desal Process

Rohm and Haas reported the development of an IX demineralization technique which is based upon the use of weak resins for treating brackish waters. The Desal process is covered by United States and foreign patents and a license may be obtained to use the invention from Rohm and Haas.

The Rohm and Haas Desal process generally consists of a series of three units. The first unit, which is the alkalization unit, contains Amberlite IRA-68 in the bicarbonate form. All anions in the influent are converted to the bicarbonate salts of sodium, calcium, and magnesium in this unit, as illustrated in the following equation:

$$(R\text{---}NH)HCO_3 + NaCl \longrightarrow (R\text{---}NH)Cl + NaHCO_3$$

The second unit, or dealkalization unit, contains Amerblite IRC-84, a weak acid cation exchange resin. In this unit the bicarbonate salts are converted to carbonic acid as denoted in the following equation:

$$R\text{---}COOH + NaHCO_3 \longrightarrow R\text{---}COONa + CO_2 + H_2O$$

The third unit, which is the carbonation unit, also contains Amberlite IRA-68, but in the free base form. In the free base form, Amberlite IRA-68 absorbs the carbonic acid from the effluent of the Amberlite IRC-84 unit, as shown in the following equation:

$$R\text{---}N + H_2O + CO_2 \longrightarrow (R\text{---}NH)HCO_3$$

Upon completion of the above exhaustion cycle, the alkalization unit is regenerated back to the free base form with ammonia, caustic, or lime, as illustrated in the following equation:

$$(R\text{---}NH)Cl +, NH_4OH \longrightarrow (R\text{---}N) + NH_4Cl + H_2O$$

The dealkalization unit is converted back to the hydrogen form by regeneration with sulfuric, hydrochloric, nitric, or sulfurous acid, as illustrated in the following equation:

$$2R\text{---}COONa + H_2OSO_4 \longrightarrow 2R\text{---}COOH + Na_2SO_4$$

Since the carbonation unit is already in the bicarbonate form, the flow pattern is reversed for the next cycle and the third unit becomes the alkalization unit and the first unit the carbonation unit.

The Office of Saline Water sponsored a field evaluation of the Desal process.[33] A mobile Desal IX unit was field tested on four different brackish waters located at Yuma, Arizona; Roswell, New Mexico; Dalpra Farm, Colorado; and Webster, South Dakota. Waters at these locations varied from 1543 mg/l TDS at Webster to 3671 mg/l TDS at the Dalpra Farm. The purpose of the tests was to evaluate the performance of the process on waters

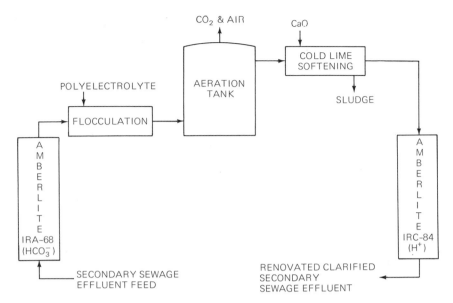

Fig. 9-11   Modified Desal process for treatment of municipal wastewater.

of different chemical characteristics and to establish the economics and design criteria for plants up to 1 mgd capacity. Various arrangements were tested in addition to the basic three bed process, including the two bed Desal system and combinations with cold lime softening. The resin performed satisfactorily on each water and produced a product of high quality.

Kunin and his co-workers at Rohm and Haas reported the results of laboratory studies on the treatment of secondary sewage effluent by a modified Desal process.[34,35] A flow diagram for the modified Desal process tested is shown in Figure 9-11. All of the processing studies were conducted on a laboratory scale using chlorinated secondary sewage effluent from the Haddonfield, N.J. activated sludge treatment plant.

The test column was approximately 36 in. long with a 1 in. ID. Using a 200 ml resin bed volume, enough headroom was available in the column to allow for at least 100 percent expansion of the bed during the upflow exhaustion cycle. The resin was regenerated with NH_4OH (hydroxylamine) by gravity. Typical results reported for the modified process treatment of the secondary effluent are shown in Table 9-19.

Santee, California.   A Desal process pilot plant was operated intermittently at Santee, California from January 1970 to October 1971.[36] Most of the operating data was collected during a two month period in August and

TABLE 9-19 **Modified Desal Process Treatment of Chlorinated Secondary Effluent.**

	CONCENTRATION, mg/l except pH	
Parameter	Secondary effluent	Modified Desal process effluent
pH	7.8	7.0
ABS	2.0	0.5
COD	90	15
Hardness (as $CaCO_3$)	98	0
NH_4^+	18	<1
Cl^-	32	3.5
NO_3^-	1.8	<1
PO_4^{3-}	21	<1
HCO_3^-	195	24

September 1971. The system was designed to operate automatically with a nominal capacity of 50,000 gpd. Each of the three resin vessels was 10 ft high and 3.5 ft in diameter. Feedwater to the ion exchange unit was activated sludge effluent treated with lime in a solids contact clarifier and then filtered in dual media filters.

A number of problems were experienced during startup; as a result, the unit was not operated as long as planned. Hydraulic loading was 0.42 gpm/ft^3 during many of the test runs, and 1 gpm/ft^3 for some of the tests. The resin manufacturer recommended a loading of 1 gpm/ft^3. The lead column (containing anion resin IRA-68) capacities in many of the runs were 5 kg/ft^3, which was below the design capacity of 14.5 kgr/ft^3. Difficulty in carbonating the anion resin was believed to be responsible for the poor performance. The anionic resin was put in the bicarbonate form by passing a carbon dioxide and water solution under pressure through the resin bed. Various difficulties were encountered in operating this carbonation system. Modifications in the carbonation system did improve the capacity of the anion resin. The highest capacity obtained in the test program was about 11 kgr/ft^3. The investigators believed that some of this reduced capacity could be accounted for by removal of organic matter by the IRA-68 resin.

The capacity of the resin in the lead anion column was found to be the limiting factor in the process. The design specifications called for a lead anion column capacity of 14.5 kgr/ft^3, and cation capacity of about 18 kgr/ft^3. Because the highest lead anion column capacity obtained during the project was 11.5 kgr/ft^3, exhaustion of the IRC-84 cation resin was never approached.

The TDS removal was as high as 86 percent, with an influent TDS ranging between 700 and 1000 mg/l as CaCO$_3$. Influent COD ranged between 25 and 35 mg/l, but COD removals were not reported.

Orange County, California. The Orange County Water District briefly evaluated the Desal process in a short test period in 1970. Secondary sewage (trickling filter) effluent was passed upflow through 700 cm^3 of IRA-68 anion resin contained in a 1.5 in. ID clear plastic column. The hydraulic loading was 2 gpm/ft^3 (16 BV/hour). The capacity of the resin, as measured by removal of chlorides, was much less than expected. The low capacity was attributed to difficulties in the carbonation system. Several changes and variations in the carbonation system were tested, but the resin could not be carbonated to produce the expected capacity, and the test program was terminated.

Elgin, Illinois

Pilot plant studies were conducted by Culligan Company under contract to the U.S. Environmental Protection Agency to evaluate treatment of municipal wastewater by several IX procedures.[37] The pilot plant was located at the Elgin Municipal Sanitary Treatment Plant, Elgin, Illinois. Supply water to the IX pilot plant was activated sludge effluent treated by lime clarification, dual media filtration, and granular activated carbon adsorption. The chemical clarification–filtration–adsorption system had a capacity of 15 gpm. The IX system was designed for flexible operation so that a number of treatment methods and modes of operation could be evaluated. It consisted of five resin containing pressure vessels, ranging from 10 to 14 in. ID and from 48 to 72 in. in height. The vessels could be operated individually or in various combinations, and the flow during the service or regeneration cycles could be either upflow or downflow.

The hydrogen form of a strong acid cation resin [IRC-120 (Rohm and Haas)] was tested and considered successful. The resin was supplied activated sludge effluent which was filtered and treated by carbon adsorption, but the chemical clarifier was not used. The performance obtained with this resin did not differ greatly from those which would be predicted on a potable water. The typical quality of water produced by the IRC-120 cation exchange resin is shown in Table 9-20. As shown in Table 9-20, the influent turbidity ranged from 2.0 to 3.4 JTU and the total organic carbon ranged from 5.4 to 10.2 mg/l. Regeneration of this cation exchange resin (IRC-120) with sulfuric acid was accomplished without difficulty. Precipitation of calcium sulfate in the resin bed was avoided; however,

TABLE 9-20 **Water Quality Summary, Strong Acid Resin[a] Ion Exchange Pilot Study, Elgin, Illinois.**

	CONCENTRATION, mg/l CaCO₃ except turbidity				Average % Removal
Constituent	Influent Range	Avg.	Effluent Range	Avg.	
Calcium	146–169	159	0–10.0	6.5	96
Magnesium	174–195	187	0–9.4	3.7	98
Sodium	164–223	185	13–71	32	83
Potassium	10.0–11.7	11.0	1.6–4.6	2.8	75
Ammonia-N	7.1–24	13.2	0.0–10.4	4.3	68
Turbidity, JTU	2.0–3.4	—	—	—	—
Total organic carbon	5.4–10.2	7.8	5.3–9.5	6.7	14

[a]Rohm and Haas IRC-120.

such precipitation occurred in the regeneration effluent about 10 minutes after leaving the column.

Weak base anion exchange resins in the free base form (Rohm and Haas IRA-93 and IRA-68) were also evaluated. Performance of the IRA-93 resin was slightly better than the IRA-68 resin. Therefore, it was believed that IRA-93 is the weak anion resin to be used in combination with, and preceded by a strong acid cation resin in the hydrogen form.

The weak acid cation resin IRC-84 (Rohm and Haas) was evaluated. Weak acid cation exchange resin will only partly demineralize wastewater, and the extent of demineralization will not exceed the content of alkalinity present in the water: effluent from this resin will contain all cations associated with nonalkaline cations. More complete cation removal requires post-treatment with a strong acid cation exchange resin. If complete demineralization is not required, a weak acid cation exchange resin may be sufficient. For example, with the sewage at Elgin, Illinois, this resin by itself reduced the TDS from about 600 mg/l to about 250 mg/l. The influent water employed in the weak acid cation resin (IRC-84) study was the sewage treated in the pretreatment section of the pilot plant (dual media filtration and activated carbon adsorption). The tests indicated that capacity was greatly affected by flow rate, which is typical for weak acid resin. At rates of 6 and 3 gpm/ft³, capacities of 17.2 and 21.5 kgr, respectively, were obtained. Effluent quality was about the same at the two flow rates. The typical quality of water produced is shown in Table 9-21.

A system utilizing both weak acid and strong acid cation exchange

TABLE 9-21   Water Quality Summary, Weak Acid Resin*
Ion Exchange Pilot Study, Elgin, Illinois.

	CONCENTRATION, mg/l CaCO₃				Average percentage removal
	Influent		Effluent		
Constituent	Range	Avg.	Range	Avg.	
Calcium	122–167	150	12–29	20	87
Magnesium	161–200	178	16–29	22	88
Sodium	212–264	238	188–200	195	18
Potassium	12–20	16	9–16	12	25
Ammonia-N	6.4–9.4	7.6	2.6–8.8	6.4	16

*Rohm and Haas IRC-84.

resins (IRC-84 and IRC-122*) was also evaluated. The advantage to using both forms of resin is the greater efficiency obtained during regeneration. Excess acid is normally required for the regeneration of a strong acid resin, and this excess acid is of sufficient quality to regenerate the weak acid resin. As a result, the ion exchange capacity obtained from the weak acid exchange resin may be obtained without increased chemical regeneration costs.

It was the general conclusion of this study that the best system for demineralizing municipal wastewater by IX consisted of weak acid, strong acid, and weak base resins in series.

SUL-biSUL Process

The SUL-biSUL process is another novel IX approach to the demineralization of water. The process is patented by the Dow Chemical Company and licensed to, and marketed by, several firms, including Culligan. The process has not been tested on municipal wastewater, but has been tested on various brackish water supplies;[38] and a 0.5 mgd plant utilizing this process is now in operation in Burgettstown, Pennsylvania.[6] This process is based upon the fact that an anion exchange resin which has been exhausted with sulfate may be in either the divalent sulfate ion form or the monovalent bisulfate form, depending upon the pH.

The process uses a conventional, strong acidic, sulfonic cationic exchange resin for cation removal; this resin is regenerated with sulfuric acid. When the water to be demineralized passes through the cation exchange resin bed, all of its salts are converted to acids. The second bed in

*Rohm and Haas Company.

the system consists of a strongly basic quaternary ammonium anion exchange resin that has been converted to the sulfate form, perhaps by waste regenerant from the cation exchanger. As the acids produced by cation exchange pass through the anion exchanger, the sulfate that occupies two exchange sites on the anion exchanger is converted to monovalent bisulfate, and one exchange site becomes available to pickup an anion from the acid mixture.

$$\begin{matrix} R \\ \\ R \end{matrix} \!\!> SO_4^{2-} + H^+A^- \longrightarrow \begin{matrix} R-HSO_4^- \\ \\ R-A^- \end{matrix}$$

The hydrogen from the acid that has come out of the cation exchange bed is removed by conversion of sulfate to bisulfate, and the strong acid anion is also removed. Thus, the strong acids are removed in toto by the anion exchange bed.

The regeneration of the two exchange units in the SUL-biSUL process, as in a conventional IX system, is performed utilizing two separate regeneration techniques. The regeneration of the cation unit is accomplished using low levels of sulfuric acid in a conventional manner. The regeneration of the anion unit, however, is unique. During exhaustion, for each molecule of acid (HA) exhausting the regenerated sulfate anion bed, one bisulfate ion is formed. If the exhausting acid is sulfuric acid, two bisulfate ions are formed.

$$\begin{matrix} R \\ \\ R \end{matrix} \!\!> SO_4^{2-} + H_2SO_4 \longrightarrow \begin{matrix} R-HSO_4^- \\ \\ R-HSO_4^- \end{matrix}$$

Then, regardless of the chemical composition of the raw water, at least 50 percent of the exhausted exchange sites are bisulfate ions.

Occuring simultaneously with bisulfate ion formation or acid adsorption is the process of normal IX. As chlorides and nitrates, present in the raw water, enter the anion bed during exhaustion they do displace some sulfates and bisulfates.

The simplest method of anion regeneration in the SUL-biSUL process is to reverse the sulfate–bisulfate equilibrium; i.e., to convert bisulfate ions back to sulfate ions, releasing the adsorbed acid, which is then discharged from the bed.

$$\begin{matrix} R-HSO_4^- \\ \\ R-HSO_4^- \end{matrix} \longrightarrow H_2SO_4 + \begin{matrix} R \\ \\ R \end{matrix} \!\!> SO_4^{2-}$$

(Exhausted site)   (Acid)   (Regenerated resin)

With water supplies containing sulfate ions it is claimed that it is possible to regenerate the anion bed by merely rinsing the unit, either upflow or downflow, with brackish water. During this operation, bisulfate ions disassociate forming sulfate ions and hydrogen ions because of the neutral or alkaline pH of the raw water. Also, during rinse regeneration, ions other than sulfate ions are eluted from the bed by a process of selective IX; the strong base anion exchange resin will selectively remove and hold sulfate ions in preference to chloride ions from dilute solutions.

The SUL-biSUL process is claimed to be economically competitive with electrodialysis in small installations, or where chemical costs are low and power costs are high. In addition, the capital costs are claimed to be lower.

Other Ion Exchange Studies

Midkiff and Weber reported the results of laboratory studies of the IX process to determine operating characteristics, design parameters, and the feasibility of IX for removal of nitrates and phosphates from waters and wastewaters containing sulfates, chlorides, and bicarbonates.[39] Factors studied included selectivity, hydraulic loading, and depth of resin bed. The investigation was carried out in a continuous flow columnar contacting system, which consisted of 1 in. diameter plexiglass column 4 ft long, with resin bed depths ranging from 3 in. to 3 ft. The particular resin used in the study was a strong base type in the chloride form. The main focus of this study was on the selective exchange of various anions, particularly phosphate and nitrate. One interesting phenomenon which was observed during this study was termed "reverse exchange", i.e., the less selective ions appeared in the effluent at a peak concentration which was considerably in excess of the feed solution concentration for that ion. The phosphate ion was one constituent which was observed to exhibit this phenomenon. This same phenomenon also occurred in the studies at Pomona, California.[30]

Another study which also utilized a strong base anion exchange resin in chloride form to treat municipal wastewater was reported by Eliassen and co-workers.[40,41] Activated sludge effluent from the City of Palo Alto, California was sand filtered and supplied continuously to a small IX column containing a strong base anion resin. This study evaluated the selective removal of phosphates, nitrates, and organics; reduction in total mineral content is not achieved with this type of resin.

Nine commercially available strong base anion exchange resins in the chloride form were used in laboratory studies for the treatment of activated sludge effluent.[42] Three of the resins were selected, and additional tests were conducted with larger pilot scale equipment. Activated sludge effluent for the tests was obtained from the Maple Lodge Works of the West Hertfordshire (England) Main Drainage Authority. The pilot studies were conducted with a column 1 m long and 5 cm ID filled with 1 l of resin, giving a bed depth of about 50 cm. The suspended solids level in the activated sludge effluent was exceptionally low, and clogging of the resin beds was not a serious problem. Significant amounts of organic matter, as measured by BOD, total organic carbon, color, and anionic detergents, were removed by the resins. During the limited number of cycles employed in the pilot work, the organic matter did not foul or interfere with the ion exchange properties of the resins. The removal of most constituents was about as expected from laboratory studies and from theory, but the behavior of phosphate was more complex. The maximum value of phosphate within the column, or in the effluent, at times was higher than the concentrations in the feedwater. All experiments conducted exhibited this same type of phosphate behavior, indicating an uptake and subsequent release of phosphate by the ion exchange resins. As discussed above, this phenomenon was also observed at Pomona[30] and by Midkiff and Weber.[39]

Although many organic materials can be removed from water by IX, a potential problem is that the organics may foul the resin, thus reducing its exchange capacity. Frisch and Kunin studied organic fouling of some of the anion resins which were available at that time.[43] However, some of the newer resins have the capacity to remove organic material and can be regenerated without marked loss of efficiency.

A laboratory study in wastewater demineralization reported scaling and fouling of the resins.[44] The principal fouling material was concluded to be detergent. Removal of the detergent by an acid–alcohol solution was required to restore the resin capacity.

The organics present in many wastewaters are high molecular weight substances which adsorb irreversibly on standard cation or anion exchangers and cannot be eluted. The search for highly porous resins not subject to these limitations has been an objective of laboratory research for many years. Several resins have been developed which are now being used commercially for the removal of organic substances similar to those present in municipal wastewaters.[45]

The removal of toxic trace elements by IX is largely unknown, but some information and speculation was presented by Abrams.[45] There are several ion exchange methods by which toxic metals can be removed from wastewaters. The least selective method is to use a column of a strong acid cation exchanger in either the acid or sodium cycle. Dissolved heavy metals are effectively removed with other cations.

Another method involves the ability of certain weakly basic resins to form complexes with the transition metals and the metals immediately following the transition elements. Moeller[46] lists the metals from which the majority of the best characterized complexes are derived as follows:

V	Cr	Mn	Fe	Co	Ni	Cu	Zn
Mo	—	Ru	Rh	Pd	Ag	Cd	
W	Re	Os	Ir	Pt	Au	Hg	

It is known, for example, that the weakly basic resins form complexes with salts of copper, cobalt, nickel, and silver even at low concentration—without the addition of acids or salts. Thus, the use of weakly basic resins for heavy metal removal has considerable advantages over the use of strong acid or strong base resins.

The resins used for this purpose are those containing the polyalkylene-polyamine groupings. Phenolic resins containing these groupings, such as Duolite A-2 and Duolite A-4 (Diamond Shamrock), have been effective. Duolite A-30B (Diamond Shamrock), and epoxy–polyamine resin removes even trace quantities of copper from aqueous solution. It may be postulated that the phenolic group of the former and the aliphatic hydroxyl group of the latter give a chelate ring closure. On the basis of laboratory tests, Duolite A-30B (Diamond Shamrock) removes about 2 lb copper/ft^3.[45]

The Sirotherm process is a unique system for the regeneration of spent resins with heat, rather than with chemicals.[47] The Sirotherm process was developed in Australia and has not been applied in the United States. There appear to be several unanswered questions about the process, and pilot scale work remains to be accomplished.

There are also several operational variations for IX units including proprietary systems patented by LA Water Treatment (CF process), Permutit (Progressive Mode process), and others. There has been some interest in layered bed units, particularly for cation exchange. Abrams and Poll anticipated that layered beds could be used advantageously in the deionization of waters in which the alkalinity is greater than about 25 percent of the total anion content.[48] It is claimed that the layered bed system offers significant advantages over conventional fixed bed concurrent cation exchange, particularly with regard to capacity delivered per unit weight of acid used for regeneration.

Abrams also believes that countercurrent operation, i.e., service in one direction and regeneration in the opposite direction (e.g., service in the downflow direction with regeneration upflow) has significant advantages over concurrent operation.[49] Countercurrent operation is principally advantageous with strong acid cation or strong base anion exchangers. In most commercial operations with concurrent systems the amount of acid required to achieve a practical regeneration level in a strong acid cation resin varies from two to five times the stoichiometric capacity. Even with these higher amounts of acids, the cation exchanger may be only 50 to 80 percent converted to the hydrogen form. Weak electrolyte resins are more easily regenerated, and countercurrent operation does not provide significant improvements over concurrent units. The advantages claimed for countercurrent over concurrent IX systems include:

1. lower leakage of ions during service,
2. lower consumption of regenerants,
3. decrease in quantity of regenerant wastes,
4. balance of regenerant wastes to give neutral effluents in deionization, and
5. lower consumption of water for rinse and backwash.

BRINE DISPOSAL

All of the methods that might be used to demineralize wastewater—RO, ED, or IX—produce a concentrated brine stream as well as the product water stream. The fraction of the feedwater which can be recovered is a function of several factors including: (1) demineralizing process utilized, (2) specific design (membrane or resin type), (3) feedwater composition, and (4) type of pretreatment. These factors are considered in a report for the Office of Saline Water.[50]

Wastewater demineralization plants will typically recover 75 to 85 percent of the feedwater resulting in a brine stream of 15–25 percent of the feed. Alternative methods of brine disposal, and the costs involved, were investigated in other studies for the Office of Saline Water.[51,52] A study was completed in May, 1970 to determine the costs for ultimate disposal of waste brines from RO, ED, and IX demineralization processes.[53]

Disposal may be accomplished by evaporation in lined ponds, underground injection, spreading on unusable arid land, ocean discharge via pipeline, stream discharge, or abandonment at the operation site. An intermediate process of evaporation to saturation or dryness may also be used in combination with the final disposal operations.

Deep well injection requires a comprehensive geologic investigation and field testing of the disposal zone and reservoir areas to determine that safe, effective underground disposal is possible, that the injected brine is compatible with fluids in the reservoir, that it cannot encroach on or pollute underground fresh water, that future natural resources will not be contaminated, and that the risk of causing unforeseen phenomena, such as earthquakes and eruptions, is minimized.

Land spreading may be feasible in certain Western U.S. locations, such as remote deserts and playas having no net annual runoff and no useful aquifers beneath, or occupying closed basins where the land has already been rendered useless for agricultural or other human purposes.

Conveyance via pipeline to the sea or to a salt lake may be feasible for those inland sites within about 150 miles of such an ultimate disposal sump.

Stream discharge is generally not feasible in the United States, and because of the local pollution problem that it introduces, it does not really qualify as an ultimate disposal method. By impoundment of brines with controlled release during periods of high flow, however, the pollution problem could be minimized.

Abandonment at the site by evaporation may be feasible in the Western United States where the annual gross evaporation rate greatly exceeds the annual rainfall. The salt residues in a lined evaporation pond remain in the pond and buildup continuously over the lifetime of the desalination plant. Alternatively, lined solar evaporation ponds may be used to preconcentrate to saturation near the desalination facility, and the concentrated brine could then be disposed of on useless land in a remote area.

Preconcentration techniques, whereby the brine is concentrated to saturation or dryness before conveyance to the ultimate disposal site, include both chemical and physical methods. The successful application of one or more of these methods in geographically feasible areas will depend entirely upon the economics of the local situation. Lime coagulation, precipitation, and ion exchange may be inexpensive chemical methods. Solar evaporation and use of multistage flash evaporators, and vertical tube evaporators are effective preconcentrating methods for concentrating waste brines to the point where they can be most economically disposed of as a liquid, a thick slurry, or a solid. For disposal as a liquid, part of the waste is evaporated, and the remaining part is transferred to the nearest nonleaching sump or ocean. For disposal as a slurry, the initial waste brine is concentrated to the point at which scaling or crystallization occurs. For disposal as a solid, the initial liquid is concentrated to saturation and then dried to solids in solar evaporation ponds.

While the costs of solar evaporation in arid areas are almost always feasible, the costs for forced evaporation by mechanical means can be expected to be high, particularly when the added costs for crystallizing and for drying to solids are considered. There are several broad categories of equipment that are commercially available:

To concentrate up to, but not exceeding, 10 percent solids, use of the conventional, single effect, multistage flash evaporator that has already been developed for seawater conversion results in a low cost per 1000 gal of water evaporated. Use of these units on wastewater instead of seawater, however, requires a reoptimization to determine minimum costs, since the existing results for seawater are not directly applicable.

For concentrations from 10 to 40 percent TDS, evaporation would have to be carried out by conventional triple or quadruple effect equipment, or by vapor compression. To prevent scaling of the heat exchange surfaces, pretreatment or cleaning methods would have to be employed and concentrations would have to be kept below saturation.

For concentrations from 40 percent to dryness, evaporation would have to be carried out by such special drying equipment as submerged combustion evaporators which are nonscaling, flash dryers, or crystallizing evaporators. Because of their low thermal efficiency and mechanical complications, these units are generally very costly to operate.

Eutectic freezing has also been suggested as a treatment method for waste brines. A bench scale investigation of eutectic freezing was conducted to determine the feasibility of the process for brine treatment.[54] It was concluded that the unit operations (freezing, separation, washing, and drying) were feasible for brine treatment. The final product of brine treatment by freezing is a slurry of salt crystals, which would then require processing to obtain dry salt.

The treatability by the activated sludge process of brine produced by an RO unit was investigated in the laboratory.[55] It was concluded that the brine resulting from treatment of municipal wastewater by RO was treatable by a conventional activated sludge system. It was also concluded that bench scale tests would have to be conducted before a design and cost of treatment by this method could be determined.

REFERENCES

1. Scheffer, S. L., "History of Desalting Operations, Maintenance, and Cost Experienced at Buckeye, Ariz.," *Journal American Water Works Assoc.*, 64, November, 1972, p. 726.

2. Moore, D. H., "Greenfield, Iowa, Reverse Osmosis Desalting Plant," *Journal American Water Works Assoc.*, 64, November, 1972, p. 781.

3. Shields, C. P., "Reverse Osmosis for Municipal Water Supply," *Water and Sewage Works*, 119, January, 1972, p. 64.

4. Hornburg, C. D., et al., "Commercial Membrane Desalting Plants Performance and Cost Experience," *Journal National Water Supply Improvement Association*, January, 1976, pp. 1–9.

5. Mattson, M. E., "Membrane Desalting Gets Big Rush," *Water and Wastes Engineering*, April, 1975, p. 35.

6. Zabban, Walter, et al., "Conversion of Coal Mine Drainage to Potable Water by Ion Exchange," *Journal American Water Works Assoc.*, November, 1972, pp. 775–780.

7. Smith, J. A., et al., "Renovation of Municipal Wastewater by Reverse Osmosis." Environmental Protection Agency, Water Pollution Control Research Series 17040, May, 1970.

8. "Pilot Wastewater Reclamation and Injection Study." Report prepared for Orange County Water District by Montgomery-Toups, December, 1967.

9. Feuerstein, D. L., "Renovation of Sewage by Reverse Osmosis." Envirogenics Company, February, 1968.

10. Dryden, F. D., "Mineral Removal by Ion Exchange, Reverse Osmosis and Electrodialysis." Proceedings of the Wastewater Reclamation and Reuse Workshop, Lake Tahoe, California, June 25–27, 1970, p. 179.

11. Cruver, J. E. and Nusbaum, I., "Application of Reverse Osmosis to Wastewater Treatment." Report by Gulf Environmental Systems Company, November 22, 1972.

12. Nusbaum, I., et al., "Reverse Osmosis—New Solutions and New Problems." Report by Gulf Environmental Systems, May 15, 1971.

13. Cruver, J. E., "Reverse Osmosis for Water Reuse." Report by Gulf Environmental Systems, June 19, 1973.

14. Cruver, J. E., et al., "Water Renovation of Municipal Effluents by Reverse Osmosis." Report for the Office of Research and Monitoring, Environmental Protection Agency, August, 1972.

15. Beckman, J. E., et al., "Control of Fouling of Reverse Osmosis Membranes When Operating on Polluted Surface Waters." Report for the Office of Saline Water, U.S. Department of Interior, February 24, 1971.

16. "Reverse Osmosis Renovation of Municipal Wastewater." Report by Aerojet-General Corporation for FWQA, Contract No. 14-12-184, December, 1969.

17. "Reverse Osmosis of Treated and Untreated Secondary Sewage Effluent." Report prepared by Eastern Municipal Water District, Hemet, California for the Office of Research and Monitoring, Environmental Protection Agency (forthcoming).

18. Feige, W. A., and Smith, J. M., "Wastewater Applications with a Tubular Reverse Osmosis Unit." Paper presented at the First AIChE Southwestern Ohio Conference on Energy and the Environment, Oxford, Ohio, October 25–26, 1973.

19. Besik, F., "Renovation of Domestic Wastewater," *Water and Sewage Works*, 120, April, 1973. p. 78.

20. Hindin, E., and Bennett, P. J., "Water Reclamation by Reverse Osmosis," *Water and Sewage Works*, 116, February, 1969, p. 66.

21. Bennett, P. J., et al., "Removal of Organic Refractories by Reverse Osmosis," in *Proceedings of 23rd Industrial Waste Conference*, Ext. Ser. 132, 1968, p. 1000.

22. Hindin, E., et al., "Organic Compounds Removed by Reverse Osmosis," *Water and Sewage Works*, 116, 1969, p. 466.

23. Lynch, M. A., Jr., and Mintz, M. S., "Membrane and Ion-Exchange Processes— A Review, *Journal American Water Works Assoc.*, 64, November, 1972, p. 711.

24. Katz, W. E., "Electrodialysis for Low TDS Waters," *Industrial Water Engineering*, June/July, 1971.

25. Smith, J. D., and Eisenmann, J. L., "Electrodialysis in Waste Water Recycle," in *Proceedings 19th Industrial Waste Conference, Purdue University*, 1964, pp. 738–760.

26. Filar, H., Jr., "Carbon Adsorption and Electrodialysis for Demineralization at Santee, California." Report for EPA (forthcoming).

27. Brunner, C. A., "Pilot-Plant Experiences in Demineralization of Secondary Effluent Using Electrodialysis," *Journal Water Pollution Control Federation*, October, 1967, pp. R1–R15.

28. Hiroi, K., "Behaviors of Some Trace Elements in the Concentration of Sea Water by Electrodialysis with Ion Exchange Membranes." Unpublished.

29. Gregg, J. C., "Nitrate Removed at Water Treatment Plant," *Civil Engineering*, 43, April, 1973, p. 45.

30. Parkhurst, J. D., et al., "Demineralization of Wastewater by Ion Exchange." Paper presented at the 5th International Water Pollution Research Conference, July–August, 1970.

31. Sanks, R. L., and Kaufman, W. J., "Partial Demineralization of Brackish Waters." Paper presented at the Sanitary Engineering Division Symposium, ASCE, Pennsylvania State University, July 28, 1965.

32. Sanks, R. L., and Kaufman, W. J., "Partial Demineralization of Saline Sewage by Ion exchange." Progress Report No. 3, Engineering Research Laboratory, School of Public Health, University of California, Berkeley, May, 1966.

33. "Ion Exchange: Field Evaluation of Desal Process." OSW, R & D Progress Report No. 631, November, 1970.

34. Pollio, F. X., and Kunin, R., "Tertiary Treatment of Municipal Sewage Effluents," *Environmental Science and Technology*, 2, January, 1968, p. 54.

35. Downing, D. G., et al., "Desal Process—Economic Ion Exchange System for Treating Brackish and Acid Mine Drainage Waters and Sewage Waste Effluents," in *Chemical Engineering Progress Symposium Series*, Vol. 64, 1968, p. 125.

36. Filar, H., Jr., "Desal Ion Exchange for Demineralization at Santee, California." Report, forthcoming.

37. Kreusch, E. and Schmidt, K., "Wastewater Demineralization by Ion Exchange." Report for EPA, Contract #14-12-599, December, 1971.

38. Schmidt, K., et al., "SUL-biSUL Ion Exchange Process: Field Evaluation on Brackish Waters." OSW Research and Development Progress Report No. 446, May, 1969.

39. Midkiff, W. S., and Weber, W. J., Jr., "Operating Characteristics of Strong-Base Anion Exchange Reactors," in *Proceedings of the 25th Industrial Waste Conference*, Purdue University Ext. Ser., Vol. 137, 1970, p. 593.

40. Eliassen, R., et al., "Ion Exchange for Reclamation of Reusable Supplies," *Journal American Water Works Assoc.,* September, 1965, pp. 1113–1122.

41. Eliassen, R., and Bennett, G. E., "Anion Exchange and Filtration Techniques for Wastewater Renovation," *Journal Water Pollution Control Federation*, 39, October, 1967, pp. R82–R91.

42. Gregory, J., and Dhond, R. V., "Wastewater Treatment by Ion Exchange," *Water Research*, 6, 1972, p. 681.

43. Frisch, N. W., and Kunin, R., "Organic Fouling of Anion-Exchange Resins," *Journal American Water Works Assoc.*, July, 1960, pp. 875–887.

44. Grantham, J., "Tertiary Effluent Demineralization," *Process Biochemistry*, January, 1970, pp. 31–38.

45. Abrams, I. M., "Ion Exchange and Resinous Adsorbents." Paper presented before the Division of Water and Waste Chemistry, American Chemical Society, Cincinnati, Ohio, January 13–18, 1963.

46. Moeller, T., *Inorganic Chemistry*. John Wiley & Sons, Inc., New York, 1952.

47. Bregman, J. I., and Shackelford, J. M., "Ion Exchange is Feasible for Desalination," *Environmental Science and Technology*, 3, April, 1969, p. 336.

48. Abrams, I. M., and Poll, R. F., "Progress in Layered-Bed Cation Exchange." Paper presented at the International Water Conference of the Engineers' Society of Western Pennsylvania, 28th Annual Meeting, Pittsburgh, Pa., December, 12, 1967.

49. Abrams, I. M., "Counter-Current Ion Exchange With Fixed Beds." 10th Annual Liberty Bell Corrosion Course, 1972.

50. "Desalting Techniques for Water Supply Quality Improvement." Report for Office of Saline Water, U. S. Department of the Interior by American Water Works Association Research Foundation, 1973.

51. Lemezis, S., et al., "Determining Feasibility of Alternative Approaches of Obtaining Solid-Dry Brine Effluent from Desalination Plants at Inland Locations." Office of Saline Water, NTIS PB 233 072, May, 1974.

52. Fluor Corporation, Ltd., "Summary of Desalination Plant Brine Disposal

Methods for Inland Locations." Office of Saline Water, NTIS, PB-233 151, February, 1971.

53. "Disposal of Brines Produced in Renovation of Municipal Wastewater." Water Pollution Control Research Series ORD-1707 ODL Y05/70, May, 1970.

54. Stepakoff, G. L., and Fraser, J. H., "Development of a Eutectic Freezing Process for Brine Disposal." Paper presented at the Industrial Water and Pollution Conference, Chicago, Illinois, March, 1973.

55. Jennett, J. C., and Patterson, C. C., "Treatability of Reverse Osmosis Raffinates by Activated Sludge," *Journal Water Pollution Control Federation*, 43, March, 1971, p. 381.

10

Land treatment of wastewaters

THE PURPOSE OF LAND TREATMENT

The basic purpose of land treatment is to recycle the nutrients and water contained in wastewaters to productive use in agricultural or silvicultural systems in such a manner as to provide simultaneously a high degree of wastewater renovation (see Figure 10-1). Land treatment is the application of effluents, usually following secondary treatment, on the land by one of several conventional methods involving irrigation or groundwater recharge. Additional treatment is provided by natural processes as the effluent moves through the natural filter provided by the soil, plants, and related ecosystem. Part of the wastewater applied is lost by evaportranspiration, while the remainder returns to the hydrologic cycle through overland flow or the groundwater system. Most of the groundwater eventually returns, directly or indirectly, to the surface water system. The treatment process, irrigation method, soil characteristics, and loading rate determines

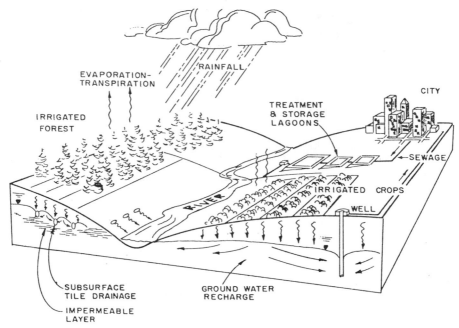

Fig. 10-1   Conceptual land application of wastewater effluents.[1]

the percentage of the flow that is lost through evaporation and transpiration, or enters the groundwater and surface streams. The degree of treatment obtained and the final destination of the water can be influenced by process, site selection, loading rate, and design.

Land application of effluent can provide moisture and nutrients necessary for crop growth. In semiarid areas, insufficient moisture for peak crop growth and limited water supplies makes water especially valuable. The primary nutrients (nitrogen, phosphorus, and potassium) are reduced only slightly in the secondary treatment process, so that most of these elements present in raw sewage are contained in secondary effluent and sludges. Soil nutrients are consumed each year by crop removal and lost by soil erosion. Fertilizer, which is highly dependent on energy input, is in short supply and prices are on the increase. Indications are that this trend will continue over the next few years. Land treatment of wastewaters can result in removal of nitrogen and phosphorus from the wastewater while simultaneously making these nutrients available to crops to reduce the need for fertilizers.

Land treatment is often considered as an alternative to advanced wastewater treatment (AWT) techniques for nitrogen removal and phosphorus removal. Land treatment systems have been used following secondary treatment even in cases where there are no requirements for AWT because of the need for supplemental irrigation water, and because of problems of excessive cost of providing treatment and outfall lines to distant points of effluent discharge into suitable surface waters.

As discussed in the next section, since July 1, 1974 EPA has required that land treatment be evaluated as an alternative method of treatment before any construction grant funds be awarded to a project. This evaluation is to be made on the basis of "application of the best practicable waste treatment technology over the life of the works," and to allow "application of technology at a later date which will provide for the reclaiming or recycling of water or otherwise eliminate the discharge of pollutants." Thus, even though immediate pollution control needs may not dictate AWT, long term planning must be based on evaluation of methods such as AWT or land treatment which will enable future water reuse.

WHAT EVALUATION IS REQUIRED BY THE EPA

The Federal Water Pollution Control Act Amendments of 1972 contain several provisions that are directly concerned with the evaluation of land treatment as an alternative waste treatment method to be evaluated for all waste treatment projects.

Sections 201 (d), (e) and (f) read as follows:

(d) The Administrator shall encourage waste treatment management which results in the construction of revenue producing facilities providing for—

1. the recycling of potential sewage pollutants through the production of agriculture, silviculture, or aquaculture products, or any combination thereof;
2. the confined and contained disposal of pollutants not recycled;
3. the reclamation of 'wastewater; and
4. the ultimate disposal of sludge in a manner that will not result in environmental hazards.

(e) The Administrator shall encourage waste treatment management which results in integrating facilities for sewage treatment and recycling with facilities to treat, dispose of, or utilize other industrial and municipal waste, including but not limited to solid waste and waste heat and thermal discharges. Such integrated facilities shall be designed and operated to produce revenues in excess of capital and operation and maintenance costs and such revenues shall be used by the designated regional management agency to aid in financing other environmental improvement programs.

(f) The Adminstrator shall encourage waste treatment management which combines 'open space' and recreational considerations with such management.

Section 201 also contains under its Subsection (g) (2) provisions that require consideration of appropriate alternate waste management techniques. They read as follows:

(g) (2) The Administrator shall not make grants from funds authorized for any fiscal year beginning after June 30, 1974, to any State, municipality, or inter-municipal or interstate agency for the erection, building, acquisition, alteration, remodeling, improvement, or extension of treatment works unless the grant applicant has satisfactorily demonstrated to the Administrator that—

A. Alternative waste management techniques have been studied and evaluated and the works proposed for grant assistance will provide for the application of the best practicable waste treatment technology over the life of the works consistent with the purposes of this title; and
B. as appropriate, the works proposed for grant assistance will take into account and allow to the extent practicable the application of technology at a later date which will provide for the reclaiming or recycling of water or otherwise eliminate the discharge of pollutants.

In commenting on this subsection, House Report No. 92-911 on the 1972 Amendments comments on pages 87–88, as follows:

The Committee believes that applicants must in the future be required to examine a much broader range of alternatives for the treatment of pollutants than they have heretofore typically done. It expects the Administrator to provide leadership and to stimulate research to assure the development and application of new treatment techniques. In arriving at the best practicable waste treatment technology consideration must be given to its full environmental impact on water, land, and air and not simply to the impact on water quality. There may be no net gain to the Nation if we adopt a technology to improve water quality without recognizing its possible adverse effect on our land and air resources. The term "best practicable waste treatment technology" covers a range of possible technologies. There are essentially three categories of alternatives available in selection of wastewater treatment and disposal techniques. These are (1) treatment and discharge to receiving waters, (2) treatment and reuse, and (3) spray-irrigation or other land disposal methods. No single treatment or disposal technique can be considered to be a panacea for all situations and selection of the best alternative can only be made after careful study.

Particular attention should be given to treatment and disposal techniques which recycle organic matter and nutrients within the ecological cycle.

In defining "best practicable waste treatment technology" for a given case, consideration must be given to new or improved treatment techniques which have been developed and are now considered to be ready for full-scale application. These include land disposal, use of pure oxygen in the activated sludge process, physical-chemical treatment as a replacement for biological treatment, phosphorus and nitrogen removal, in collection line treatment, and activated carbon adsorption for removal of organics. Planners must also give consideration, however, to future use of new techniques that are now being developed and plan facilities to adapt to new techniques.

The EPA's interpretation of the law is reflected in the following quotation from *Engineering News Record* (2/7/74):

> EPA administrator Russell E. Train said . . . "Under proper conditions, land application offers a viable alternative to advanced or tertiary treatment processes, and the technique deserves serious consideration by many communities and industries throughout the U.S."

> An EPA spokesman points out that beginning with fiscal year 1975, any community applying for a waste treatment construction grant must show that it has considered all possible alternative methods of treatment. This would include land disposal, he said.

Thus, it is clear that the evaluation of the feasibility of land treatment of wastewaters as an alternative wastewater treatment method is now an essential part of all wastewater treatment studies. EPA has published specific guidelines for such evaluations.[64]

ALTERNATIVE LAND TREATMENT SYSTEMS

Land treatment systems are in widespread use. The results of a survey in 1964 indicated that there were 2192 land disposal systems in the United States, including 1278 industrial systems and 914 municipal systems (379 surface applications and 535 subsurface septic tank systems). More recently EPA published a 1972 Municipal Wastewater Facilities Inventory in which 571 surface application systems were identified.

Land application approaches can be classified into three main groups: overland flow, irrigation, and infiltration–percolation. These approaches are illustrated in Figure 10-2 and their characteristics compared in Table 10-1.

Overland Flow

Overland flow is the controlled discharge, by spraying or other means, of effluent onto the land with a large portion of the wastewater appearing as runoff. The rate of application is measured in inches per week, and the wastewater travels in a sheet flow down the grade or slope.

Overland flow is essentially a biological treatment process in which wastewater is applied over the upper reaches of sloped terraces and allowed to flow across the vegetated surface to runoff collection ditches. Renovation is accomplished by physical, chemical, and biological means as the wastewater flows in a thin sheet down the relatively impervious slope.[4]

TABLE 10-1   Characteristics for Four Land Application Processes for Secondary Treated Wastewater[3]

	Annual Loading	Land Area Requirement for One mgd Flow	Objective	Suitable Soils	Dispersal of Applied Water	Impact on Quality of Applied Water
Overland flow	5–25 ft/yr	45–225 Ac. plus buffer areas, etc.	Maximize water treatment. Crop is incidental.	Slow permeability and/or high water table.	Most to surface runoff. Some to evapotranspiration and groundwater.	BOD and SS greatly reduced. Nutrients reduced by fixation and crop growth. TDS increases some.
Irrigation	<1 to >5 ft/yr	<225 to >1,100 Ac. plus buffer areas, etc.	Maximize agricultural production.	Suitable for irrigated agriculture.	Most to evapotranspiration. Some to groundwater; little or no runoff.	BOD and SS removed. Most nutrients consumed in crop or fixed. TDS greatly increased.
High rate irrigation	1 to >10 ft/yr	<110 to 1,100 Ac. plus buffer areas, etc.	Maximize water and treatment by evapotranspiration and percolation with crop production as a side benefit.	More permeable soils suitable for irrigated agriculture; may use soils marginal because of coarse texture.	Evapotranspiration and groundwater; little or no runoff.	BOD and SS mostly removed. Nutrients reduced. TDS substantially increased.
Infiltration-recharge	11 to 500 ft/yr	2 to 100 AC. plus buffer areas, etc.	Recharge water or filter water; crop may be grown with little or no benefit.	Highly permeable sands and gravels.	To groundwater some evapotranspiration; no runoff.	BOD and SS reduced. Little change in TDS.

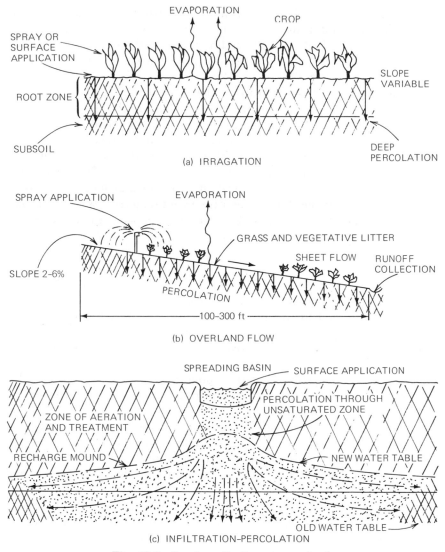

Fig. 10-2 Land application approaches.[2]

Soils suited to overland flow are clays and clay loams with limited drainability. The land for an overland flow treatment site should have a moderate slope—between 2 and 6 percent. The surface should be evenly graded with essentially no mounds or depressions. The smooth grading and ground slope make possible sheet flow of water over the ground without ponding or stagnation. Grass is usually planted to provide a habitat for biota and to prevent erosion. As the effluent flows down the slope, a portion infiltrates into the soil, a small amount evaporates, and the remainder flows to collection channels. As the effluent flows through the grass, the suspended solids are filtered out and the organic matter is oxidized by the bacteria living in the vegetative litter.

The overland flow treatment process has been used in the U.S. primarily for treatment of high strength wastewater, such as that from canneries. In Australia overland flow or grass filtration has been used for municipal waste treatment for many years. Overland flow can be used as a secondary treatment process where discharge of a nitrified effluent low in BOD is acceptable or as an advanced wastewater treatment process. The latter objective will allow higher rates of application (5 in./week or more), depending on the degree of advanced wastewater treatment required.[4] Where a surface discharge is prohibited, runoff can be recycled or applied to the land in irrigation or infiltration–percolation systems.

Irrigation

Irrigation is the controlled discharge of effluent, by spraying or surface spreading, onto land to support plant growth. The wastewater is "lost" to plant uptake, to air by evapotranspiration, and to groundwater by percolation. Application rates are measured either in inches per day or week, or in gallons per acre per day. The method of application depends upon the soil, the type of crop, the climate, and the topography. Sloping land is acceptable for irrigation provided that application rates are modified to prevent excessive erosion and runoff.

Renovation of the wastewater occurs generally after passage through the first 2–4 ft of soil. The use of irrigation as a treatment and disposal technique has been developed for municipal wastewater and a variety of industrial wastewaters. Crops grown have ranged from vegetables to grasses and cereals.

As noted in Table 10-1, irrigation may be practiced so as to maximize crop production and nutrient removal ("irrigation"), or it may be practiced so as to maximize the amount of water disposed of per acre while sacrificing the optimization of crop production and nutrient removal ("high rate irrigation"). The high rate approach is often not compatible with local farming practices because the crop becomes secondary in importance to the wastewater disposal.

There are a number of different ways to apply wastewater to the land in irrigation systems. Each site will have its own physical characteristics that will influence the choice of the method of application. The three that are most commonly used are spraying, ridge and furrow, and flooding. Each of these methods is illustrated on Figure 10-3.

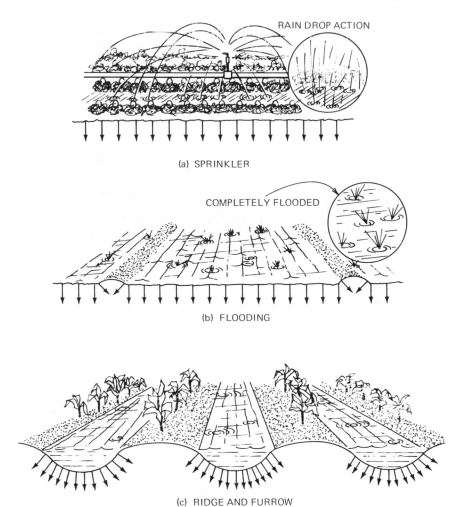

RAIN DROP ACTION

(a) SPRINKLER

COMPLETELY FLOODED

(b) FLOODING

(c) RIDGE AND FURROW

Fig. 10-3   Basic methods of application.[2]

In the spraying method, effluent is applied above the ground surface in a way similar to rainfall. The spray is developed by the flow of effluent under pressure through nozzles or sprinkler heads. By adjusting the pressure and nozzle aperture size, the rate of discharge can be varied. The elements of a spray system are the pump or source of pressure, a supply main, laterals, risers, and nozzles or sprinkler heads. Since the system operates under pressure, there is a wide variety of ground configurations suitable for this type of disposal. The spray system can be portable or permanent, moving or stationary. (See Figures 10-4 and 10-5). The capital cost of a spray system is relatively high because of pump and piping costs and pump operating

Fig. 10-4   Moving center pivot spray irrigation equipment. (*Courtesy Lockwood Company*)

Fig. 10-5   Fixed, solid set spray irrigation equipment.

costs. The effluent used in a spray disposal system cannot have solids that are large enough to plug the nozzles. Sprinkling is the most efficient method of irrigation with respect to uniform distribution.

The ridge and furrow method is accomplished by gravity flow. The effluent flows in the furrows and seeps into the ground. Ground for this type of operation must be relatively flat. The ground is groomed into alternating ridges and furrows, the width and depth varying with the amount of effluent to be disposed and the type of soil. (Figure 10-6). The rate of infiltration into the ground will control the amount of effluent used. If crops are to be irrigated with effluent, the width of the ridge where the crop is planted will vary with the type of crop. The furrows must be allowed to dry out after application of sewage effluent so that the soil pores do not become clogged.

The third type of application is flooding. Flooding, as the term implies, is the inundation of the land with a certain depth of effluent. The depth is determined by the choice of vegetation and the type of soil. The land has to be level or nearly level so that a uniform depth can be maintained. The land does need "drying out" so that soil clogging does not occur. The type of crop grown has to be able to withstand the periodic flooding. The flooding technique is not as widely applicable as the spray or surface techniques.

The irrigation method used depends on many factors, including soil

Fig. 10-6   Ridge and furrow irrigation.

characteristics, crop, labor requirements, maintenance, topography, costs, water supply, and control required. Each method has distinct advantages for a specific set of existing conditions. Surface irrigation depends on soil permeability and uniformity for uniform distribution of water. For sprinkler irrigation, the distribution of water is controlled by the selection and design of the equipment used. If soil conditions are suitable, surface irrigation normally offers the advantages of savings in power and hardware requirements. The degree of complexity needed to meet the requirements of the regulatory agencies and the public should be considered in the selection of the irrigation method.

Important considerations in irrigation economics are automation of the irrigation system and labor requirements. Until recently, equipment was not available for automation of surface irrigation systems. New automation equipment makes surface irrigation comparable with sprinkler equipment for application of wastewater effluents.

Infiltration–Percolation

This method of treatment is similar to intermittent sand filtration in that application rates are measured in feet per week or gallons per day per square foot. The major portion of the wastewater enters the groundwater, although there is some loss to evaporation. The spreading basins are generally dosed on an intermittent basis to maintain high infiltration rates. Soils are usually coarse textured sands, loamy sands, or sandy loams.

This process has been developed for groundwater recharge of municipal effluents, municipal wastewater disposal, and industrial wastewater treatment and disposal. The distinction between treatment and disposal for this process is quite fine. Wastewater applied to the land for the purpose of disposal is also undergoing treatment by infiltration and percolation, whether or not monitoring for detection of renovation is being practiced.

The infiltration–percolation approach is primarily a groundwater recharge system and does not attempt to recycle the nutrients through crops and is not comparable to the irrigation techniques in this regard. Where groundwater quality is being degraded by salinity intrusion, groundwater recharge can reverse the hydraulic gradient and protect the existing groundwater. Where existing groundwater quality is not compatible with expected renovated quality, or where existing water rights control the discharge location, a return of renovated water to surface water can be designed using pumped withdrawal, underdrains, or natural drainage. At Phoenix, Arizona, for example, the native groundwater quality is poor, and the renovated water is to be withdrawn by pumping with discharge into an irrigation canal.[4]

REVIEW OF EXISTING MAJOR LAND TREATMENT PROJECTS

It is the purpose of this section to review some existing, major land treatment projects. These projects will illustrate actual installations of all of the various types of systems described in the previous section. More extensive details on these and other projects may be found in references 2 and 5-9. Data from existing projects also discussed in the subsequent section, "Major Factors Requiring Evaluation." A comprehensive review of available literature has also been published.[11]

EPA-APWA Study

The American Public Works Association has published a field survey (funded by EPA) of about 100 facilities using land application of wastewaters.[5] Also, data were gathered from many other existing installations in the U.S. and in foreign countries. They found the spray irrigation technique to be the predominant form of irrigation technique in existing practice as well as in recently proposed large land application projects. They found that 90 percent of the communities using land treatment plan to continue its use, indicating that this approach is acceptable to most of its current users, which were found to be predominantly small communities. The choice of land treatment was found to be based on such factors as: need for supplemental irrigation water; augmentation of groundwater resources; simplicity and economy of providing required degrees of treatment; problems of excessive cost of providing treatment and outfall lines to distant points of effluent discharge into suitable surface water sources; and merely "to get rid of the wastewater" in a convenient, trouble free manner.

The following points are borne out by the report: Existing practice stresses land application of treated effluents, not raw wastewaters; the percentage of land application acreage frequently represents only a portion of the land reserved by the owners for their systems; application periods may vary from one month to twelve months a year, and from one to seven days a week, depending on climatic conditions, need for land application for excess flows, seasonal industrial processing, such as in the food industry, and other local factors; land values are relatively low, zoned for either agriculture or residential uses, often in undeveloped areas, and subject to minimal or no degradation of value due to use for irrigation purposes; all types of soil are utilized, with sand, clay and silt most favored; groundwater interference problems influence choice of sites, and when appropriate sites are selected, cause minimal difficulties with land application methods.

Use of the irrigated land varies with the owners' needs and dictates, from

Fig. 10-7   Surface application of wastewater. (*Courtesy U.S. E.P.A.*)

no ground cover, to grass cover, to cultivated crops, to forested areas; grass is the most common ground cover in community systems. It is evident that the cropping value of supplemental irrigation with wastewaters and their nutrient components is not universally utilized.

Rates of application of wastewater effluents to the land, and duration of uninterrupted application vary from 0.1 in. per day to more than 1 in. per day, with varying periods of irrigation and resting. The most commonly used application rate is 2 in./week, but the report cautions that the suitable application rate must be carefully tailored to the specific site.

Relatively little need was found for providing special environmental protection measures in land application areas. Rather, such facilities were often considered to enhance the environment. Security provisions are not universally used to protect against intrusion of outsiders. Fencing and patrolling are not universally practiced. Buffer zones to isolate land application areas and impede dispersal of aerosol sprays are used, but no common practice is in effect. Monitoring of groundwater, surface water sources, soils, crops, animals, and insects is practiced in some locations and minimally used in others, often dependent solely on the requirements of health authorities.

Specific findings of this survey related to design considerations will receive further discussion in the subsequent section, "Major Factors Requiring Evaluation." The rest of this section will be devoted to discussions of existing major land treatment projects.

Muskegon County, Michigan

At the time of this writing, the Muskegon County project is the largest spray irrigation project in operation in the United States.[8,10,12-15] The project is a countywide system treating municipal and industrial effluents from throughout the county. It replaced three existing municipal plants and four industrial discharges. The design capacity is 43.4 mgd, with the wastewater made up of domestic sewage and industrial pulp and paper mill effluents. Land area is 230 acres/mgd of capacity for disposal. The 10,000 acre site is composed of treatment lagoons, storage lagoons, roads, buffer areas, and approximately 5400 acres of irrigated area (Figure 10-8). The system calls for secondary treatment in the aerated lagoon system (3 day detention), storage during winter months (in two 850 acre lagoons), application on the land with a center pivot sprinkler irrigation system, a subsurface drainage system, and control and monitoring of all surface runoff and subsurface drainage. The design loading rate is 2.5 million gal/acre/year or 7.7 ft/year over the irrigated area. The topography at Muskegon is quite flat and the soils very sandy. The irrigation equipment includes 54 irrigation rigs of the center pivot type, with each covering 40–160 acres. The subsurface drainage system is conprised primarily of drainage pipes which discharge to drainage ditches, with a small portion of the site drained by wells or by natural drainage.

Construction of the project began in 1971 (Figure 10-9). Pumping of wastewater to the site began in May of 1973. Because the irrigation machines were not yet ready, no irrigation took place in 1973. By the spring of 1974, over one year's wastewater was biologically treated and stored in the storage lagoons. Irrigation began in late May of 1974. (Figures 10-10, 10-11, 10-12). At that time, only 20 of the 54 irrigation rigs were available. Wastewater was irrigated wherever rigs were available. During this first year a wet

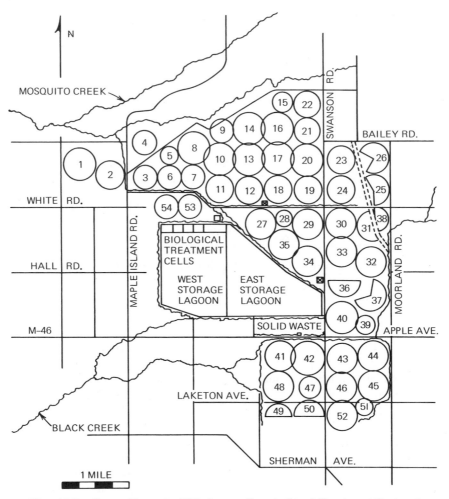

Fig. 10-8   General layout of Muskegon County Land Treatment System.[8]

spring and maximum applied water caused corn crop damages. Under these adverse conditions in 1974, there was a crop yield of approximately 28 bushels/acre. Yields from the 1975 irrigation season were much improved and averaged about 60 bushels/acre, with yields in some portions of 100 bushels/acre.[10]

The treatment performance data for 1974 and 1975 show that the Muskegon County Wastewater System treated wastewater very effectively. Table 10-2 summarizes the data. The treated wastewater quality changed substantially in the storage lagoons. In 1974 in the west storage lagoon where the

Fig. 10-9   The Muskegon County System under construction. (*Courtesy Muskegon County Wastewater Management System*)

wastewater had been in storage and impounded for over a year, the water quality changed dramatically. BOD was reduced to 5 mg/l. Oxygen levels were at saturation. The pH level was stabilized. Nitrogen levels were reduced. Bacteriological data show reductions below 200 organisms/100 ml for fecal coliform.

The drain tile water quality analysis reflects the effluent quality after passage through the soil. The trend of the physical parameters indicates a reduction in the specific conductivity, pH, color, turbidity, etc. The bacteriological results show a mixed pattern and no trends are indicated. There is a reduction in Na^+ and C^- concentrations by a factor of 50 percent. Sulfate and iron concentrations are either the same or increased by the high iron and sulfate content in the farm site soil. There was a substantial reduction in phosphate levels from 5 ppm to 0.07 ppm or less. The nitrogen picture is very difficult to evaluate at this time because the leachate in instances contained more nitrogen than the irrigation water. This could be attributed

Fig. 10-10   Aerial view of completed Muskegon County System. Aerated lagoons and storage lagoons are in the center of the picture. (*Courtesy Muskegon County Wastewater Management System*)

Fig. 10-11   Aerated and outlet lagoons at Muskegon. (*Courtesy Muskegon County Wastewater Management System*)

Fig. 10-12   Crops and center pivot equipment at Muskegon. (*Courtesy Muskegon County Wastewater Management System*)

to the nitrogen levels in the soil from prior years, and the heavy application rate of effluent throughout the first season causing excessive amounts of nitrogen to leach out. There is about a 50 percent reduction in the nitrogen concentration levels from the irrigated water by the soil and crop uptake. The water quality in the receiving streams show the same general trend as in the drain tiles, but most concentrations are somewhat lower because of groundwater dilution of the drain tile water. The quality of final discharges from the project relative to the NPDES permit requirements were as follows:

	System Design	30-Day NPDES Limit	1974 Discharge	1975 Discharge
BOD_5	4 mg/l	4 mg/l	3.7	2.7 mg/l
Suspended solids	4 mg/l	10 mg/l	8	7–31 mg/l
Total P	0.5 mg/l	0.5 mg/l	0.009	0.02 mg/l
Ammonium-N	0.5 mg/l	—	0.7	0.6 mg/l
Nitrate-N	5.0 mg/l	—	1.3	1.44 mg/l
Fecal coliform	0	200/100 ml	238	$1–5 \times 10^3$ $1–5 \times 10^2$

TABLE 10-2 Summary of Treatment Performance, Muskegon, Michigan.

Average Results

Parameter	Influent 1974	Influent 1975	Storage Lagoon East 1974	East 1975	West 1974	West 1975	Drain Tiles 1974	Drain Tiles 1975	Mosquito Creek 1974	Mosquito Creek 1975	Black Creek 1974	Black Creek 1975
BOD, mg/l	220	205	20	12	5	13	2.2	1.2	2	3.3	2	2.3
DO, mg/l	0	0	3	6	8.5	5.4	2–9	—	9.5	6.2	1.6	3.4
Temp., °C	24	23.5	1–26	1–27	1–26	1–27	—	—	1–5	9.9	12	9.7
pH	7.5	7.3	7.6	7.9	8.2	7.7	7	—	7.2	7.5	6.8	7.0
Sp. Cond. μmhos	1300	1049	1200	872	750	777	600	599	750	574	800	670
DS, mg/l	1050	1093	750	728	550	654	—	—	375	466	700	691
MLVSS, mg/l	500	460	300	228	200	222	—	—	160	172	160	205
SS, mg/l	325	249	20	23	10	18	—	—	10	7	30	31
COD, mg/l	550	545	140	131	70	104	—	—	30	33	25	23
TOC, mg/l	140	107	30	43	20	32	5	11.6	10	15	10	11
NH$_4^+$, mg/l	9.0	6.1	2.5	2.5	0.2	2.3	0.40	0.29	0.45	0.6	0.5	0.5
Total Kjeldahl-N, mg/l	—	8.2	—	4.6	—	4.5	—	—	—	—	—	—
NO$_3^-$/NO$_2^-$, mg/l	0.0	—	2.5	1.3	0.8	0.8	2.8	2.2	1.9	1.9	1.4	1.0
PO$_4^{3-}$, mg/l	6.5	4.9	5	3.9	0.7	3.0	0.05	0.04	0.1	0.09	0.05	0.06
SO$_4^{2-}$, mg/l	85	75	95	101	70	78	140	142	80	81	320	284
Cl$^-$, mg/l	175	182	160	169	90	138	50	60	60	78	18	32
Na, mg/l	150	166	145	163	85	125	40	42	40	66	7	21
Ca, mg/l	70	73	65	60	60	56	70	72	60	61	110	107
Mg, mg/l	16	14	16	16	16	16	25	23	20	18	40	38
K, mg/l	11	11	11	10	6	7.5	2.8	2.6	5	4	2.5	3
Fe, mg/l	1.25	0.8	1.0	1.2	0.7	0.9	4.0	7.7	0.08	1	0.4	17
Zn, mg/l	0.9	0.6	0.25	0.12	0.15	0.09	0.06	0.01	0.01	0.07	0.2	0.11
Mn, mg/l	0.25	0.28	0.25	0.22	0.08	0.09	0.15	0.20	0.08	0.11	0.4	0.38
Color, units	100	—				—	20–150	2–750	130	—		
Total coliform (colonies/100 ml), 1974					0–1.3×10^5		10–1000		40–1.5×10^4			
1975						100–1.2×10^8		<1–170		<1–9.6×10^4		
Fecal coliform (colonies/100 ml), 1974					0–2400		0–440		1–1500			
1975						4–1.2×10^6		<1–32		<1–4800		
Fecal strep (colonies/100 ml), 1974					0–2300		2–700		7–5500			
1975						2–3800		<1–47		—		

The high suspended solids of 31 mg/l occurred in the southern portions of the project, where large amounts of iron were leached from the soil.

During the 1974 season, the first year of full farm operation, approximately 4500 acres were planted to corn. Air planters were used to plant all acreage. Plant populations ranged from 13,500 to 21,600 plants/acre. Some commercial nitrogen fertilizer was applied by broadcast methods prior to planting and by dissolution in irrigation water intermittently throughout the growing season. Irrigation began the last week in May and ceased around

TABLE 10-3   County of Muskegon—Wastewater System, Schedule of Construction Costs Through 1974[8]

Description	Costs Incurred
Land acquisition	$ 5,371,614.17
Relocation, dislocation, moving expense	1,165,660.70
Clearing, paving, and project fencing	1,915,027.14
6″, 8″, 10″ Interior drainage pipe—perforated plastic (6″, 8″) and concrete drain tile (10″)	2,636,155.10
Ditches, channels, and related structures	1,285,889.98
Irrigation pressure and well discharge piping, electrical distribution system	2,000,502.05
Administration building, chlorine building, and related structures	350,661.03
Observation wells and drainage wells	385,538.50
Force mains and sewer lines	7,881,434.75
Pumping stations, lift stations, main station, and irrigation, etc.	3,495,116.35
Lagoons, treatment facilities, and related structures	9,965,305.50
Irrigation spray rigs and pads	1,353,372.30
Utility relocation	295,468.37
Eggleston Township access point	137,173.95
Engineering feasibility study, project engineer, operations manual, and legal and fiscal	2,266,342.08
Capitalized interest	880,947.67
Machinery and equipment	595,969.05
Laboratory equipment	
Farm machinery and equipment	
Office furnishings and equipment	
Radio equipment	
Wastewater treatment equipment	
Miscellaneous construction costs	84,302.80
Inspection related	
Pipes, pump station and ditches	
Mechanical electrical	
Buildings	
Total	$42,066,561.56

TABLE 10-4   1975 Actual and 1976 Budget Operation and Maintenance Costs, Muskegon, Michigan

	1975 Actual	1976 Budget
O & M expenses, less capital outlay	$1,822,497	$2,406,996
Revenues		
Crop	706,109	703,000
Grants	207,794	128,134
Laboratory services	7,687	18,866
Miscellaneous	967	0
Subtotal, revenues	$ 922,557	$ 850,000
Net expenses	899,940	1,556,996
Net expense/gallon $\times 10^6$		
26.79 mgd	92.04	
29 mgd		146.69 (+59 percent)

the 20th of November. A soil insecticide was used in addition to broadleaf and grass herbicides. Harvest began during the last week of September and continued until the middle of November. Approximately 5000 acres were planted to corn in 1975 and yields increased to 60 bushels/acre.

The cost of the Muskegon land treatment system was $42,066,561.56. Table 10-3 shows a detailed breakdown of construction costs. The cost per acre of the total project is $42,066,561.56 divided by 10,800 acres = $3,895.05/ acre. Table 10-4 summarizes the actual 1975 costs and budgeted 1976 costs for operation and maintenance of the system. The net expenses are budgeted to be 73 percent greater in 1976 than actually experienced in 1975 or, recognizing the higher projected flow rate, 59 percent more per gallon treated. The chief factor in increased costs was a 50 percent increase in the electrical rates. The composition of the 1976 budgeted costs may be summarized as follows:

	(%)
Salaries and fringes	29.3
Electricity	34.9
Other utilities	0.9
Consumable farm supplies	9.9
Miscellaneous supplies	9.9
Outside services (custom harvest, equipt. repair, etc.)	6.0
Miscellaneous	1.1
Administrative and contingency	8.0
	100.0

Melbourne, Australia[16-18]

The Werribee Board of Works sewage farm was established in 1892 in an area adjacent to Melbourne. The farm, which once was a barren, windswept plain, is now one of the showplaces of Australia. The 26,809 acre farm is composed of 10,376 acres of land filtration, 3472 acres of grass filtration, and 3393 acres of lagoons, with the remaining 9568 acres devoted to irrigation and drainage canals, roads, residences, and dryland pasture. The farm produces cattle. During a portion of the year sheep are grazed.

The farm was established over a period of several years. Extensive preparations were required before irrigation was attempted. The soil of the original farm area is described as having a dense clay layer about 12 in. thick, 9–12 in. below the surface. It was necessary to break up the soil to a depth of at least 2 ft. The soil surface was accurately graded to provide for uniform distribution of sewage by surface irrigation. Finally an effective drainage system was provided. By the end of 1908, there were 3282 acres under irrigation. As the city has grown, so has the farm. New land was purchased and prepared for the border check method of irrigation (water is run over the soil surface in a wide, flat sheet). A bank is provided at the lower end of the field to prevent runoff of raw sewage into the drain.

Nearly all soil irrigated is alluvial delta soil of the ancient Werribee River. Some of the new area added was basaltic clay with a thin capping of loam and very impermeable. These soils are used mostly for dry pasture. More recently exposed alluvium along the lower reaches of the river and the bay are used for grass filtration and lagoons.

The treatment procedure is as follows:

Land filtration   for the period of high evaporation between September and April.

Grass filtration   for the period of low evaporation between May and August.

Lagooning   for peak daily flows and wet weather flows.

Both the land filtration and grass filtration treatment processes are accomplished by the border method of surface irrigation.

During late spring, summer, and autumn most of the flow is applied in the land filtration system where approximately two-thirds of the water is consumed by evaporation. The irrigations average approximately 4 in. and the interval between irrigations averages about 18 days. During 1970–71, approximately 9020 million gal were applied to the land filtration area, which is approximately 2.7 ft/year over the area. One third of the water applied in land filtration is collected by the subsurface drainage system and transported to Port Phillip Bay.

Since irrigation of pastures is not practicable during the winter and early spring, an alternative method was developed for treatment of sewage. The grass filtration system utilizes a sedimentation tank to remove the settleable materials. The effluent then flows continuously over borders of dense vegetation, where suspended matter is filtered out and oxygen demanding material is oxidized. This effluent then flows directly into collecting drains. 12,886 million gal were treated by the grass filtration system in 1970–71, which is about 11.4 ft/year over the area involved.

Distribution channels and the irrigation system are designed to handle a maximum of 76 mgd. All flows in excess are overflowed to the lagoon treatment system. On an average day, peak flows may reach 150 mgd. Thus, 76 mgd would be distributed for irrigation and 74 mgd would be overflowed to the lagoons. The quantity treated by lagoons during 1970–71 was 21,049 million gallons, which was approximately 49 percent of total flow.

The discharge standard of 10 ppm for BOD and suspended solids from lagoons and drains to Port Phillip Bay is the same as recommended for effluents discharged to inland streams in Great Britain. Frequent and regular monitoring insures that these standards are maintained.

The average rainfall is approximately 20 in. and the evaporation is approximately 45 in., which results in a net deficit in moisture of 25 in. This amounts to a possible consumption of about 30 percent of the total yearly flow on the area irrigated or in ditches. Normally, 15,000 head of cattle are on the farm and annual sales are about 5000 head. In addition, 35,000–50,000 sheep are fattened each summer.

Table 10-5 summarizes the removal efficiencies for the three treatment methods used at Melbourne. Soil analyses[18] showed that irrigation over 48–73 years resulted in increased soil contents of N, organic C, and P. As C increased, cation exchange capacity was found to increase 1.4 meq/percent C. Soil C:N ratios decreased with sewage irrigation while Ca:Mg ratios in-

TABLE 10-5   Estimated Performance by Treatment Processes on an Annual Basis at Melbourne, Australia.[17]

Characteristics	Method of Treatment		
	Crop Irrigation	Overland Flow	Lagoon System
Percent of total flow treated	20	30	50
Percent removal			
BOD	98	96	94
Suspended solids	97	95	87
Total nitrogen	95	60	40
Total phosphorous	95	35	30
Detergent	80	50	30
E. Coli	98	99.5	99.8

creased, indicating greater accumulations of N and Ca relative to C and Mg, respectively. Soil K content was found to remain unchanged or declined with sewage irrigation. With some exceptions, total soil content of all trace elements included in the soil analysis were increased by sewage irrigation.

Sewage irrigation of pastureland was found to have resulted in increased or unchanged forage contents of all the chemical elements determined, except Mn. This one element was decreased, but not to the extent that it would be considered to be at a deficiency level in grasses and legumes. In terms of animal health, there were no indications that forage from irrigated sites contain trace elements in either excessive or deficient amounts.[18]

Flushing Meadow, Phoenix, Arizona[19-26]

In the Phoenix area, one-third of the agricultural water comes from groundwater. The remaining two-thirds of the irrigation water and the municipal supplies are obtained from surface reservoirs on the Salt and Verde Rivers. In recent years the groundwater table in the Phoenix area has been dropping 10 ft/year. In 1971, the water table dropped 20 ft. The depth to groundwater varies from 400 ft in the Mesa area to 50 ft near the Salt River in the Phoenix area.

In 1967, the Flushing Meadows Project was begun. The objectives were to study the treatment of secondary effluent by rapid infiltration and determine infiltration rates. Specifically, the removal of BOD, suspended solids, nitrogen, fluoride, and pathogenic organisms was important. It was desired to obtain renovated water of a quality sufficiently high to permit unrestricted irrigation.

A site was located west of Phoenix within the floodplain of the Salt River. The 2 acre site was divided into six basins that are 20 ft wide by 700 ft long. Secondary effluent is pumped into these parallel basins. Usually 1 ft of water is held in the basins by an overflow structure at their lower end. The infiltration rate for the 2 acre system is approximately 0.5 mgd. Most of the soil in the basins is fine, loamy sand to a depth of about 3 ft. Then coarse sand and gravel layers extend to about 250 ft, where a clay layer forms the lower boundary of the aquifer. The static water table is at a depth of about 10 ft. Observation wells for sampling groundwater and renovated sewage water were installed in a line midway across the basin area. Infiltration rates of 1 ft/day or 300 ft/year are regularly achieved by flooding for 14 days and resting 10–20 days. During the two weeks of inundation (surcharge is about 1 ft), the infiltration rate drops from 2.5 ft/day to 1.5 ft/day with an average of 2 ft/day. During the summer, 10 days are sufficient for drying, re-aeration, and biological oxidation, which restore the infiltration capacity, but winter operation requires 20 days. If the suspended solids content of the effluent can be kept below 20 mg/l, little or no sludge accumulates on the bottom of the basins, and they can be operated for several years without having to be cleaned. On the other hand, annual or more frequent removal of accumulated solids is necessary if the effluent has more than 20 mg/l of suspended solids.[23] The permeability of the soil using well water is 4 ft/day. Removals have been: BOD, fecal coliform, and suspended solids—essentially complete; phosphorus and fluoride—70 percent; nitrogen—30 percent; boron, lead, and cadmium—essentially zero; total salts—no change. These data indicate that such a system would produce renovated water of more than sufficient quality to permit unrestricted irrigation and recreation.

For this type of infiltration–percolation system, it was found that non-vegetated basins can yield similar or even higher infiltration rates than vegetated basins. Based on a hydraulic loading of 300 ft/year, one acre of infiltration basins can handle 300 acre-ft/year, or about 0.27 mgd. A series of wells in the center of the riverbed are used to pump the infiltrated wastewater to prevent the renovated water from moving into aquifers outside the water reclamation system (Figure 10-13). This project is an excellent illustration of the infiltration–percolation approach. The total cost of the spreading operation and to pump the renovated water out of the Salt River bed was estimated at about $5.00/acre-ft, or about $15.00/million

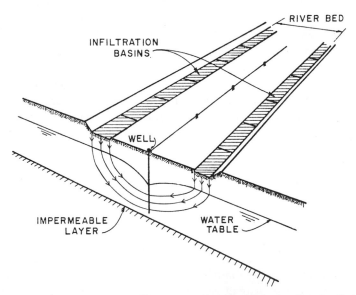

Fig. 10-13 System of infiltration basins on both sides of river bed and wells in center for pumping renovated water at Phoenix, Arizona.[21]

gal in May, 1973.[9] This is much less than the cost of equivalent tertiary treatment at a conventional sewage plant.

Pennsylvania State University[9, 27-30]

This research project began in 1962 and consists of irrigation of farm and forest lands with 0.5 mgd of secondary effluent. Based upon the success of the pilot program, the project is now being expanded to handle the entire flow of 4 mgd from the treatment plant serving the University and much of the borough of State College.

Spray irrigation has been successful in disposing of secondary effluent, increasing reed canary grass hay yields and increasing wood growth in mixed hardwoods and other tree stands. The groundwater quality has not been adversely affected; effects on wildlife, birds, and insect life are now being quantitatively evaluated.

In all, approximately 80 acres of land, both crop and forest areas, have been spray irrigated. The split between the crop and forest areas has been varied; now it is about 50 percent to each such area.

Soil conditions are variable. In the Gameland Areas, the soil is a deep, sandy loam, with 20–160 ft of residual overlay above recurring beds of sandstone, quartzite, and dolomite. In the Agronomy–Forestry Area, soil overlay ranges from 5–80 ft in depth, with a dolomite bedrock; the soil is somewhat less permeable, of more clay–loam character. The forest mulch cover was not disturbed and undercover weeds and grass are not cut in the nonfarm areas. The mulch improves the ability of the area to absorb winter spraying. Groundwater tables are deep—from 100 to 300 ft below the spray areas, which are in upland areas.

The efficiency of the land treatment process has been monitored by several monitoring wells to groundwater level and by several clusters of lysimeter stations to varying depths to determine the character of the capillary water in the soil.

Phosphorus removals are excellent in both the farming and forest areas, with the applied phosphorus concentration of 4–9 mg/l being generally reduced to less than 0.1 mg/l in the first 2 ft of soil. Nitrogen removals are poorer and more variable than phosphorus removals. Applied nitrogen concentration varied from 8 to 18 mg/l over the 7 year period of 1963–1970. In corn rotation plots, nitrate-nitrogen concentrations at 4 ft depths varied from 3.4 to 13.5 mg/l. In reed canary grass plots, receiving 2 in. of wastewater weekly, the nitrate-nitrogen concentrations varied from 2.4 to 3.7 mg/l at 4 ft. In the forested areas, concentrations in 1969–1970 varied from 2.3 to 42.8 mg/l, depending on the vegatative cover. Under red pine, less nitrogen was assimilated in annual growth than under the herbaceous annuals and perennials. Substantial increases in sodium and chloride were observed. Boron, manganese, and potassium were reduced substantially. The nitrogen loading was the most critical parameter in terms of meeting drinking water standards.

At Penn State the perennial grasses were the most suitable crops for lands receiving wastewater effluent because of several factors. They have fibrous root systems and form sod that helps control erosion and still allows a high rate of infiltration. The grasses are also tolerant of a wide range of ecological conditions. They have a high uptake of nutrients over a long period of growth. In 6 years 2127 lb of nitrogen were applied to the reed canary grass in 536 in. of sewage effluent and sludge. Of this, 2071 lb were removed in the harvested crop, a 93 percent renovation efficiency. During the same period of time, 797 lb of phosphorus were applied in the wastewater, and 279 lb were removed when the crop was harvested. The overall crop renovation efficiency was 35 percent for phosphorus.

Hardwood forests were found to be not as efficient as agronomic crops in removing the nutrients. For example, a corn silage crop removed 145 percent of the nitrogen applied in the sewage effluent. In contrast, the trees removed only 39 percent, and most of it was returned to the soil in falling leaves. The silage corn crop also removed 143 percent of the phosphorus from the sewage effluent, while the hardwoods removed only 19 percent.[28,29]

Tallahassee, Florida[9, 31-35]

Tallahassee's two treatment plants, the Lake Bradford Plant (4.5 mgd) and the Dale Mabry Plant (0.5 mgd), were placed in operation during the 1940s. Their effluent was discharged to a natural drainage stream which flows into Lake Munson. Since this stream also receives most of the storm runoff water from the city, during the past thirty years Lake Munson has become heavily laden with silt and shows the typical signs of accelerating eutrophication. In 1961, the city placed in operation a 60,000 gpd high rate trickling filter plant to serve the municipal airport. Over a 6 month period during 1961–1962, field experiments at this plant demonstrated that the effluent could be satisfactorily disposed of on land by irrigating at the rate of 4 in./day over 8 hours.

When Tallahassee's two plants reached their planned capacity in 1965, a new southwest wastewater treatment plant (high rate trickling filter) was constructed near the airport, where soil and groundwater conditions were similar to the experimental irrigation plot. In order to protect Lake Munson, the City is in the process of developing an alternative 850 acre effluent disposal site 1.5 miles north of the lake where land irrigation of the entire combined flow of 11 mgd has been shown to be feasible. The system has

been operating continuously since the initial flow of 0.25 mgd began in
the summer of 1966. Daily flows were gradually increased to 1 mgd by the
summer of 1969. Plant effluent BOD and suspended solids averaged
15–20 ppm during this period.

The soil is mostly Lakeland fine quartz sand, with a depth to water table
and limestone aquifer of approximately 50 ft. This soil typically has an
infiltration capacity of 3–4 in./hour. Usually it has 1–2 percent organic
matter and less than 5 percent clay. The low natural fertility of the soil is
reflected in the native vegetation. Besides scrub oak and similar plants,
slash pine grow rather slowly in the area. Furthermore, the soil has a poor
moisture holding capacity, with an available water content of about 1
in./ft of soil. Sieve analyses on samples collected 1 ft below ground surface
show an average effective size of 0.15 mm and a uniformity coefficient
of 2.3. These characteristics provide an almost unlimited hydraulic absorption capacity. This assures that flooding and the attendant runoff will
not be a problem.

The irrigation system is composed of 6–8 in. aluminum main lines and
2 in. aluminum lateral lines. The sprinklers are spaced on 100 ft centers,
and each one delivers 45 gpm at an application rate of 0.45 in./hour. The
piping system is valved so that sixteen sprinklers can be operated at one
time to apply effluent at the rate of approximately 1 mgd to 4 acres. Altogether
16 acres are under irrigation in four 4 acre tracts. One or a combination
of the four plots can be sprinkled at any time. It was soon established that
only 8 acres was necessary for the 1 mgd flow, except to observe how grasses
responded to municipal wastewater irrigation. To determine which grasses
responded best, the four plots were seeded with Pensacola Bahia, Argentina
Bahia, Centipede, and mixed wild grasses.

In the spring of 1971 four big sprinkler guns were installed on 400 ft
centers in a rectangular plot to irrigate with flow bypassed from the Lake
Bradford plant. Each gun delivers 1060 gpm in a 555 ft circle. Each pair
of sprinklers applied 2 million gal daily.

Under the initial experimental design, the aluminum farm irrigation
pipe was laid on the surface so that it could be rerouted with a minimum of
effort. The aluminum pipe has proven not to be completely satisfactory because the exposed lines have been bent, broken, or corroded by external
mechanical damage and internal wear from abrasion. Therefore, underground cast iron pipe will be used when the irrigation field is expanded
Alternative pumps are also being installed to eliminate downtime for pump
repair unless both pumps in a pair fail simultaneously. Sprinkler heads
have been satisfactorily protected from stoppages by placing a self cleaning
traveling screen with $\frac{1}{4}$ in. openings in front of the pumps to remove suspended debris.

Selected chemical and biological parameters are monitored throughout the system. At 300,000 gal/acre/day loading rates, the concentration of orthophosphate dropped from 25 mg/l at the surface to 0.04 mg/l at a depth of 10 ft. Most of the ammonium-nitrogen was converted to nitrate-nitrogen in the upper 24 in. of soil. There was no clear evidence of denitrification or that the plants absorbed any appreciable amount of nitrate nitrogen as the effluent percolated downward through the soil. While density of fecal coliform bacteria in the influent was normally in the range of $10^5-10^6/100$ ml. the density in water from monitoring wells was usually zero or occasionally one or two bacteria per 100 ml.

According to available data,[31,32] a crop rotation schedule of coastal bermuda grass in the summer and rye grass in the winter can potentially remove about 600 lb of nitrogen per acre per year. Assuming an irrigation rate of 3 in./week and 25 mg/l nitrogen, about 900 lb of nitrogen would be applied to the soil. The combination would have a 67 percent recovery efficiency. This crop rotation would leave about 7.5 mg/l of residual nitrogen. Some of this remains in the root system of the plants and is released when the roots decay. Carbon is also released during microbial decomposition. It appears likely that some carbon moves down to the water table where it can be metabolized by denitrifying bacteria to convert nitrate ion to nitrogen gas, which then escapes. If 10 percent of the original nitrogen were lost by this process, then approximately 2.5 mg/l would be denitrified. Under these conditions, no more than 5 mg/l nitrate-nitrogen would remain in groundwater. The degree of denitrification at Tallahassee is not yet known.

Other Projects

Experiences at many other projects or general conclusions drawn from projects are reviewed in available literature[5,11, 36-49] and are drawn upon in the following sections of this chapter, which deal with specific design factors.

MAJOR FACTORS REQUIRING EVALUATION

Pretreatment Requirements

The APWA–EPA study[5] found that virtually all U.S. municipal land treatment installations utilize secondary treatment prior to land application. Most also use chlorination prior to land application. The use of secondary treatment and disinfection minimizes health hazards and odors in the land treatment system. Although many existing municipal systems use conventional secondary plants, the Muskegon County, Michigan system provides secondary treatment in a 3 day aerated lagoon followed by settling lagoons. Also, in some cases, oxidation pond effluent has been irrigated.

Treatment of wastewater prior to land application may be necessary for a variety of reasons, including (1) maintaining a reliable distribution system, (2) allowing storage of wastewater without nuisance conditions, (3) maintaining high infiltration rates into the soil, or (4) allowing the irrigation of crops that will be used for human consumption.[4] In the following paragraphs, minimum preapplication treatment levels are described for irrigation, infiltration–percolation, and overland flow, and wastewater quality is briefly discussed.

Irrigation. Reduction of BOD and suspended solids in municipal wastewaters is not generally necessary from the standpoint of loadings on the soil. Reductions may be necessary, however, where clogging of the distribution system may occur, where disinfection is required, or where considerable storage is required. The bacteriological quality of municipal wastewater is usually limiting where food crops or landscape areas (parks, golf courses, etc.) are to be irrigated or where aerosol generation by sprinkling is anticipated. Generally, the state public health agencies place limitations on the quality of municipal wastewater that can be used for irrigation. In California, for example, biological treatment plus disinfection to a level of 23 total coliform organisms per 100 ml is required for irrigation of golf courses, parks, freeway landscapes, and pastures grazed by milking animals. For direct irrigation of food crops, California requires an oxidized, coagulated, filtered wastewater disinfected to a level of 2.2 total coliform organisms per 100 ml. Primary effluent is acceptable for surface irrigation of orchards, vineyards, and fodder, fiber, and seed crops in California.

Infiltration–Percolation. Reduction of suspended solids is the most important preapplication treatment criterion for infiltration–percolation systems, so that soil clogging and nuisance conditions from odors are minimized. Biological treatment is often provided for this purpose.

Overland Flow. This process is often used in place of conventional secondary processes[50]. When used as a secondary treatment process, the minimum preapplication treatment is screening and possibly grit and grease removal to avoid clogging the distribution system. No food crops are grown, and sprinkling systems can be designed to minimize the generation of mists by using rotating boom sprays or by sprinkling at low pressures. In a pilot system at Ada, Oklahoma, raw comminuted wastewater, which was settled for 10 minutes for grease and grit removal, was applied successfully using a 30 ft diameter rotating boom with sprays discharging slightly downward at 15 psi.[50]

When overland flow is used as an advanced wastewater treatment process, lagoon or conventional secondary effluent can be used as feed. Nitrified effluent can be treated successfully for nitrogen removal. Disinfection prior to application may avoid postdisinfection and allow sprinkling at higher pressures.

Inorganic Loading Limitations

Suggested values for major inorganic constituents in water applied to the land are shown in Table 10-6. In arid portions of the country, total dissolved solids may present a hazard in irrigating certain crops. Crops vary in their tolerance to salinity and boron.[52,53]

Sodium can be toxic to crops; however, the effects of sodium on permeability usually occur first. The sodium adsorption ratio should be maintained below 9 to prevent deflocculation of the soil structure or sealing of the soil.[2] The sodium adsorption ratio is of special concern when the soil has a high clay content, and it can be reduced by increasing the wastewater concentrations of calcium and magnesium through the addition of gypsum or other amendments.[52] The effects of other wastewater constituents, such as heavy metals, are discussed later in this chapter, as is removal of the two key algal nutrients, nitrogen and phosphorus.

Phosphorus and Nitrogen Removal

Phosphorus is removed by adsorption on the cation exchange complex, by precipitation, and by sorption with iron and aluminum oxides. The removal of phosphorus is therefore dependent on the soil textures, the cation exchange capacity, the amount of iron and aluminum oxides, and the uptake of phosphorus by the crop. Because of these removal mechanisms, little movement of phosphorus through the soil system with the drainage water occurs in most soils. Phosphorus concentrations in groundwater and from subsurface drainage systems are seldom over 0.2 mg/l and seldom as low as 0.01 mg/l. Common concentrations of phosphorus in the water which has moved through the soil are expected to be in the range of 0.01–0.1 mg/l, with midpoints of this range most common. In acid soils, Al and Fe phosphates are precipitated, while in soils of higher pH, Ca phosphates predominate. In cases where the soil is nearly all sand, excellent hydraulics can result but phosphorus removal suffers and phosphorus may actually leak readily. However, this is the exception. Most soils will remove virtually all of the phosphorus in the upper 1–2 ft and will have a tremendous capacity for phosphorus removal. Phosphorus removal capacity is finite, and over a long time period or with high loadings the phosphorus additions may exceed the soil's capacity for removal.

TABLE 10-6   Suggested Values for Major Inorganic Constituents in Water Applied to the Land.[51]

Problem and Related Constituent	No Problem	Increasing Problems	Severe
Salinity[a]			
EC of irrigation of water, mmho/cm	<0.75	0.75–3.0	>3.0
Permeability			
EC of irrigation water, in mmho/cm	>0.5	<0.5	<0.2
SAR (sodium adsorption ratio)[b]	<6.0	6.0–9.0	>9.0
Specific ion toxicity[c]			
From root absorption			
Sodium (evaluate by SAR)	<3	3.0–9.0	>9.0
Chloride, meq/l	<4	4.0–10	>10
Chloride, mg/l	<142	142–355	>355
Boron, mg/l	<0.5	0.5–2.0	2.0–10.0
From foliar absorption[d] (sprinklers)			
Sodium, meq/l	<3.0	>3.0	—
Sodium, mg/l	<69	>69	—
Chloride, meq/l	>3.0	>3.0	—
Chloride, mg/l	<106	>106	—
Miscellaneous[e]			
NH_4^+-N			
mg/l for sensitive crops	<5	5–30	>30
NO_3^--N			
HCO_3^-, meq/l only with over-	<1.5	1.5–8.5	>8.5
HCO_3^-, mg/l head sprinklers	<90	90–520	>520
pH	Normal range =	6.5–8.4	—

[a]Assumes water for crop plus needed water for leaching requirement (μ) will be applied. Crops vary in tolerance to salinity. mmho/cm $\times$ 640 = approximate total dissolved solids (TDS) in mg/l or ppm; mmho $\times$ 1000 = μmho.
[b]SAR = $Na/\sqrt{(Ca + Mg)/2}$, where Na = sodium, meq/l, Ca = calcium, Mg = magnesium.
[c]Most tree crops and woody ornamentals are sensitive to sodium and chloride (use values shown). Most annual crops are not sensitive.
[d]Leaf areas wet by sprinklers (rotating heads) may show a leaf burn due to sodium or chloride absorption under low humidity, high evaporation conditions. (Evaporation increases ion concentration in water films on leaves between rotations of sprinkler heads.)
[e]Excess N may affect production or quality of certain crops, e.g., sugar beets, citrus, grapes, avocados, apricots, etc. (1 mg/l NO_3^--N = 2.72 lb N/acre-ft of applied water.) HCO_3^- with overhead sprinkler irrigation may cause a white carbonate deposit to form on fruit and leaves.

Note: Interpretations are based on possible effects of constituents on crops and/or soils. Suggested values are flexible and should be modified when warranted by local experience or special conditions of crop, soil, and method of irrigation.

Phosphorus removal by adsorption can be estimated by Langmuir adsorption isotherms.[54] The adsorption capacity has been estimated for several soils with a range of 77 to over 900 lb/acre-ft of soil profile.[54] It appears[54] that the adsorption capacity of the soil for phosphorus can be restored in a few months. Apparently this saturation and restoration cycle can occur many times. Restoration occurs when adsorbed phosphorus is precipitated or removed by crops.

For many soils the total removal capacity is large and would greatly exceed the planning life of the land application project. For coarse textured soils with little calcium, iron, or aluminum, the removal capacity may be limited. If phosphorus removal is critical under these conditions, a soil chemist should be consulted regarding the storage capacity for phosphorus. A detailed discussion of phosphorus adsorption isotherms and methods for estimating the concentration in percolate is available in the literature.[55]

Phosphorus accumulating in soil over a long period of time may potentially interfere with crop growth.[1] This interference generally would occur as a nutrient imbalance in the plant; that is, high phosphorus levels in the soil may reduce the availability of some crop micronutrients. Thus, the plant may develop low levels of other required nutrients rather than toxic levels of phosphorus. Precipitation of the phosphorus would be expected to minimize any toxicity problem; otherwise, corrective measures may be required. In cases where nitrogen removal is required, nitrogen loadings are likely to be the limiting factor in determining the area requirements and suitable crops.

Nitrogen contained in wastewater applied to the land may be in any of four forms: organic, ammonium, nitrate, and nitrite. Nitrite-nitrogen is easily oxidized to nitrate in the presence of oxygen so that concentrations above 1.0 mg/l for nitrite are rare. All nitrogen, regardless of form when applied, is mineralized to the nitrate form unless lost by volatilization. Nitrogen in the nitrate form is subject to movement with the water through the root zone and thus may not be retained by the soil. The ammonia and organic forms of nitrogen are retained in the soil by adsorption and ion exchange until mineralized to the nitrate form or removed by crop uptake. Nitrification, mineralization, and denitrification are biological processes that can occur in soil, as shown in Figure 10-14.

Generally, organic and ammonium nitrogen are the principal forms applied to land. Organic nitrogen, being suspended instead of dissolved, is filtered out in the soil matrix and mineralized into ammonium nitrogen. Ammonium exists in equilibrium with ammonia gas; at a pH between 7.5 and 8.0 about 10 percent of the nitrogen will be in the gaseous ammonia form. Volatilization of significant quantities of ammonia requires not only a high pH but also considerable air–water contact. Therefore, this mechanism is not expected to provide significant nitrogen removal in land application systems. Nitrate-nitrogen is not retained in soil by adsorption or ion exchange, but instead leaches readily with applied water. Nitrate-nitrogen can be removed only by growing and removing from the area a crop which takes up nitrogen, or by denitrification. The conditions necessary for significant denitrification do not normally occur in soil systems used

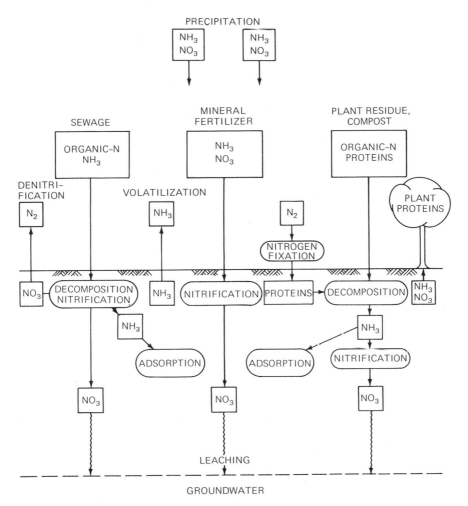

Fig. 10-14   The nitrogen cycle in soil.[2]

for wastewater application in irrigation systems. The denitrification rate is highest for stratified soils with saturated or near saturated conditions at the bottom of the root zone. The least denitrification occurs in uniform soils with moderate to rapid permeabilities. Because the root zone must be well aerated for most crops, significant anaerobic areas will not exist and high rates of denitrification are not likely to occur in the root zone. Only nitrates which move below the root zone will be subject to a significant amount of denitrification. Denitrification has been shown to vary from 10 percent to 80 percent of the nitrate-nitrogen moving below the root zone.

An elaborate system to create anaerobic conditions and supply a carbon energy source for the denitrifying bacteria is required to accomplish predictable high rates of denitrification in soil systems. Denitrification is dominant and crop uptake negligible in the infiltration–percolation process under proper wet–dry cycles. Both crop uptake and denitrification are significant nitrogen removal mechanisms for the overland flow process method.

Nitrogen removal by crops is dependent on the length of growing season and crop type as well as nitrogen availability. Crop requirements for nitrogen during the growing season approximately parallel the evapotranspiration demand. Thus, applications paralleling the seasonal changes in evapotranspiration may be more beneficial to the crop than a constant application.

Crops normally used with land application can be divided into three broad groups and removal rates. A forage crop will remove 150–600 lb/acre or more, field crops will remove 75–150 lb/acre, and forests will remove 20–100 lb/acre, as shown in Table 10-7. Wide variation occurs within each group, and some crops do not fit this generalization.

Some crops show luxurious uptake of nutrients with total removal being as much as twice the noted values. Nitrogen removal is generally dependent on the amount applied. However, removal efficiency (nitrogen removed by crop divided by nitrogen applied times 100) decreases as the loading increases.[1] Removal efficiencies of 84 and 68 percent were reported for reed canary grass with 421 and 524 lb/acre, respectively, of total nitrogen applied.[56] Figures 10-15 and 10-16 show nitrogen removal and removal efficiency for corn grain and corn silage. Removal efficiency may be greater than 100 percent when loadings are low because of release of nitrogen stored in the soil. As nitrogen removal efficiency decreases, nitrate-nitrogen concentrations in the percolate may increase significantly, as shown on Figure 10-17 for a corn crop. However, the increase in concentration will depend on the crop grown. For example, the nitrate-nitrogen concentration in percolate at a 4 ft depth beneath reed canary grass averaged 3.1 mg/l for 4 years and showed no trend of increasing as the total nitrogen applied increased.[56] Crops vary widely in efficiency for nitrogen removal and total amount removed depending on both the crop species and nitrogen loading. Where strict limits are placed on the nitrogen concentration of percolate from land application, a crop with high removal efficiency may be required together with controlled loadings.

High levels of nitrogen can interfere with productivity in some crops. Sugar beets, for example, will not obtain high sugar levels if excess nitrogen is available late in the season. Flavor of grapes as well as their sugar content and pH, factors controlling wine quality are dependent in part on nitrogen availability.

TABLE 10-7   Reported Nutrient Removal
by Forage Crops, Field
Crops, and Forest Crops.[1]

	Nitrogen uptake (lb/acre/year)
Forage Crops	
Costal bermuda grass	480–600
Reed canary grass	226–359
Fescue	275
Alfalfa	155–220[a]
Sweet clover	158[a]
Red clover	77–126[a]
Lespedeza hay	130
Field Crops	
Corn	155
Soy beans	94–113[a]
Irish potatoes	108
Cotton	66–100
Milo maize	81
Wheat	50–76
Sweet potatoes	75
Sugar beets	73
Barley	63
Oats	53
Forest Crops	
Young deciduous (up to 5 years)	100[b]
Young evergreen (up to 5 years)	60[b]
Medium and mature deciduous	30–50[b]
Medium and mature evergreen	20–30[b]

[a]Legumes remove substantial nitrogen requirements from the air.
[b]Estimated.

There is little hazard of nitrogen toxicity of crops where typical municipal effluent is applied to land. When loadings are low, the total quantities applied are usually comparable to normal fertilizer applications. When loadings are high, the nitrogen is leached through the soil profile and thus should be little problem to the plant.

A potential toxicity problem occurs when applications continue through the winter when nitrogen is not nitrified.[1] The organic nitrogen and ammonia–nitrogen are stored in the soil during winter. When the weather warms, nitrification occurs over a short time and concentrations may reach high

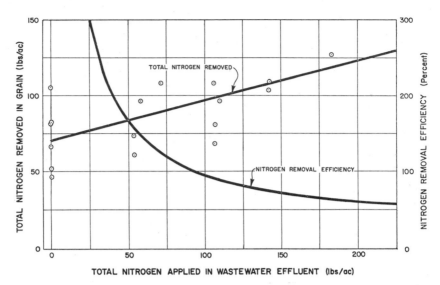

Fig. 10-15   Total nitrogen removed and removal efficiency for grain corn.[1]

levels. With excess nitrogen available some crops, particularly forages, can accumulate high levels of nitrate, which may be toxic to livestock and people. The potential nitrate toxicity from crops would normally occur only during early spring, probably before the first harvest. The potential

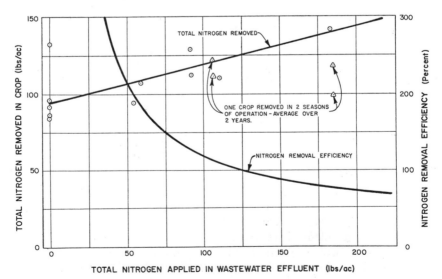

Fig. 10-16   Total nitrogen removed and removal efficiency for corn silage.[1]

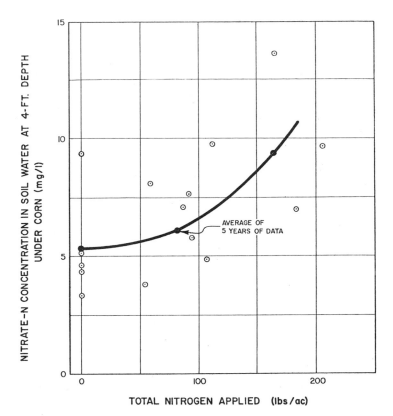

Fig. 10-17   Nitrate-nitrogen concentration of percolate beneath a corn crop vs. total nitrogen applied in wastewater.[1]

hazard should not persist long because leaching through the soil should reduce soil nitrate levels. However, pollution of the groundwater may then be a problem. Monitoring of vegetation can detect potential toxicity problems before they cause damage.

Under long term operation (steady state conditions), the removal of nitrogen by the crop and denitrification will control the nitrogen concentration in the percolate. Initially, higher rates of removal may be obtained with an associated accumulation of organic nitrogen (humus) in the soil. After a few years, the rate of accumulation will decline and will reach a steady state condition with little or no additional accumulation.

Typical steady state nitrogen balance estimates for two effluent application rates and two crop groups are shown in Table 10-8 to illustrate a sample nitrogen budget. With steady state conditions, the wastewater nitrogen additions of 1300, 300, and 135 lb/acre produce 10, 5, and 5 mg/l nitrogen concentrations, respectively, in the percolate water as indicated. The balances in Table 10-8 are based on an assumed removal by crops and denitrification. Increasing the loadings above those shown will probably increase the nitrogen concentration and the pounds of nitrogen in the percolate, unless the crop is changed. Requirements for a lower nitrogen concentration in the percolate will necessitate reductions in the loading rate. Thus, nitrogen limitations for discharge may control the loading rate.

TABLE 10-8   Approximate Nitrogen Balance for Two Land Application Processes Under Steady State Conditions.[1] (Values in lbs/acre except as noted.)

	Infiltration–percolation	High rate irrigation
Forage Crop		
Application		
Effluent (acre-ft/acre)	24	6.9
Nitrogen in effluent	1300	300
Nitrogen added in precipitation		
and fixation less losses to		
ammonia volatilization, etc.	0	0
Removed by crop	300	200
Leached below root zone	1000	100
Lost to denitrification (35 and 25 percent)	350	25
Returned to groundwater		
Treated effluent (acre-ft/acre)	24	5.6
Nitrogen in treated effluent	650	75
Concentration in return flows (mg/l)	10	5
Cultivated Field Crop		
Application		
Effluent (acre-ft/acre)		3.1
Nitrogen in effluent		135
Nitrogen added in precipitation		
and fixation less losses to		
ammonia volatilization, etc.		0
Removed by crop		100
Leached below root zone		35
Lost to denitrification (25 percent)		9
Returned to groundwater		
Treated effluent (acre-ft/acre)		1.8
Nitrogen in treated effluent		26
Concentration in return flows (mg/l)		5

Except for legume plants, nitrogen additions to the root zone from rainfall and fixation (atmospheric nitrogen made available by bacteria living in the soil) are approximately equal to the incidental losses from volatilization of ammonia and other removal mechanisms. Bacteria living on the roots of legume plants fix (convert to an available form) large quantities of nitrogen from the atmosphere. Indications are that with effluent application, where high nitrogen loads are available, this fixation process is greatly reduced.

If there is no nitrogen removal requirement, then other factors—probably hydraulic loading limitations—will determine the area requirements.

Loading Rates

The permissable loading rate of wastewater is affected by conditions of the soil, climate, and crop. The liquid loading rate must be adjusted to the crop use and the infiltration rate of the soil so that ponding does not occur. A hydraulic application of 1–5 ft/year has been considered the normal range for the classification of systems under the irrigation approach, 1–10 ft/year for high rate irrigation, 5–25 ft/year for overland flow, and 10–500 ft/year for infiltration–percolation. For each of these processes, the effluent loading rate may be limited by any of several constraints. The process of water moving into and through the soil depends on the soil's capacity for infiltration and percolation. Thus, these soil properties can restrict loading rates. Limits placed on composition of deep percolation to groundwater or of return flow to surface streams may also result in a constraint on the loading rate or require higher levels of pretreatment. Another possible loading rate constraint is the finite capacity of the soil–crop system to remove various pollutants. The following parameters may be limiting for a situation where nitrogen loading is not the controlling factor:

Infiltration capacity of the soil.

Permeability of the root zone and underlying geologic materials.

Soil capacity to remove and assimilate inorganic chemicals (heavy metals, specific ions, salts, etc.).

Discharge requirements to groundwater and surface water.

Climatic influences such as precipitation, evapotranspiration, and growing season.

Infiltration. The infiltration capacity of a soil limits the rate at which water can be applied without runoff. Steeper slopes, previous erosion, and lack of a dense vegetative cover will reduce the infiltration capacity and necessitate a corresponding reduction in application rates. The rate of infiltration is dependent on soil properties, crop cover, and slope. The higher the initial soil moisture content, the lower the infiltration capacity, and infiltration rates decrease over time as the water application continues. A rest period from application allows the soil to dry and restores the soil's infiltration capacity. Also, infiltration may be only vertical (downward), as with most irrigation designs, or it may be both vertical and lateral, as with ridge and furrow irrigation. The infiltration rate limits the rate of application (i.e., in./hour), but only in rare instances or in infiltration–percolation systems will it limit the total seasonal application (i.e., ft/year). Once the water has infiltrated the surface of the soil, it must move through the root zone to the groundwater. This movement depends on vertical permeability and the lateral movement of groundwater. The permeabilities in these two directions may be quite different for some soils. The permeability of the soil determines the rate of drainage and amount of aeration that occurs between the applications. The permeability of the soil in a vertical direction will determine the total precipitation and effluent application that can be transmitted to the groundwater. In a wet year, the amount of effluent which can be applied will be reduced.

General ranges of sprinkler application rates for various soil types are given in Table 10-9. The infiltration rates in Table 10-9 are usually adequate for pre-

TABLE 10-9   Typical Ranges of Infiltration Rates and Available Soil Moisture by Soil Type.[1]

Soil Type	Available Soil Moisture Storage,[a] in./ft.		Good Condition Base Soil Basic Infiltration Rates,[b] in./hr.		
	Range	Avg.	0–3%	3–9%	9+%
Very coarse textured sands and fine sands	0.50–1.00	0.75	1.0–	0.7+	0.5+
Coarse textured loamy sands and loamy fine sands	0.75–1.25	1.00	0.7–1.5	0.50–1.00	0.40–0.70
Moderately coarse textured sandy loams and fine sandy loams	1.25–1.75	1.50	0.5–1.0	0.40–0.70	0.30–0.50
Medium textured, very fine sandy loams, loams, and silt loams	1.50–2.30	2.00	0.3–0.7	0.20–0.50	0.15–0.30
Moderately fine textured sandy clay loams and silty clay loams	1.75–2.50	2.20	0.2–0.4	0.15–0.25	0.10–0.15
Fine textured sandy clays, silty clays and clays	1.60–2.50	2.30	0.1–0.2	0.10–0.15	<0.10

[a]Storage between field capacity (1/10 to 1/3 atm) and wilting point (15 atm).
[b]For good vegetative cover these rates may increase by 25–50 percent. For poor surface soil conditions these rates may decrease by as much as 50 percent.

liminary planning in the absence of field measurements. However, in unusual circumstances or more detailed planning, it may be necessary to obtain specific field data on infiltration. When detailed information is required, cylinder, furrow, or sprinkler infiltration tests can be made. Because of inherent differences between these test methods, it is necessary to use the test appropriate to the irrigation method to be used. Prior to final design, detailed field investigations should be made.

Permeabilities of Soil and Geologic Materials.   When water has infiltrated the soil, movement through the root zone to the groundwater depends on vertical permeability, and the movement of groundwater depends on lateral permeability. The permeability in these two directions may be quite different for some soils. The permeability of the soil in a vertical direction will determine the total excess water that can percolate through the soil to the groundwater. The permeability of the soil and surface geology in a horizontal direction will determine the extent of the groundwater mound beneath a land application system.

Aeration of the root zone is important to most agricultural crops and thus is required in a well operated irrigation or high rate irrigation process. Most agricultural crops show adverse effects from inadequate oxygen at air contents less than 10 percent of soil volume. Virtually no oxygen diffusion occurs during and following the water application until the soil has drained. Drainage requires a few hours for coarse textured soil to several days for fine textured soil.

Soil permeability is the major factor determining the time required for the soil to re-aerate after water is applied. Soil with low permeability requires a much longer rest period to allow aeration of the soil profile. The rest periods must be long enough to allow CO_2 to escape and O_2 to diffuse into the soil profile. The length of rest time necessary is also dependent primarily on the soil profile, climate, crop growth, and biological activity.

For the irrigation and high rate irrigation processes at least 3–4 ft of aerated soil is needed to provide sufficient soil material to treat the applied effluent and allow crop growth. The layer above the water table saturated by capillary action also restricts aeration. Thus, the water table should be at least 5 ft below the soil surface for the irrigation process.

The overland flow process, which utilizes a daily loading and soils of low permeability, will produce anaerobic soil conditions most of the time. The anaerobic soil will result in a lower permeability and require selection of a plant tolerant to wet anaerobic soil conditions.

The infiltration–percolation process will develop an anaerobic condition a few days after loading is started. This anaerobic condition will promote high rates of denitrification in the soil. During the rest period, the soil will drain and aerobic conditions will be restored. If a crop is grown, it should be tolerant to standing water and periodic anaerobic soil conditions.

Soil permeabilities are less for wastewater than for clear water. The average permeability for wastewater at Flushing Meadows has been about one-half that for clear water. Permeabilities for wastewater under other conditions have been reported as low as 10 percent of the clear water value. Effluent should be used for field measurements of permeability if at all possible. If effluent cannot be used, the design should be based on a reduction factor which would vary with water quality and soil conditions. Once the permeability is known, the excess water and effluent loading can be established.

Soil Classifications.   Deep soils that have a large total surface area and contain much clay and organic matter adsorb and filter wastes more effectively than shallow, sandy soils. But soil must also have the ability to accept the wastes. Liquid wastes must be able to infiltrate and percolate the soil. This requirement is best met by relatively coarse, sandy soil that, unfortunately, has only limited capacity to adsorb and filter. Hence, the requirements for soil as a medium for waste disposal are conflicting, and even the best site, found after a long search, will be a compromise. Soil surveys prepared by the Soil Conservation Service of the United States Department of Agriculture in cooperation with agricultural experiment stations and units of local government can be of help in this search.

Soil drainability depends primarily on the mechanical properties of texture and structure. These properties are largely influenced by the percentages of the three mechanical classes of soil—sand, silt, and clay. Coarse sand particles are from 0.25 mm to 0.05 mm; silt particles, from 0.05 mm to 0.005 mm; and particles smaller than 0.005 mm are clay.

Clay soils do not drain well. Soils with a relatively high content of clay are fine textured and often described as heavy. They retain large percentages of water for long periods of time. As a result, crop management is difficult but not impossible. These soils expand or swell with moisture increases, and at such times the soil structure is susceptible to being destroyed by compaction or cultivation. When clay soils dry there is often considerable shrinkage and the ground becomes cracked and very hard.

Sandy soils, in contrast, do not retain moisture very long, which is important for crops that cannot withstand prolonged submergence or saturated root zones. Soils are considered to be well drained if the infiltration rate (entrance velocity of water into soil) is 2 in./day or more as measured in full scale tests. The drainability should be determined for a large area, not for a localized test pit.

Drainage also depends upon the absence of lateral and subsurface constraints to the flow of water. An example of lateral constraint would be a sandy soil in a narrow valley with impermeable clay or rock on all sides. The lateral transmissibility and percolation rate must be equal to or higher than the infiltration rate to avoid ponding and waterlogging of soil.

A guide (see Table 10-10) for assessing the suitability of kinds of soil as media for disposal of wastes has been prepared. Six soil properties—permeability of the most limiting subsoil layer, infiltration rate, soil drainage, runoff, flooding, and available water holding capacity—are used to rate kinds of soil as having slight, moderate, or severe limitations for use for waste treatment. Most of these criteria are self explanatory, except for available water holding capacity. This criterion, measured in inches for

TABLE 10-10   Soil Limitations for Accepting Nontoxic Biodegradable Liquid Waste.[7]

| Item[b] | Degree of Soil Limitation | | |
	Slight	Moderate	Severe
Permeability of the most restricting subsoil horizon to 60 in.	Moderately rapid and moderate 0.6–6.0 in./hr	Rapid and moderately slow 6–20 and 0.2–0.6 in./hr	Very rapid, slow, and very slow >20 and <0.2 in./hr
Infiltration	Very rapid, rapid, moderately rapid, and moderate >0.6 in./hr	Moderately slow 0.2–0.6 in./hr	Slow and very slow <0.2 in./hr
Soil drainage	Well drained and moderately well drained	Somewhat excessively drained and somewhat poorly drained	Excessively drained, poorly drained, and very poorly drained
Runoff	None, very slow, and slow	Medium	Rapid and very rapid
Flooding	Soil not flooded during any part of the year	Soil flooded only during nongrowing season	Soil flooded during growing season
Available water capacity to 60 in. or to a limiting layer T[c] >7.8 in.		3–7.8 in.	<3 in.
P[d] >3 in.			<3 in.

[a]Modified from a draft guide for use in the Soil Conservation Service, USDA.
[b]For definitions see the *Soil Survey Manual*, U.S. Department of Agriculture Handbook No. 18, 1951.
[c]Temporary Installation.
[d]Permanent Installation.

the whole soil, is the depth of the layer of water that would be formed if all water in the soil that can be used by plants were concentrated at the soil surface. The equivalent volume measure would be the acre-inch (27,000 gal) or the acre-foot (326,000 gal). Available water holding capacity used in this way integrates the effects of soil texture and depth and is a convenient measure for effective soil volume. It should be noted that in this guide severe limitations reflect limiting capacity of the soil to accept wastes and limited capability of the soil to render wastes harmless to the environment. Rapid and slow permeability, for example, are both considered severe limitations.

Although soil drainage cannot be predicted from soil type alone, a general indication of soil type vs. ranges of liquid loading rates is shown in Figure 10-18. It should be noted that the ranges in Figure 10-18 denote typical practice and not recommended loadings. Loamy soils are preferred for irrigation systems; however, most soils from sandy to clay loams are acceptable. Infiltration–percolation systems require well drained soils, such as sands, sandy loams, loamy sands, and gravels. A depth to existing groundwater of 10–15 ft is preferred.[2] Lesser depths may be acceptable where underdrainage is provided. Soils with limited drainability, such as clays and clay loams, are best suited for overland flow systems. Soil depth should be sufficient to form the necessary slopes and maintain a vegetative cover.

In addition to physical factors, the balance of chemical constituents in the soil is important to plant growth and wastewater renovation. Factors such as salinity, alkalinity, and nutrient level of the soil should be determined prior to planting and should be monitored during irrigation to determine the rate and extent of any buildup. Some of the indicators of adverse soil conditions are pH, conductivity, and SAR (sodium adsorption ratio). Most crops grow best in a soil with a neutral or slightly acid pH. Both highly acid and alkaline conditions can produce sterile soil. Additions of calcium sulfate (gypsum) will aid acid soils. The salinity or TDS of the soil is commonly measured as electrical conductivity. In arid regions salts will accumulate in nearly all soils unless leaching is done. According to the University of California Committee on Irrigation Water Quality Standards, there is a definite hazard to permeability from using water having a SAR of 8 or more on certain soils. The adverse factors of high salinity, pH, and SAR may occur in the same soil producing a saline–alkaline soil. Saline soils are those with conductivities of saturation extracts greater than 4,000 μmhos/cm. It has been found that the conductivity of the saturation extract of a soil, in the absence of salt accumulation from groundwater, usually ranges from 2 to 10 times as high as the conductivity of the applied irrigation water.

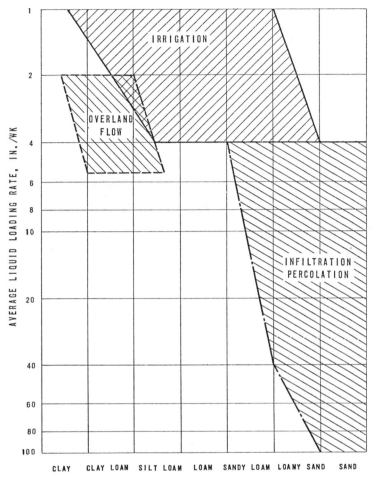

Fig. 10-18 Soil type vs. liquid loading rates for different land application approaches.[2]

When effluent is applied to a land disposal area, changes in the soil system may result. Unfortunately, these changes are difficult to predict quantitatively, but some qualitative comments may be made. The hydraulic properties of medium to fine texture soils decrease when moist. Under continual infiltration the rate of infiltration drops. Moist subsoils can have permeabilities that are a fraction of permeability they would have when partially dried by drainage and by root activity. These changes are due to the hydration and swelling of the natural peds under continually moist conditions, which reduces the large pores. If organic matter is present, oxidation–reduction potential can drop and cause further structural deterioration. Upon drying the peds shrink, the structure is stabilized, and the infiltration rates and permeabilities recover. Land treatment systems therefore will have to include drying cycles for crop and soil manipulations, and some soils may need drying cycles for the recovery of desirable physical soil conditions. Another potential problem is high sodium content of an effluent. Sodium can cause the clay in the soil to disperse, the structure to degrade, and the surface soil to seal. This causes the infiltration rate and permeability to drop markedly.

Climate. Factors such as temperature range, annual precipitation, humidity, and wind velocity have a direct effect on the amount of water that can be disposed of at a certain location. These factors also have an effect on the type of crop that can be grown successfully in that area.

The consumptive use by plants is in direct relation to the climate of the area. Consumptive use or evapotranspiration is the total water used in transpiration, stored in plant tissue, and evaporated from adjacent soil. The consumptive use varies with the type of crop, humidity, air temperature, length of growing season, and wind velocity. The amount of water lost by evapotranspiration can be estimated from the pan evaporation data supplied by the U.S. Weather Bureau in the vicinity of the site. The amount of evapotranspiration is equal to a crop factor times the amount of pan evaporation. The crop factor varies with the type of crop and the location. Presently available crop factors have been derived for use under natural moisture conditions or prevalent irrigation practices and may not be applicable to wastewater irrigation at high rates. Figure 10-19 shows the excess of mean annual potential evapotranspiration over mean annual precipitation as estimated for the United States. The values are in inches.

The hydraulic loading for the irrigation process which attempts to maximize the crop production and wastewater treatment is a function of the precipitation in a given month and the potential evapotranspiration (ET) rate of the crop in the same month. The monthly irrigation requirement is the difference between the potential ET and the study area effective precipitation divided by the irrigation efficiency. An irrigation efficiency of 70 percent is consistant with good irrigation design and local practices. The ET rates must be selected specifically for the crop in the specific study area, as must the precipitation rates. A recent study for Boulder, Colorado[3] may be used as an illustrative example (see Table 10-11). A precipitation rate of 15.4 in./year was used. Only a portion of the annual precipitation is effective for crop use because of runoff and percolation below the root zone. The average annual effective precipitation for this study area is 10.0 in./year, as estimated by Soil Conservation Service methods. The effective precipitation is used to estimate the irrigation requirements of crops.

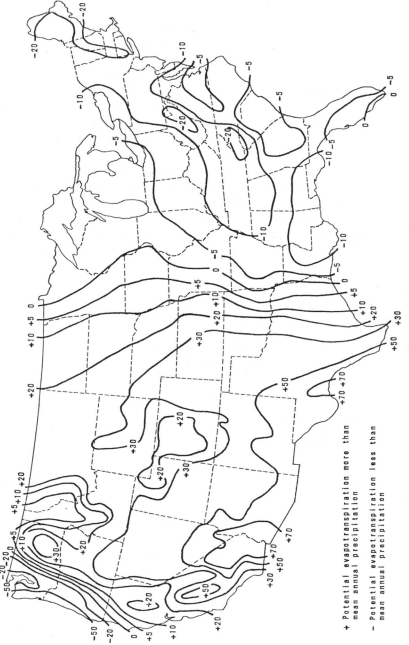

+ Potential evapotranspiration more than
  mean annual precipitation

− Potential evapotranspiration less than
  mean annual precipitation

Fig. 10-19 Potential evapotranspiration vs. mean annual precipitation (in.).[4]

The potential evapotranspiration (ET) for several crops was then computed as shown in Table 10-11. The crop ET requirements range from 21.0 to 37.7 in./year. An assumed distribution of these crop plantings was made for the land treatment area. A mean ET requirement of 25.8 in./year was calculated for use in estimating the average irrigation requirement. The average amount of effluent that would be applied for the irrigation process of land application is estimated to be 22.6 in./year. Effluent is applied only during those months of the year when the ET exceeds the precipitation. Also, soil temperature data indicate that the soil may be frozen during parts of some years from November through March. The time during which the ground is frozen varies from year to year. Because applications on frozen soil would result in runoff, applications should be limited to the time of year when soil is not frozen. The 22.6 in./year corresponds to an irrigated area of about 600 acres/million gal.

If the effluent quality goals do not require optimum treatment, and if less than optimum crop production is satisfactory, then "high rate irrigation" (see Table 10-1) is a viable alternative which will reduce the area requirements from those required by the lower rate irrigation approach.

The allowable hydraulic loading for the high rate irrigation process is dependent upon the soil's capacity for transmitting water rather than on the crop irrigation requirement. The maximum hydraulic loading is the

TABLE 10-11   Crop Evapotranspiration and Irrigation Requirement[3] (inches).

Month	Study Area Mean Effective Precipitation	Mean Potential ET[a]	Average Supplemental Irrigation Requirement[b]	ET Corn (Silage)	ET Pasture	ET Alfalfa	ET Spring Grain
Jan.	0.2	0.2	0.0	0.1	0.3	0.4	0.1
Feb.	0.3	0.3	0.0	0.2	0.5	0.6	0.2
Mar.	0.6	0.6	0.0	0.3	1.0	1.2	0.3
April	1.2	1.2	0.0	0.5	2.0	2.3	1.1
May	2.1	3.0	1.3	1.9	3.8	4.5	4.4
June	1.4	5.5	5.9	4.8	5.3	6.5	7.3
July	1.1	7.0	8.4	7.6	6.5	7.9	4.3
Aug.	1.0	4.5	5.0	4.0	5.8	6.8	1.8
Sept.	0.8	1.8	1.4	0.8	3.4	3.9	0.8
Oct.	0.7	1.1	0.6	0.5	2.0	2.3	0.5
Nov.	0.4	0.4	0.0	0.2	0.7	0.9	0.2
Dec.	0.2	0.2	0.0	0.1	0.4	0.4	0.1
Total	10.0	25.8	22.6	21.0	31.7	37.7	21.1

[a]Based on an assumed cropping pattern of $\frac{1}{2}$ corn, $\frac{1}{4}$ pasture, $\frac{1}{8}$ alfalfa, and $\frac{1}{8}$ spring grain.
[b](Mean potential ET minus mean effective precipitation) divided by 70% irrigation efficiency.

sum of soil moisture depletion by ET plus that quantity which can be transmitted through the root zone. The actual quantities are dependent on the soil permeability and soil depth. The soils in the potential land disposal sites must be inventoried and classified as to their suitability for high rate irrigation, as discussed in the preceding subsection.

Variations in climate from year to year must also be considered. A system that may work well in an average year may be seriously overloaded during an exceptionally wet year. The disadvantages of erratic precipitation can be partly compensated for by proper system design and by restricting disposal sites to the most favorable soils.

The length of the growing season affects the amount of water used by the crop. The growing season for perennial crops is generally the period beginning when the maximum temperature stays well above the freezing point for an extended period of days; the season continues uninterrupted despite later freezes. Beside consideration of runoff from frozen soils, one must consider the period when active crop growth and excess ET will occur. As indicated in Table 10-11, the six month period of May–October represents the period for the irrigation approach in the Boulder example. The high rate irrigation approach could be placed in operation a month earlier. Even if nonfreezing conditions occurred during the winter, effluent application would result in nitrate movement through the soil into the groundwater, since there would be no crop uptake of nitrogen. Sewage flows in the nonirrigation months must be stored. Generally, large lagoons are provided for off season storage.

Topography. Site topography will have an effect on the method of irrigation selected. Generally, sites with slopes greater than 15 percent are not suitable for any of the wastewater irrigation systems because of the difficulty of erosion control. Of course, techniques such as leveling and terracing can be used to modify the existing topography for spray irrigation or ridge and furrow irrigation. The flooding technique requires slopes of less than 1 percent. Center pivot spray irrigation equipment has been used on slopes up to 15 percent.

Steeper slopes will promote overland flow and subsurface lateral flow more than will gentle slopes, all other factors being equal. Also, subsurface seepage problems are more likely to occur on steep slopes. Slopes of 2–8 percent are preferred for overland flow.[2]

Vegetation

Selection of the vegetative cover to be utilized or established and maintained on a spray irrigation site depends upon many factors. The following are some of the criteria which should be considered:

1. Water requirements and tolerance.
2. Nutrient requirements and tolerance.
3. Optimum soil conditions for growth.
4. Season of growth and dormancy requirements.
5. Sensitivity to toxic heavy metals and salts.
6. Nutrient utilization and renovation efficiency.
7. Ecosystem stability.
8. Length of harvesting rotation.
9. Insect and disease problems.
10. Natural range.
11. Demand or market for the product.
12. Maintenance requirements.

The high uptake crops such as grass, which require little maintenance during the growing season, represent good selections for cover crops. Another factor is the need for annual planting. With perennials, such as grasses, planting has to be done only once and the crop is established for years. With vegetable crops, the planting has to be done every year, increasing the operational cost. Also, most vegetable crops require care during the growing period in order to produce a marketable product.

Salt tolerance can be an important parameter in crop selection. A salt buildup in the soil can be toxic to plants or can stunt their growth and produce a poor crop. A listing of forage and field crops and their tolerance to salts and boron is given in Table 10-12. Bermuda grass can tolerate

TABLE 10-12   Relative Tolerances of Field and Forage Crops to Salt or Electrical Conductivity.[2]

Tolerant (partial listing)	Semitolerant (partial listing)	Sensitive (complete listing)
Barley	Alfalfa	Alsike clover
Bermuda grass	Corn (field)	Burnet
Birdsfoot trefoil	Flax	Field beans
Canada wildrye	Oats	Ladino clover
Cotton	Orchard grass	Meadow foxtail
Rape	Reed canary grass	Red clover
Rescue grass	Rice	White Dutch clover
Rhodes grass	Rye	
Sugar beet	Sorghum (grain)	
	Sudan grass	
	Tall fescue	

18,000 μmho/cm, while ladino clover suffers a 50 percent decrease in yield with 2000 μmho/cm.

Adequate time for harvesting of crops must be scheduled into the design of the irrigation system. If farm machinery is used, the ground must be able to support the vehicles. Some states have regulations that require a period of 30 days between the last irrigation with wastewater and harvesting of many edible crops. The harvesting of grasses requires drying times after cutting unless silage is being produced.

The nutrients available from the wastewater are fairly constant throughout the year; however, crop demands can vary. A crop such as grass has a fairly uniform nutrient requirement during the growing season; however, corn and cotton need nutrients only at certain times. Since nearly constant amounts of nutrients are added to the soil, nutrients can build up and lead to future groundwater contamination when there is little or no crop uptake.

The application of wastewater to crops is very beneficial because of the natural fertilizers and nutrients in the liquid. Virtually all essential plant nutrients are found in wastewater. Measurements made at Pennsylvania State University show that the crop yield increases when wastewater rather than ordinary water is used for irrigation. Hay fields increased as much as 300 percent, corn grain increased 50 percent. The nutrients derived from wastewater are nitrogen, phosphorus, potassium, lime, trace elements, and humus.

Calcium in the form of lime is an indirect fertilizer that neutralizes acidity and checks some plant diseases. Soils high in organic matter, such as muck and peat, are generally deficient in calcium, as are clayey soils. Calcium in sewage exists in the form of carbonate, which is favorable to important soil organisms. Trace elements in wastewater are sulfur, magnesium, iron, iodine, sodium, boron, manganese, copper, and zinc. These elements can be helpful in plant development; however, in high concentrations, they can be toxic. Toxic elements can be toxic either to the plants themselves or to the animal that consumes the crop. Analysis of the soil and of the crop itself will give the levels of concentration of any toxic elements so that proper crops can be selected. Certain crops have a higher tolerance for toxic substances than others. Examples are oats and flax with respect to nickel: oats have a high tolerance, at 100 mg/l, while flax has a low tolerance, at 0.5 mg/l.

Water Rights

The evaluation of the water rights aspects of land treatment of wastewaters may be an important aspect in areas where water rights are long established, and often present a complex problem when new diversions of water are proposed. The complexity of the problem will be dependent upon how extensively downstream water users are now using the streams to which wastewaters, which would be diverted to land treatment sites, are now being discharged.

In the United States, water laws governing the right to withdraw surface water fall into three general classifications: (1) land ownership, common law, or riparian rights, (2) appropriative rights, and (3) a combination of the two.[4] These distinct divisions in water law have developed approximately in relationship to the surplus and deficiency of water, as shown in Figure 10-20. Areas of water surplus are those where the precipitation exceeds the evapotranspiration and results in more runoff and stream flow. Water deficient areas are those where evapotranspiration exceeds precipitation and little annual runoff occurs.

Simply stated, the land ownership or riparian doctrine says that the owner of land along a stream is entitled to unreduced flow and undiminished quality. Thus, in general, water may be taken from the stream and used but must be returned. In areas of surplus water, only a small portion of wastewater would be lost to evapotranspiration by land application. Thus, there would probably not be any water rights problems associated with land application of effluents where riparian rights are in force.

The appropriate doctrine dedicates the waters to the public. State laws differ, but in general a right to the use of water is obtained by putting the water to use, usually after or while filing for the right. The first person

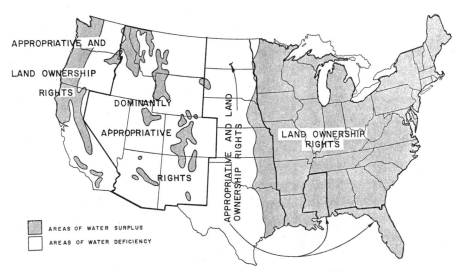

Fig. 10-20   Dominant water rights doctrines.[4]

who puts water to beneficial use receives senior rights. Subsequent appropriators are called junior appropriators and must not damage senior rights in times of short water supply. Thus, appropriated rights have a priority value based on the principle of "first in time, first in right." Appropriated water rights are property, and usually may be bought, sold, exchanged, or transferred.

Where water use is highly developed in deficient areas, water rights would probably be required for a land application wastewater treatment system. Land application on presently unirrigated land would result in increased evapotranspiration (consumptive use), which would decrease streamflows. Downstream water rights would be damaged by this action, and it may be necessary to appropriate or purchase water rights to ensure a continued land application system.

In water deficient areas, the total loss of wastewater to evapotranspiration would be inversely proportional to the loading rate. If the irrigation process is used to supply water according to crop needs, losses to evapotranspiration would typically be greater than 50 percent of the wastewater. Evapotranspiration losses for the overland flow, high rate irrigation and infiltration-percolation processes would likely be, respectively, 10–20, 15–30, and less than 10 percent of the applied water.[4]

The status of water rights in the local area should be investigated to avoid potentially serious legal problems later. If water rights are required, the expense and quantity of the right can be controlled to some extent in the process and site selection. For instance, if land presently irrigated is selected as the application site, it would be possible to purchase the water right with the land and make any necessary transfers. If a water right is necessary for only the portion of water consumptively used, it may be desirable to select the process which would minimize the evapotranspiration. A consultant familiar with water law and water resources can determine the legal implication of land application and extent of replacement required.

Farming Management

The management of an irrigation system is as important as the site selection and system design. It is vital that management personnel have a working knowledge of farming practices as well as of principles of wastewater treatment. Important items in management include seasonal (often weekly) variation in operation to respond to changing crop requirements for nutrients and water, monitoring to establish removal efficiencies and to forecast buildups of toxic compounds, and ongoing observation of the system to avoid problems of ponding, runoff, or mechanical breakdowns.

Options for land treatment system management include:[1]

Managed and operated by the implementing agency.

Managed by the implementing agency and operated by a private party through contact or crop sharing.

Managed and operated by contractual agreement with the same private party.

Managed by contractual agreement with a private party and operated by a subcontractual agreement with another private party.

Close cooperation between the treatment system management and the farm operation is required in all cases. Scheduling or irrigation with farm operations such as planting, tilling, spraying, and harvesting is vital to successful management. If adequate consideration is not given to the system operation, the design may be inappropriate. For instance, if crops are to be harvested, flexibility is needed in the irrigation system for sufficient drying and harvesting time. Farm management specialists can be helpful in setting up the management of the crops, soil, and irrigation portions of the operation.

Land Availability

There are several alternative approaches to obtaining the land required. The irrigated land could be some already in the public domain, such as a park, a publicly owned greenbelt, or some other public land. Undeveloped land may be purchased and developed into irrigated land. Contracts for sale of the effluent may be signed with farmers currently using irrigation. Currently irrigated land could be purchased. The wastewater effluent could be discharged directly to existing irrigation ditches for use in existing irrigation systems. Each approach has its advantages and disadvantages.

Those approaches calling for purchase of the land provide the most positive control of the land treatment system, since the ownership and operation can be in the hands of those responsible for wastewater treatment. The cost of these approaches will be significantly affected by the land costs. Also, they may involve condemnation proceedings and associated legal and public relations problems. Development of new irrigated land may also involve significant water rights problems. Leasing of land may retain the advantages of positive control of the operation and keep the land on the tax rolls, but adds the risk of lease renewal rights, the cost of removal of equipment if the site is abandoned, and the problem of finding enough contiguous land for which suitable leases can be arranged.

Irrigation by contract minimizes the initial costs but results in the loss of positive control of the irrigation system and, for this reason, is often not a realistic approach.

Irrigation of land already in the public domain eliminates the initial land purchase costs, but will still entail the initial irrigation system development costs. Also, it would be an extremely fortuitous situation if adequate quantities and types of publicly owned land are already available in a suitable location.

Groundwater Conditions

The groundwater table is the level where free water is present in the soil. The soil is saturated, and bacteria and dissolved solids can travel freely in the water between soil particles.

Before a site is selected a great deal about the groundwater should be known, including depth, variation of depth throughout the site, seasonal variation of depth, direction of groundwater flow, and groundwater quality. If, for example, one selects a natural groundwater recharge area for irrigation, rather serious negative results could occur. Poor drainage conditions and runoff would soon result, and soils may become waterlogged. This could cause the biologically active zone to shift from an aerobic to anaerobic state, cause a destruction or alteration of the physical–chemical–biological treatment processes, and ultimately a degradation of soil, groundwater, and surface water quality. Also, following installation of a major land treatment system, areas that were formerly well drained recharge areas may become poorly drained discharge areas. These areas can occur several hundred to thousands of feet beyond stream channels or appear as discontinuous upland patches remote from nearby streams. Also, as noted earlier, to ensure an aerobic root zone, the groundwater level should be maintained at least 5 ft below the ground surface.

In order to control the resulting changes in groundwater level, artificial draining systems are often a part of land treatment systems. In addition to water level control, such drainage systems may be needed for quality control, since the quality of the renovated water will usually be inferior to that of the native groundwater. Where there is a concentrated source of renovated water entering the groundwater, as with high rate systems, provisions may be needed to restrict the spread of renovated water into the groundwater basin or to control the level of the groundwater. This can be accomplished in some cases by taking the renovated water out of the aquifer at some distance from where it entered the groundwater, as happens naturally when the groundwater drains to a stream or a lake. If the renovated water does not adequately leave the aquifer in a natural manner, it should be collected by drains (for shallow aquifers) or wells (for deep aquifers). Figure 10-21 illustrates these drainage systems.

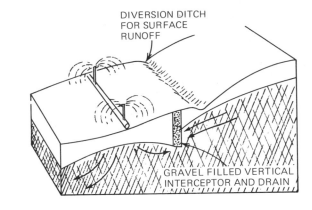

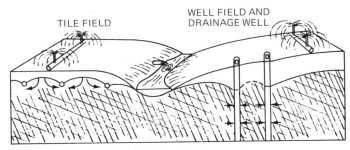

Fig. 10-21   Control of surface runoff by diversion ditches, and groundwater levels by gravel filled interceptor drains on steep slopes; and by tile fields, drainage wells, or water supply wells.[57]

The decision to drain a site artificially using wells, tile fields, or ditches immediately limits the concern for the position of the water table because its position will be controlled. Even given the will to use such a method, however, artificial drainage cannot always be achieved at a reasonable expense. Some soils may be so nearly impervious that leaching and artificial drainage are impossible using conventional means. Sites must be examined carefully with this point in mind.

Wells may be used in five different manners: (1) a wastewater renovation facility can be located adjacent to an existing well field where site conditions are suitable for irrigation, and effluent used is free of toxic or harmful nondegradable substances; (2) a well field can be placed in close proximity to the site after it has been established that adequate renovation is being achieved; (3) reclaimed water discharged to the surface can be recollected in wells relying upon induced streambed infiltration; or (4) drainage wells can be placed within the irrigation area with no beneficial use planned for the water except to insure adequate drainage; (5) where there is doubt that adequate renovation can be achieved without relying upon dilution, the irrigation site should be located remote from areas of existing or potential groundwater development. For cases 1 through 4 to be effective, unconfined aquifers or at least semi-unconfined aquifers should underlie the irrigation site.

If the water table and the impermeable layer are relatively close to the surface, wells may not be effective and the renovated water is better collected by open or closed drains. For example, a study in Montgomery, Co., Maryland,[58] where shallow (5–20 ft) soils were underlain by impervious rock, found that wells would be ineffective in controlling drainage. A subsurface drainage system composed of 4 in. polyethylene pipe in an envelope of gravel filter material with the pipe drains spaced at 100 ft intervals was recommended. The depth of the drains varied from 4 to 6 ft, depending on soil conditions.

Spacing of the drainage system will be controlled by permeabilities and depth of wetted materials. Standard procedures are available for designing drainage systems under nearly all conditions. Where high loadings occur, as in the infiltration–percolation process, or where permeabilities are low, drain spacing may be as close as 50–100 ft. Where higher permeabilities occur and loadings are lower, drains may be spaced much further apart, up to 500 ft or more. Careful consideration of drainage is a necessity, and normally some field investigation is required to specify spacings and depth even in preliminary studies.

Heavy Metals

When sewage effluent is applied to soil, toxic elements are retained by the soil. These elements will accumulate and persist, and are a potential long term environmental hazard in land treatment systems. Elements in sludge and effluent that are potential hazards to plants or the food chain are: B, Cd, Co, Cr, Cu, Hg, Ni, Pb, and Zn. Table 10-13 shows recommended concentration limits for heavy metals and other specific elements for irrigation water based on an application of 3 ft/year, a typical application rate of the irrigation process. Estimated limits are also shown for loadings of 8 and 80 ft/year, as might be used with high rate irrigation or infiltration–percolation. If these criteria are met, there should be little concern about toxic effects on plants or excessive accumulation in soils. Many of the elements are micronutrients or trace elements required by plants in small quantities.

From the standpoint of plant toxicity, chromium appears to be noninjurious because it occurs as Cr^{3+} and is not available to injure plants; chromate is rapidly reduced to Cr^{3+} in soil. Mercury forms insoluble compounds in the soil and does not injure plants at the low levels typically added with

TABLE 10-13 Recommended and Estimated Maximum Concentrations of Specific Ions in Irrigation Waters.[a]

Element	Removal Mechanism[b]	For Waters Used Continuously on All soil — 3 ft/yr Application Recommended Limit[c]	For Waters Used Up to 20 Years on Fine Textured Soils of pH 6.0 to 8.5		
			3 ft/yr Application Recommended Limit[c]	8 ft/yr Application Estimated Limit	80 ft/yr Application Estimated Limit
Aluminum	PR, S	5.0	20.0	8.0	0.8
Arsenic	AD, S	0.10	2.0	0.8	0.08
Beryllium	PR	0.10	0.50	0.2	0.02
Boron	AD, W	0.75	2.0–10.0	2.0	2.0
Cadmium	AD, CE, S	0.010	0.050	0.02	0.002
Chromium	AD, CE, S	0.10	1.0	0.4	0.04
Cobalt	AD, CE, S	0.050	5.0	2.0	0.2
Copper	AD, CE, S	0.20	5.0	2.0	0.2
Fluoride	AD, S	1.0	15.0	6.0	0.6
Iron	PR, CE, S	5.0	20.0	8.0	0.8
Lead	AD, CE, S	5.0	10.0	4.0	0.4
Lithium	CE, W	2.5[d]	2.5[d]	2.5	2.5
Manganese	PR, CE, S	0.20	10.0	4.0	0.4
Mercury	AD, CE, S	—	—	—	—
Molybdenum	AD, S	0.010	0.050[e]	0.02[e]	0.002[e]
Nickel	AD, CE, S	0.20	2.0	0.8	0.08
Selenium	AE, W	0.020	0.020	0.02	0.02
Silver	AD, CE, S	—	—	—	—
Zinc	AD, CE, S	2.0	10.0	4.0	0.4

[a] These levels will normally not adversely affect plants or soils. No data are available for mercury, silver, tin, titanium, or tungsten.

[b] AD = adsorption with iron or aluminum hydroxide, pH dependent; AE = anion exchange; CE = cation exchange; PR = precipitate, pH dependent—iron and manganese are also subject to changes by oxidation reduction reaction; S = strong strength of removal; W = weak strength of removal.

[c] EPA Water Quality Criteria, 1972.

[d] Recommended maximum concentration for irrigating citrus is 0.075 mg/l.

[e] For only acid fine textured soils or acid soils with relatively high iron oxide contents.

sludge and effluent. Cadmium can be toxic to plants, but as explained below, this element is considered hazardous to the food chain at low levels and thus dare not reach phytotoxic levels. Lead can injure plants in low phosphate, acid soils. Lead added with sludge and effluent appears to be nontoxic to plants because the large amount of phosphate also present ties up the lead and prevents injury. Sludge or effluent seldom contains cobalt, which is a product of specific industrial pollution. Boron in excess can injure plants; boron in effluents could become phytotoxic in soils there the boron content of normal irrigation water is already of concern. Zinc, copper, and nickel are toxic to plants and commonly occur in sewage sludge and effluent. These divalent metal ions are normally not found in soils in large quantities; their presence in wastewater may increase the total content of these elements in soils significantly.

Among the factors that affect the toxicity of metals to plants are:

The amount of toxic metals present in the soil.

The toxic metals present. Zn, Cu, and Ni differ in their toxicity to specific plants and in specific soils.

The pH of the amended soil. The toxic metals are much more available pH values below 6.5–7.0. A soil toxic metal content safe at pH 7 can easily be lethal to most crops at pH 5.5. Land disposal may lead to a lowering of the soil pH due to nitrification of the NH_4^+-N added.

The organic content of the amended soil. Organic matter chelates the toxic metals and makes them less available to injure plants.

The phosphate content of the amended soil. Phosphate is well known for reducing Zn availability to plants and decreasing the stunting injury caused by excessive levels of toxic metals.

The cation exchange capacity (CEC). The CEC of the soil is important in binding all cations, including the toxic metal cations. A soil with high CEC is inherently safer for disposal of sludge and effluent than a soil with low CEC.

The plants grown on sludge or effluent treated soil. Plant species vary widely in tolerance to toxic metals, and varieties within a species can vary three to tenfold. Vegetable crops very sensitive to toxic metals are the beet family (chard, spinach, redbeet, and sugarbeet), turnip, kale, mustard, and tomatoes. Beans, cabbage, and collards, and other vegetables are less sensitive. Many general farm crops (corn, small grains, and soybeans) are moderately tolerant. Most grasses (fescue, lovegrass, bermuda grass, orchard grass, and perennial ryegrass) are tolerant to high amounts of metals.

Toxic metals added to soils are not a hazard to the food chain until they have entered an edible part of a plant. The elements that are a significant potential hazard to the food chain through plant accumulation are Cd, Co, and Zn. Very high amounts of Cr^{3+} added to soil do not increase the Cr contents of crops appreciably and constitute no hazard. B, Co, and Ni at levels that severely injure the plant present no threat to the food chain. Mercury from sludge will increase soil Hg levels, but the increase in plant Hg will be small. Hg does not appear to constitute the food chain accumulator in agriculture that it does in the oceans. Pb is not translocated readily to plant tops, and is especially excluded from grain, fruits, and edible roots. The lack of Pb accumulation appears to be related to the presence of the high amount of PO_4^{3-} typically found in effluent. Because yield usually is reduced at lower plant Zn levels than those that injure the animal that consumes the plant, the food chain appears to be protected from Zn. Increases in soil Cd from sludge or effluent can lead to increased food chain Cd. The Food and Drug Administration expects eventually to specify the permissible level of Cd in foods in the marketplace. The relative quantities of Cd and Zn present appear to affect the injurious effect of Cd. There is much yet unknown about this relationship, however. Also, little or no information is available on the movement of Cd from feed grains or pasture into beefsteak. It appears that the permissible Cd content of foods in the marketplace may be established at the current natural background levels. One study[59] found that 35 years of sludge application resulted in large accumulations of Cd and other trace elements in the soil but no significant accumulation in the grain of corn plants. Concentrations in the leaves and roots were significantly higher than normal.

Some precipitants, such as iron and manganese, become more soluble under anaerobic conditions and are then subject to leaching and potential pollution of the groundwater. Maintaining an aerobic soil may be important to effective removal of some heavy metals. In soluble forms the heavy metals are available to plants and subject to movement to the groundwater.

In the overland flow process, water does not move through the soil, so there is little chance of removal of heavy metals in the soil matrix. For the irrigation and high rate irrigation processes, the soil will be very similar and heavy metal removal will be similar. Because of the higher loadings for infiltration-percolation, the soil capacity for removal of heavy metals will be reached sooner. The limitation for heavy metals concentration should be a function of site life and loading rate. The higher the loading rate and the longer the site life desired, the lower the concentration limitation should be.

Trace elements, salt, or other pollutants toxic to plants may have toxic effects sooner when high concentrations directly contact the vegetation than when the same concentrations are applied to the soil. Plants may absorb some pollutants directly through contacted exposed parts. Sprinkler and flood irrigation produce direct contact of water on the plant. If high concentrations of pollutants are of concern, furrow irrigation might be used to prevent foliage contact.

Pathogens

The removal of pathogenic organisms from effluent before discharge to the receiving water is a nearly universal requirement. In land disposal by sprinkler irrigation these organisms are subject to movement through the air as aerosols. One study[60] reports aerosolization of about 0.3 percent of the material being sprayed. The potential travel distance of the aerosols is dependent on wind velocity, sprinkler pressure, humidity, solar radiation, nozzle size, height of nozzle, and roughness of the vegetation or ground surface. Trees in the buffer strip, for instance, will substantially reduce the potential travel distance.

Under normal circumstances, pathogenic organism contamination by air transport is not expected to be a problem if the wastewater is disinfected prior to spraying and an adequate buffer strip is provided around the irrigated area. To prevent possible contamination of flowing streams or adjacent private and public property, a buffer strip or "green belt" is provided for protection. The width of this buffer strip is dependent on vegetation conditions. Trees are about 4 times better than short vegetation at preventing wind movement of particles across an area. There is considerable variation around the U.S. in regard to regulatory requirements on buffer areas (50–400 ft spans most requirements). One study[60] concluded that buffer zones of a practical size were of limited value in controlling aerosols, since distances of 700–1800 m were required to approach background concentrations of bacteria. The study concluded that disinfection is far more effective. Also, the use of sprinklers which direct the spray downward can be used to limit aerosol travel.

The effectiveness of the soil in removing pathogens has been demonstrated. A removal of 95 percent of these organisms in the surface layer of 0.5 in. of soil would be expected, and 3–5 ft of vertical movement above the water table has been shown to be sufficient for nearly 100 percent removal. The two mechanisms involved in removing pathogens are mechanical filtration (especially at the soil surface) and sorption on the soil. Sorption is the important removal process. Fine textured soils have a larger sorption capacity, dry or slightly moist soils are considered more effective in removal of pathogens. Rainfall can cause a sudden downward movement of viruses in the soil by release of previously adsorbed virus.[61] The underlying soil should not containchannels or fractures which may permit pathogens to move long distances.

The presence of pathogens even in a chlorinated secondary effluent places a restriction on the type of crops which are suited to wastewater irrigation. There appears to be a danger of pathogen contamination of vegetables to be eaten raw if wastewater is applied to the fields within 2–3 months of harvesting. This limits the usefulness of wastewater irrigation for crops such as lettuce and radishes which have short growing seasons.

Surface Runoff Control

Requirements for control of surface runoff resulting from both applied effluent and stormwater depend mainly on the expected quality of the runoff— for which few data exist.[4] Surface runoff control concerns are limited to irrigation and overland flow systems because infiltration–percolation systems are designed so that no runoff is allowed. Surface runoff control considerations for irrigation systems center on tailwater return, storm runoff, and system protection.

Tailwater Return.   Surface runoff of applied effluent is usually designed into surface application systems, such as ridge and furrow and flooding, because it is difficult to maintain even distribution across the field, and some excess water may accumulate at the end of furrows or strips. Generally, this tailwater is returned by means of a series of collection ditches, a small reservoir, a pumping station, and a force main to the main storage reservoir or distribution system.

The most common range of flows for tailwater is between 5 and 25 percent of applied flows (depending on the management provided, the type of soil, and the rate of application). In humid climates, the tailwater system design may be controlled by stormwater runoff flows, as discussed below.

Storm Runoff.   For intense storms, some form of runoff control may be required for irrigation systems, except those with well drained soils, relatively flat sites, or where the quality of the runoff is acceptable for discharge. Where runoff control is necessary, it generally consists of the collection and return of the runoff from a storm of specified intensity, with a provision for the overflow of a portion of larger flows. The amount of runoff to be expected as a result of precipitation will depend on (1) the infiltration capability of the soil, (2) the existing moisture condition of the soil, (3) slope, (4) type of vegetation, and (5) temperature of both the air and the soil. The relationships between runoff and these factors are common to many other hydrologic problems and are covered in a number of texts.

How much runoff should be contained or reapplied depends on the quality of runoff, water quality of the stream during storms, and downstream uses of the water. Requirements may vary from capturing and recycling the first flush from the field, as a minimum, to capturing all of the flow resulting from a specified frequency storm event.

System Protection. An additional requirement for runoff control often results from the need for protection from runoff caused by system failures. System failures may include ruptured sprinkler lines, inadvertent overapplication, or soil sealing. This requirement would normally be satisfied by a storm runoff control system.

More extensive surface runoff control features are normally required for overland flow systems, which are designed principally for runoff of applied effluent rather than percolation. Typically 40–80 percent of the applied effluent runs off. In most cases, the runoff is collected in ditches at the toe of each terrace and then conveyed to a discharge point, where it is monitored and in some cases disinfected.

Nonirrigated Area Requirements

In addition to the areas required for the irrigated land (as discussed earlier), there will be additional area requirements for such things as secondary treatment facilities, buffer strips along existing roads and other existing developments within or next to the irrigated area, storage lagoons for retaining the wastewater flow during winter months, small zones within the irrigated zone which are not suitable for irrigation, etc. Winter storage lagoons must be sized to retain several months of winter sewage flow in most areas, as well as to retain any storm flow which may drain into the storage lagoon. A rule of thumb for preliminary calculations of gross area requirements is that the total land needs, including all of the above factors, will be about 1.4 times the irrigated area needed. A detailed analysis of specific potential sites is, of course, needed to determine the total space needs more accurately.

Effects on Effluent

Unfortunately, many of the existing land treatment systems have not been adequately monitored to determine the effluent quality produced, but rather have been operated so as to avoid nuisance conditions. The previous discussions of nutrient and hydraulic loadings contain data on the removal of some constituents in wastewater as it passes through a land treatment system. The expected treatment efficiencies of well designed and well operated systems are summarized in Table 10-14. Through selection of appropriate design criteria for application rates, application cycles, or type of crop grown, improvements in removal efficiencies may be achieved. Because treatment

TABLE 10-14   Reported Removal Efficiencies After Biological Treatment.[53]

Constituent	Removal Efficiency, Percent		
	Irrigation	Infiltration–Percolation	Overland Flow
BOD	98+	85–99	92+
COD	80+	50+	80+
Suspended solids	98+	98+	92+
Nitrogen (total as N)	85+[a]	0–50[b]	70–90[a,b]
Phosphorus (total as P)	80–99	60–95	40–80[c]
Metals	95+	50–95[d]	50+
Microorganisms	98+	98+	98+[e]
TDS	+30–0[f]	+10–0[f]	+30–0[f]

[a]Depends on crop uptake.
[b]Depends on denitrification.
[c]May be limiting.
[d]Ion exchange capacities may be limited.
[e]Chlorination of runoff may be needed.
[f]May increase.

efficiencies are site specific, depending on loading rates, soils, crops, climate, design objectives, and operation, they are difficult to generalize.

COST CONSIDERATIONS

An extensive guide to estimating the costs of land treatment systems has been published by EPA.[62] It is not practical to duplicate this lengthy report as a part of this chapter. Thus, the discussion here will be limited to a review of those major factors that influence the costs of land treatment systems and to make some general observations on the relative economics of AWT and land treatment. The reader is referred to the EPA document for detailed cost information.

It is virtually impossible to accurately generalize the costs of land treatment because of the tremendous impact that local conditions may have on costs. This section will review those factors that influence costs and the range of costs. Table 10-15 summarizes a typical range of costs for conditions ranging from favorable to unfavorable for land treatment.

1. *Land costs.*   Obviously, land costs will vary substantially from one locale to another. Local costs must be determined. Land costs when land is used as part of the treatment process are eligible for 75 percent EPA grants.

TABLE 10-15   Effect of Conditions on Land Treatment (Spray Irrigation) Costs per Acre (1974 Costs).[63]

Item	Very Favorable Conditions	Moderately Favorable Conditions	Unfavorable Conditions
Land preparation	Little earthwork and clearing, $50	$ 150	Extensive earthwork and clearing, $350
Surface runoff control	Relatively level site and moderate rain, $200	$ 500	Rolling topography and intense storms, $1000
Subsurface drainage	None required	$ 400	Extensive underdrain system needed, $1000
Irrigation system			
Pumping station and distribution main	$ 400	$ 550	$ 700
Laterals and sprinklers	Center pivot $ 300	Solid set $1400	Solid set $2000
Land costs	$ 500	$1000	$2000
Relocation costs	$ 10	$ 30	$ 50
Totals[a]	$1460	$4030	$7100

[a]Not including transmission, storage lagoon, or pretreatment costs.

2. *Transmission to site.* This cost will depend on the location of the land disposal site relative to the sources of wastewater. It will include any pumping and transmission costs. This is site dependent and is not included in Table 10-15.
3. *Site development costs.* These costs can be extremely variable and have varied from $140 to $5100/acre and include the following:
   a. *Relocation costs.* This will depend upon the number of families displaced from the site. There are Federal relocation laws which prescribe the reimbursement costs. They are typically a very minor fraction of the total costs (less than 1 percent).
   b. *Land preparation.* This will be a function of the earthwork and clearing required to make the site compatible with the selected irrigation system. The ridge and furrow system generally requires more land preparation than spray systems.
   c. *Surface runoff control system.* This includes the needed dikes, ponds, and pumping stations to collect surface runoff and return it to the irrigation system. The flatter the terrain, the lower the costs.
   d. *Subsurface drainage system.* This cost depends upon the soils and groundwater conditions. Wells are generally lower in cost than underdrains if the site is suited to wells.

e. *Distribution and irrigation system.* These costs are largely affected by the type of irrigation system used. Center pivot rigs may cost as little as $300/acre, while solid set systems may cost as much as $2000/acre. The solid set systems are suited to a wider range of conditions than are center pivot rigs, a fact which accounts for their more frequent use. However, if the site is suited to use of the center pivot rigs, their cost is substantially less. Regardless of the irrigation equipment selected, a pumping station and distribution main must be installed to convey the wastewater from the storage lagoon to the irrigation laterals and sprinkler system.

f. *Storage lagoons.* The storage lagoon must be sized for storage of winter wastewater flows as well as for any stormwaters which drain into the lagoon site. In extremely flat sites (such as Muskegon), the lagoon may be formed by diking. This is so variable depending upon climate and site conditions that it is not included in Table 10-15.

g. *Pretreatment costs.* Unless the effluent from an existing treatment facility is to be applied to the land, the costs for pretreatment facilities must be included. As discussed earlier, the degree of pretreatment required varies substantially.

In evaluating the cost effectiveness of land treatment, one must consider some significant items which are credits against costs. These items are the income received from the sale of crops grown and the salvage value of the land. EPA's cost effectiveness analysis guidelines specify that land used as part of the treatment process shall be assumed to have a salvage value at the end of the planning period equal to its prevailing market value at the time of the analysis.

In order to make some general observations on the relative economics of AWT and land treatment, generalized cost curves have been prepared to reflect the general nature of the effects of the conditions, ranging from relatively favorable to relatively unfavorable (see Table 10-15), on land treatment costs. These curves are for sprinkler systems, since they are more widely applicable. According to data in the EPA cost document,[62] the total costs for overland flow systems will approach those shown for very favorable conditions for spray irrigation in this chapter.

The cost curves (Figures 10-22 through 10-24) *do not* include pretreatment costs, the costs to deliver wastewater to the irrigation site, or revenue from (or costs to dispose of) crops. The purpose of the curves is to compare costs of land *treatment* with AWT techniques; thus, inclusion of costs to transport wastewater to the land treatment site (which are totally site specific in any case) would be unfavorably biased against land treatment. Cost curves are plotted for irrigated area requirements from 100 to 500 acres/million gal, based upon the capital costs shown in Table 10-15 and estimated operation and maintenance costs. It was assumed that total area requirements were 130 percent of the irrigated area to provide for buffer zones and to account

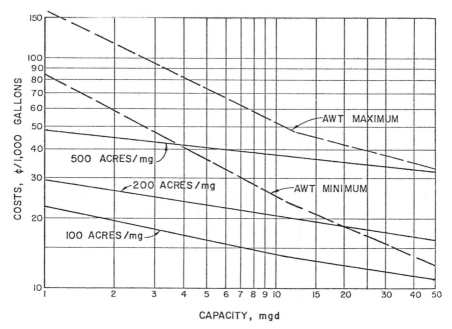

Fig. 10-22   Cost comparison—very favorable conditions for land treatment.[63]

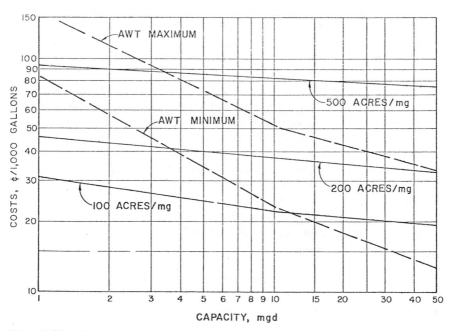

Fig. 10-23   Cost comparison—moderately favorable conditions for land treatment.[63]

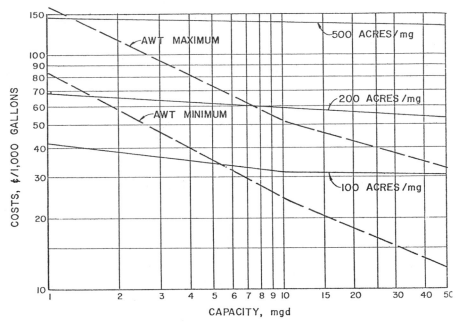

Fig. 10-24   Cost comparison—unfavorable conditions for land treatment.[63]

for unusable areas within the irrigation site. Storage for 5 months' flow was assumed. Engineering, legal, and contingency costs were applied to the non-land costs only. Capital costs were amortized at 7 percent over 20 years.

In order to span a range of AWT alternates for comparison's sake, two levels of treatment were assumed. "AWT minimum" consists of coagulation, sedimentation, and filtration. This would reduce phosphorus, BOD, suspended solids, and coliform to levels comparable to that achieved by a land treatment system where nitrogen removal is not of concern. "AWT maximum" adds biological nitrogen removal and activated carbon adsorption and regeneration to the AWT minimum approach. As with land treatment, secondary treatment and raw sewage transport costs are not included. It was assumed that the chemical sludges would be lime sludges, which would be dewatered and recalcined. It is probable that AWT costs can be reduced if dewatering and burial of lime sludges near the plant site is practical for a given locale. Recalcining costs are included, however, to insure an adequately high AWT cost estimate. AWT costs are also expressed in 1974 cost levels.

The AWT costs and land treatment costs are plotted in Figures 10-22, 10-23, and 10-24. Although such generalized curves have limitations, they do indicate general trends in the relative costs of AWT and land treatment.

Analysis of these curves shows that *increases in the degree of treatment required and decreases in plant size improve the competitive economic position of land treatment with conventional AWT processes.* Even under moderately favorable conditions with a land requirement of 200 acres/million gal, land treatment would be expected to have an economic edge over minimum AWT for plants with 4 mgd capacity or less—a capacity encompassing over 90 percent of the plants in the U.S.

REFERENCES

1. Powell, G. M., "Land Treatment of Municipal Wastewater Effluents, Design Factors—Part II." Paper presented at U.S. EPA Technology Transfer Seminars (1975).

2. Pound, C. E., and Crites, R. W., "Wastewater Treatment and Reuse by Land Application—Volumes I and II." U.S. Environmental Protection Agency Report 660/2-73-006a & b, August, 1973.

3. "Comparative Study of Sewage Treatment for Domestic Sewage Effluent and Sludge—Boulder, Colorado." CH2M/Hill Engineers, March, 1974.

4. Pound, C. E., and Crites, R. W., "Land Treatment of Municipal Wastewater Effluents, Design Factors—Part I." Paper presented at U.S. EPA Technology Transfer Seminars, 1975.

5. Sullivan, R. H.; Cohn, M. M.; and Baxter, S. S., "Survey of Facilities Using Land Application of Wastewater." U.S. Environmental Protection Agency Report 430/9-73-006, July, 1973.

6. Sopper, W. E., and Kardos, L. T., *Recycling Treated Municipal Wastewater and Sludge Through Forest and Cropland.* The Pennsylvania State University Press, 1973.

7. "Recycling Municipal Sludges and Effluents on Land," Proceedings of a Joint Conference in Champaign, Illinois, July 9–13, 1973.

8. Demirjian, Y. A., "Muskegon County Wastewater Management System." Paper presented at U.S. EPA Technology Transfer Seminars, 1975.

9. D'Itri, F. M.; Smith, T. P.; Bouwer, H.; and Myers, E. A., "An Overview of Four Selected Facilities That Apply Municipal Wastewater to the Land." Paper presented at U.S. EPA Technology Transfer Seminars, 1975.

10. "Record Harvest Set by Effluent–Treated Cornfield," *Engineering News Record*, October 2, 1975, p. 22.

11. "Land Application of Sewage Effluents and Sludges: Selected Abstracts." U.S. EPA Environmental Protection Technology Series, EPA-660/2-74-042, June, 1974.

12. Stevens, L. A., "The County That Reclaims Its Sewage," *Readers Digest*, July, 1975, p. 39.

13. "The Muskegon County Plan of Wastewater Reuse," *Public Works*, October, 1973, p. 77.

14. Forestell, W. L., "Sewage Farming Takes a Giant Step Forward," *The American City*, October, 1973, p. 73.

15. Chaiken, E. I.; Poloncsik, S.; and Wilson, C. D., "Muskegon Sprays Sewage Effluents on Land," *Civil Engineering*, May, 1973, p. 49.

16. Searle, S. S., and Kirby, C. F., "Waste Into Wealth," *Water Spectrum*, Fall, 9172, p. 15.

17. Seabrook, B. L., "Land Application of Wastewater in Australia—The Werribee Farm System." U.S.E.P.A. Report, April, 1975.

18. "Selected Chemical Characteristics of Soils, Foragees, and Drainage Water from the Sewage Farm Serving Melbourne, Australia." U.S. Army Corps of Engineers Report, January, 1974.

19. "Wastewater Renovation by Ground-Water Recharge Through Surface Spreading in the Salt River Bed, Phoenix, Arizona—The Flushing Meadows Project." U.S. Dept. of Agriculture, Phoenix, Arizona, Annual Report for the Year 1968.

20. Bouwer, H., "Renovating Secondary Effluent by Groundwater Recharge with Infiltration Basins," in *Recycling Treated Municipal Wastewater and Sludge Through Forest and Cropland*, The Pennsylvania State University Press, 1973.

21. Bouwer, H.; Rice, R. C.; Escarcega, E. D.; and Riggs, M. S., "Renovating Secondary Sewage by Ground Water Recharge with Infiltration Basins." U.S. Environmental Protection Agency, Water Pollution Control Research Series, Project No. 16060 DRV. U.S. Government Printing Office, Washington, D.C., March, 1972.

22. Bouwer, H., "Ground-Water Recharge Design for Renovating Wastewater," *Journal Sanitary Engineering Division, American Society of Civil Engineers Proceedings*, 59, 1970.

23. Rice, R. C., "Soil Clogging During Infiltration with Secondary Sewage Effluent," *Journal Water Pollution Control Federation*, in press.

24. Lance, J. C., and Whisler, F. D., "Nitrogen Balance in Soil Columns Intermittently Flooded with Sewage Water," *Journal Environmental Quality*, 1972, p. 180.

25. Lance, J. C.; Whisler, F. D.; and Bouwer, H., "Oxygen Utilization in Soils Flooded with Sewage Water," *Journal Environmental Quality*, in press.

26. Bouwer, H., "Salt Balance, Irrigation Efficiency, and Drainage Design," *Journal of the Irrigation and Drainage Division, American Society of Civil Engineers Proceedings*, 1969, p. 153.

27. Sopper, W. E., and Kardos, L. T., "Vegetation Responses to Irrigation with Treated Municipal Wastewater," in *Recycling Treated Municipal Wastewater and Sludge Through Forest and Cropland*, The Pennsylvania State University Press, 1973.

28. Sopper, W. E., "Crop Selection and Management Alternatives—Perennials," in Proceedings of the Joint Recycling Municipal Sludges and Effluents on Land, National Associations of State Universities and Land-Grant Colleges, Champaign, Illinois, July 9–13, 1973.

29. Murphey, W. K.; Brisbin, R. L.; Young, W. J.; and Cutter, B. E., "Anatomical and Physical Properties of Red Oak and Red Pine Irrigated with Municipal Wastewater," in *Recycling Treated Municipal Wastewater and Sludge Through Forest and Cropland*, The Pennsylvania State University Press, 1973.

30. Wood, G. W.; Simpson, D. W.; and Dressler, R. L., "Effects of Spray Irrigation of Forests with Chlorinated Sewage Effluent on Deer and Rabbits," in *Recycling Treated Municipal Wastewater and Sludge Through Forest and Cropland*, The Pennsylvania State University Press, 1973.

31. Overman, A. R., "Research Experience in Florida." Proceedings Workshop of Land Renovation of Municipal Effluent in Florida, University of Florida, 1972.

32. Burton, G. W., "Wastes—Forages—Animals—Man." Proceedings Workshop on Land Renovation of Municipal Effluent and Sludge in Florida, University of Florida, 1973.

33. Adams, W. E.; Stelly, M.; Morris, H. D.; and Elkins, C. B., "A Comparison of Coastal and Common Bermuda Grasses in the Piedmont Region," *Agronomy Journal*, 1967, p. 281.

34. Parks, W. L., and Fisher, W. B., Jr., "Influence of Soil Temperature and Nitrogen on Rye Grass Growth and Chemical Composition," *Soil Science Society of America, Proceedings*, 257, 1958.

35. Fiskell, J. G. A., and Ballard, R., "Prediction of Phosphate Retention and Mobility in Florida Soils." Proceedings Workshop on Land Renovation of Municipal Effluent and Sludge in Florida, University of Florida, 1973.

36. "Spray Irrigation Manual." Pennsylvania Bureau of Water Quality Management Publication No. 31, 1972.

37. "Soil Systems for Municipal Effluents—A Workshop and Selected References.' U.S. Environmental Protection Agency Report 16080 GWF, February, 1972.

38. "Agricultural Utilization of Sewage Effluent and Sludge—An annotated Bibliography." Federal Water Pollution Control Administration Report CWR-2, January, 1968.

39. Godfrey, K. A., "Land Treatment of Municipal Sewage," *Civil Engineering*, September, 1973, p. 103.

40. Hinsley, T. D., "Water Renovation for Unrestricted Re-Use," *Water Spectrum*, p. 1, published by the Corps of Engineers, 1973.

41. White, W. T., "Reusing Wastes is St. Petersburg Goal," *Water and Wastes Engineering*, March, 1975, p. 66.

42. Crites, R. W., "Wastewater Irrigation: This City Can Show You How," *Water and Wastes Engineering*, July, 1975, p. 49.

43. Applegate, C. H., and Gray, D. V., "Land Spreading Effluent From A Secondary Facility," *Water and Sewage Works*, July, 1975, p. 85.

44. Dove, L. A., "We Don't Waste Wastes," *The American City*, November, 1972, p. 68.

45. Dove, L. A., "Recycling Allows Zero Wastewater Discharge," *Civil Engineering*, February, 1975, p. 48.

46. Bogedain, F. O.; Adamczyk, A.; and Tofflemire, T. J., "Land Disposal of Waste-Water in New York State." New York State Dept. of Environmental Conservation, 1974.

47. Malhotro, S. K., and Myers, E. A., "Design, Operation, and Monitoring of Municipal Land Irrigation Systems in Michigan." Paper presented at the WPCF Conference, Denver, Colorado, October, 1974.

48. Dugan, G. L.; Young, R. H. F.; Lau, L. S.; Ekern, P. C.; and Loh, P. C. S., "Land Disposal of Wastewater in Hawaii," *Journal Water Pollution Control Federation*, 1975, 2067.

49. Bendixen, T. W.; Hill, R. D.; Schwartz, W. A.; and Robeck, G. G., "Ridge and Furrow Liquid Waste Disposal in a Northern Latitude," *Journal of the Sanitary Engineering Division, ASCE*, No. SA1, February, 1968, p. 147.

50. Thomas, R. E.; Jackson, K.; and Penrod, L., "Feasibility of Overland Flows for Treatment of Raw Domestic Wastewater." US EPA Environmental Protection Technology Series EPA-660/2-2-74-087, December, 1974.

51. Ayers, R. S., "Water Quality Criteria for Agriculture," UC Committee of Consultants, California Water Resources Control Board, April, 1973.

52. "Diagnosis and Improvement of Saline and Alkali Soils," U.S. Salinity Laboratory, Agriculture Handbook No. 60, U.S. Department of Agriculture, 1954.

53. "Evaluation of Land Application Systems," EPA-430/9-75-001, Office of Water Program Operations, Environmental Protection Agency, March, 1975.

54. Ellis, B. G., "The Soil as a Chemical Filter," in *Recycling Treated Municipal Wastewater and Sludge Through Forest and Cropland*. The Pennsylvania State University Press, 1973.

55. Taylor, A. W., and Kunishi, H. M., "Soil Adsorption of Phosphates from Wastewater," in *Factors Involved in Land Application of Agricultural and Municipal Wastes*, USDA, Agricultural Research Service, July, 1974.

56. Kardos, L. T.; Sooper, W. E.; Myers, E. A.; Parizek, R. R.; and Nesbitt, J. B., "Renovation of Secondary Effluent for Use as a Water Resource." EPA-660/2-74-016, February, 1974.

57. Parizek, R. R., "Site Selection Criteria for Wastewater Disposal—Soils and Hydrogeologic Considerations," in *Recycling Treated Municipal Wastewater and Sludge Through Forest and Cropland.* The Pennsylvania State University Press, 1973.

58. "Wastewater Treatment Study—Montgomery County, Maryland." CH2M/Hill Engineers, November, 1972.

59. Kirkham, M. B., "Trace Elements in Corn Grown on Long-Term Sludge Disposal Site," *Environmental Science and Technology*, August, 1975, p. 765.

60. Sorber, C. A.; Baysum, H. T.; Schaub, S. A.; and Small, M. J., "A Study of Bacterial Aerosols of a Wastewater Irrigation Site." Paper presented at the 48th WPCF Conference, Miami Beach, Florida, 1975.

61. Wellings, F. M.; Lewis, A. L.; and Mountain, C. W., "Pathogenic Viruses May Thwart Land Disposal," *Water and Wastes Engineering*, March, 1975, p. 70.

62. Pound, C. E.; Crites, R. W.; and Griffes, D. A., "Costs of Wastewater Treatment by Land Application." EPA-430/9-75-003, June, 1975.

63. Culp, G. L., "Example Comparisons of Land Treatment and Advanced Waste Treatment," Paper presented at EPA Technology Transfer Seminars, 1975.

64. "Evaluation of Land Application Systems," EPA-430/9-75-001, March, 1975.

11

Estimating the costs of wastewater treatment facilities

The purpose of this chapter is to present a guide for estimating the costs of various methods of wastewater treatment. No attempt is made to judge the relative merits of alternative treatment approaches; rather, the intention is to provide a basis for *preliminary estimates* of capital and operating costs of wastewater treatment processes which the reader has selected as potentially applicable to a given problem. There is no question that the most accurate approach to such estimates is to prepare cost curves for each treatment unit and then to combine the specific treatment units involved in a given plant to determine the estimate. This approach was used in an earlier, comprehensive report for EPA.[1] However, merely to cover comprehensively the chemical clarification of secondary effluent required 81 separate capital and operation and maintenance curves.[6] It is simply not practical completely to cover all unit processes of interest at that level of detail within this textbook. Thus, the approach taken here is to combine many separate cost curves into a simple, single curve for specific unit processes. The assumed design conditions are spelled out. The resulting curves will provide a convenient means of quickly estimating the general level of costs associated with a specific process.

It must be recognized that any generalized cost curves will not yield a precise estimate for any given plant. Local conditions may cause significant variations from such generalized curves. The limitations of such generalized cost curves might best be reinforced by quoting from a cost estimating report prepared for EPA:[1] "It must be recognized that the estimating data and methods presented in this report cannot in any way be used as a substitute for cost estimating based on detailed knowledge of a particular wastewater treatment plant situation. Thorough construction cost estimates must recognize not only the particular plant design characteristics, but also such items as current and projected labor costs, attitudes of proposed contractors regarding their need for work at the time of construction bidding, availability of materials, climate and seasonal factors, local site conditions, and many other variables which affect actual construction costs." However, generalized cost curves are often useful for considering the relative economics of alternative general approaches in such studies as regional water quality management studies or in preliminary evaluations of the general cost levels expected for a potential project.

COST INDICES

The use of any method of cost estimating requires careful consideration of inflation. This has been especially true in recent years since inflation of construction costs averaged about 9 percent per year in the period 1970–1975. Such rapid change in costs effects both the use of previous cost estimates to predict project costs and the planning for cost on a project which may be six months to a year away from implementation at the time the estimate is prepared.

The basis for cost indices used in the construction industry is specific construction material and labor costs. These costs are proportioned by a predetermined factor to derive an index. Among the most frequently used indices in the wastewater field are probably the *Engineering News Record* (*ENR*) Construction Cost Index and Building Cost Index.

The *ENR* indices were started in 1921 and intended for general construction cost monitoring. The *ENR* Construction Cost Index consists of 200 hours of common labor, 2500 lb of structural steel shapes, 1.128 tons of Portland Cement and 1.088 mbfm of 2 × 4 lumber. The *ENR* Building Cost Index consists of 68.38 hours of skilled labor and the same materials as are included in the construction cost index. The large amount of labor included in the construction cost index was appropriate prior to World

War II; however, on almost all contemporary construction, the labor component is far in excess of this amount. In fact, there should be little, if any, application of the Construction Cost Index to wastewater plant projects. This index does not include mechanical equipment, pipes, and valves, which are normally associated with such plant construction, and the proportional mix of materials and labor are not specific to wastewater plant construction.

To provide a more specific index the Environmental Protection Agency developed the Sewage Treatment Cost Index. This index was based on the cost components of a hypothetical 1 mgd trickling filter plant. The quantities of labor, materials, construction equipment and contractor's overhead and profit remain constant and the unit prices and price changes as derived from the U.S. Bureau of Labor Statistics and *Engineering News Record* are applied to the constant quantities to derive the index.

Values for the *ENR* Building Cost Index and the EPA Sewage Treatment Plant Index are plotted in Figure 11-1 for 1967 until early 1975, compared to Bureau of Labor Statistics (BLS) indices for concrete, fabricated steel

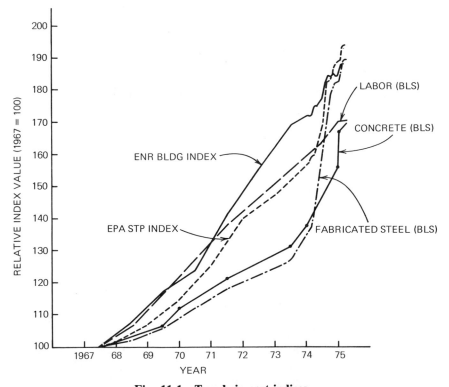

Fig. 11-1   Trends in cost indices.

products, and labor. A common base of 100 for 1967 was used. The EPA index did fairly well in reflecting the cost increase in concrete and steel; however, it lagged about 9 months behind the major price changes which took place in mid-1973. Labor increased steadily during this period, but concrete and steel jumped radically during 1973–1974. Obviously, the more specific an index is, the more accurately it will track cost change. The variation in inflation of various cost components cannot be accurately monitored by a single component index. If an index is based on an improper mixture of several single component indices, it will fail.

Indices are available which should accurately reflect changes in the major cost components of wastewater plants, which can be summarized as follows:

Equipment: estimated purchase cost of pumps, drives, process equipment, and other items which are factory made and sold as equipment.

Concrete: delivered cost of ready mix concrete.

Steel: reinforcing steel and miscellaneous steel required for weirs, launders, handrails, etc.

Electrical and instrumentation: the cost of electrical service and instrumentation associated with the process equipment.

The appropriate BLS wholesale price index for each of these categories is:

	Price category	*Index number*	*June, 1975 value*
Equipment	BLS general purpose machinery and equipment	114	180.1
Concrete	BLS concrete ingredients	132	174.0
Steel	BLS steel mill products	1013	195.0
Labor	ENR wage index (skilled labor)	—	213[a]
Electrical and instrumentation	BLS electrical machinery and equipment	117	140.9

[a]Kansas City, Missouri

The better to reflect the cost components of the types of plant being constructed today, EPA has recently developed a new index based on the components of a hypothetical 5 mgd activated sludge plant and a 50 mgd activated sludge plant followed by chemical clarification and filtration.[2] Each index is calculated quarterly for 25 cities around the country by an extensive computer program. Input to the programs consists of local wage rates, material prices, and some additional factors to account for productivity variations, vendor profit, and overhead. The plant programs are called Small City Conventional Treatment (SCCT) and Large City Advanced Treatment (LCAT). A representative design for each of the cases is converted into a list of labor and materials quantities for that case. The computer then applies the unit labor and materials costs for each of the 25 cities to the above quantities and computes a total cost. The total cost is then compared with a similar total cost for the same city in a datum period, which is the third quarter of 1973. The design basis for each of the cases may be summarized as follows:

Small City Conventional Treatment (SCCT)
Nominal plant size: 5 mgd
Unit processes: bar screen, grit chamber, primary clarification, conventional activated sludge, chlorination, gravity thickening, and vacuum filtration.

Large City Advanced Treatment (LCAT)
Nominal plant size: 50 mgd
Unit processes: bar screen, grit chamber, primary clarification, conventional activated sludge, lime clarification, recalcination, multimedia gravity filtration, chlorination, gravity thickening, vacuum filtration, and multiple hearth incineration.

The values of these indices on a national basis in the fourth quarter of 1975 were: SCCT = 110; LCAT = 120 (base value is 100 in third quarter of 1973). There were substantial regional variations with the range of indices as follows: SCCT = 77–139; LCAT = 75–142. Thus, it is prudent to determine the index for the specific locale involved. The cost curves which follow in this chapter are based upon fourth quarter 1975 cost levels.

BASIS OF CURVES

The cost curves presented in this chapter each include the basic construction costs, costs for engineering, costs for legal and administrative services, a contingency allowance, and costs for financing during construction. A factor of 35 percent of the basic construction costs was used for engineering, legal, administrative, and contingency costs. Financing during construction was estimated at 12 percent of the project costs based upon an average interest rate of 8 percent over a three year construction period. *Each curve includes all of the above factors.* Factors which are not included in the curves and which may be significant in a specific instance include raw sewage pumping, effluent disposal systems, any unusual site development problems (representative site development costs are incorporated in the curves), land costs, effluent storage systems, or standby treatment units.

The following sections describe the assumptions and basis of each of the process curves which follow. Throughout the various processes, a labor cost (total, including fringes) of $10/hour and a power cost of $0.02/kwh were used in calculating operation and maintenance (O & M) costs. The annual O & M costs shown do not include capital amortization but include labor, power, fuel, maintenance materials, and any chemical purchase required.

Activated Sludge (Figure 11-2)

The activated sludge plant costs are based upon the data presented in the report "Estimating Costs and Manpower Requirements for Conventional

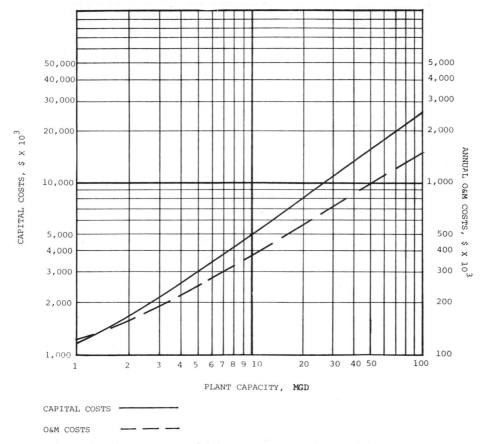

Fig. 11-2   Activated sludge plant costs, including primary treatment but not including sludge treatment.

Wastewater Treatment Facilities."[1] The following curves were integrated into a single curve for activated sludge plants: preliminary treatment, primary sedimentation, aeration basins and aeration equipment, secondary clarifiers, chlorine feed equipment, chlorine contact basins, primary sludge pumping, garage and shop, and administration and laboratory. In addition, site preparation costs were included in the total capital cost curve. These site preparation costs are those typically associated with the above treatment units and would not reflect any unusual foundation or excavation problems. A 1.14 factor was used to include costs of general yardwork and yard piping. A 1.35 multiplier was used to incorporate engineering costs, legal costs, administrative costs, and contingencies.

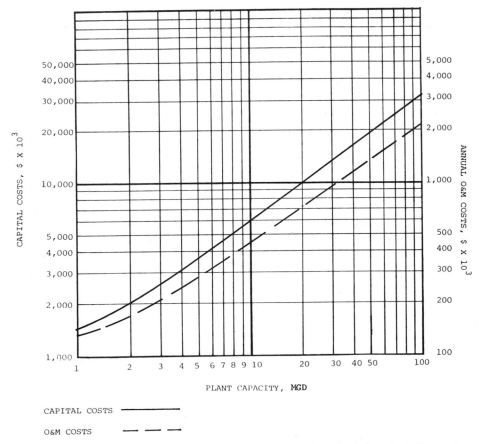

PLANT CAPACITY, MGD

CAPITAL COSTS ——————

O&M COSTS — — —

Fig. 11-3 Cost of nitrification with single stage activated sludge, including primary treatment by not including sludge treatment.

The basic design criteria used in estimating the activated sludge costs include the following: average overflow rate in the primary clarifier—900 gpd/ft^2; average detention time in the aeration basin—6 hours; average amount of air supplied—700 ft^3/lb BOD removed; secondary clarifiers— average overflow rate of 600 gpd/ft^2; average chlorination feed capacity— 5.0 mg/l; and, chlorine contact time—30 minutes.

The operation and maintenance costs were calculated by using the basic man hour requirements contained in the previously referenced report.[1] A total, average (including administrative, managerial, and operational labor) labor cost, including all overhead costs, of $10/hour was used. The resulting numbers for the unit processes described above were then calcu-

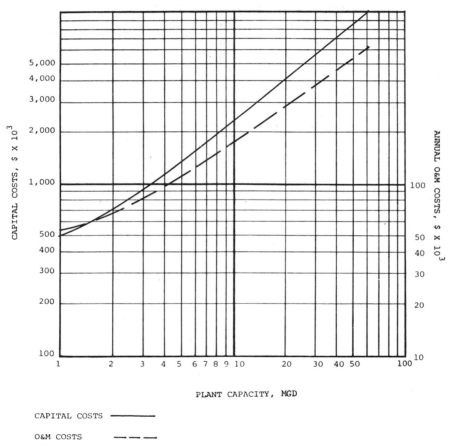

CAPITAL COSTS   ——————

O&M COSTS   — — —

Fig. 11-4   Cost of biological nitrification downstream of activated sludge in mixed reactor.

lated. In addition, laboratory work, yardwork, administration, and general costs were calculated and included in the total operation and maintenance costs for activated sludge plants.

Biological Nitrogen Removal (Figures 11-3 through 11-6)

Costs are presented for both mixed reactors and fixed film denitrifiers. Costs for nitrification in a single stage activated sludge plant are based on an average aeration time of 9.3 hours, an average F/M ratio of 0.19, and 2480 lb oxygen/mgd/day. Other costs for biological nitrogen removal are based

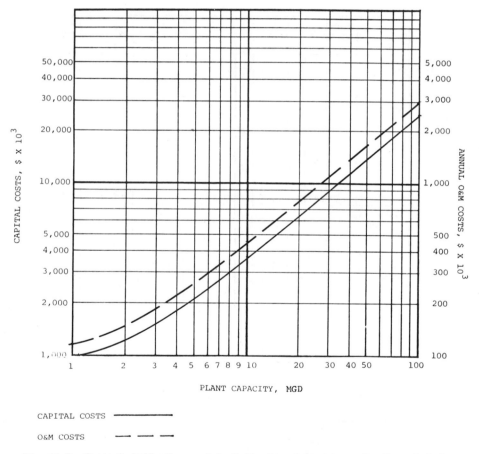

CAPITAL COSTS ————————

O&M COSTS — — — —

Fig. 11-5   Cost of nitrification and denitrification downstream of activated sludge in mixed reactors.

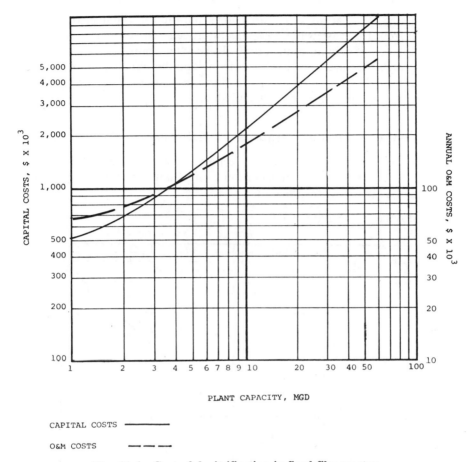

PLANT CAPACITY, MGD

CAPITAL COSTS ─────────

O&M COSTS   ─ ─ ─

Fig. 11-6   Cost of denitrification in fixed film reactor.

on data developed by the U.S. Environmental Protection Agency, Cincinnati laboratory.[3-5] Mixed reactor costs are presented for separate mixed reactors for nitrification and denitrification downstream from the conventional activated sludge plant described earlier. Nitrificaton is achieved in a 4 hour detention time reactor followed by a clarifier operating at 630 gpd/ft^2. Denitrification occurs in a 4 hour detention time mixed reactor, followed by a 21 minute aeration tank, followed by a final clarifier operating at 740 gpd/ft^2. Recycle pumps with a capacity of 100 percent of the average flow are provided for each set of mixed reactors and clarifiers.

The costs of fixed film denitrification are based on the use of $\frac{1}{4}$ in. rock as the media for the attached biological growth used to accomplish the denitrification step. No clarifier is required for separation of the denitrifying organisms from the wastewater with this approach. The estimates are based on the use of a 15 minute contact time. The rock is housed in 12 ft diameter by 21 ft high pressure vessels. The costs include the influent pump station, backwash and surface wash pumps, and the methanol feed system. An average methanol dose of 60 mg/l is included.

Chemical Clarification (Figures 11-7 through 11-9)

Figure 11-7 presents the costs for tertiary chemical clarification, rapid mixing, flocculation, and sedimentation using alum. The following criteria were used as the basis for the costs:

Alum dosage   200 mg/l

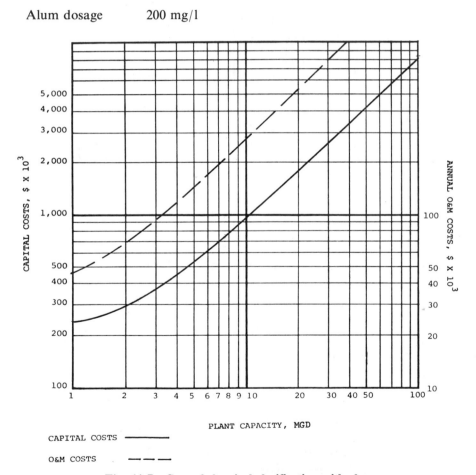

CAPITAL COSTS ─────────

O&M COSTS   ─ ─ ─

Fig. 11-7   Cost of chemical clarification with alum.

Alum cost	$73/ton
Polymer dosage	1 mg/l
Polymer cost	$1/pound
Rapid mixing	30 seconds, $G = 600$
Flocculation	15 minutes, $G = 70$
Sedimentation	800 gpd/ft^2
Sludge quantity	710 lb/million gal
	(0.5% solids in clarifier underflow

The cost curve includes chemical feed and storage and chemical sludge pumping but does not include chemical sludge processing or disposal. The annual O & M costs include the cost of chemical purchase.

Figure 11-8 presents the costs for single stage lime clarification (chemical

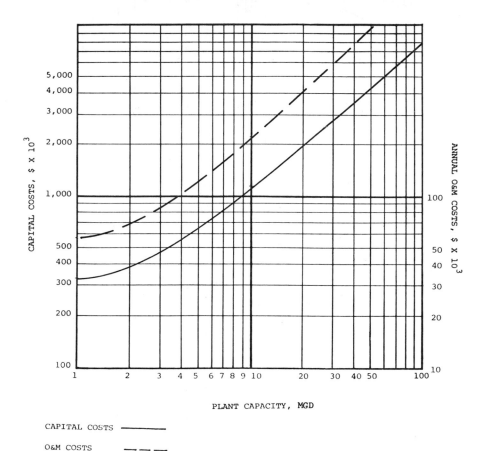

PLANT CAPACITY, MGD

CAPITAL COSTS ———————

O&M COSTS — — —

Fig. 11-8   Cost of chemical clarification in a single stage lime system.

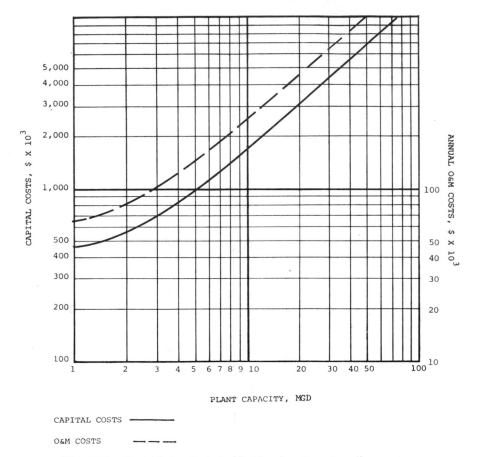

PLANT CAPACITY, MGD

CAPITAL COSTS ————

O&M COSTS — — —

Fig. 11-9   Cost of chemical clarification in a two stage lime system.

storage and feed, rapid mixing, flocculation, sedimentation, and recarbonation). The recarbonation *basin* costs are included but the cost of the CO_2 feed system is not included. Costs for various sources of CO_2 (liquid, submerged burners, and stack gas) are given in Chapter 3 and the appropriate costs should be added to the values obtained from Figure 11-8. The design criteria used were:

Lime dosage	300 mg/l as CaO (80% recycle from recalcining)
Lime cost	$28/ton
Polymer dosage	1 mg/l
Polymer cost	$1/lb
Rapid mixing	30 seconds, $G = 600$

Flocculation	15 minutes, $G = 70$
Sedimentation	1000 gpd/ft^2
Chemical sludge pumping	24 gpm/mgd
Recarbonation basin	5 minute detention

No sludge handling or disposal costs are included. Annual O & M costs include chemical purchase, assuming all lime is purchased (e.g., no recalcining and lime recycle). Should lime recalcining and recycling be practiced, the O & M costs shown in Figure 11-8 should be corrected by deducting the lime costs ($12,800/year/mgd) from Figure 11-8 and then adding makeup lime costs. Lime recalcining costs are presented in Figure 11-18.

Figure 11-9 presents costs for lime clarification followed by two stage recarbonation (e.g., chemical storage feed, rapid mixing, flocculation, sedi-

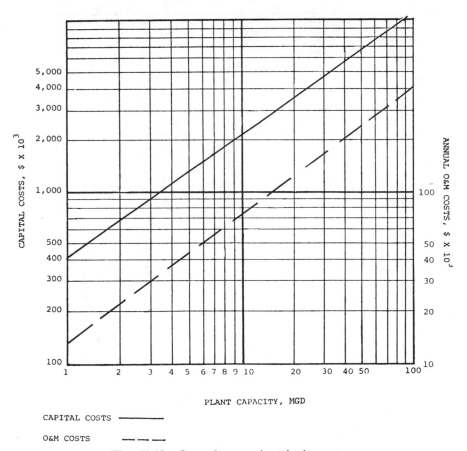

CAPITAL COSTS ———

O&M COSTS — — —

Fig. 11-10   Cost of ammonia stripping system.

mentation, first stage recarbonation basin, sedimentation, second stage recarbonation). As with the single stage system, the CO_2 feed costs are not included and must be determined from the curves in Chapter 3. The recarbonation sedimentation basin is sized at 1000 gpd/ft^2 and both recarbonation basins are sized for 5 minutes detention. The same adjustments described for the single stage system costs would be required if recalcining and lime recycle are used.

Ammonia Stripping (Figure 11-10)

The cost curves are based on a hydraulic loading rate of 2 gpm/ft^2 and an air flow of 400 cf/gallon. The curves do not include the cost of pH elevation for the process nor any wastewater pumping costs.

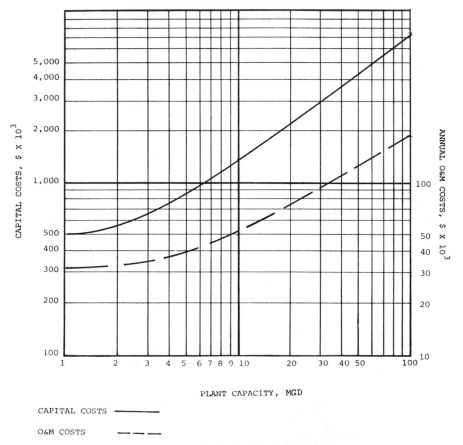

Fig. 11-11   Cost of effluent filtration.

Effluent Filtration (Figure 11-11)

The costs are based on gravity filtration at 5 gpm/ft^2 with hydraulic back-wash-surface wash. A minimum of 4 filters is provided at all capacities to provide flexibility in operation should one filter be out of service. Power costs are based on each filter backwashing once per 12 hours. Capital costs include the basic filter structures, media, backwash, and surface wash systems, and polymer feeding (filter aid) equipment. The curves do not include any costs associated with filter influent or effluent pumping.

Activated Carbon Adsorption and Regeneration (Figures 11-12 and 11-13)

These costs are based on cost curves presented in the 1973 edition of the EPA Technology Transfer Manual on carbon adsorption. A 20 percent

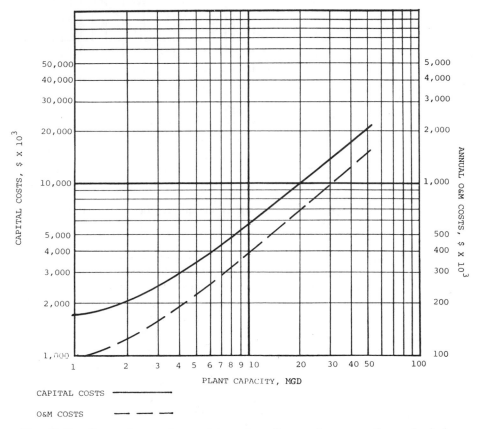

Fig. 11-12   Cost of granular carbon adsorption and regeneration—physical-chemical system (1500 lb carbon/million gal).

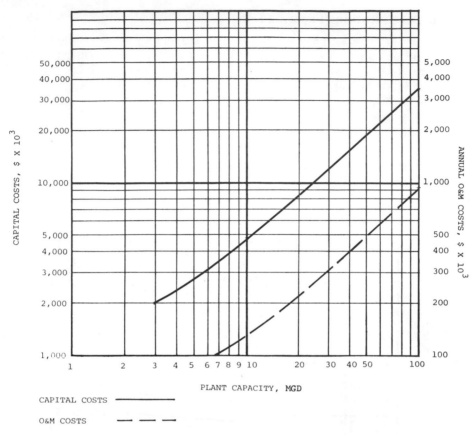

CAPITAL COSTS ——————

O&M COSTS — — —

Fig. 11-13   Cost of granular carbon adsorption and regeneration—tertiary system (200 lb carbon/million gal).

allowance for yardwork and 12 percent for financing during construction was made in adjusting these curves. Two curves are presented. One is based upon tertiary use of activated carbon adsorption based on a carbon dosage of 200 lb/million gal. The second curve is based upon a carbon dosage of 1500 lb/million gal to reflect conditions if the carbon is used in a purely physical-chemical system without any upstream biological treatment. Carbon contact time provided is 30 minutes in an upflow, countercurrent, expanded bed. Carbon regeneration is provided in a multiple hearth furnace loaded at 40 lb carbon/ft^2/day (allowing 30 percent downtime) with 8 percent carbon loss per cycle (makeup carbon costs are included at $0.50/lb). Costs include pumping wastewater (40 ft total head) through the columns. Fuel requirements were based on 4250 Btu/lb carbon at a cost of $1.50/10^6 Btu.

Ammonia Removal by Selective Ion Exchange (Figure 11-14)

The following criteria were used in estimating the system costs:

Design flow rate	15 BV/hour
Ammonia exchange cap.	6.6 eq/ft^3
Zeolite volume	430 ft^3/mgd
Ion exchange vessels	6 ft deep, reinforced concrete gravity flow units
Cycle time	13 hr including an 11 hr exhaustion cycle and 2 hr for regeneration, rinsing, and reprocessing of rinse water
Throughput	165 BV/cycle
Regenerant recovery	Closed loop stripping tower (see Figures 7-9 and 7-22)

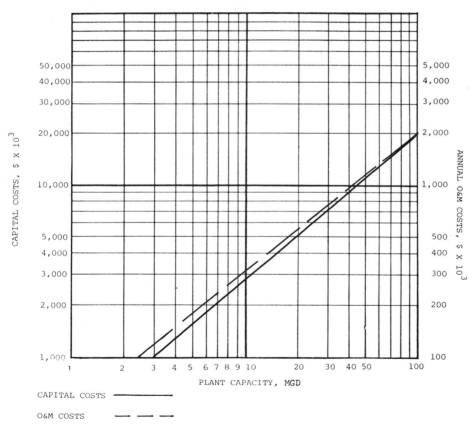

CAPITAL COSTS ————————

O&M COSTS — —— ——

Fig. 11-14   Cost of ammonia removal by selective ion exchange.

No credit for income from the potential sale of ammonium sulfate is incorporated in the cost curve.

Organic Sludge Handling and Disposal

There are so many combinations of sludge characteristics, mixtures, and unit processes that can be used for organic sludge handling and disposal that it is impractical to estimate the costs of all combinations in the space available. Detailed estimates for various system components may be found in reference 1. The process used as the basis for the general cost curve presented herein utilizes gravity thickening of the primary sludge, dissolved

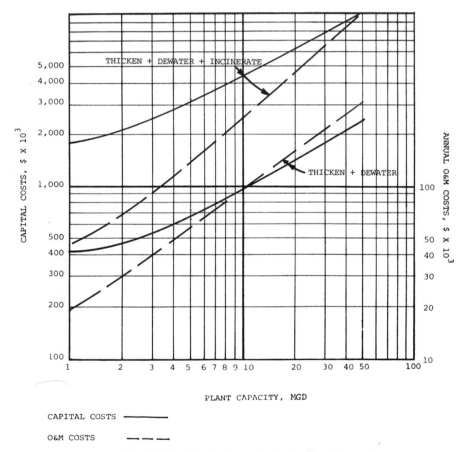

Fig. 11-15   Cost of organic sludge handling.

air flotation of waste activated sludge, blending of the thickened sludges, dewatering by vacuum filtration, and multiple hearth incineration.

The following criteria were used:

Gravity Thickener (Primary Sludge)

Solids load:	20 lb/ft^2/day
Solids concentration in,	5.0%
Solids concentration out,	10.0%

Dissolved Air Flotation Thickener (Waste Activated Sludge)

Solids loading:	20 lb/ft^2/day
Solids concentration in,	1%
Solids concentration out,	3%

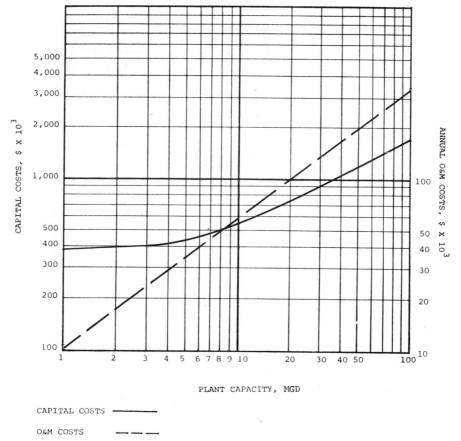

PLANT CAPACITY, MGD

CAPITAL COSTS ———————

O&M COSTS   — — —

Fig. 11-16   Cost of alum sludge thickening and dewatering.

Vacuum Filtration

Solids loading	3 lb/ft^2/hr
Polymer dosage	18 lb/ton
Solids concentration out	16%
Run time	20 hr/day

Multiple Hearth Incineration

Loading	6 lb/ft^2/hour
Downtime	30%

Sludge quantities used were 1080 lbs primary sludge/million gal and 875 lb waste activated sludge/million gal. Figure 11-15 shows the costs associated with the thickening and dewatering portion of the process as well as the costs of the total system of thickening, dewatering, and incineration.

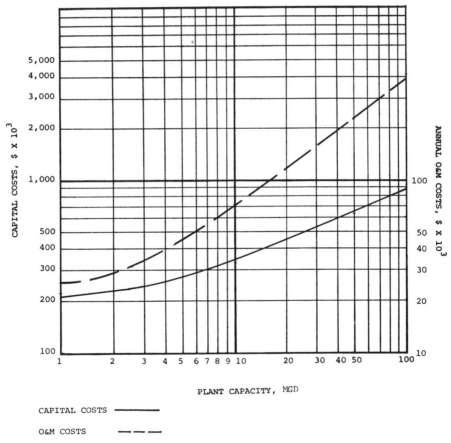

CAPITAL COSTS   —————

O&M COSTS   — — —

Fig. 11-17   Cost of lime sludge thickening and dewatering.

Chemical Sludge Handling (Figures 11-16, 11-17, 11-18)

Curves are presented for alum sludge thickening and dewatering (Figure 11-16); lime sludge thickening and dewatering (Figure 11-17); and lime sludge thickening, dewatering, and recalcining (Figure 11-18).

The alum sludge costs are based on a tertiary alum sludge which would be subjected to flotation thickening at 2 lb/hour/ft^2 and then vacuum filtered at 2 lb/hour/ft^2. The estimated alum sludge quantity is 710 lb/million gal for an alum dose of 200 mg/l. Lime sludge costs are based on gravity thickening at 200 lb/day/ft^2, centrifugation (4.8 gpm/mgd), and recalcining at 7 lb/hour/ft^2 (wet basis), with 40 percent allowance for downtime. Lime sludge quantity was estimated at 2700 lb/million gal for a dose of 300 mg/l as CaO.

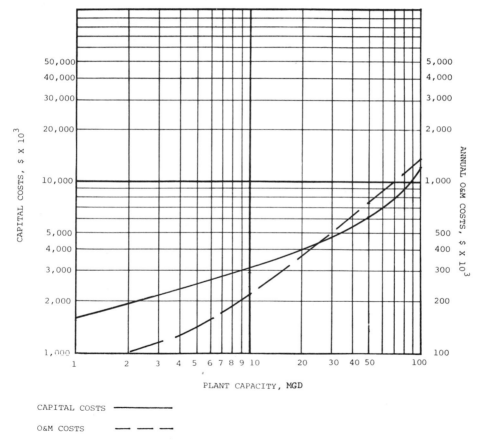

Fig. 11-18   Cost of lime sludge thickening, dewatering, and recalcining.

TABLE 11-1   Unit Process Costs at 10 mgd.

Process (design criteria as defined in Chapter 11)	Capital Costs[a]	Annual O & M costs	Total costs,[b] ¢/1000 gal
Activated sludge	5,000,000	380,000	23.3
Organic sludge dewatering and incineration	4,500,000	250,000	18.5
Nitrification—single stage activated sludge	6,000,000	450,000	27.8
Nitrification—second stage mixed reactor	2,300,000	180,000	10.9
Nitrification—denitrification downstream of activated sludge in mixed reactors	3,800,000	450,000	22.2
Denitrification—fixed film	2,100,000	180,000	10.7
Alum clarification	1,000,000	290,000	10.5
Lime clarification—single stage	1,200,000	220,000	9.1
Lime clarification—two stage	1,800,000	260,000	11.8
Ammonia stripping	2,200,000	73,000	7.7
Effluent filtration	1,400,000	52,000	5.0
Carbon adsorption and regeneration			
physical-chemical	5,800,000	400,000	25.6
tertiary	4,700,000	140,000	16.0
Selective ion exchange—NH_4 removal	2,900,000	310,000	16.0
Alum sludge dewatering	550,000	60,000	3.1
Lime sludge dewatering	350,000	70,000	2.8
Lime sludge dewatering and recalcining	3,100,000	220,000	14.0

[a]Fourth Quarter, 1975. Includes engineering, legal, fiscal, administrative, contingencies, and financing during construction. Does not include effluent disposal, unusual site development problems, land costs, reserve funds.
[b]Capital amortized at 7 percent over 20 years.

Illustrative Unit Process Costs

Table 11-1 presents the estimated costs for various unit processes for a 10 mgd design capacity operating at capacity. Because of the generalized nature of the estimates and the inherent assumptions related to design criteria, the costs in Table 11-1 cannot be used to determine the most cost effective solution for a specific project, but rather the estimates serve to put the overall costs in perspective. A cost of 10¢/1000 gal is equivalent to a cost of $1.05/month/home at an average flow of 350 gpd/home.

REFERENCES

1. "Estimating Costs and Manpower Requirements for Conventional Wastewater Treatment Facilities." Black and Veatch Engineers, EPA Report 17090 DAN 10/71, 1971.

2. "EPA Switches to New Cost Indexes," *Journal Water Pollution Control Federation*, February, 1976, p. 233.

3. Smith, R., "Costs of Wastewater Renovation." EPA memo, November, 1971.

4. Smith, R., "Updated Cost of Dispersed Floc Nitrification and Denitrification for Removal of Nitrogen From Wastewater." EPA memo, April 13, 1973.

5. Smith, R., "The Cost of Columnar Denitrification for Removal of Nitrogen From Wastewater." Internal Report, EPA, January, 1972.

6. "Costs of Chemical Clarification of Wastewater," prepared by Culp/Wesner/ Culp, EPA Contract 68-03-2186, 1976.

12

Selecting and combining unit processes

Table 1-2 provides information on typical effluent qualities expected from various unit process combinations. It is the purpose of this chapter to provide a summary of the general factors to be considered in selecting and combining processes to achieve specific effluent quality goals. It is impractical to discuss all possible combinations of unit processes. Thus, selected examples are offered to illustrate the considerations involved in process selection.

GENERAL CONSIDERATIONS

Nature of the Raw Wastes

Current or future industrial wastes could have a significant effect on the capacity and performance of the treatment facility. Industrial pretreatment requirements may minimize these effects but the process should be flexible enough to accommodate variations in pretreatment efficiency.

TABLE 12-1   Process Capabilities.

	BOD, mg/l	COD, mg/l	Turbidity, JTU	PO_4^{3-}, mg/l	N, mg/l	TDS, mg/l	Coliform, MPN/100 ml	Approximate Cost @ 10 mgd, ¢/1000 gal[a] Per Process	Cum.
A. Overall Process Capabilities Without Chemical Coagulation									
Secondary effluent	20–40	30–80	10–40	20–30	15–30	400–600	$1-3 \times 10^6$	23	23
Carbon adsorption and regeneration	3–6	10–12	1–2	20–30	18–28	400–600	$1-3 \times 10^6$	16	39
Disinfection	3–6	10–12	1–2	20–30	18–28	450–650	2–100	1	40
Demineralization (R—O)	1	0.8–2	1	0.4	3–5	100	2	40	80
B. Overall Process Capabilities with Chemical Coagulation									
Secondary effluent	20–40	30–80	10–40	20–30	15–30	400–600	$1-3 \times 10^6$	23	23
Chemical coagulation (lime)	3–7	30–50	2–7	0.5–2	15–30	400–600	2000	9	32
Filtration	1–2	25–45	0.1–1	0.1	15–30	400–600	50	5	37
Nitrogen removal (ion exchange)	1–2	25–45	0.1–1	0.1	1–3	400–600	50	16	53
Carbon adsorption and regeneration	0–1	3–10	0.1–1	0.1	1–3	400–600	50	16	69
Disinfection	0	3–10	0.1	0.1	1–3	450–650	0	1	70
Demineralization (R—O)	0	0.8–1.5	0.1	0.1	0.3–1	100	0	40	110

[a]Includes capital, operation, and maintenance costs. EPA STP Construction Cost Index = 200. Excludes sludge handling costs.

Effluent Quality Required

One of the first considerations in the selection of unit processes is the degree of treatment required at present and that which may be required in the future. Once the required water quality is established, one must then select unit processes to accomplish the desired result. Obviously, for almost every situation there are a number of combinations of unit processes which will satisfy conditions. Known and possible future water quality requirements must be kept in mind so that unit processes selected for the original plant can be expanded or supplemented readily to meet future needs.

Reliability

The simplicity and reliability of unit processes and combinations thereof are top priority factors. While simplicity and reliability are not synonymous, they are very closely related. The simpler a process is to operate and control, the more likely it is to operate successfully and continuously. Regardless of how highly efficient a process may be when operating at its very best under rigidly controlled conditions, if it is sensitive to minor changes in flow, temperature, or applied water quality, or if it requires constant expert control and supervision, then it is not likely to operate as intended all of the time in practical plant application. Processes of this type may make excellent research or laboratory exercises and may be instrumental in developing useful new treatment concepts, but they may be ill suited to direct adaption to plant scale use. Simplicity also adds to reliability by making it easier to design and engineer the structures and equipment needed to carry out the process.

The state of development of a process, particularly the number of fullscale plants from which engineering design and equipment information, actual operating results, and cost data can be obtained, has a great influence on the degree of reliability which may be expected of it. Obviously, if new approaches are not tried on a plant scale, little progress will be made. However, the engineer and client involved should enter such a project aware of the degree of uncertainty involved. The acceptable risk level must be determined individually for each project.

The operation of conventional secondary treatment plants is often notably erratic. Advanced wastewater treatment processes which follow secondary ones must be designed to cope with this situation. They must be able to handle these variations in performance and still produce an effluent of uniform quality.

The first tests of any new treatment method are usually made in the laboratory. The next step is pilot plant operation on a continuous flow basis. Pilot plant work gives a good insight into the results to be expected from liquid processing. Unfortunately, it does not always adequately evaluate sludge problems or problems associated with recycled process streams or return solids. It is only at plant scale that the real engineering design problems and the practical operation problems are faced for the first time. The transformation of pilot plant data gathered by biologists, engineers, chemists, and laboratory technicians into workable engineering designs is a difficult and demanding task, and one not to be taken lightly or for granted. It is here, in the hands of the sanitary engineering designer, that the project either fails or succeeds. The success depends not only on how well he selects and combines the unit processes, but more importantly how well he selects and specifies the performance of equipment, and how well the excruciating details of design are executed. Actually, the development of this engineering design knowhow forms the great gap in the development and application of new and better methods for wastewater treatment. There is a great backlog and wealth of information and scientific research which cannot be used until it is applied to actual working plants. This transformation requires the design of new structures and equipment, and the provision of sufficient safety and flexibility in the design to allow for any contingencies which may not be apparent or solvable at pilot scale.

Sludge Handling

Of major importance in process selection are the circumstances related to the handling and ultimate disposal of the sludge produced. In locations where there are available large, remote areas of land, almost any kind of sludge, wet or dry, stable or decomposing, can be, and is, disposed of by hauling or pumping to these land disposal sites. In many places this method for sludge disposal probably will not be tolerated indefinitely, but might suffice for the time being. Ocen dumping has long been an easy way to evade the knotty problems involved in proper sludge disposal. However, the nuisances which have been created and the damages wrought to beaches and coastal waters have aroused the public to the point where this method is now in almost universal public disfavor.

Many methods of sludge disposal involve dewatering of the sludge. As pointed out previously, the ease with which sludge may be dewatered is a prime factor in unit process selection. There are many alternative ways to process the liquid component of wastewater to secure the desired result at about equal costs, but there may be very few ways to satisfactorily and economically dewater sludge. In certain wastewaters, dewatering of mixtures of organic and chemical sludges may be satisfactory, but care must be taken to check this out before designing a fullscale plant. Favorable pilot plant tests are a prerequisite.

Another approach which has been used successfully is to keep all organic sludges entirely separate from all chemical ones. Then conventional equipment used for either of these types of sludge can be installed. Pilot tests are still highly desirable, even with this approach.

Process Compatibility

Another consideration in process selection is compatibility with other unit processes used in the overall treatment scheme. The possible effects of waste streams or recycled solids is also very important.

The most favorable point for addition of chemical coagulants and flocculants may be influenced by the form in which phosphorus is present. Raw sewage in primary tanks, for example, usually contains polyphosphates which are later broken down to orthophosphate by biological treatment. Polyphosphates in relatively high concentrations (more than 1 mg/l of P_2O_5) are capable of interfering with coagulation and sedimentation. Orthophosphate compounds produce no interference with normal coagulation and sedimentation. If the amount of polyphosphate interference in the primary stage is great, then, obviously chemical addition should follow reversion of this material to orthophosphate.

The optimum pH for various unit processes is a factor which influences their compatibility, and may affect placement of processes in the plant flow sequence. There are many examples of this point. Various chemical coagulants operate best within certain pH ranges. In this case, there may be a choice—either change the pH or change the chemical used. It is also possible to broaden the effective pH range of chemicals by use of coagulant aids, such as activated silica or polymers. Biological processes may be adversely affected by pH values above 9.3, although it is well to keep in mind that activated sludge cultures have great buffer capacity and it may be possible to allow the pH of the influent to activated sludge basins to exceed 9.3 so long as the pH in the aeration tank itself does not exceed this value. Phosphorus removal by lime requires raising the pH to values of 9-11. Ammonia stripping also is most efficient at pH values of 10.5–11. Calcium reaches minimum solubility at a pH near 9.3. Carbon adsorption of organics is best at low pH values, and is very poor at values above 9.0. Chlorination is more effective at low pH values, at low turbidities, and in waters with low chlorine demand. Granular activated carbon adsorption is favored by good clarification of the applied water. Granular activated carbon will adsorb chlorine. With information on process interrelationships and pH effects, then, a logical process sequence for maximum reliability and effluent quality would be biological treatment followed by advanced treatment. Although the costs may exceed those of approaches combining certain steps, there are at least three advantages to this arrangement which must be given due consideration: the pH is favorable for biological treatment; the organic sludges are kept separate from the chemical sludges, an arrangement which usually minimizes sludge dewatering problems; and the polyphosphates in the raw wastewater are converted to orthophosphate before chemical coagulants are applied, thus avoiding potential interference with chemical coagulation.

In considering compatibility, attention must be given to the point in the overall process at which centrate, filtrate, backwash water, scrubber water, and other plant process water and recycle streams are introduced. It is wise to provide more than one point of return for some of these streams, particularly to take care of those times when certain plant units are temporarily out of service.

Air Pollution

A careful evaluation of the ultimate fate of pollutants removed from the wastewater must be made to ensure that water pollution control has not been achieved at the expense of air pollution.

Resource Consumption

High degrees of wastewater treatment cannot be achieved without the expenditure of resources such as power and chemicals. It is obviously desirable to minimize the consumption of these resources and the relative consumption by alternative processes is a factor for consideration. Potential beneficial reuse of the resources in the wastewater should also be considered carefully.

Space Requirements

The relative space requirements of alternative processes is another factor in process selection.

Safety Considerations

Any potential hazards within the plant boundaries or those that could affect the surrounding area due to plant malfunction or transport of materials to or from the plant must be considered.

Costs

The final factor in process selection is indeed important, and that is capital and operating costs. It is obviously and patently false economy to use a process which will not do the job because it is cheap to install and operate. In those cases where only a low risk level is acceptable, if the designer has a choice between a proven process with a high degree of reliability and a process which may be cheaper, but is of unknown or questionable reliability, then prudence is on the side of the more costly but workable process. One special word of caution is appropriate with regard to the attractive and tempting possibility of combining various processes into a single process or into a single structure. By reducing the total numer of structures, there are obvious savings in construction costs. Actually, this initial cost saving is usually the only advantage of combining processes, and there may be many disadvantages. The history of water and wastewater treatment is replete with failures to satisfactorily combine two or more processes. Yet some of these mistakes are perpetuated in present day designs, and new unworkable combinations may be devised. This is not to say that some of these efforts may not end in success, but only to inject a word of caution and to point out some of the inherent weaknesses in making one process out of two or more. First of all, there is invariably a loss of control or flexibility in operation by combining two processes. A change which benefits one of the processes may interfere with or reduce the efficiency of the other. Also, it is much more difficult to relate cause and effect, which is necessary to make treatment adjustments, because of the greater complexity introduced by the combination. The following sections discuss some illustrative consideration related to specific process needs.

COAGULATION-SEDIMENTATION

Upflow solids contact basins of the sludge blanket type combine the four processes of rapid mixing, flocculation, settling, and solids recycling in a single unit. They have been applied successfully to treatment of a constant flow of water (principally well waters) which are of constant physical and chemical quality, at a considerable savings over the use of four separate operations. However, these sludge blanket basins are often not stable when operated at varying flow rates, or with waters which have significant variations in chemical composition or physical characteristics, such as temperature or organic content. It is difficult to change the point of chemical addition in this type of basin if it becomes necessary. Also with changes in flow rate or chemical composition or physical character of the applied water, or with the accumulation of organics in the sludge blanket, the chemical dosages required to maintain a sludge blanket of the proper density change so frequently and rapidly as to present an impossible operating condition, and the basin fails to function as a settling device. Wastewaters possess properties which make operation of the sludge blanket clarifiers difficult. Experience indicates that this type of equipment can be used satisfactorily for wastewater treatment if the overflow rates are the same as for a conventional horizontal flow clarifier, and if the sludge is removed from the basin as rapidly as it reaches the bottom of the tank. Separate rapid mixing, flocculation, and settling basins with provisions for external sludge recirculation are much more satisfactory for treatment of wastewater. Mutual interference is avoided, and complete flexibility of operation and control is afforded.

PHOSPHORUS REMOVAL

One method demonstrated upon a plant scale to provide efficient, consistently reliable phosphorus removal is the unit process of chemical coagulation. In an integrated biological-chemical process approach, the coagulant may be added either ahead of the primary clarifier or directly to the aeration tank of an activated sludge system. When added ahead of the primary clarifier, alum, iron or lime may be used. When added directly to the aeration tank, iron or alum salts are used. In the approach which separates the physical-chemical processes from biological treatment, any of the above coagulants may be added to the secondary effluent with subsequent settling downstream. The following major points of comparison should be evaluated.

Effluent Quality Achievable

The literature indicates that typically the systems in which chemical coagulation is integrated with the biological process provide overall phosphorus removals of 75–85 percent, corresponding to effluent phosphorus concentrations of 1–2 mg/l. Subsequent filtration of effluents from such an integrated process may reduce the phosphorus concentrations to 0.5 mg/l if the coagulant dose is proper. In systems using chemical coagulation of secondary effluent as a separate process, lower phosphorus concentrations may be achieved. For example, the South Tahoe project produced an average phosphorus concentration of 0.05 mg/l for an entire year. By separating the processes of chemical treatment and biologic treatment, it is possible to employ efficient rapid mixing and flocculation devices. Also, it is possible to optimize the coagulant dose solely for the phosphorus removal desired without concern for the effects of such dosages on the biological treatment system. The same features which permit lower phosphorus concentrations to be achieved also result in the ability to achieve lower suspended solids in the final effluent when the physical-chemical processes are downstream of the biological process.

Flexibility of Operation

Separation of the chemical coagulation process from the biological process keeps the chemical and biological sludges separate. This provides the designer with more options in terms of coagulant choice and sludge handling. For example, in many instances, use of lime as the coagulant in a tertiary step has an advantage in that coagulant recovery may be achieved. This reduces the quantity of sludge requiring ultimate disposal and may, in some instances, provide a more economical coagulation process than the use of alum or iron salts. Separation of the chemical step also provides more operational flexibility to insure *maximum* removal of phosphorus, if this is a goal. Recycle of concentrated chemical sludges to the rapid mix or flocculation basins can be accomplished to maximize phosphorus removal when the chemical coagulation is carried out in a separate step. The coagulant dose can be selected solely on the basis of phosphorus removal rather than on the maximum dose compatible with the operation of the biological system.

Reliability

The addition of a chemical coagulation and settling step downstream of a secondary biological plant provides a more reliable system. Any engineer who has worked in an activated sludge plant realizes that the settling nature of the biological floc may vary from day to day. These variations may be dampened by the additions of coagulants for phosphorus removal. However, the variability in settling nature and in dewatering characteristics will be greater in dealing with a mixture of biological and chemical floc than in dealing with a chemical floc alone. The presence of a clarifier and coagulation facility downstream of the secondary clarifier provides an important tool for controlling the amount of solids discharged to the receiving stream. Variations in solids carryover from the secondary clarifier can be easily removed by proper operation of the chemical clarifier. If still more units, such as filtration or activated carbon, are downstream, the chemical clarification step then serves to increase the reliability of the downstream processes by reducing the opportunity for high solids carryover which could blind any downstream filtration processes.

Cost

Certainly one of the incentives for considering combining chemical and biologic treatment processes into common treatment units is the potential cost savings. However, if one examines the total cost involved in removal of phosphorus by chemical coagulation, one soon discovers that about 75 percent (this percentage may vary somewhat with the size of the plant) are involved in operation costs such as raw chemical cost and labor, which are largely independent of the point at which chemicals are added. There is obviously a savings in capital cost in that the separate chemical clarifier is eliminated by the integrated biological-chemical approach. However, if one examines the total cost over the life of the project, do the economic incentives justify the reduced flexibility, reduced efficiency, and reduced reliability resulting from the integration of the process? In evaluating the economics, one must also consider the relative cost of handling the resulting sludges. The ability to recover and reuse coagulants when the chemical step is separated from the biological step may provide savings which offset the higher capital cost associated with the tertiary sequence.

NITROGEN REMOVAL

The biological systems for providing nitrogen removal are all based upon the principle of biological nitrification and denitrification. There are a variety of approaches for applying these basic concepts. Some involve three activated sludge systems in series, where the first provides oxidation of the carbonaceous materials, the second provides oxidation of the nitrogenous material, and the third is an anoxic system. In all approaches, some supplemental oxygen demand source (such as methanol) is added to the anoxic system to make the reaction times required for denitrification practical. Alternative approaches to applying these biological reactions include the use of fixed film reactors for the denitrification step and, in some cases, also for the nitrification step. Also, it may be possible to combine the first two aerobic biological processes into one mixed reactor or one fixed film reactor, although some loss of reliability in nitrification may result. Although there are a variety of approaches available for applying the biological concept, they all are dependent upon the successful series operation of more than one biological treatment system. The biological organisms involved are sensitive to pH, temperature, and toxic materials. If reliable nitrogen removal is to be achieved on a year round basis, the biological systems must be designed for the coldest operation conditions. The fact that the nitrifying reactions depress the wastewater alkalinity, which may result in a depression in pH in some waters while the optimum range for biological nitrifiers is about 8.5, may result in variations in efficiency unless proper pH control is provided. The Contra Costa approach (Figure 7-48) of applying lime to the raw wastewater prior to entering the biological nitrogen removal system has merit in that it may reduce the concentration of potentially toxic heavy metals by forming insoluble metal hydroxide salts at high pH values while providing alkalinity to offset the acidic effect of the nitrification step. The Contra Costa system is an excellent example of combining unit processes to their mutual advantage.

The nitrifying organisms are slow growing organisms. If some mechanical or hydraulic upset occurs within the plant resulting in the loss of these organisms, a prolonged period where nitrogen removal efficiency is depressed will result. The three sludge biological nitrogen removal systems consist of three biological processes in series, each of which is dependent upon successful operation of all the processes upstream. The use of fixed film reactors provides a more positive means of retaining the critical biological growths required for the successful operation of these systems. In many instances, removing most of the nitrogen most of the time by the biological approach may provide adequate protection of the receiving watercourse. However, for cases *requiring* removal of essentially all of the nitrogen all of the time, use of physical-chemical nitrogen removal process as backup to (or in place of) the biological process may be required.

Chapter 7 reviewed three major physical-chemical alternates for nitrogen removal: ammonia stripping, selective ion exchange, and breakpoint chlorination. Many engineers hearing only partial reports of the full scale installation of ammonia stripping at South Tahoe tend to discount this approach as an alternative worthy of consideration. The severe winter weather and problems of scaling of the stripping tower packing have presented problems at Tahoe. However, the recent developments on this process and its inherent merits certainly justify consideration. One of the major merits is its simplicity. The process requires little operator control. The major limitations of the process are its reduced efficiency in cold weather and the maintenance problems which can potentially occur from calcium carbonate deposition within a stripping tower. Where nitrogen removal is required for 12 months a year and cold winter weather occurs, the stripping process alone will be incapable of providing low nitrogen concentrations during the winter. In such instances, breakpoint chlorination as a supplemental means of nitrogen removal during the cold weather season should be considered as an alternative.

Neither of the other two physical-chemical approaches considered (ion exchange and breakpoint chlorination) suffer from the cold weather limitation of ammonia stripping. However, this advantage is offset by the disadvantages of more complex operation and higher cost. There are several plants now under construction which plan to utilize the breakpoint chlorination process for removal of all of the ammonia or the ammonia residual escaping some other upstream nitrogen removal system. The process has the disadvantage of the addition of large quantities of dissolved solids to the wastewater as a result of the addition of chlorine or sodium hypochlorite required to remove the ammonia. Careful control of the process is required to prevent either the formation of excessive quantities of nitrate at high pH or the formation of nitrogen trichloride at low pH values. Both the breakpoint process and the ion exchange process are capable of providing very low nitrogen concentrations. The breakpoint process can remove all of the ammonia nitrogen, while the degree of removal provided by the ion exchange columns will vary somewhat with the mode of operation, i.e., the frequency of regeneration.

In comparing the alternative of one or more of the physical-chemical approaches to nitrogen removal with the biological approach, there are several major points to consider. One of major concern is the controlability of the biological approach as compared to that offered by the physical-chemical approaches. Should, for some reason, the nitrifying population be lost from the biological system, there is little the operator can do to accelerate the recovery of the system, and substantial time periods with reduced nitrogen removal can occur. Each of the physical-chemical processes respond rapidly to changes in operating conditions. That is, one can make a change in operating conditions and immediately obtain a predictable response in the process. As in any biological system, the operator has a limited degree of flexibility in controlling the biological nitrogen removal process in light of changing operation conditions. Certainly, each of the physical-chemical approaches has its own advantages and disadvantages. In warm climates, the stripping process simplicity and reliability are major advantages. There is little room for operator error or any inherent process instability to result in any malfunction. The cold weather limitations reduce the applicability of this process as a sole means of nitrogen removal on a year round basis in many climates. The ion exchange approach provides consistent nitrogen removal under a variety of conditions without the drawback of adding dissolved solids as occurs with the breakpoint chlorination process. However, the ion exchange approach involves a higher capital investment and operating costs than the other approaches. It is obvious that the physical-chemical alternatives and the biological alternatives each have advantages and disadvantages which must be weighed by the designer in light of his own given circumstances.

OVERALL CAPABILITY AND COSTS OF TREATMENT PROCESS COMBINATIONS

Because of variations in plant influent municipal wastewater composition from city to city, it is impossible to present predictions of effluent quality for various unit process combinations which would be universally applicable Table 1-2 provides some estimates of achievable effluent qualities. Based on the preceding chapters, it is possible to make some general predictions of typical quality and related costs. Table 12-1 summarizes two general approaches. There are of course a great many other possible combinations of processes which can be envisioned, but this table illustrates two commonly used approaches and the approximate costs associated with improved effluent quality.

Table 12-1A presents a minimal system where secondary effluent is subjected directly to carbon adsorption. This system removes neither nitrogen nor phosphorus. As an optional step, demineralization could also be provided. Table 12-1B shows several other unit processes. Again, demineralization is shown as an optional step. As previously discussed, the overall sequence given in Table 12-1B (without demineralization) would produce acceptable concentrations of heavy metals and viruses even for water reuse. Although this is not fully reflected by the data in the table, the effluent will have characteristics which are aesthetically pleasing—it will be odorless, colorless, and very clear. The greater reliability (see section on reliability considerations) and level of treatment offered by the process in Table 12-1B as compared to 12-1A is achieved at an incremental cost of 30¢/1000 gal, which is equivalent to a total incremental cost of $3.60/month/home. (Receipt of construction grants would reduce this increment to about $1.80. The cases compared in Table 12-1 offer a severe difference in effluent quality produced (i.e., no nitrogen or phosphorus removal in one case vs. very high removals of nitrogen and phosphorus in the other). In cases where one is comparing alternatives all of which provide comparable removals of the same pollutants, the economic differences will be much, much less—probably on the order of a few cents per month per home. The value of reliability and efficiency must be carefully weighed against such small differences in cost.

REFERENCES

1. Culp, G. L., and Hamann, C. C., "Advanced Waste Treatment Process Selection," *Public Works*, 3 parts, March, April, May, 1974.
2. Culp, G. L., panel discussion, "Advanced Methods of Wastewater Treatment," ASCE National Environmental Engineering Division Conference on Wastewater Effluent Limits, Ann Arbor, Michigan, July, 1973.

Index

Index

625